Review of Physiological Chemistry

SEVENTH EDITION

1959

HAROLD A. HARPER, Ph.D.

Associate Professor of Biochemistry
University of California School of Medicine
San Francisco
Biochemist Consultant to the Clinical Investigation Center,
U. S. Naval Hospital, Oakland
Biochemist Consultant to St. Mary's Hospital
San Francisco

Lange Medical Publications
LOS ALTOS, CALIFORNIA

Spanish Edition: Ediciones Morata, Madrid, Spain

Library of Congress Catalog Card Number: 59-9598

Lithographed in U. S. A.

Lange Medical Publications

LOS ALTOS, CALIFORNIA

PREFACE

The author's intention in writing this book has been to present a critical and concise summary of that which is considered essential in the expanding universe of physiological chemistry. To achieve this objective, a judicious selection of material has been necessary. The Review may thus serve as a supplement and as a companion volume to the larger standard texts and reference books. Much of the material which has been included is drawn from the author's basic science lectures in connection with the postgraduate training of physicians. It is therefore anticipated that the Review will not only meet the needs of the physician who is preparing for state and specialty boards but that it will also serve as one means of maintaining the practicing physician's perspective of the constantly increasing dimensions of this basic clinical discipline.

The author is indebted to many of his colleagues and to his readers, whose suggestions and criticisms have been of inestimable value in the preparation of this edition. He also gratefully acknowledges the excellent artwork of Mr. Henry Symmes, Mrs. Laurel V. Gilliland, and Professor Ralph Sweet. As in past editions, the editorial skills of Mr. James Ransom and Dr. Jack Lange have been an invaluable contribution to the success this book has enjoyed.

Harold A. Harper

San Francisco
May, 1959

TABLE OF CONTENTS

ILLUSTRATIONS

CHARTS

TABLES

1...

Carbohydrates

The carbohydrates, often termed starches or sugars, are widely distributed both in animal and in plant tissues. In plants they are produced by photosynthesis and include the starches of the plant framework as well as of the plant cells. In animal cells carbohydrate serves as an important source of energy for vital activities. Some carbohydrates have highly specific functions in vital processes (e.g., ribose in the nucleoprotein of the cells, galactose in certain fats, and the lactose of milk).

Carbohydrates are defined chemically as aldehyde or ketone derivatives of the higher polyhydric (more than one OH group) alcohols or as compounds which yield these derivatives on hydrolysis.

Classification:

Carbohydrates are divided into four major groups as follows: (1) Monosaccharides (often called "simple sugars") are those which cannot be hydrolyzed into a simpler form. The general formula is $C_nH_{2n}O_n$. The simple sugars may be subdivided as trioses, tetroses, pentoses, hexoses, or heptoses, depending upon the number of carbon atoms they possess; and as aldoses or ketoses, depending upon whether the aldehyde or ketone groups are present. Examples are:

		Aldo Sugars	Keto Sugars
Trioses -	($C_3H_6O_3$)	glycerose	dihydroxyacetone
Tetroses -	($C_4H_8O_4$)	erythrose	erythrulose
Pentoses -	($C_5H_{10}O_5$)	ribose	ribulose
Hexoses -	($C_6H_{12}O_6$)	glucose	fructose

(2) Disaccharides are carbohydrates which yield two molecules of the same or of different monosaccharides when hydrolyzed. The general formula is $C_n(H_2O)_{n-1}$. Examples are sucrose, lactose and maltose.

(3) Oligosaccharides are those which yield two to ten monosaccharide units on hydrolysis.

(4) Polysaccharides yield more than ten molecules of monosaccharides on hydrolysis. The general formula is $(C_6H_{10}O_5)_x$. Examples of polysaccharides are the starches and dextrins. These are sometimes designated as hexosans, pentosans, or mixed polysaccharides, depending upon the nature of the monosaccharides which they yield on hydrolysis.

Asymmetry:

In the formulas for glucose shown on the following page it will be noted that a different group is attached to each of the four bonds on the central carbon atom. For example, the four groups attached to carbon No. 2 are $C{\overset{O}{\diagup\!\!\diagup}}\!-\!H$, H, OH, and HO—C—H. A carbon atom to which four different atoms or groups of atoms are attached is said to be **asymmetric**.

Geometric (Spatial) Isomerism:

The presence of asymmetric carbon atoms in a compound makes possible the formation of **isomers** of that compound. Such compounds, which are identical in composition and differ only in spatial configuration, are called **stereoisomers** or **geometric isomers.** Two such isomers of glucose, one of which is the mirror image of the other, are shown below.

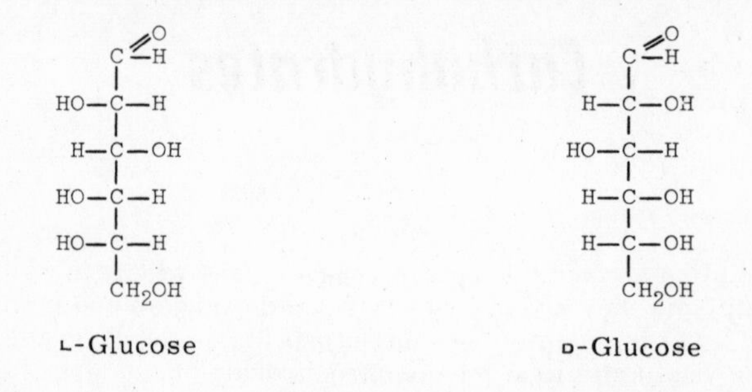

L-Glucose D-Glucose

The number of possible isomers of any given compound depends upon the number of asymmetric carbon atoms in the molecule. According to the **rule of n** (where "n" represents the number of asymmetric carbon atoms in a compound), 2^n equals the number of possible isomers of that compound. Thus glucose, with four asymmetric carbon atoms, would be expected to have 2^4 or 16 isomers. Eight would belong to the D series, and the mirror images of each of these would comprise the L series.

Two sugars which differ from one another only in the configuration around a single carbon atom are termed **epimers.** Galactose and glucose are examples of an epimeric pair which differ with respect to carbon 4; mannose and glucose are epimers with respect to carbon 2. Epimerisation or interconversion of epimers in the tissues is illustrated by the conversion in the liver of galactose to glucose. The conversion is catalyzed by a specific enzyme, designated as an epimerase (see p. 195).

The designation of an isomer as a D form or of its mirror image as an L form is determined by its spatial relationship to the parent substance of the carbohydrate family, the three-carbon sugar, glycerose. The L and D forms of this sugar are shown below.

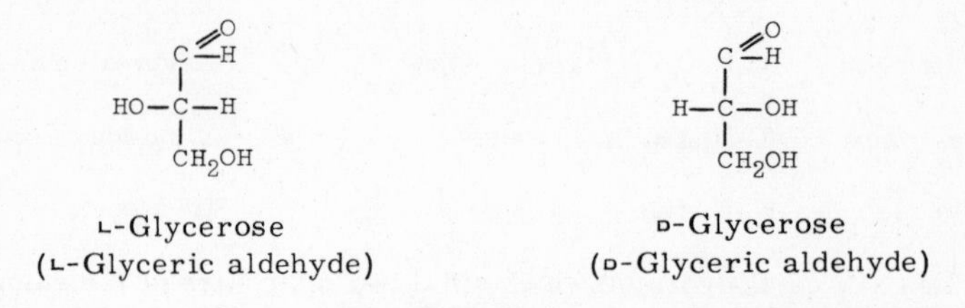

L-Glycerose
(L-Glyceric aldehyde)

D-Glycerose
(D-Glyceric aldehyde)

The orientation of the H and OH groups around the carbon atom just adjacent to the terminal primary alcohol carbon (e.g., carbon atom No. 5 in glucose) determines the family to which the sugar belongs. When the OH group on this carbon is on the right, the sugar is a member of the D series; when it is on the left, it is a member of the L series. The distribution of the H and OH groups on the other carbon atoms in the molecule is of no importance in this connection.

The majority of the monosaccharides occurring in the body are of the D configuration.

THE KILIANI SYNTHESIS

The structural relationships of the monosaccharides may be readily visualized by the formulas shown on p. 4. A method for the synthesis of these sugars was first proposed by Kiliani. It is based upon the addition of HCN to the carbonyl group of aldehydes or ketones. The application of the Kiliani synthesis to the production of the two tetroses, **erythrose** and **threose**, from glycerose is shown on the following page.

This process can be repeated and the four isomeric D-pentoses would be formed; from these, the eight isomeric hexoses would result.

If the D- and L-aldoses be conceived as thus evolving from the trioses, the aldehyde carbon becomes the one most recently added. It is for this reason that the carbon adjacent to the terminal primary alcohol group (on the opposite end of the molecule) is the true indicator of the original family of origin, i. e., of the D series or of the L series, since this carbon is the original and the only asymmetric carbon of the parent sugars.

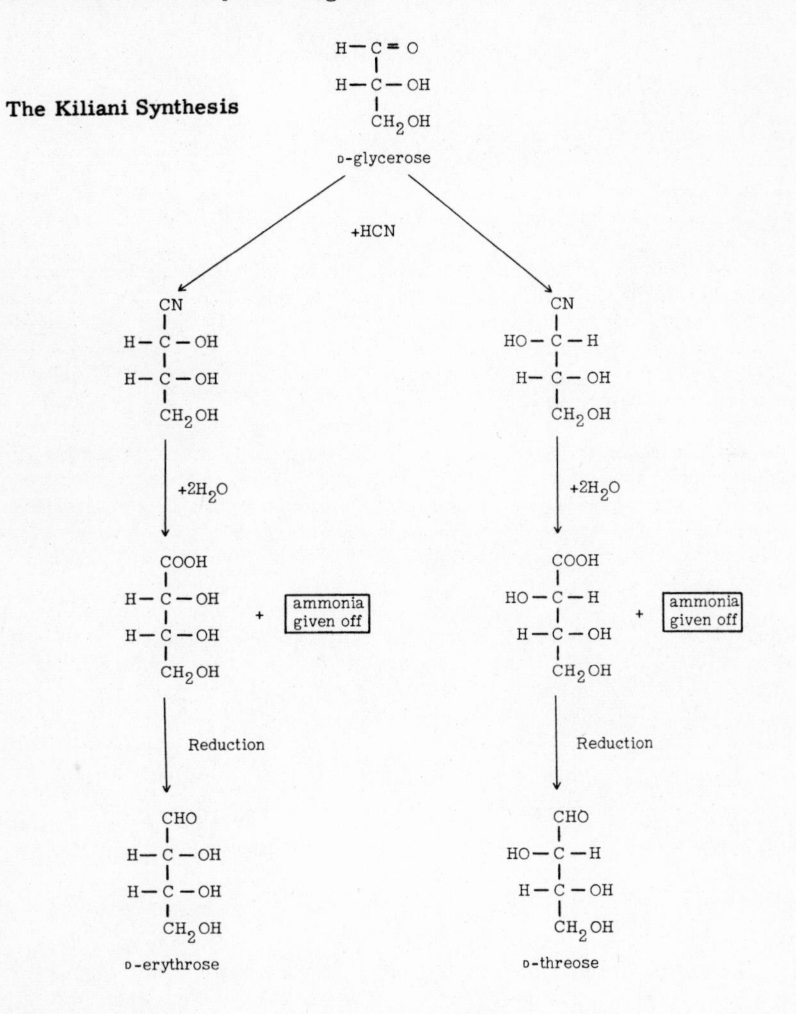

Optical Isomerism:

The presence of asymmetric carbon atoms also confers **optical activity** on the compound. When a beam of polarized light is passed through a solution exhibiting optical activity, it will be rotated to the right or left in accordance with the type of compound, i. e., the **optical isomer** which is present. A compound which causes rotation of polarized light to the right is said to be dextrorotary and a plus (+) sign is used to designate this fact. Rotation of the beam to the left (levorotary action) is designated by a minus (-) sign.

When equal amounts of dextro- and levorotary isomers are present, the resulting mixture has no optical activity since the activities of each isomer cancel one another. Such a mixture is said to be a **racemic**, or a DL mixture. Synthetically produced compounds are necessarily racemic because the opportunities for the formation of each optical isomer are identical. The separation of optically active isomers from a racemic mixture is called resolution, i. e., the racemic mixture is said to be "resolved" into its optically active components.

THE STRUCTURAL RELATIONS OF THE ALDOSES

D-Series

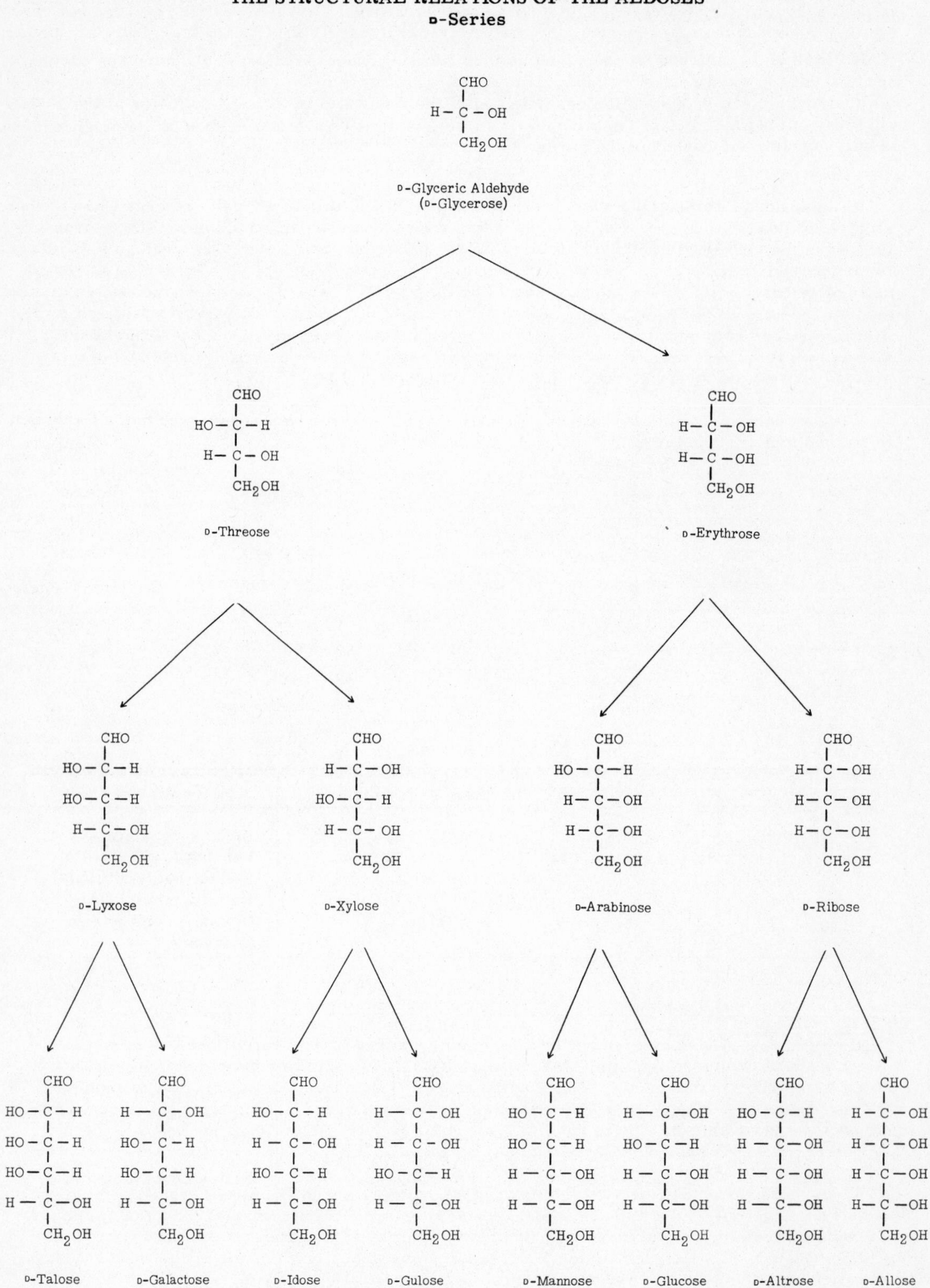

Geometric isomerism and optical isomerism are independent properties. Thus a compound might be designated D (-) or L (+), indicating structural relationship to D or L glycerose but exhibiting the opposite rotary power. The naturally-occurring form of lactic acid, the L (+) isomer, is an example.

MONOSACCHARIDES

The monosaccharides include trioses (containing three carbon atoms), tetroses (four carbon atoms), pentoses, hexoses, and heptoses (five, six, and seven carbon atoms). The trioses are formed in the course of the metabolic breakdown of the hexoses (see p. 175); pentose sugars are important constituents of nucleic acids and many coenzymes; they are also formed in the breakdown of glucose by the direct oxidative pathway (see p. 186), and the hexoses glucose, galactose, and fructose are physiologically the most important of the monosaccharides (see table on p. 7). A seven-carbon keto sugar, sedoheptulose, was first discovered in 1917 in the sedum plant. It also occurs in animal tissues, where it is formed as a phosphate ester in the metabolism of pentose phosphates by the direct oxidative pathway (see p. 187).

The structures of the aldo sugars are shown on p. 4. Five keto sugars which are important in metabolism are shown below:

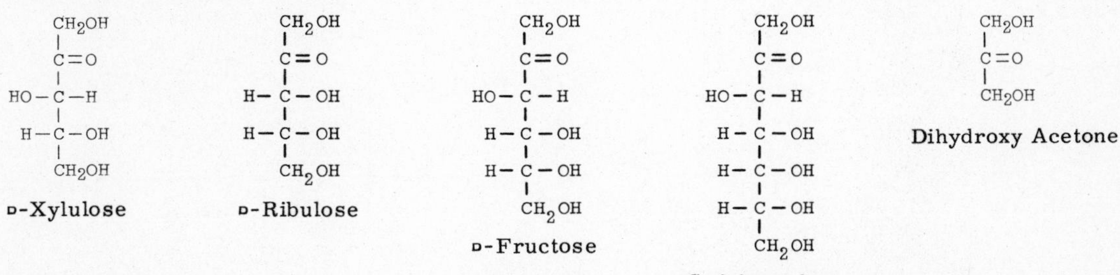

Pentoses

Sugar	Source	Importance	Reactions
D-Ribose	Nucleic acids.	Structural elements of nucleic acids and coenzymes, e.g., ATP, DPN, TPN (Co I-II), flavoproteins.	Reduce Benedict's, Fehling's, Barfoed's, and Haynes' solutions. Form distinctive osazones with phenylhydrazine.
D-Ribulose	Formed in metabolic processes.	Intermediates in direct oxidative pathway of glucose breakdown (see p. 186).	Keto sugars.
D-Arabinose	Gum arabic. Plum and cherry gums.	These sugars are used in studies of bacterial metabolism, as in fermentation tests for identification of bacteria. They have no known physiological function in man.	With orcinol-HCl reagent gives colors: violet, blue, red, and green.
D-Xylose	Wood gums.		With phloroglucinol-HCl gives a red color.
D-Lyxose	Heart muscle.	A constituent of a lyxoflavin (see p. 79) isolated from human heart muscle.	

HEXOSES

The hexoses are more important physiologically than all the other monosaccharides combined (see table on p. 7). Examples are ᴅ-glucose, ᴅ-fructose, ᴅ-galactose, and ᴅ-mannose.

Ring Structures:

The formulas on pp. 4 and 5 illustrate the so-called ''straight chain'' structure of hexose sugars, but this does not satisfactorily explain many carbohydrate reactions. On the basis of the two ring structures known to exist in the glycosides (see below), Haworth in 1929 proposed similar structures for the sugars themselves. The terminology for such structures was based on the simplest organic compounds exhibiting a similar ring structure (i. e., pyran and furan).

The ring structures shown below represent **hemiacetal** formation, i. e., a condensation between the aldehyde group and a hydroxy group of the same compound. Because of the asymmetry present in the terminal carbon, two forms for each ring structure can exist. These are designated as α and β. For ᴅ-glucose, for example:

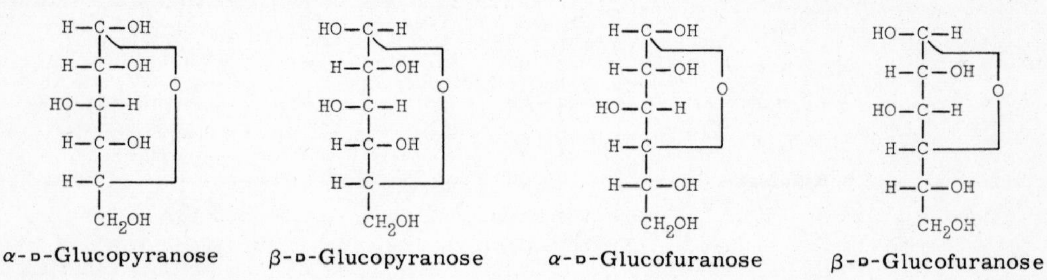

α-ᴅ-Glucopyranose β-ᴅ-Glucopyranose α-ᴅ-Glucofuranose β-ᴅ-Glucofuranose

Ketoses may also show ring formation. Thus one may find α-ᴅ-fructofuranose, β-ᴅ-fructofuranose, or the corresponding pyranoses. Other ring forms (e. g., between C atoms 1 and 2 or 1 and 3) may exist but are so unstable that when liberated from their glycosides as sugars they tend to mutate readily to the pyranose form, as do the freed furanoses also.

GLYCOSIDES

Glycosides are compounds containing a carbohydrate and a noncarbohydrate residue in the same molecule. In these compounds the carbohydrate residue is attached by an acetal linkage at carbon atom No. 1 to a noncarbohydrate residue or **aglycone.** If the carbohydrate portion is glucose, the resulting compound is a **glucoside.**

A simple example is the methyl glucoside formed when a solution of glucose in boiling methyl alcohol is treated with 0.5% hydrogen chloride as a catalyst. The reaction proceeds as follows:

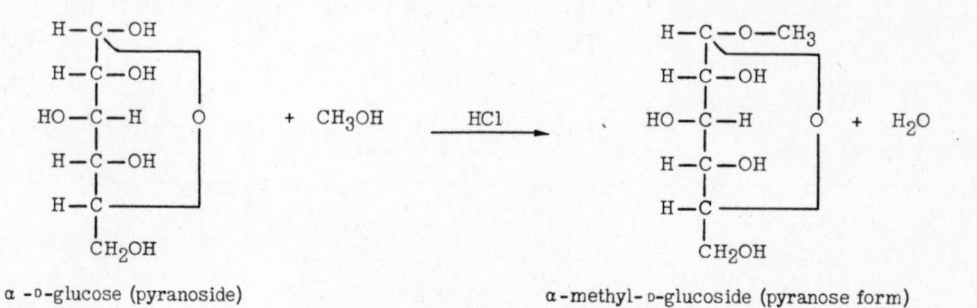

α-ᴅ-glucose (pyranoside) α-methyl-ᴅ-glucoside (pyranose form)

From β-D-glucose, β-methyl-glucoside would be formed.

Glycosides are found in many drugs, spices, and in the constituents of animal tissues. The aglycone may be methyl alcohol, glycerol, a sterol, phenols, etc. The glycosides which are important in medicine because of their action on the heart all contain steroids as the aglycone component. These include derivatives of digitalis and strophanthus. Other examples of glycosides which contain steroids as the aglycone are phlorhizin, salicin, amygdalin, and saponin.

Mutarotation:

When a sugar is first dissolved in water, its optical rotation gradually changes until a constant rotation characteristic of the sugar is reached. This phenomenon, known as mutarotation, appears to be due to changes of α to β forms and vice versa, probably via the straight chain aldo or keto form (see p. 10). When equilibrium is reached, the characteristic rotation is observed.

Hexoses

Sugar	Source	Importance	Reactions
D-Glucose	Fruit juices. Hydrolysis of starch and of cane sugar.	The "sugar" of the body. The sugar carried by the blood, and the principal one used by the tissues. Glucose is usually the "sugar" of the urine when glycosuria occurs.	Reduces Benedict's, Haynes', Barfoed's reagents (a reducing sugar). Gives osazone with phenylhydrazine. Fermented by yeast. Forms soluble saccharic acid with HNO_3.
D-Fructose	Fruit juices. Honey. Hydrolysis of cane sugar and of inulin (from the Jerusalem artichoke).	Can be changed to glucose in the liver and intestine and so used in the body.	Reduces Benedict's, Haynes', Barfoed's reagents (a reducing sugar). Forms osazone identical with that of glucose. Fermented by yeast. Cherry-red color with Seliwanoff's resorcinol-HCl reagent.
D-Galactose	Hydrolysis of lactose.	Can be changed to glucose in the liver and so used. Synthesized in the body to make lactose of mother's milk. A constituent of glycolipids (see p. 24).	Reduces Benedict's, Haynes', Barfoed's reagents (a reducing sugar). Forms osazone, distinct from above. Phloroglucinol-HCl reagent gives red color. Forms insoluble mucic acid with HNO_3. Not fermented by yeast.
D-Mannose	Hydrolysis of plant mannosans and gums.	A constituent of prosthetic polysaccharide of albumins, globulins, mucoids. Also the prosthetic polysaccharide of tuberculoprotein.	Reduces Benedict's, Haynes', Barfoed's reagents (a reducing sugar). Convertible to glucose in the body. Forms same osazone as glucose.

IMPORTANT CHEMICAL REACTIONS OF MONOSACCHARIDES

Several reactions are of importance as proof of the structure of a typical monosaccharide such as glucose. These include the following:

Iodo Compounds:

An aldose heated with concentrated hydriodic acid (HI) loses all of its oxygen and is converted into an iodo compound (glucose to iodohexane, $C_6H_{13}I$). Since the resulting derivative is a straight chain compound related to normal hexane, this is evidence of the lack of any branched chains in the structure of the sugar.

Acetylation:

The ability to form sugar esters (e.g., acetylation with $CH_3CO.Cl$) indicates the presence of alcohol groups. The total number of acyl groups which can thus be taken up by a molecule of the sugar is a measure of the number of such alcohol groups. Because of its five OH groups, the acetylation of glucose, for example, results in a penta acetate.

Other Reactions:

Various reactions dependent upon the presence of aldehyde or ketone groups are particularly important because they form the basis for most analytical tests for the sugars. The best-known tests involve **reduction** of metallic hydroxides together with oxidation of the sugar. The alkaline metal is kept in solution with sodium potassium tartrate (Fehling's solution) or sodium citrate (Benedict's solution). Various modifications permit quantitative detection of the copper reduced as a measurement of the sugar content. Other metallic hydroxides may be used (bismuth, Nylander test, ammoniacal silver, Tollens' test). Barfoed's test distinguishes between monosaccharides and disaccharides, since copper acetate in dilute acid is reduced by the former in 30 seconds but only after several minutes' boiling (to produce hydrolysis) of the disaccharides.

Osazone formation is a useful means of preparing crystalline derivatives of the sugars. These compounds have characteristic crystal structures, melting points, and precipitation times, and are valuable in the identification of sugars. They are obtained by adding a mixture of phenylhydrazine hydrochloride and sodium acetate to the sugar solution and heating in a water bath at 100°C. The reaction involves only the carbonyl carbon (i.e., aldehyde or ketone group) and the next adjacent carbon. For example, with an aldose the following reaction occurs:

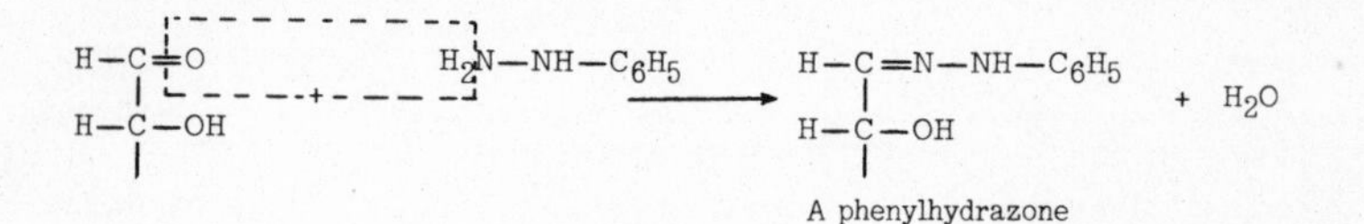

A phenylhydrazone

The hydrazone then reacts with two additional molecules of phenylhydrazine to form the osazone:

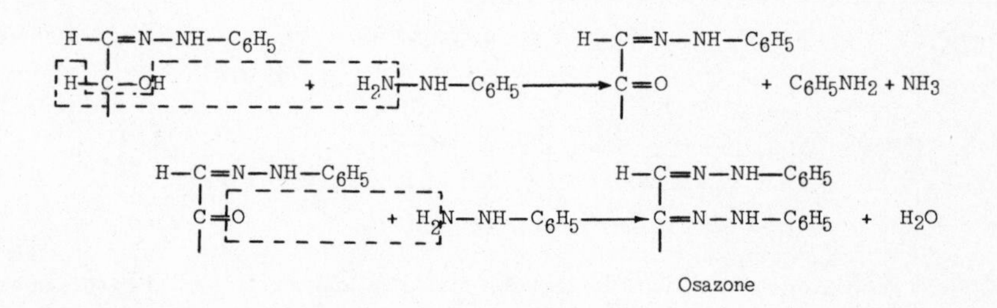

Osazone

The reaction with a ketose is similar.

It will be noted from a comparison of their structures that glucose, fructose, and mannose have the same osazones; but since the structure of galactose differs in that part of the molecule unaffected in osazone formation, it would form a different osazone.

Interconversion. Glucose, fructose, and mannose are interconvertible in solutions of weak alkalinity such as $Ba(OH)_2$ or $Ca(OH)_2$. Presumably, the physiologic interconversion which takes place in the body is similarly explained. These changes are easily visualized structurally through an enediol form common to all three sugars:

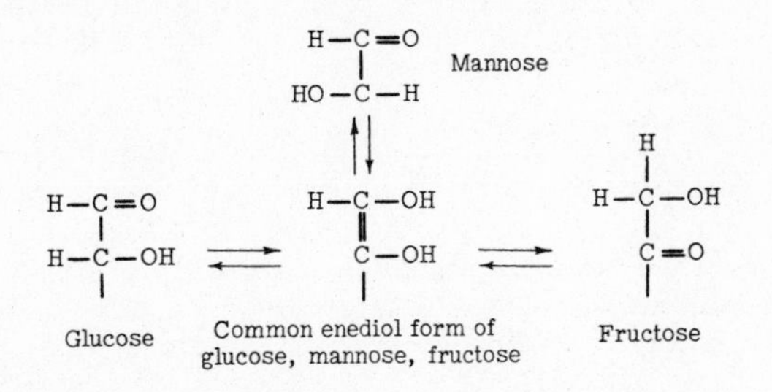

Oxidation may take place in aldoses, forming acids as end products. Oxidation of the aldehyde group forms "aldonic acids." If the aldehyde group remains intact the primary alcohol group at the opposite end of the molecule is oxidized and "uronic acids" are formed instead.

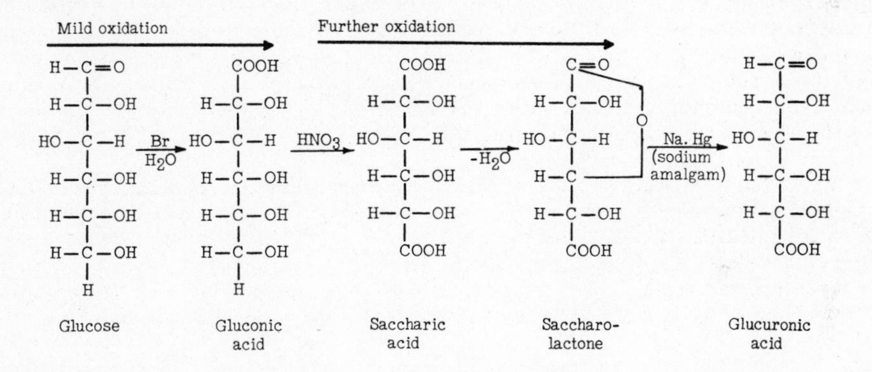

Note that glucuronic acid exerts "reducing" activity because of the free aldehyde group. These so-called hexuronic acids are important in connection with conjugation reactions (see pp. 168 and 273).

Oxidation of galactose with concentrated HNO_3 yields the dicarboxylic mucic acid. This compound crystallizes out readily and is useful as an identifying test. Galacturonic acid is found in natural products (e.g., pectins).

Reduction. The monosaccharides may be reduced to their corresponding alcohols.

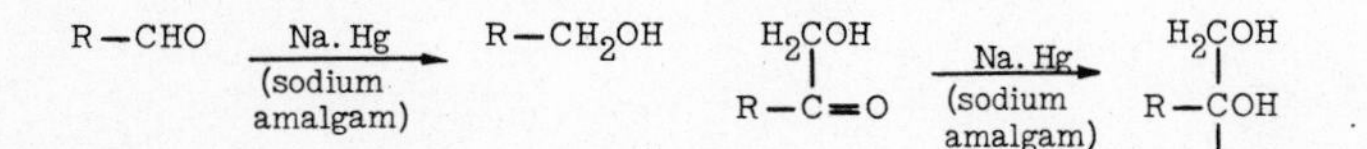

Thus, glucose yields sorbitol; galactose yields dulcitol; mannose yields mannitol; and fructose yields mannitol and sorbitol.

With **strong mineral acids** there is a shift of hydroxyl groups toward and of hydrogen away from the aldehyde end of the chain.

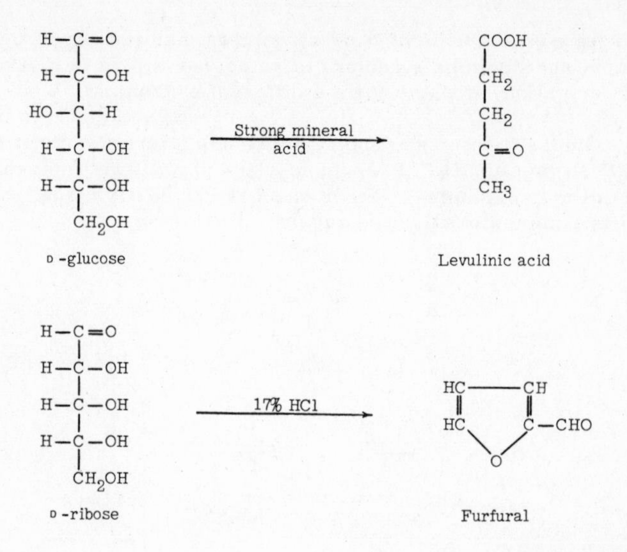

Reaction products with acid (furfural or one of its derivatives) will condense with certain organic phenols to form characteristically-colored products. Color tests for the various sugars are based on such reactions.

Heating of gluconic acid produces lactones. These are cyclic structures resembling the pyranoses and furanoses described on p. 6.

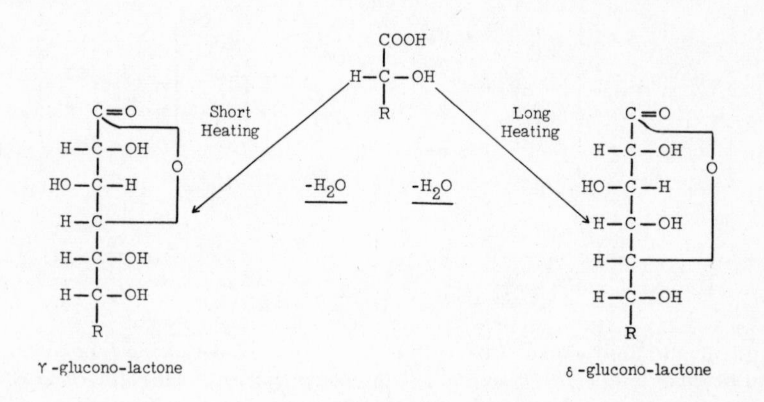

With **alkali** monosaccharides react in various ways. (1) In dilute alkali the sugar will change to the cyclic α and β structures, with an equilibrium between the two isomeric forms. (See Mutarotation, p. 7.)

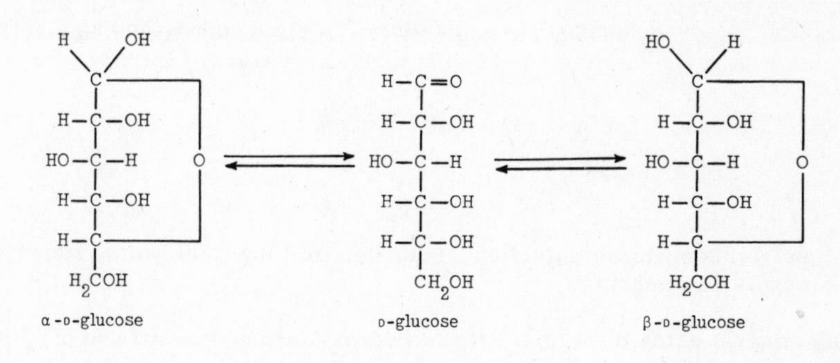

On standing, a rearrangement will occur which produces an equilibrium mixture of glucose, fructose, and mannose through the enediol form. (See p. 9.)

(2) If the mixture is heated to 37° C., the acidity increases and a series of enols are formed in which double bonds shift from the oxygen-carbon link to positions between various carbon atoms.

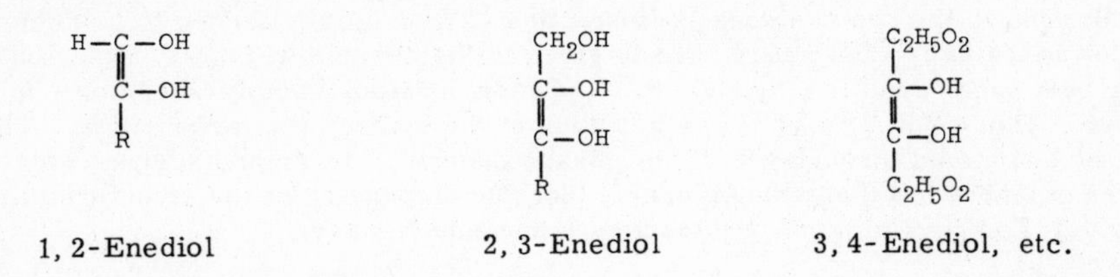

1, 2-Enediol 2, 3-Enediol 3, 4-Enediol, etc.

(3) In concentrated alkali, sugar caramelizes and produces a series of decomposition products. Yellow and brown pigments develop, salts may form, many double bonds between carbon atoms are formed, and chains may rupture.

DEOXY SUGARS

Deoxy sugars are those containing fewer atoms of oxygen than of carbon. They are obtained on hydrolysis of certain substances which are important in biologic processes. An example is the deoxyribose occurring in nucleic acids (DNA, see p. 46).

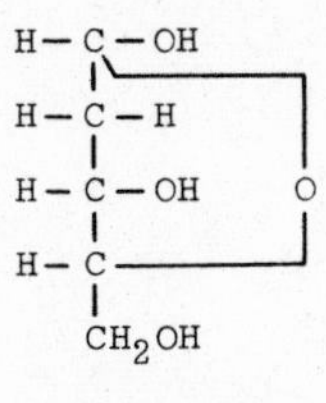

2-Deoxy-D-ribo-furanose (α form)

AMINO SUGARS (HEXOSAMINES)

Sugars containing an amino group are called **amino sugars**. Examples are D-glucosamine and D-galactosamine, both of which have been identified in nature. Glucosamine (also called chitosamine) is a constituent of the chitin in the exoskeleton of crustaceans. Galactosamine (chondrosamine) is a constituent of chondroitin (see p. 382). Amino sugars are also important constituents of mucoprotein (see p. 132).

Several antibiotics (erythromycin, carbomycin) contain amino sugars. Erythromycin contains a dimethylamino sugar. Carbomycin contains the first known 3-amino sugar, 3-amino-D-ribose. The amino sugars are believed to be related to the antibiotic activity of these drugs.

```
 H—C—OH
    |\_________
 H—C—NH2       |
    |          O
HO—C—H         |
    |          |
 H—C———————————
    |
 H—C—OH
    |
   CH2OH
```

Glucosamine
(2-Amino-D-glucofuranose) (α form)

DISACCHARIDES

The disaccharides are sugars which can be hydrolyzed into two monosaccharides. The two sugars are united by a glycosidic linkage (see p. 6). They are named chemically according to the structures of their component monosaccharides. The suffix **-id(e)** is given to that residue in which the oxygen common to the two residues is linked to a carbon atom attached to another oxygen which is not common to the two. The suffix **-ose** is given to the residue without this peculiarity. Thus sucrose (see next page) might be considered a pyranoglucosido-fructofuranose or a furanofructosido-glucopyranose. The α and β refer to the grouping at the starred (*) carbon atom. The **numbers** refer to the carbon atoms through which the linkage occurs. **-furan** and **-pyran** refer to the structural resemblances to these substances. (See the discussion of the structure of monosaccharides, p. 6.) The physiologically important disaccharides are:

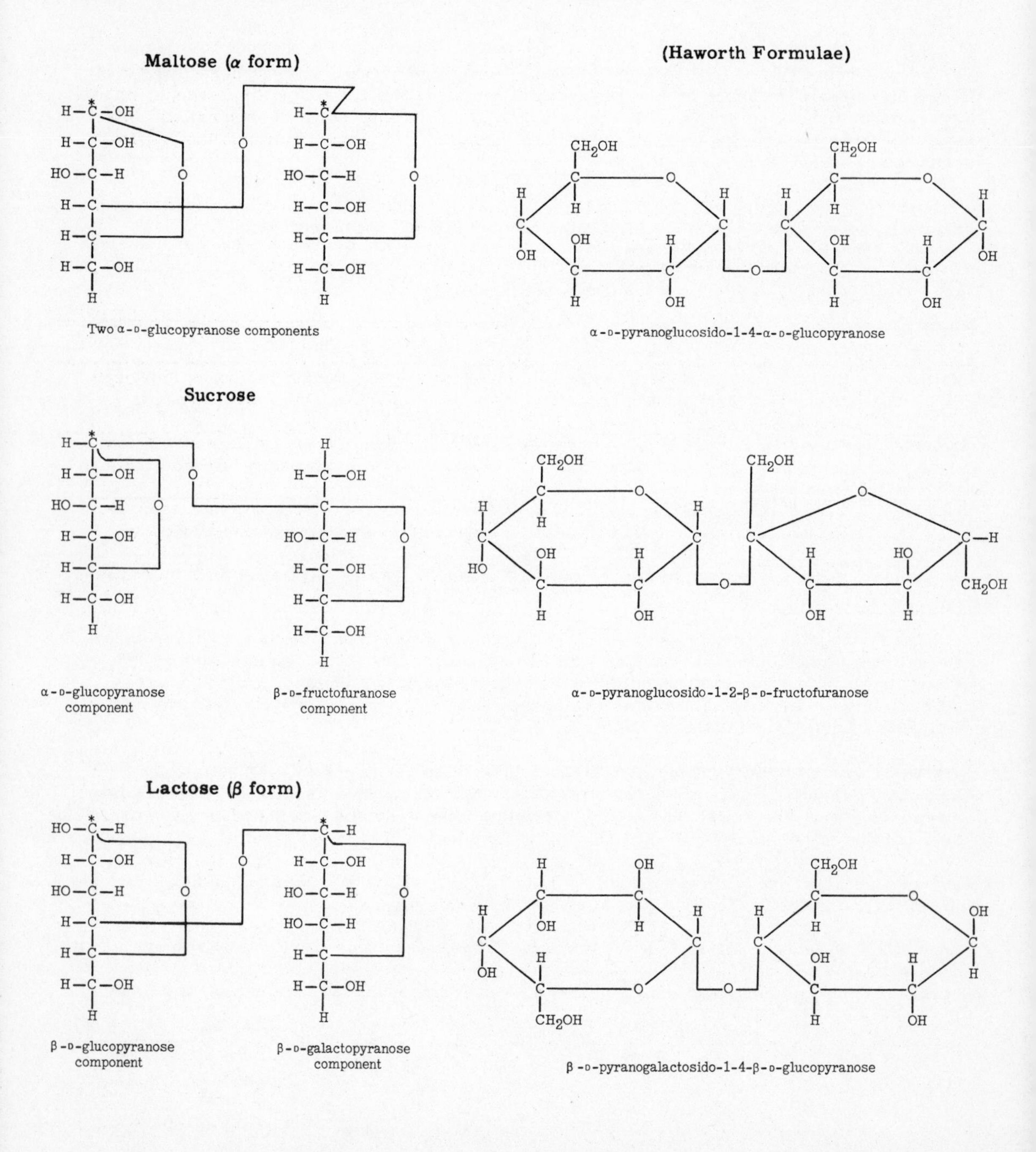

Maltose (α form)

(Haworth Formulae)

Two α-D-glucopyranose components

α-D-pyranoglucosido-1-4-α-D-glucopyranose

Sucrose

α-D-glucopyranose component

β-D-fructofuranose component

α-D-pyranoglucosido-1-2-β-D-fructofuranose

Lactose (β form)

β-D-glucopyranose component

β-D-galactopyranose component

β-D-pyranogalactosido-1-4-β-D-glucopyranose

Trehalose (α form)

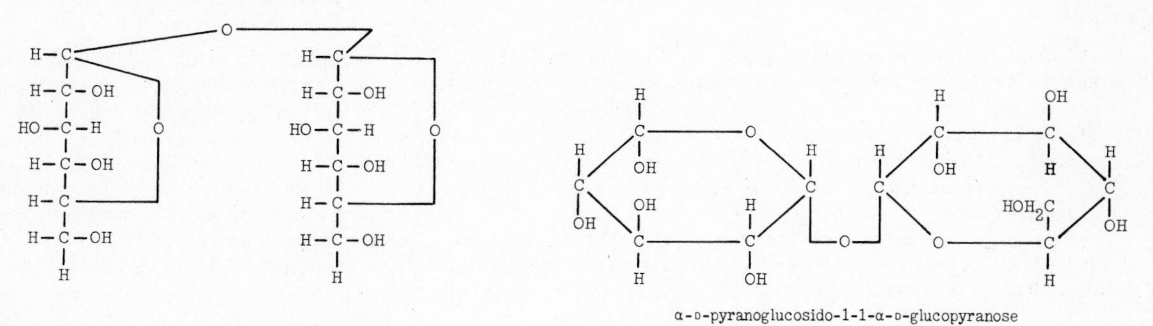

α-D-pyranoglucosido-1-1-α-D-glucopyranose

Since sucrose has no free carbonyl group, it gives none of the reactions characteristic of "reducing" sugars. Thus it fails to reduce alkaline copper solutions, form an osazone, or exhibit mutarotation. Hydrolysis of sucrose yields a crude mixture which is often called "invert sugar" because the strongly levorotary fructose thus produced changes (inverts) the previous dextrorotary action of the sucrose.

Lactose gives rise to mucic acid on prolonged boiling with HNO_3. This is derived from the galactose produced as a result of hydrolysis of lactose by the boiling with acid.

Disaccharides

Sugar	Source	Reactions
Maltose	Diastatic digestion or hydrolysis of starch. Germinating cereals and malt.	Reducing sugar. Forms osazone with phenylhydrazine. Fermentable. Hydrolyzed to D-glucose.
Lactose	Milk. May occur in urine during pregnancy. Formed in the body from glucose.	Reducing sugar. Forms osazone with phenylhydrazine. Not fermentable. Hydrolyzed to glucose and galactose.
Sucrose	Cane and beet sugar. Sorghum. Pineapple. Carrot roots.	Nonreducing sugar. Does not form osazone. Fermentable. Hydrolyzed to fructose and glucose.
Trehalose	Fungi and yeasts.	Nonreducing sugar. Does not form an osazone. Hydrolyzed to glucose.

TECHNIC OF TESTS FOR CARBOHYDRATES*

(See p. 14.)

Anthrone Test: To 2 ml. of anthrone test solution (0.2% in concentrated H_2SO_4) add 0.2 ml. of unknown. A green or blue-green color indicates presence of carbohydrate. The test is very sensitive; it will give a positive reaction with filter paper (cellulose). The anthrone reaction has been adapted to the quantitative colorimetric determination of glycogen, inulin, and sugar of blood.

Barfoed's Test (copper acetate and acetic acid): To 5 ml. of reagent add 1 ml. of unknown. Place in boiling water bath. (See chart on p. 14 for interpretation.)

Benedict's Test (copper sulfate, sodium citrate, sodium carbonate): To 5 ml. of reagent in test tube add 8 drops of unknown. Place in boiling water bath for 5 minutes. A green, yellow, or orange-red precipitate gives a semi-quantitative estimate of the amounts of reducing sugar present.

*Chromatographic technics (see p. 36) are also now used for separation and identification of carbohydrates.

QUALITATIVE TESTS FOR IDENTIFICATION OF CARBOHYDRATES

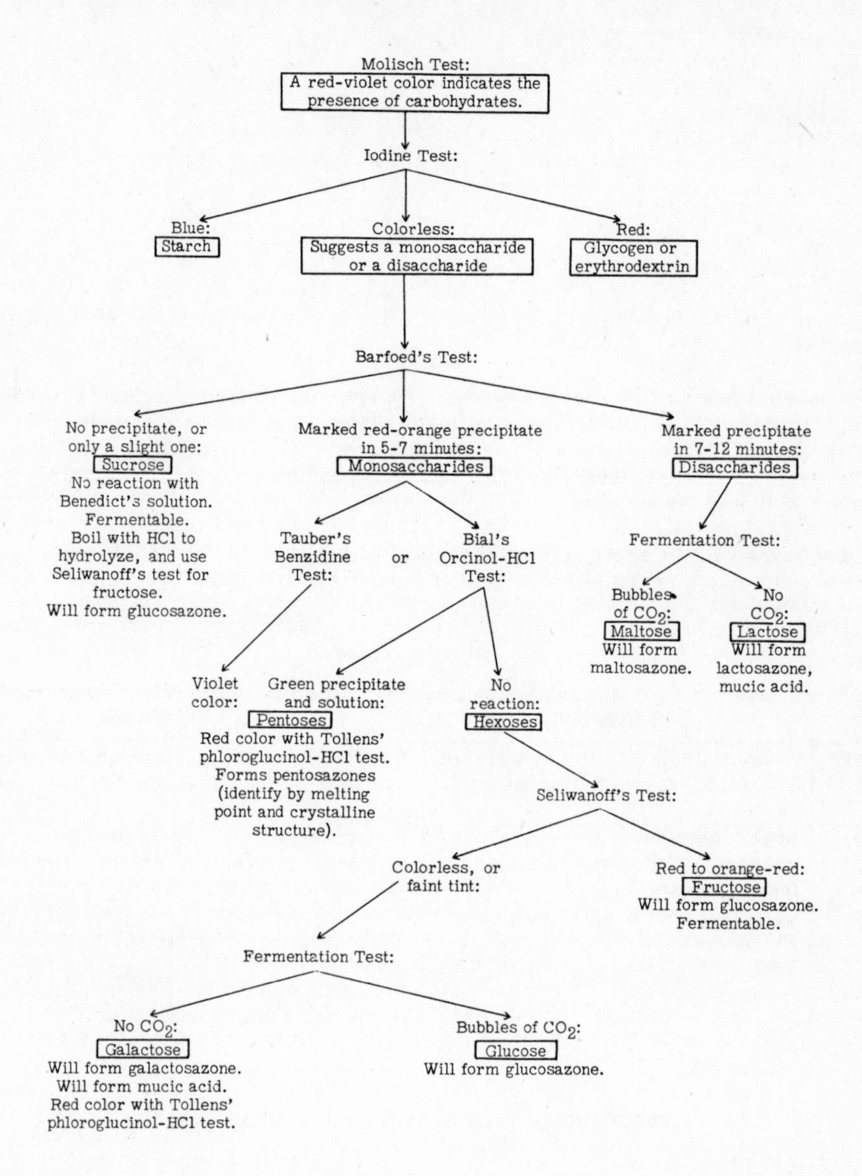

Bial's Orcinol-HCl Test: To 5 ml. of reagent add 2 to 3 ml. of unknown and heat until bubbles of gas rise to the surface. Green solution and precipitate indicate pentose.

Fermentation Test: To 5 ml. of a 20% suspension of ordinary baker's yeast add about 5 ml. of unknown solution and 5 ml. of phosphate buffer (pH 6.4 to 6.8). Place in a fermentation tube or test tube and let stand one hour. Bubbles of CO_2 indicate fermentation.

Haynes' Test (Rochelle salt or sodium potassium tartrate, glycerol, copper sulfate): Performed similarly to Benedict's.

Iodine Test: Acidify the unknown solution with HCl and add one drop of the mixture to a solution of I_2 in KI. The formation of a blue color indicates the presence of starch; a red color indicates the presence of glycogen or erythrodextrin.

Molisch Test: To 2 ml. of unknown add 2 drops of fresh 10% α-naphthol reagent and mix. Pour 2 ml. of concentrated H_2SO_4 so as to form a layer below the mixture. A red-violet ring indicates the presence of carbohydrate.

Pavy's Test (Rochelle salt, ammonium hydroxide, copper sulfate): Similar to Benedict's test.

Phenylhydrazine Reaction (osazone formation): Heat phenylhydrazine reagent with 2 ml. of a solution of the sugar in a test tube in a boiling water bath for 30 minutes; cool, and examine crystals with a microscope. Compare with diagrams in laboratory manuals or with crystals prepared from known solutions.

Seliwanoff's Resorcinol Test: To 1 ml. of unknown add 5 ml. of freshly-prepared reagent. This is made by adding 3.5 ml. of 0.5% resorcinol to 12 ml. of concentrated HCl and diluting to 35 ml. with distilled water. Place in boiling water bath for 10 minutes. Cherry-red color indicates fructose.

Tauber's Benzidine Test: To 1 ml. of benzidine solution add 2 drops of unknown sugar; boil and cool quickly. A violet color indicates the presence of pentose.

Tollens' Naphthoresorcinol Reaction: To 5 ml. of unknown in a test tube add 1 ml. of 1% alcoholic solution of naphthoresorcinol. Heat gradually to boiling; boil for one minute with shaking; let stand for four minutes, then cool under tap. Then prepare ether extract. A violet-red color in the ether extract indicates presence of hexuronic acids and eliminates pentoses.

Tollens' Phloroglucinol-HCl Test: To equal volumes of the unknown solution and HCl add phloroglucinol. Glucuronates may be distinguished from pentoses or galactose by the naphthoresorcinol test (see above).

POLYSACCHARIDES

Polysaccharides include the following physiologically important substances:

Starch $(C_6H_{10}O_5)_x$ is formed of an α-glucosidic chain. Such a compound, yielding only glucose on hydrolysis, is called a **glucosan**. It is the most important food source of carbohydrate and is found in cereals, potatoes, legumes, and other vegetables. Natural starch is insoluble and gives a blue color with iodine solution. The microscopic form of the granules is characteristic of the source of the starch. The two chief constituents are amylose (98%), which is linear in structure, and amylopectin, which consists of highly branched chains. Each is composed of a number of α-glucosidic chains having 24 to 30 molecules of glucose apiece. The glucose residues are united by one to four linkages.

Glycogen is the polysaccharide of the animal body. It is often called animal starch. It is a branched structure with straight chain units of 11-18-α-D-glucopyranose (in α[1-4]-glucosidic linkage) cross-linked by means of α(1-6) glucosidic bonds (see p. 177). It is nonreducing and gives a red color with iodine.

Inulin is a starch found in tubers and roots of dahlias, artichokes, and dandelions. It is hydrolyzable to fructose and hence it is a fructosan. No color is given when iodine is added to inulin solutions. This starch is easily soluble in warm water. It is used in physiologic investigation for determination of the rate of glomerular filtration (see p. 297).

Dextrins are substances which are formed in the course of the hydrolytic breakdown of starch. The partially digested starches are amorphous. Dextrins which give a red color when tested with iodine are first formed. These are called **erythrodextrins.** As hydrolysis proceeds the iodine color is no longer produced. These are the so-called **achroodextrins.** Finally, reducing sugars will appear.

Cellulose is the chief constituent of the framework of plants. It gives no color with iodine and is not soluble in ordinary solvents. Since it is not subject to attack by the digestive enzymes of man, it is an important source of "bulk" in the diet.

Chitin is an important structural polysaccharide of invertebrates. It is found, for example, in the shells of crustaceans. Structurally, chitin apparently consists of N-acetyl-D-glucosamine units joined by β(1-4) glucosidic linkages.

Polysaccharides which are associated with the structure of animal tissues are analogous to the cellulose of the plant cells. Examples are **hyaluronic acid** and the **chondroitin sulfates** (see p. 383). These substances are members of a group of carbohydrates, the **mucopolysaccharides,** which are characterized by their content of amino sugars (see p. 11) and uronic acids (see p. 9). Heparin (see p. 126), which occurs in several animal tissues, is also a mucopolysaccharide because on hydrolysis heparin yields glucuronic acid and glucosamine, as well as acetic and sulfuric acids.

Many of the mucopolysaccharides occur in the tissues as prosthetic groups of conjugated proteins, the **mucoproteins** or **glycoproteins** (see p. 132). Examples are found among the α_1 and α_2 globulins of plasma (see p. 129). The mucoproteins of the plasma are characterized by the presence of an acetyl hexosamine (N-acetyl glucosamine?) and a hexose (mannose or galactose) in their polysaccharide portion. In addition, a methyl pentose (**L-fucose**) and **sialic acid** commonly occur in these conjugated proteins.

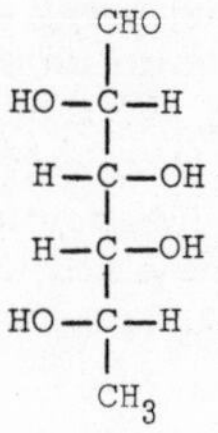

```
     CHO
      |
  HO—C—H
      |
   H—C—OH
      |
   H—C—OH
      |
  HO—C—H
      |
     CH3
```

L-Fucose

```
        OH          OH   NH.CO.CH3
        |           |    |
HOOC — C — CH2 — CH — CH — CH — (CH.OH)2 — CH2OH
        |_____________O______|
```

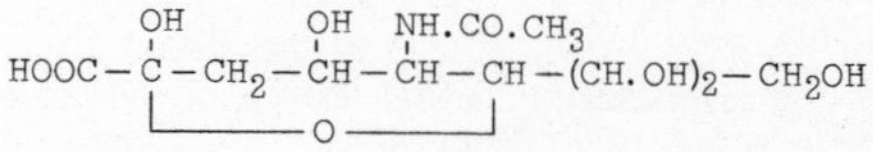

Sialic Acid
(N-Acetyl Neuraminic Acid)

The "blood group" substances of the erythrocytes (isoagglutinogens) which are responsible for the major immunological reactions of blood (blood types) are also mucoproteins. L-fucose is an important constituent of human blood group substances (19% in blood group B substance). Other examples of mucopolysaccharides which produce specific immune reactions are found among the bacteria. The capsular polysaccharides (haptenes) of pneumococci have been the most extensively studied in this connection. Preparations of capsular polysaccharide from Type I pneumococci yield, on hydrolysis, glucosamine and glucuronic acid.

Some of the pituitary hormones, although mainly proteins (e.g., the gonadotropins and thyrotropic hormone) also contain carbohydrate. This suggests that they may also be mucoproteins or glycoproteins.

• • •

Bibliography:

Fieser, L.F., and Fieser, M.: Organic Chemistry, 3rd Ed. Reinhold, 1956.
Kabat, E.A.: Blood Group Substances. Academic, 1956.
Meyer, K.: Advances in Protein Chemistry 2:249. Academic, 1945.
Percival, E.G.V.: Structural Carbohydrate Chemistry. Miller, 1953.
Pigman, W.W., and Goepp, R.M., Jr.: Chemistry of the Carbohydrates. Academic, 1948.
Stacey, M.: Advances in Carbohydrate Chemistry 2:161. Academic, 1946.

2...
Lipids

The lipids are a group of organic substances of fatty nature which are (1) insoluble in water; (2) soluble in fat solvents, such as ether, chloroform, and benzene; (3) related to the fatty acids as esters, either actually or potentially; and (4) utilizable in metabolism by living organisms. The lipids thus include fats, oils, waxes, and related compounds.

A lipoid is a fat-like substance which may not actually be related to the fatty acids. Occasionally the terms "lipid" and "lipoid" are synonymous.

An oil is a fat which is liquid at ordinary temperatures.

Lipids are important dietary constituents, not only because of their high energy value but also because of the vitamins and the essential fatty acids (see p. 203) which are associated with the fat of natural foods. In the body, fat serves as an efficient form of energy (both for direct use and for storage) and as an insulating material in the subcutaneous tissues and around certain organs. The fat content of nervous tissue is particularly high. Combinations of fat and protein (lipoproteins) are important cellular constituents, occurring both in the cell membrane and in the mitochondria within the cytoplasm.

Classification:

The following classification of lipids, proposed by Bloor, is generally accepted:

A. Simple Lipids: Esters of fatty acids with various alcohols.
   1. Fats - Esters of fatty acids with glycerol.
   2. Waxes - Esters of fatty acids with alcohols other than glycerol.

B. Compound Lipids: Esters of fatty acids containing groups in addition to an alcohol and the fatty acid.
   1. Phospholipids - Substituted fats containing, in addition to fatty acids and glycerol, a phosphoric acid residue, nitrogen-containing compounds, and other substituents. These lipids include:
      a. Lecithins (phosphatidyl choline).
      b. Cephalins (phosphatidyl ethanolamine).
      c. Lipositols (phosphatidyl inositol).
      d. Phosphatidyl serine.
      e. Plasmalogens (acetal phosphatide).
      f. Sphingomyelins (phosphatidyl sphingosides).
   2. Cerebrosides (glycolipids) - Compounds of the fatty acids with carbohydrate, containing nitrogen but no phosphoric acid.
   3. Other compound lipids, such as sulfolipids and aminolipids. Lipoproteins (see p. 132) may also be placed in this category.

C. Derived Lipids: Substances derived from the above groups by hydrolysis.
   1. Fatty acids, both saturated and unsaturated.
   2. Glycerol.
   3. Sterols and other sterids.
   4. Alcohols other than glycerol and sterols.
   5. Fatty aldehydes.
   6. Proteins of lipoproteins.

FATTY ACIDS

Fatty acids are obtained from the hydrolysis of fats. Fatty acids which occur in natural fats usually contain an even number of carbon atoms and are straight-chain derivatives. The chain may be saturated or unsaturated.

Saturated Fatty Acids (Those Containing No Double Bonds):

Saturated fatty acids are theoretically built up on acetic acid as the first member of the series. The general formula is $C_nH_{2n+1}COOH$. Examples of the acids in this series are as follows (the most important are in bold type):

Acid	Formula	Occurrence
Butyric	C_3H_7COOH	Present in certain fats in small amounts (especially butter).
Caproic	$C_5H_{11}COOH$	
Caprylic (octanoic)	$C_7H_{15}COOH$	In small amounts in many fats, especially those of plant origin.
Decanoic (capric)	$C_9H_{19}COOH$	
Lauric	$C_{11}H_{23}COOH$	Spermaceti, laurels.
Myristic	$C_{13}H_{27}COOH$	Nutmeg, wax myrtles.
Palmitic	$C_{15}H_{31}COOH$	Common in all animal and plant fats.
Stearic	$C_{17}H_{35}COOH$	
Arachidic	$C_{19}H_{39}COOH$	Peanut oil.

Other higher members of the series are known to occur, particularly in waxes. A few branched-chain fatty acids have also been isolated from both plant and animal sources.

Unsaturated Fatty Acids (Those Containing One or More Double Bonds):

These may be further subdivided as to degree of unsaturation.

A. Oleic Series: (One double bond. General formula: $C_nH_{2n-1}COOH$.)

Acid	Position	Formula	Occurrence
Crotonic	Δ-2*	$CH_3-CH=CH-COOH$	Croton oil.
Palmitoleic	Δ-9 (cis)	$CH_3(CH_2)_5CH=CH(CH_2)_7COOH$	Plant and animal fats and oils.
Oleic	Δ-9 (cis)	$CH_3(CH_2)_7CH=CH(CH_2)_7COOH$	
Erucic	Δ-13	$CH_3(CH_2)_7CH=CH(CH_2)_{11}COOH$	Rapeseed oil, nasturtium seed oil.

B. Linoleic Series: [Two double bonds between carbon atoms 9 and 10 and 12 and 13 (Δ9;12). General formula: $C_nH_{2n-3}COOH$.]

Acid	Position	Formula	Occurrence
Linoleic†	Δ9;12	$C_{17}H_{31}COOH$	Linseed oil, various plant and animal fats.

C. Linolenic Series: [Three double bonds (Δ9;12;15). General formula: $C_nH_{2n-5}COOH$.]

Acid	Position	Formula	Occurrence
Linolenic†	Δ9;12;15	$C_{17}H_{29}COOH$	Occurs with linoleic.

D. Fatty acids with four double bonds, $C_nH_{2n-7}COOH$, e.g., arachidonic† (C-20;Δ5;8;11;14); and **five double bonds,** $C_nH_{2n-9}COOH$, e.g., clupanodonic, are also known. Various other irregular structures, such as hydroxy groups (ricinoleic acid) or cyclic groups, have been found in nature. Examples of the latter are the chaulmoogric acids, which have been used in the treatment of leprosy.

---

*This refers to the position of the double bond. The carbon atoms are numbered to the left of the carboxyl carbon.

†Linoleic, linolenic, and arachidonic are the so-called "essential" fatty acids (EFA). See p. 203.

Chaulmoogric Acid

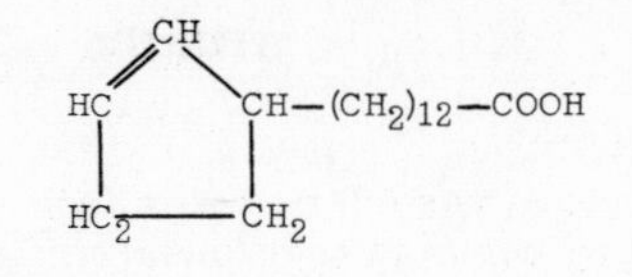

Isomerism in Unsaturated Fatty Acids:

A variation in the location of the double bonds along the unsaturated fatty acid chain produces an isomer of that compound. Oleic acid might thus form 16 different isomers.

Geometric isomerism also may exist, depending on the orientation of the molecules about the axis of the double bonds. The presence of double bonds in a molecule forms, in a sense, a rigid axis about which the rest of the molecule may be oriented. Some compounds may differ from one another only in the orientation of their parts about this axis. This is noteworthy in the chemistry of sterids. If the radicals which are being considered are on the same side of the bond, the compound is called a ''cis'' form; if on opposite sides, a ''trans'' form. This can be illustrated with oleic and elaidic acids or with fumaric and maleic acids.

$CH_3(CH_2)_7CH$ ‖ $HOOC(CH_2)_7CH$	$CH_3(CH_2)_7CH$ ‖ $CH(CH_2)_7COOH$	COOH | CH ‖ CH | COOH	COOH | HC ‖ CH | COOH
Cis Form (Oleic Acid)	Trans Form (Elaidic Acid)	Cis Form (Maleic Acid)	Trans Form (Fumaric Acid)

In acids with a greater degree of unsaturation there are, of course, more geometric isomers.

Alcohols:

Alcohols contained in the lipid molecule include glycerol, cholesterol, and the higher alcohols (e.g., cetyl alcohol, $C_{16}H_{33}OH$), usually found in the waxes. The presence of glycerol is indicated by the acrolein test:

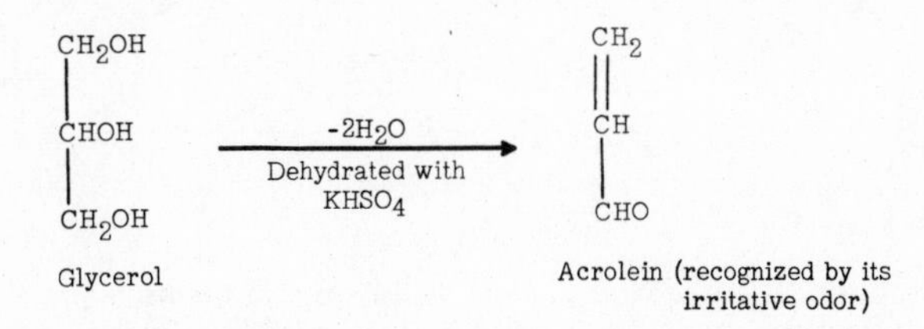

Among the unsaturated alcohols found in fats are a number of important pigments. These include phytol (phytyl alcohol), which is also a constituent of chlorophyll; and lycopin, the pigment of tomatoes. A hydrocarbon, carotene, is related to lycopin. Carotene is easily split in the body at the central point in the chain to form two molecules of an alcohol, vitamin A alcohol. (See p. 68.)

Fatty Aldehydes:

The fatty acids may be reduced to fatty aldehydes. These compounds are found either combined or free in natural fats.

STERIDS

The sterids are often found in association with fat. They may be separated from the fat after the fat is saponified, since they occur in the "unsaponifiable residue" (see p. 25). All of the sterids have a similar cyclic nucleus resembling phenanthrene. However, since the rings are not uniformly unsaturated, the parent (completely saturated) substance is better designated as cyclopentanoperhydrophenanthrene. The positions on the sterid nucleus are numbered as follows:

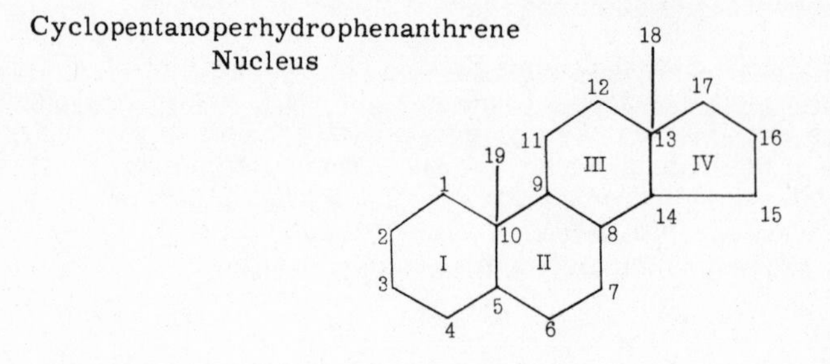

Methyl groups are frequently attached at positions 10 and 13 (constituting C atoms 19 and 18). A side chain at position 17 is usual (as in cholesterol; see below). If the compound has one or more hydroxyl groups and no carbonyl or carboxyl groups, it is a **sterol**, and the name terminates in -ol. If it has one or more carbonyl or carboxyl groups, it is a **steroid**.

Cholesterol:

Cholesterol is widely distributed in all cells of the body, but particularly in nervous tissue. It occurs in animal fats but not in plant fats. The metabolism of cholesterol is discussed on p. 213. Cholesterol is designated as 3-hydroxy-5, 6-cholestene.

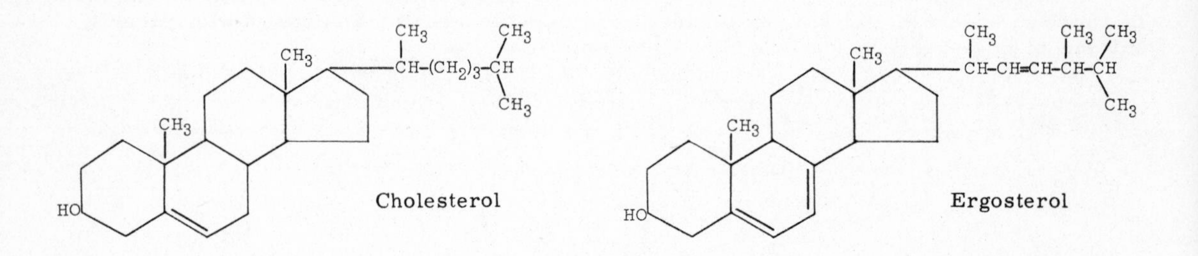

Ergosterol:

Ergosterol occurs in ergot and yeast. It is important as a precursor of vitamin D. When irradiated with ultraviolet light, it acquires antirachitic properties consequent to the opening of ring II. Ergosterol resembles cholesterol except that there are double bonds between 5 and 6 and between 7 and 8 and that the side chain attached at position 17 has a different structure (see above).

Coprosterol:

Coprosterol (coprostanol) occurs in feces as a result of the reduction of cholesterol by bacteria in the intestine. The double bond between 5 and 6 is saturated, and the orientation of rings I and II (between carbon atoms 5 and 10), which is **trans** in cholesterol, is **cis** in coprosterol.

Other Important Sterids:

These include the bile acids, sex hormones, adrenocortical hormones, D vitamins, cardiac glycosides, and the sitosterols of the plant kingdom.

Color Reactions to Detect Sterols:

Saturated sterols, like coprosterol, do not give these color tests.

The Liebermann-Burchard reaction. A chloroform solution of a sterol, when treated with acetic anhydride and sulfuric acid, gives a green color. The usefulness of this reaction is limited by the fact that various sterols give the same or a similar color. This reaction is the basis of the colorimetric test for blood cholesterol.

Salkowski test. A red to purple color appears when a chloroform solution of the sterol is treated with an equal volume of concentrated sulfuric acid.

Digitonin, $C_{56}H_{92}O$ (a glycoside occurring in digitalis leaves and seeds), precipitates cholesterol as the digitonide if the hydroxyl group in position 3 is free. This reaction serves as a method for the separation of free cholesterol and cholesterol esters.

NEUTRAL FATS (TRIGLYCERIDES) AND WAXES

Structurally, the lipids are esters. In the so-called neutral fats (triglycerides) the alcohol is glycerol, which is esterified with fatty acids. The esterified fatty acids may be alike or different.

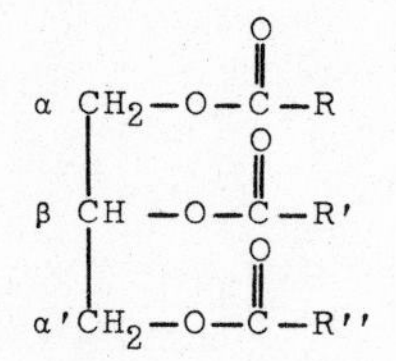

In the above example, if all three fatty acids were the same and if R was $C_{17}H_{35}$, the fat would be known as tristearin, since it consists of three stearic acid residues esterified with glycerol. In a mixed glyceride more than one fatty acid is involved, e.g.:

$CH_2-O-\overset{O}{\overset{\|}{C}}-C_{17}H_{35}$
$CH-O-\overset{O}{\overset{\|}{C}}-C_{15}H_{31}$
$CH_2-O-\overset{O}{\overset{\|}{C}}-C_{17}H_{35}$

α,α′-Distearopalmitin (Symmetrical)

$CH_2-O-\overset{O}{\overset{\|}{C}}-C_{17}H_{35}$
$CH-O-\overset{O}{\overset{\|}{C}}-C_{17}H_{35}$
$CH_2-O-\overset{O}{\overset{\|}{C}}-C_{15}H_{31}$

α,β- Distearopalmitin (Asymmetrical)

Waxes:

If the fatty acid is esterified with an alcohol of high molecular weight instead of with glycerol, the resulting compound is called a **wax**.

PHOSPHOLIPIDS[1]

The phospholipids include six groups: (1) lecithins, (2) cephalins, (3) lipositols, (4) phosphatidyl serine, (5) plasmalogens, and (6) sphingomyelins.

Lecithins (Phosphatidyl Choline):

The lecithins contain **glycerol** and **fatty acids**, as do the simple fats, but they also contain

phosphoric acid and **choline**. The lecithins are widely distributed in the cells of the body but are particularly important in the metabolism of fat by the liver (see p. 205).

$$HOCH_2-CH_2-\overset{+}{N}(CH_3)_3 \quad OH^-$$

Choline
(One of the Constituents of Lecithins)

Example of an α lecithin:

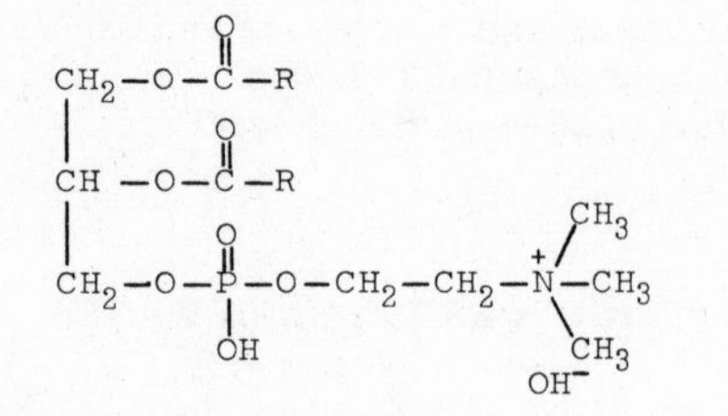

The β lecithin contains the phosphoric acid–choline complex on the center or β carbon of glycerol.

<u>Cephalins</u> (Phosphatidyl Ethanolamine):

The cephalins differ from lecithins only in that ethanolamine replaces choline. Both α and β cephalins are also known.

Example of an α cephalin:

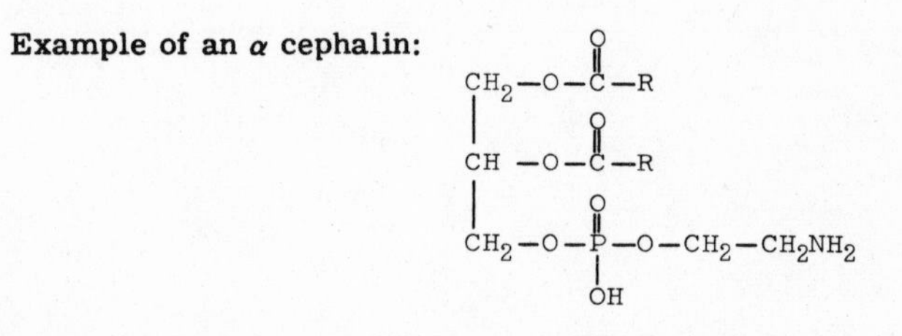

<u>Lipositols</u> (Phosphatidyl Inositol):

Inositol (see p. 95) as a constituent of lipids was first discovered in acid-fast bacteria. Later it was found to occur in phospholipids of soybean and of brain tissue as well as in other plant phospholipids. Hawthorne and Chargaff[2] prepared and analyzed the inositol phosphatides of soybean and of ox brain. The hydrolysis products of the soybean preparation included, in addition to inositol monophosphate, a mixture of other phosphoric acid esters thought to be the galactoside and arabinoside of inositol monophosphate.

The structure of an inositol phospholipid obtained from commercial soybean lecithin is as follows:

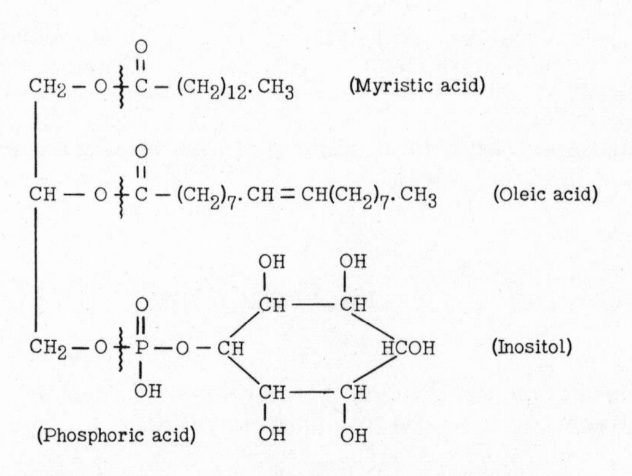

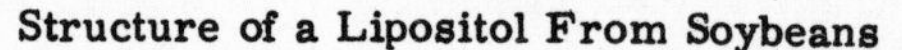

Structure of a Lipositol From Soybeans

Phosphatidyl Serine:

A cephalin-like phospholipid which contains the amino acid serine rather than ethanolamine has been found in brain tissue. Traces of this phospholipid, phosphatidyl serine, have also been detected in the blood[3].

```
                  O
                  ‖
CH2 — O — C — R
 |
 |                O
 |                ‖
CH —— O — C — R'
 |
 |                O
 |                ‖
CH2 —— O — P — O —|— CH2.CHNH2COOH
                  |
                  OH        (Serine)
```

Plasmalogens:

These compounds constitute as much as 10% of the phospholipids of the brain and muscle. Structurally, the plasmalogens resemble lecithins and cephalins but give a positive reaction when tested for aldehydes with Schiff's reagent (fuchsin-sulfuric acid) after pretreatment of the phospholipid with mercuric chloride. It was originally suggested (Thannhauser) that a long chain "fatty aldehyde" (designated >CH.R in formula at left below) was attached by an acetal linkage to the α and β hydroxy groups of glycerol. Hence the term "acetal phosphatide" was used to describe these compounds.

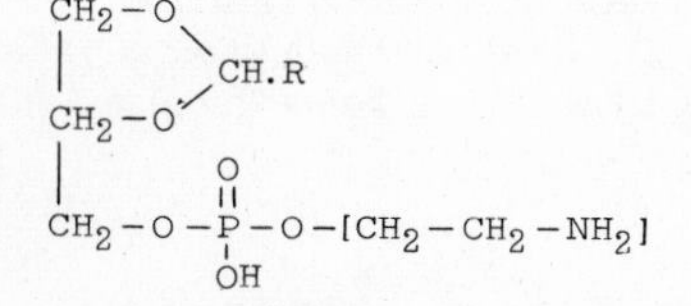

Structure of Acetal Phosphatide

```
                      OH
                      |
CH2 — O — CH — R'
 |
 |                    O
 |                    ‖
CH —— O — C — R''
 |
 |                    O
 |                    ‖
CH2 — O — P — O — [CH2 — CH2 — NH2]
                      |
                      OH
```

Structure of Plasmalogen

The portion of the above formulas shown in brackets is an ethanolamine residue. In some cases choline may be substituted.

More recent work has suggested that the acetal structure is an artefact, produced during isolation of the compound. According to these newer investigations two alkyl (R) groups are present, one in a fatty acyl group linked in an ester, the other as part of an ether linkage rather than a hemi acetal (see formula at right, above).

Sphingomyelins (Phosphatidyl Sphingosides):

Sphingomyelins are found in large quantities in brain and nervous tissue (see Chapter 22). There is no glycerol present. On hydrolysis they yield a fatty acid, phosphoric acid, choline, and a complex amino alcohol, sphingol (or sphingosine). The structure of sphingol:

$$CH_3(CH_2)_{12}CH{=}CHCHOH{-}CHNH_2{-}CH_2OH$$

The structure of a sphingomyelin:

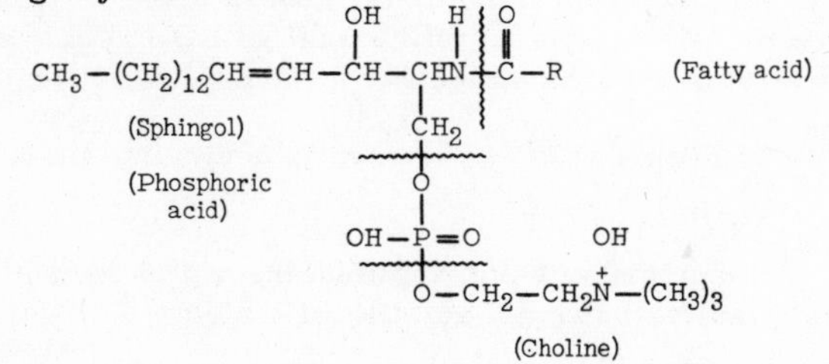

CEREBROSIDES (GLYCOLIPIDS)

Cerebrosides contain galactose, a high molecular weight fatty acid, and sphingol. Individual cerebrosides are differentiated by the kind of fatty acid in the molecule. Four have been isolated and their fatty acids identified. These are **kerasin**, containing normal lignoceric acid ($C_{23}H_{47}COOH$); **cerebron** (phrenosin), with a hydroxy lignoceric acid (cerebronic acid); **nervon**, containing an unsaturated homologue of lignoceric acid ($C_{23}H_{45}COOH$) called nervonic acid; and **oxynervon**, having apparently the hydroxy derivative of nervonic acid as its constituent fatty acid.

The cerebrosides are found in many tissues besides brain. In Gaucher's disease (a hepatosplenomegaly) the kerasin content of the spleen is very high (see p. 219). The cerebrosides are in much higher concentration in medullated than in nonmedullated nerve fibers.

The structure of cerebroside sulfuric ester (beef brain) is as follows:

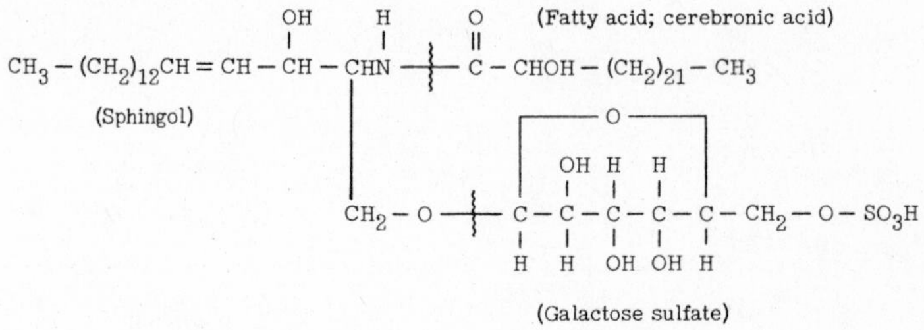

```
                         OH     H         O      (Fatty acid; cerebronic acid)
                         |      |    }    ||
CH3 - (CH2)12CH = CH - CH - CHN --+-- C - CHOH - (CH2)21 - CH3
                                |
          (Sphingol)            |       +--------- O ---------+
                                |       |                     |
                                |       |  OH  H   H          |
                                |       |  |   |   |          |
                               CH2 - O --+-- C - C - C - C - C - CH2 - O - SO3H
                                         |   |   |    |    |
                                         H   H   OH   OH   H

                                           (Galactose sulfate)
```

Gangliosides are glycolipids, occurring in the brain, which contain **neuraminic acid** in addition to C_{24} or C_{22} fatty acids, sphingosine, and three molecules of a hexose (glucose and galactose). Neuraminic acid is a substance structurally similar to the sialic acid of mucopolysaccharides (see p. 16). A hexosamine may be substituted for neuraminic acid in some gangliosides or, in other instances, both may be present.

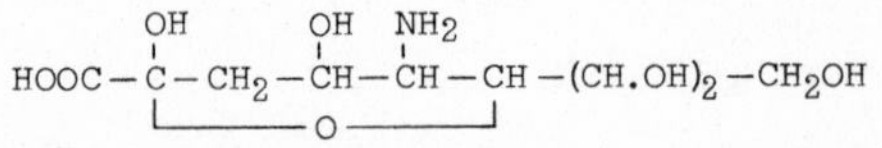

```
        OH          OH   NH2
        |           |    |
HOOC - C - CH2 - CH - CH - CH - (CH.OH)2 - CH2OH        Neuraminic Acid
        |_________ O _________|
```

CHARACTERISTIC CHEMICAL REACTIONS AND PROPERTIES OF THE LIPIDS

Hydrolysis:

Hydrolysis of a lipid may be accomplished enzymatically through the action of **lipases**, yielding fatty acids and glycerol. Lecithinases and cerebrosidases attack ester linkages in lecithins and cerebrosides.

Saponification:

Hydrolysis of a fat by alkali is called **saponification**. The resultant products are glycerol and the alkali salt of the fatty acids, which are called **soaps**. Acid hydrolysis of a fat yields the free fatty acids and glycerol. Soaps are cleansing agents because of their emulsifying action. Some soaps of high molecular weight and a considerable degree of unsaturation are selective germicides. Others, such as sodium ricinoleate, have detoxifying activity against diphtheria and tetanus toxins.

Analytic Methods for the Characterization of Fats:

These include a determination of melting point, solidification temperature, and refractive index, as well as certain **chemical determinations**. These are:

A. Saponification Number: The number of milligrams of KOH required to saponify 1 Gm. of fat or oil. It varies inversely with the molecular weight of the fat or oil.

B. Acid Number: The milligrams of KOH required to neutralize the free fatty acid of 1 Gm. of fat.

C. Polenske Number: The number of cubic centimeters of 0.1 normal KOH required to neutralize the insoluble fatty acids (those not volatile with steam distillation) from 5 Gm. of fat.

D. Reichert-Meissl Number: This is the same as the Polenske number except that after a 5 Gm. sample of the fat has been saponified, the **soluble** fatty acids are measured by titration of the distillate obtained by steam distillation of the saponification mixture.

E. Iodine Number: In the presence of iodine monobromide (Hanus method) or of iodine monochloride (Wijs method), unsaturated lipids will take up iodine. The iodine number is the amount (in grams) of iodine absorbed by 100 Gm. of fat. This is a measure of the degree of unsaturation of a fat. Oils like linseed or cottonseed oil have higher iodine numbers than solid fats like tallow or beef fat because the former contain more unsaturated fatty acids in the fat molecule.

F. Acetyl Number: The amount in milligrams of KOH required to neutralize the acetic acid obtained by saponification of 1 Gm. of fat after it has been acetylated. This is a measure of the number of hydroxy-acid groups in the fat. Castor oil, because of its high ricinoleic acid content, a fatty acid containing one OH group, has a high acetyl number (about 146).

Chromatographic technics (see p. 36) have recently been utilized very successfully to separate and identify the fatty acids of the lipids. These methods include the application of gas-liquid chromatography to the separation of the methyl esters of fatty acids[4a,b]. The countercurrent distribution technic of Craig (see p. 39) has also been used to separate and estimate individual fatty acids quantitatively.

Unsaponifiable Matter:

Unsaponifiable matter includes substances in natural fats which cannot be saponified by alkali but are soluble in ether or petroleum ether. Consequently they may be separated from lipid mixtures by extraction with these solvents following saponification of the fat since soaps are not ether-soluble. Ketones, hydrocarbons, high-molecular-weight alcohols, and the sterids are examples of unsaponifiable residues of natural fats.

Hydrogenation:

Hydrogenation of unsaturated fats in the presence of a catalyst (nickel) is known as "hardening." It is commercially valuable as a method of converting these fats, usually of plant origin, into solid fats as lard substitutes or margarines.

Rancidity:

Rancidity is a chemical change which results in unpleasant odors and taste in a fat. The oxygen of the air is believed to attack the double bond to form a peroxide linkage. The iodine number is thus reduced, although little free fatty acid and glycerol are released. Lead or copper catalyze rancidity; exclusion of oxygen delays the process.

Spontaneous Oxidation:

Oils which contain highly unsaturated fatty acids (e.g., linseed oil) are spontaneously oxidized by atmospheric oxygen at ordinary temperatures and form a hard, waterproof material. Such oils are added for this purpose to paints and shellacs. They are then known as "drying oils."

• • •

References:

1. Celmer, W.D., and Carter, H.E.: Physiol. Rev. **32**:167, 1952.
2. Hawthorne, J.N., and Chargaff, E.: J. Biol. Chem. **206**:27, 1954.
3. Artom, C.: J. Biol. Chem. **157**:595, 1945.
4. a. James, A.T., and Martin, A.J.P.: Biochem. J. **63**:144, 1956.
   b. James, A.T., and Webb, J.: Ibid. **66**:515, 1957.

Bibliography:

Bloor, W.R.: Biochemistry of the Fatty Acids. Reinhold, 1943.
Bull, H.B.: Biochemistry of the Lipids. Wiley, 1937.
Deuel, H.J., Jr.: The Lipids, Their Chemistry and Biochemistry. 3 Vols. Interscience, 1951-57.
Ralston, A.W.: Fatty Acids and Their Derivatives. Wiley, 1948.
Sobotka, H.: Chemistry of the Sterids. Williams and Wilkins, 1938.

3...
Proteins

Proteins are organic substances of high molecular weight formed by a number of amino acids united by a peptide linkage. Non-amino groups may be present. Like the fats and carbohydrates, the proteins are composed of carbon, hydrogen, and oxygen but differ from typical examples of these foodstuffs in that they also contain nitrogen and often sulfur. On complete hydrolysis they yield various crystalline alpha-amino acids. Proteins are amphoteric, i.e., they possess both acid and basic properties, depending upon the reaction of the solution. They usually form colloidal solutions (emulsoids) in water. Separation from solution is accomplished by precipitation in the presence of electrolytes or at low temperatures with alcohol at varying H^+ concentrations. Physical technics using centrifugal methods or electrophoresis may also be used.

The proteins are the basic components of protoplasm in the cells of animals and plants. For this reason they are probably the most important of all biologic materials since they serve as the only source of the "building blocks" required for the synthesis of the structural constituents of the body. Many hormones and all of the known enzymes are proteins.

Classification:

The system of classification of proteins given below is based mainly on solubility reactions and only in part on composition. In fact, the system is now falling into disuse because present knowledge of the composition of proteins, while still inadequate, nevertheless indicates that many of the proteins have been erroneously categorized. For example, albumins and globulins cannot be rigorously delineated by solubility in water or in salt solutions, respectively, because one group of proteins (pseudoglobulins) possesses some of the other characteristics of globulins but is soluble in water (whereas the euglobulins do not dissolve in salt-free water). Many of the so-called "simple proteins" have been found to be associated with nonprotein substances such as the carbohydrate component of egg albumin. The histones and protamines might more properly be classified as "derived" rather than "simple" proteins. However, until much more information on protein composition and structure has been obtained, it is likely that the terminology given below will continue to be used.

A. Simple Proteins: These contain only alpha-amino acids or their derivatives and occur as such in nature. Simple proteins include the following groups:
   1. Albumins - Soluble in water, coagulated by heat, precipitated by saturated salt solutions. Examples: lactalbumin, serum albumin.
   2. Globulins - Soluble in dilute salt solutions of the strong acids and bases; insoluble in pure water or in moderately concentrated salt solutions. Examples: serum globulin, ovoglobulin.
   3. Glutelins - Soluble in dilute acids and alkalies; insoluble in neutral solvents. Example: glutenin from wheat.
   4. Prolamines - Soluble in 70 to 80% alcohol; insoluble in absolute alcohol, water, and other neutral solvents. Examples: zein (corn) and gliadin (wheat).
   5. Albuminoids (scleroproteins) - Essentially the same characteristics as other simple proteins, but insoluble in all neutral solvents and in dilute acids and alkalies. These are the proteins of supportive tissue. Examples: keratin, collagen. Gelatin is a degenerated albuminoid.
   6. Histones - Soluble in water and very dilute acids; insoluble in very dilute NH_4OH; not coagulated by heat. Basic amino acids predominate in the hydrolytic products. Example: globins.
   7. Protamines - Basic polypeptides, soluble in water or in NH_4OH, not coagulated by heat; predominate in basic amino acids and precipitate other proteins. Found principally in cells. Examples: salmine (salmon) and sturine (sturgeon).

B. Conjugated Proteins: Those which are united with some nonprotein substance (the prosthetic group) by other than a salt linkage.
   1. Nucleoproteins - Compounds of one or of several molecules of proteins with nucleic acid. Examples: nuclein, nucleohistone from nuclei-rich material (glands).

2. Glycoproteins and mucoproteins - Compounds with carbohydrate prosthetic groups (mucopolysaccharides) which on hydrolysis yield amino sugars (hexosamines) and uronic acids. The distinction between glycoproteins and mucoproteins may be based on the amount of carbohydrate: glycoproteins contain less than 4% of carbohydrate in the molecule, and mucoproteins contain more than 4%. Examples include the protein, mucin, and proteins in the plasma which migrate electrophoretically with the α_1 and α_2 fractions.
3. Phosphoproteins - Compounds with a phosphorus-containing radical other than a phospholipid or nucleic acid. Example: casein.
4. Chromoproteins - Compounds with a chromophoric group in conjugation. Examples: hemoglobin, hemocyanin, cytochrome, flavoproteins.
5. Lipoproteins - Conjugated to neutral fats (triglycerides) or other lipids such as phospholipid and cholesterol.

C. Derived Proteins: This category was originally devised to include the artificially synthesized protein-like compounds and those resulting from the decomposition of proteins. Those proteins which may be isolated after removal of the nonprotein prosthetic groups of conjugated proteins might also be included here rather than among the simple proteins (e.g., protamines and histones from nucleoproteins or from hemoglobin).
1. Primary derivatives - Coagulated proteins and those formerly called proteans and metaproteins. Most of these compounds are denaturation products of proteins resulting from the action of heat, alcohol, or acids and alkalies. Their chemical nature is poorly understood, and the tendency at present is not to classify them as proteins.
2. Secondary derivatives - Products of partial to complete hydrolysis of proteins.
   a. Proteoses - Products of the hydrolytic decomposition of proteins. Soluble in water, precipitated in saturated salt solution, not coagulated by heat.
   b. Peptones - Products of further hydrolytic decomposition. Soluble in water and saturated salt solutions, not coagulated by heat.
   c. Peptides - Compounds of two or more amino acids, either synthesized or resulting from hydrolysis of proteins. Soluble in water and salt solutions, not coagulated by heat, may not give biuret reaction (see p. 42).
   d. Diketopiperazines - Cyclic anhydrides of two amino acids.

AMINO ACIDS[1]

The alpha-amino acids are the structural elements of the proteins and are obtained from them when they are hydrolyzed. All of the known amino acids are capable of forming isomers except glycine, which has no asymmetric carbon atom*. The majority of the amino acids found in nature within the intracellular proteins are of the L configuration. However, D-amino acids have been identified in natural materials, particularly in the products of the activity of microorganisms (e.g., antibiotics).

Serine is an amino acid which serves as the reference compound for the assignment of an amino acid to the D or to the L series; the position of the alpha-amino group is the point of reference, as was the penultimate OH group on the carbohydrate molecule (see p. 2). The symbols (+) or (-) are used to designate direction of rotation of polarized light and hence optical isomerism (see p. 3).

*Threonine has two asymmetric carbon atoms, resulting in the formation of four isomers, two of which are referred to as forms of "allothreonine."

Neutral Amino Acids - Monoamino-monocarboxylic Amino Acids:

A. Aliphatic:

1. Glycine	$C_2H_5O_2N$	aminoacetic acid	$NH_2CH_2.COOH$
2. L-Alanine	$C_3H_7O_2N$	α-aminopropionic acid	$CH_3.CHNH_2.COOH$
3. L-Serine	$C_3H_7O_3N$	β-hydroxy-α-aminopropionic acid	$CH_2OH.CHNH_2.COOH$
4. L-Threonine	$C_4H_9O_3N$	β-hydroxy-α-aminobutyric acid	$CH_3.CHOH.CHNH_2.COOH$
5. L-Valine	$C_5H_{11}O_2N$	α-aminoisovaleric acid	$(H_3C)_2$ CH. $CHNH_2$. COOH
6. L-Leucine	$C_6H_{13}O_2N$	α-aminoisocaproic acid	$(H_3C)_2$ CH. CH_2. $CHNH_2$. COOH
7. L-Isoleucine	$C_6H_{13}O_2N$	β-methyl-α-aminovaleric acid	$CH_3.CH_2.CH(CH_3).CHNH_2.COOH$

B. Aromatic:

1. L-Phenylalanine $C_9H_{11}O_2N$ β-phenyl-α-aminopropionic acid

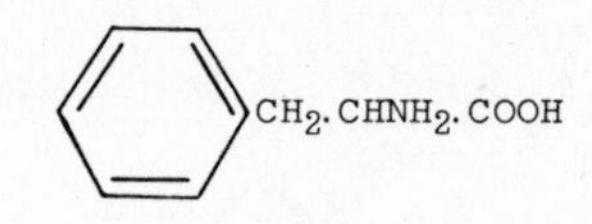

2. L-Tyrosine $C_9H_{11}O_3N$ β-parahydroxyphenyl-α-aminopropionic acid

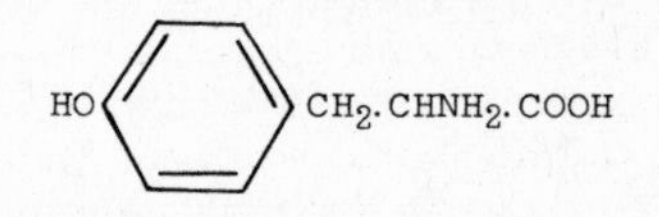

C. Sulfur-containing:

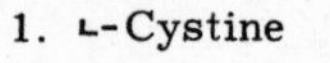

1. L-Cystine $C_6H_{12}O_4N_2S_2$ di-(β-thio-α-aminopropionic acid)

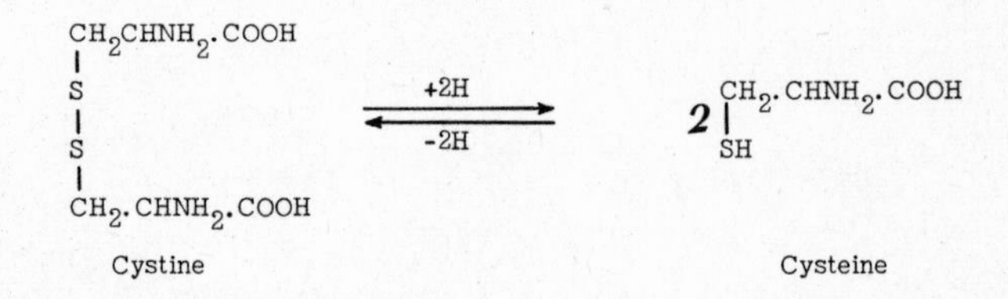

Cysteine, $C_3H_7O_2NS$ (β-thio-α-aminopropionic acid), is often referred to as "half cystine." It is readily produced by the reduction of cystine.

2. L-Methionine $C_5H_{11}O_2NS$ γ-methylthiol-α-amino-n-butyric acid

$CH_3.S.CH_2.CH_2.CHNH_2.COOH$

D. Iodine-containing:

1. L-Iodogorgoic acid $C_9H_9O_3NI_2$ 3, 5-diiodotyrosine

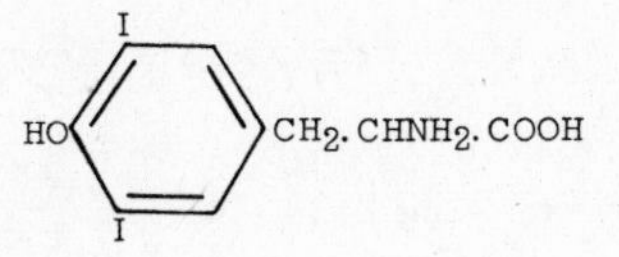

2. L-Triiodothyronine $C_{15}H_{12}O_4NI_3$ α-[3, 5-diiodo-4-(3′-iodo-4′-hydroxyphenoxy) phenyl-alanine]

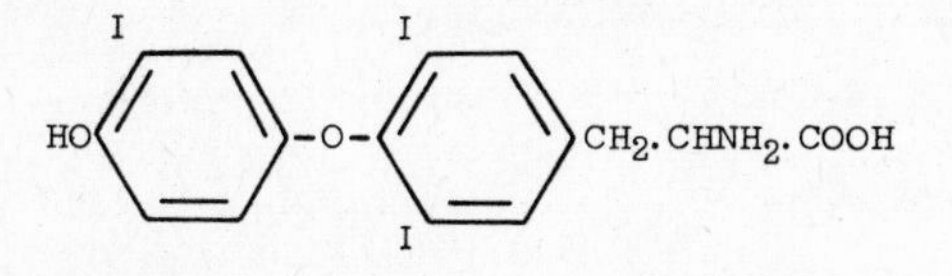

3. L-Thyroxine* $C_{15}H_{11}O_4NI_4$ α-[3, 5-diiodo-4-(3′, 5′, -diiodo-4′-hydroxyphenoxy) phenyl-alanine]

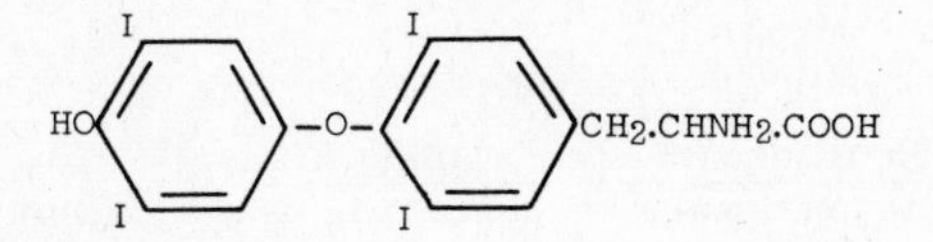

E. Heterocyclic:

1. L-Tryptophan 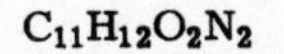 $C_{11}H_{12}O_2N_2$ β, 3-indol-α-aminopropionic acid

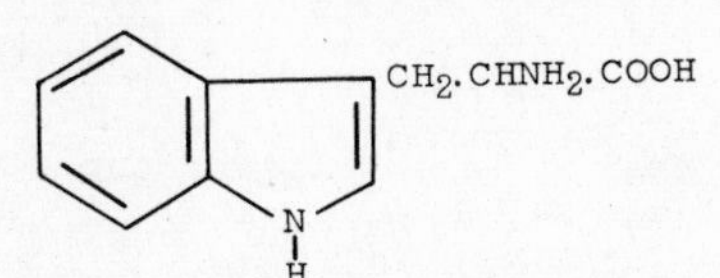

2. L-Proline 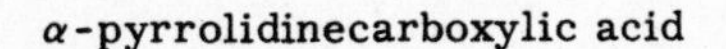 $C_5H_9O_2N$ α-pyrrolidinecarboxylic acid

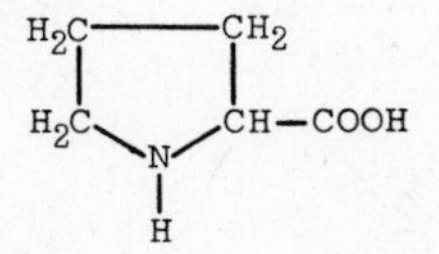

*The completely de-iodinated thyroxine is referred to as a **thyronine**. Thus thyroxine itself might be termed tetraiodothyronine.

3. L-Hydroxyproline $C_5H_9O_3N$ γ-hydroxy-α-pyrrolidinecarboxylic acid

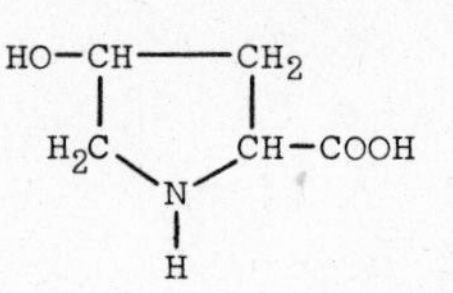

4. L-Histidine $C_6H_9O_2N_3$ β, 4-imidazolyl-α-aminopropionic acid

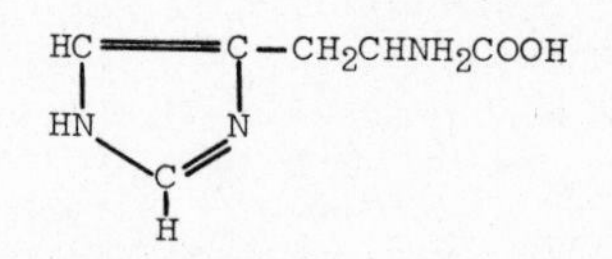

Acid Amino Acids - Monoamino-dicarboxylic Amino Acids:

A. L-Aspartic acid $C_4H_7O_4N$ α-amino succinic acid $HOOC.CH_2.CHNH_2.COOH$

B. L-Glutamic acid $C_5H_9O_4N$ α-amino glutaric acid $HOOC.CH_2.CH_2.CHNH_2.COOH$

These amino acids also occur in proteins as the mono amides, asparagine and glutamine.

$$H_2N.\overset{O}{\overset{\|}{C}}.CH_2.CHNH_2.COOH$$

Asparagine

$$H_2N.\overset{O}{\overset{\|}{C}}.CH_2.CH_2.CHNH_2.COOH$$

Glutamine

Basic Amino Acids - Diamino-monocarboxylic Amino Acids:

The first three are the hexone bases, each containing six carbon atoms.

A. L-Arginine $C_6H_{14}O_2N_4$ δ-guanidyl-α-aminovaleric acid

$$H_2N.\underset{\underset{NH}{\|}}{C}.NH.CH_2.CH_2.CH_2CHNH_2COOH$$

B. L-Lysine $C_6H_{14}O_2N_2$ α-ε-diaminocaproic acid

$$H_2NCH_2.CH_2.CH_2.CH_2.CHNH_2.COOH$$

C. L-Hydroxylysine $C_6H_{14}O_3N_2$ α-ε-diamino-5-hydroxycaproic acid

$$H_2NCH_2.CHOH.CH_2.CH_2.CHNH_2.COOH$$

D. L-Ornithine* $C_5H_{12}O_2N_2$ α-δ-diaminovaleric acid $\underset{NH_2}{\underset{|}{CH_2}}.CH_2.CH_2.\underset{NH_2}{\underset{|}{CH}}.COOH$

*These amino acids are probably not constituents of the protein molecule. They are, however, important in metabolism as compounds involved in the urea cycle. (See p. 232.)

E. L-Citrulline* $C_6H_{13}O_3N_3$ δ-carbamido-α-aminovaleric acid

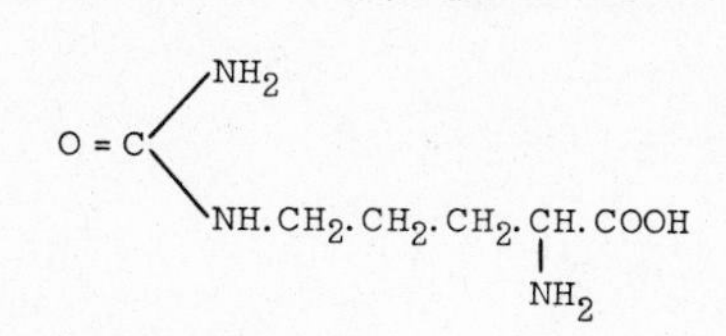

Amphoteric Properties of Amino Acids:

The properties of amino acids include ionization by the formation of "zwitterions," or ampholytes (amphoteric electrolytes). They dissociate both as acids and as bases, depending upon the pH of the solution. If an amino acid in solution is placed in an electrical field, the molecules will also migrate to one pole or the other in accordance with the pH of the solution. At a given pH the amino acid behaves neither as an acid nor as a base and does not migrate to anode or cathode. This is called the **iso-electric point.** Whole proteins also exhibit an iso-electric point (see Precipitation Reactions of Proteins, p. 41). It is usually expressed in terms of the pH of the solution at which this occurs. In the iso-electric state the amino acid is thought to be dissociated both as an acid and as a base and fails to appear electrically charged because its positive and negative charges are equal. When acid is added, an amino acid behaves as a base; when alkali is added, it behaves as an acid. Thus amino acids may have more than one dissociation constant (pK) to express the relative dissociation power of the acidic and basic groups.

Example: Glycine: $pK_1 = 2.42$
$pK_2 = 9.47$

Iso-electric point: pH = 6.10

CHEMISTRY OF THE AMINO ACIDS

IMPORTANT GENERAL REACTIONS OF AMINO ACIDS

Formation of a Salt with Acids and Bases:

$$R-CHNH_2COOH + HCl \longrightarrow RCH\overset{HCl}{\overset{|}{N}}H_2COOH$$

$$R-CHNH_2COOH + NaOH \longrightarrow R-CHNH_2COONa + H_2O$$

Formation of Esters with Alcohols:

$$RCHNH_2COOH + R'OH \longrightarrow RCHNH_2COOR' + H_2O$$

Methylation:

$$R-\overset{NH-CH_3}{\overset{|}{CH}}-COOH \quad \text{or} \quad R-\overset{N(CH_3)_2}{\overset{|}{CH}}-COOH$$

*These amino acids are probably not constituents of the protein molecule. They are, however, important in metabolism as compounds involved in the urea cycle. (See p. 232.)

Acetylation with Acetyl Chloride or Acetic Anhydride:

$$\begin{array}{l} R-CH-COOH \\ \quad\;\; | \\ \quad\;\; NH-\underset{\underset{O}{\|}}{C}-CH_3 \end{array}$$

Acetylated or chloracetylated racemic amino acids are asymmetrically attacked by various pancreatic, bile, and kidney enzymes. This has been utilized by Greenstein as an effective means for resolution of racemic (synthetic) amino acids into their optically active components.

Acid Chloride Formation: (The amino group is first masked by acetylation.)

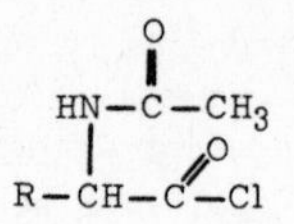

Primary Amine Formation:
By heating with $Ba(OH)_2$ decarboxylation occurs.

$$RCHNH_2\boxed{COO}H \longrightarrow RCH_2NH_2 + CO_2$$

Dehydration to Form Diketopiperazines:
These may then be hydrolyzed to dipeptides.

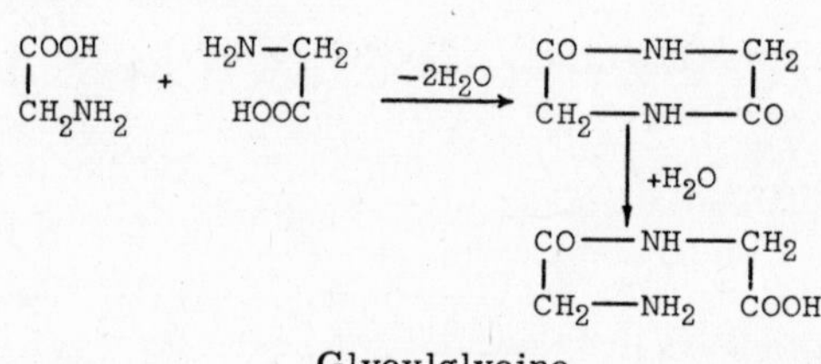

Glycylglycine

Formation of DNP (Dinitrophenyl) Derivatives: (See p. 39.)
In mildly alkaline solutions, fluorodinitrobenzene reacts with amino acids to form yellow derivatives which can be extracted from a mixture by the use of ether or chloroform. This technic is useful for chromatographic separation of amino acids.

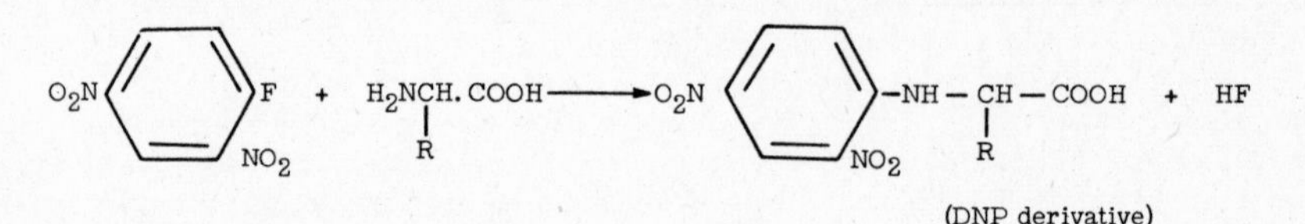

Van Slyke Nitrous Acid Reaction:
The liberated nitrogen may be measured to estimate free amino groups in proteins or in amino acids themselves.

$$RCHNH_2COOH + HONO \longrightarrow RCHOHCOOH + N_2 + H_2O$$

Sørensen Formol Titration:

Formaldehyde masks the amino group by forming an amino acid–formaldehyde complex and thus permits titration of free carboxyl groups. This reaction is useful in following the course of hydrolysis in proteins.

Ninhydrin Reaction:

A blue compound is formed when ninhydrin is heated with alpha-amino acids or with peptides or proteins which contain at least one free amino and carboxyl group. This color reaction is used to identify such compounds in chromatographic analysis.

Proline and hydroxyproline which contain the **imino** (-NH-) group rather than the amino group form condensation products with ninhydrin, which have a red color that rapidly changes to yellow. The red colors may be used for a specific colorimetric determination of these two amino acids. The reaction with ninhydrin can be stopped after the formation of the red color complexes by extraction with benzene, in which the other amino acid–ninhydrin complexes are insoluble.

The reaction with ninhydrin is believed to be more specific than the nitrous acid reaction. It is now applied to blood, urine, and other fluids for measurement of alpha-amino nitrogen[2]. When amino acids are heated with ninhydrin, carbon dioxide is given off from any carboxyl group adjacent to a free amino group. The liberated carbon dioxide is quantitatively measured in the manometric apparatus of Van Slyke and Neill.

The reactions of an amino acid with ninhydrin are illustrated on p. 35.

CHEMISTRY OF THE PROTEINS

Structure of Proteins:

Several methods have been used to determine the external structure of a protein molecule:

Streaming birefringence or double refraction of flow. A beam of light is passed through a polarizing lens. The polarized light is then passed through a solution of a protein and, finally, through a second polarizing lens which is oriented at right angles to the first lens. No light will emerge from the second lens if the protein solution does not affect the polarized light. This is the case with spherical protein molecules. It is also true of fibrous molecules when the solution is at rest because the protein molecules are then randomly distributed. However, when a solution of a fibrous protein is put in motion, the elongated molecules arrange themselves lengthwise in the axis of the stream and thus act as of another polarizer were added to the system. As a result the polarized light does pass through the second lens. This phenomenon is referred to as "streaming birefringence" or "double refraction of flow." It has been used to calculate the "axial ratio" of a fibrous protein, i.e., the ratio of the length of the long axis to that of the short axis. An example is fibrinogen, which has a calculated axial ratio of 20:1 (see p. 128).

X-ray diffraction. A single crystal of a protein or layers of protein or protein fibers will deflect x-rays, and the resultant image on a photographic plate can be analyzed to yield information on the crystal or on the structure of the fiber.

Electron photomicrography. An actual picture of very small objects can be obtained with the electron microscope. Magnifications as high as 100,000 diameters can be obtained with this instrument. This permits the visualization of proteins of high molecular weight, such as virus particles.

Studies on the shape of proteins indicate that there are two general types in nature: globular proteins, which have an axial ratio (length:width) of less than 10; and fibrous proteins, with axial ratios greater than 10. **Keratin,** the protein of hair, wool, and skin, is a typical fibrous protein. It consists of a long peptide chain (see below) or groups of such chains. The peptide chains may be coiled in a spiral or helix formation and cross-linked by S-S bonds (see p. 35). The coiled form is referred to as α-keratin. It changes to β-keratin by unfolding.

Myosin, the major protein of muscle, is also a fibrous protein which undergoes a change in its structure during muscle contraction and relaxation. The following diagrams illustrate these types of structures.

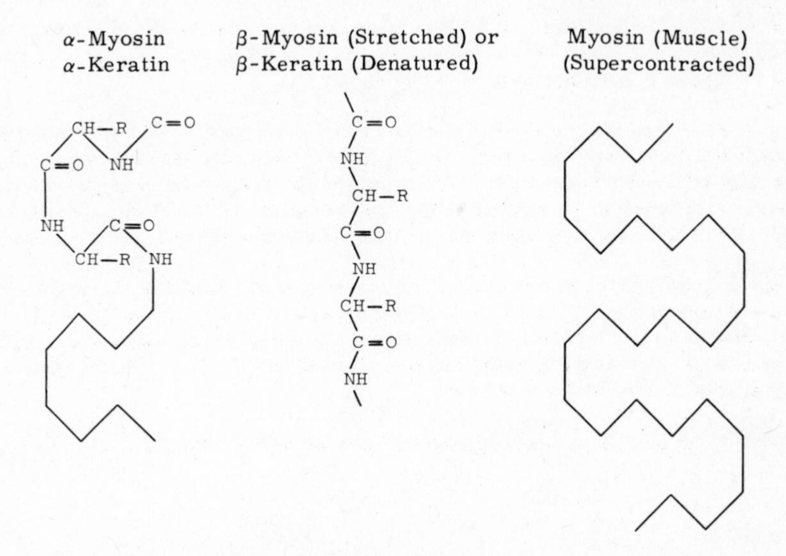

Globular proteins are characterized by the presence of peptide chains which are folded or coiled in a very compact manner. Axial ratios are usually not over 3 or 4. Examples of globular proteins are found among the fractions of the albumins and globulins in the plasma. **Insulin** is another globular protein.

The Peptide Linkage:

The basic structure of the protein molecule is the peptide linkage, i. e., a union between the carboxyl group of one amino acid and the α-amino group of another. An example of a tetrapeptide in which four amino acids are united by a peptide linkage is shown below.

$$H_2N-\underset{}{\overset{R}{CH}}-\underset{O}{\overset{}{C}}\,\}\,NH-\underset{R}{CH}-\overset{O}{C}\,\}\,NH-\underset{H}{\overset{R}{C}}-\underset{O}{C}\,\}\,NH-\underset{R}{\overset{COOH}{CH}}$$

The remainder of the amino acid molecules (designated above as R) form side chains. The presence of unbound carboxyl or amino groups (from dicarboxylic amino acids such as aspartic or glutamic, or basic amino acids such as arginine or lysine) makes the molecule more hydrophilic than one without such groups on the side chains.

Long chains of amino acids linked by the peptide bond are called **polypeptides.** The polypeptide chains are maintained by various mechanisms within the protein molecule. An important example is the disulfide bond, which may be used to interconnect two parallel chains through cysteine residues within each polypeptide. This bond is relatively stable and thus is not readily broken under the usual conditions of denaturation. Performic acid treatment oxidizes the S-S bonds. This reagent is used, for example, to oxidize insulin in order to separate the protein molecule into its constituent polypeptide chains without affecting the other parts of the molecule (see p. 39). The union of two parallel peptide chains by an S-S linkage is illustrated on p. 35.

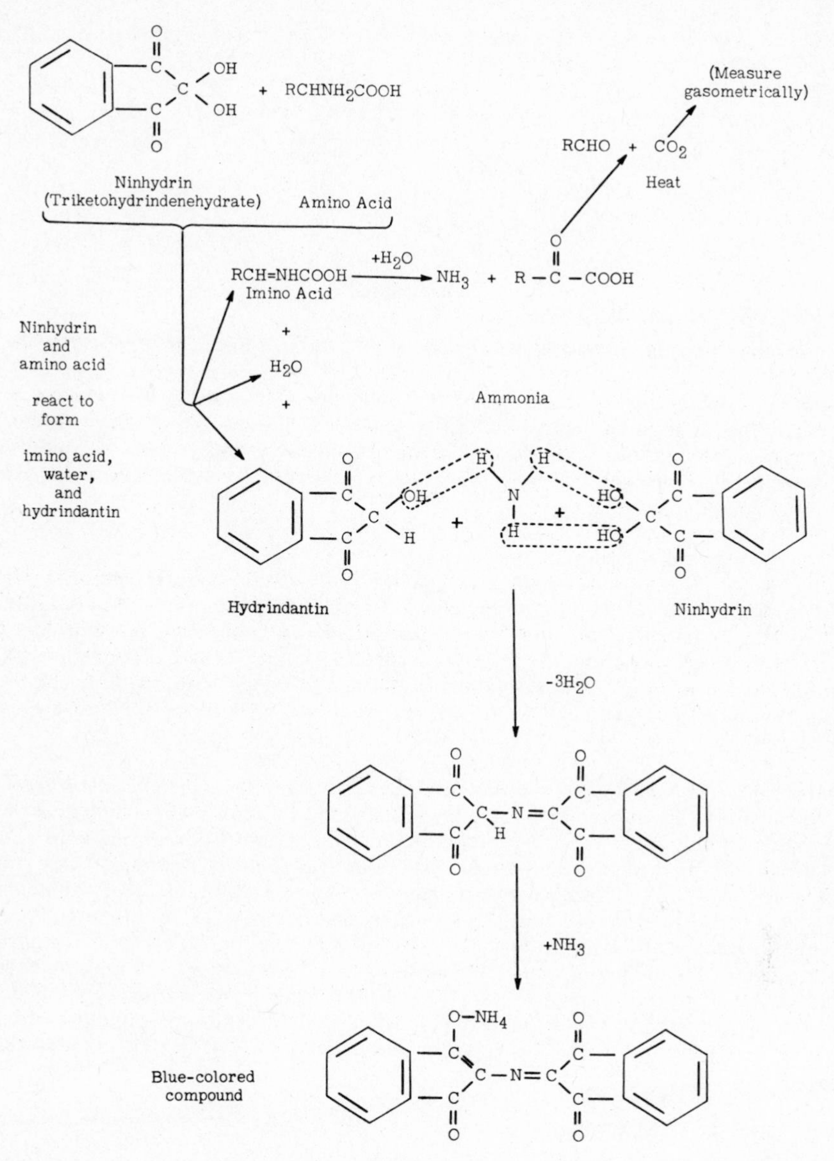

In addition to S-S bonds, one other major force is involved in the preservation of the structure of a protein molecule. This is the **hydrogen bond,** which is produced by the sharing of hydrogen atoms between the nitrogen and the carbonyl oxygen of the same or of different peptide chains.

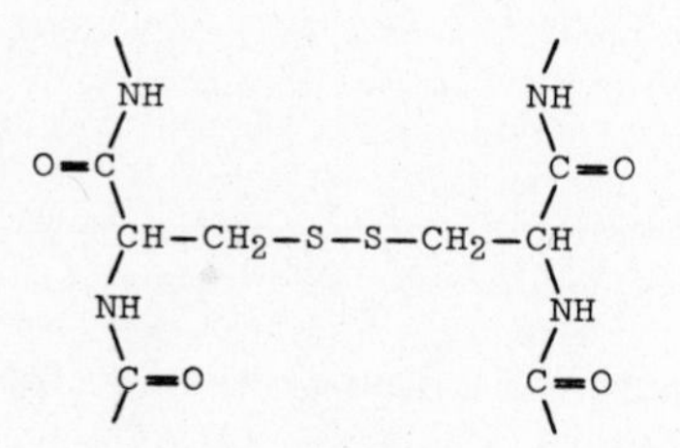

Two Peptide Chains United by a Disulfide Linkage

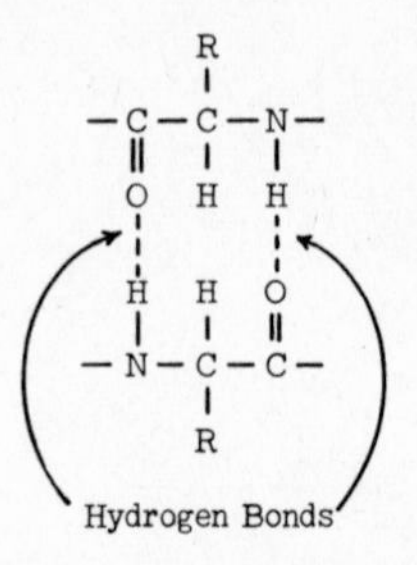

The concept that peptide chains are folded in the form of a helix assumes that the coiled structure is maintained by the hydrogen bonds between the $-\overset{O}{\overset{\|}{C}}-$ and $-\overset{H}{\overset{|}{N}}-$ groups of a single peptide chain and that it is these bonds which are broken in the unfolding process which occurs during denaturation of proteins. Individual hydrogen bonds are very weak, but the reinforcing action of a large number of such bonds in the protein molecule produces a stable structure.

The Amino Acid Content of Proteins:

Determination of the amino acid content of a purified protein usually requires first its hydrolysis to the free amino acids. This is accomplished by heating with acid in a sealed tube. (Acid hydrolysis destroys tryptophan, and hydrolysis with alkali is used when this amino acid is to be determined.) The content of amino acids in the hydrolysate may be measured by chemical or biological methods. Analyses for some amino acids can be made with specific chemical tests (e.g., Millon reaction for tyrosine; see p. 42), but such methods usually require relatively large amounts of the sample.

Chromatographic procedures are now widely used for the separation and determination of a variety of compounds of biochemical interest, including fatty acids, carbohydrates, and amino acids. Chromatographic procedures have been particularly useful for amino acid analyses such as those required for the measurement of the amino acid content of proteins. One method in common use employs strips of filter paper suspended in a sealed cylindrical jar or sealed cabinet (see below). One end of the paper dips into an organic solvent–water mixture such as collidine, n-butyl alcohol, n-propyl alcohol, or phenol. The solvent may be placed in a trough from which the paper strip hangs ("descending paper chromatography"), or the strip may be suspended from the top of the jar and dip into a trough at the bottom of the jar ("ascending paper chromatography"). The procedures are carried out in a closed system in order that the atmosphere will remain saturated with both water vapor and the vapor of the organic solvent which is being used.

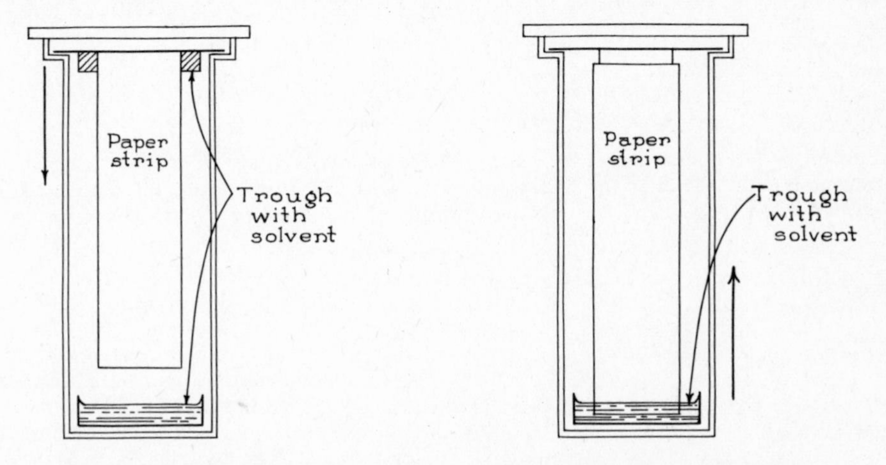

Cross Section of Apparatus for Descending (Left) and Ascending (Right) Paper Chromatography. (Redrawn from Fruton and Simmonds: General Biochemistry. John Wiley and Sons, Inc., New York, 1953.)

An extremely small amount (about 0.005 ml.) of an amino acid solution containing about 0.01 mg. of amino acids is applied at a marked point 5 cm. from that end of the paper which is to be dipped into the solvent. The system is allowed to operate for several hours, and the distance which the solvent has travelled from the marked point at which the amino acid solution was added is then measured. The paper is allowed to dry and sprayed with a solution of ninhydrin in butyl alcohol. This compound produces colored derivatives of amino acids (see p. 35), so that the position of the amino acids on the paper strip can be seen. The amino acids in the mixture will have moved along the paper strip at varying rates so that they will be separated from one another. The ratio of the distance travelled by an amino acid to the distance of the solvent front, both measured from the marked point of application of the amino acid mixture, is called the R_f value for that amino acid. These values for a given amino acid differ according to the experimental conditions, e.g., the solvents used. It is possible to identify an amino acid by a comparison of its R_f value with that obtained by running known solutions of each amino acid separately under the same experimental conditions. Alternatively, solutions whose amino acid composition is known may be chromatographed on an adjoining paper strip simultaneously with the unknown mixture, so that the spots on the test strip may be readily identified by comparison with the corresponding known counterparts on the standard strip. Quantitation of the amino acids may be accomplished by cutting out each separate spot, removing (eluting) the compound with a suitable solvent, and performing a direct colorimetric (ninhydrin) or chemical (nitrogen) analysis.

Certain modifications of this simple paper chromatographic technic have been introduced in an effort to obtain better separation of the components of the mixture and to improve their quantitation. One such modification makes use of a square sheet rather than a strip of filter paper. The sample is added at the upper left corner and chromatographed for several hours with one solvent mixture (e.g., phenol-ammonia-water), after which the paper is turned through 90° and run for an additional period with another mixture (e.g., collidine-water). This is called "two-dimensional paper chromatography" to distinguish it from the one-dimensional type described above. It permits more efficient separation of the components of the mixture. Diagrammatic representations of a one- and of a two-dimensional chromatogram are shown on p. 38.

A further increase in the resolving power of paper chromatography is achieved by **paper electrophoresis**, that is, by passing an electric current through the paper. This technic is particularly useful for analyses of the components of a protein mixture as well as peptide and amino acid mixtures (see p. 130).

As a substitute for filter paper, other substances have been used for the supporting medium in chromatographic analyses. In one method, starch is packed into a cylindrical glass column and equilibrated with a solvent such as a butyl alcohol–water mixture. The protein hydrolysate sample (equivalent to about 3 mg. of protein), dissolved in a butyl alcohol–water mixture, is added at the top of the column and solvent is added by a continuous drip method. As the solvent emerges at the bottom of the column, it is collected in small aliquots or fractions with the aid of a "fraction collector," a device which operates automatically to introduce small tubes under the outlet of the column at frequent regular intervals, depending upon the number of fractions desired. The amino acids in the mixture are adsorbed onto the starch and move slowly through the column at differing rates. Consequently, under ideal conditions each amino acid can be expected to emerge from the column at a different time. For example, leucine and isoleucine might be found in the tubes representing the first 14 ml. of the solvent (eluate) which comes from the column, whereas proline would not emerge until approximately 50 ml. of eluate had been collected. Further resolution may be obtained by varying the solvents used to elute the amino acids.

Stein and Moore[3] utilized an ion exchange resin (Dowex 50®) and various solvents as well as alterations in temperature to accomplish separation on a chromatographic column of all of the amino acids in a protein hydrolysate. Modifications in this procedure recently introduced by Moore, Spackman, and Stein[4] permit a complete amino acid analysis of a peptide or of a protein hydrolysate in 24 to 48 hours. The modified system can be used either with a fraction collector or with automatic recording equipment. In this new method, columns of finely pulverized 8% cross-linked sulfonated polystyrene resin (Amberlite IR-120®) are used. The method has recently been used for analysis of histones, hemoglobin, and ribonuclease; with modifications it can be used for determination of amino acids and related compounds in plasma, urine, and animal tissues.

Microbiologic analyses[5]. Many microorganisms, particularly the lactic acid bacilli, require a number of nutrient factors for optimal growth. It is possible to grow these organisms on a medium which is entirely composed of chemically defined nutrients. If any one essential compound

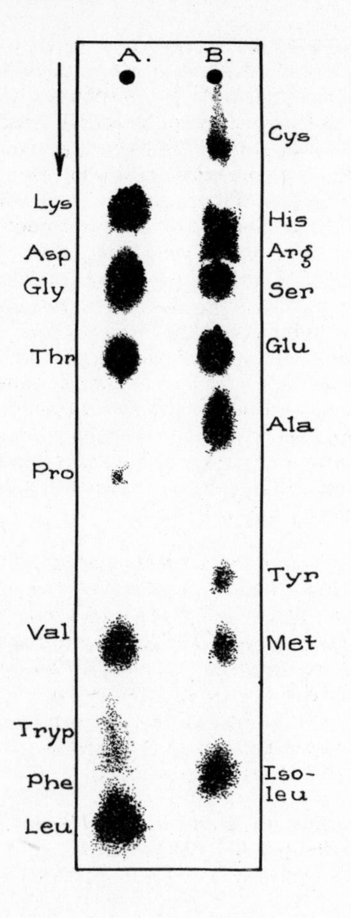

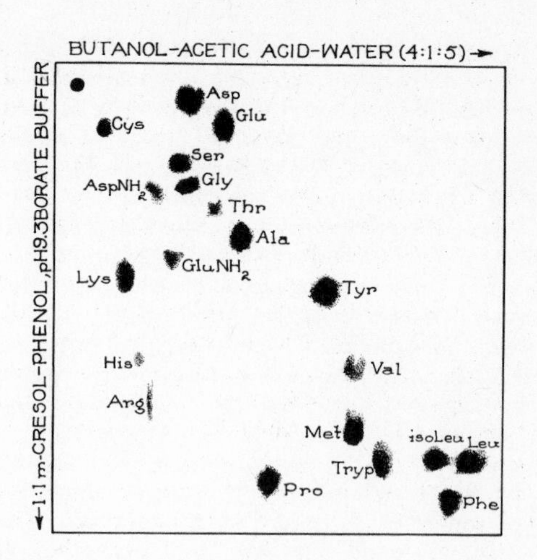

Single-dimensional (Left) and Two-dimensional (Above) Chromatogram of Natural Amino Acids. (Redrawn from Levy and Chung, "Two-dimensional Chromatography of Amino Acids on Buffered Papers," Analytical Chemistry **25**:396, March, 1953. Copyright 1953 by American Chemical Society; reproduced with permission.)

is eliminated, the organism will not grow. Restoration of the limiting compound will restore the growth of the bacteria in proportion to its concentration. For example, a broth may be mixed which contains all of the essential nutrients for the organism except tryptophan. The addition of very small amounts of tryptophan permits growth of the lactobacillus as measured by production of lactic acid from glucose or by the turbidity of the culture. In practice, measured amounts of the amino acid are added to tubes of the tryptophan-free broth, and these tubes are then inoculated with a suspension of the test organism and incubated for two to three days. Lactic acid production is measured by titration with sodium hydroxide, and the acid production, when plotted against the amino acid concentration, gives a typical growth curve. The tryptophan content of an unknown sample may be measured by adding a known amount of the sample to an assay tube and relating the lactic acid production found in that tube to the growth curve obtained with the standard solutions of tryptophan. By the use of appropriate basal media and test bacteria the majority of the naturally-occurring amino acids can be measured. These methods are reasonably specific and require only small amounts of material for analysis. Thus they have found wide application in analyses of the amino acid content of blood and other body fluids as well as protein hydrolysates.

Microbiologic assays for vitamins and certain other growth factors are also commonly used. The technic is similar to that described for the amino acids.

The Sequence of Amino Acids in the Peptide Chain; End-group Analysis:

Some knowledge of the arrangement of the constituent amino acids in the protein molecule has been obtained by the partial hydrolysis of the protein followed by isolation and analysis of small peptide fragments. A more useful technic, however, is the use of a reagent which combines strongly with the free amino groups at the end of the polypeptide chain; when the chain is subsequently hydrolyzed, the compound remains attached to the amino acid. Such a compound is fluorodinitrobenzene. It reacts with terminal α-amino groups and with the free ε-amino groups of lysine. Its reaction with a tripeptide to form a dinitrophenyl (DNP) derivative is shown below.*

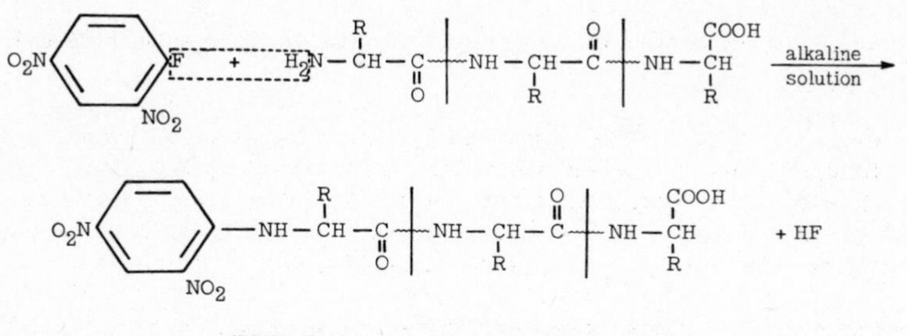

DNP Derivative of the Tripeptide

In practice, the reagent is allowed to react with a protein under mildly alkaline conditions. After removal of excess reagent, the protein is hydrolyzed with acid and the yellow DNP-amino acids are extracted with ether or chloroform and separated by chromatographic methods.

Separation of Peptides by Counter-current Distribution:

Craig[5a,b] has devised a method for the fractionation of mixtures of closely related substances which depends on the same principle commonly employed when a substance is extracted from one solvent such as water by a second solvent such as ether. Partition chromatography (see p. 36) is another example of this same principle. The method of Craig, which is termed **counter-current distribution,** permits the performance of up to 100 extractions in one operation, as well as determinations of the distribution of the components of a mixture. Counter-current distribution has been employed in the purification of polypeptides of biologic importance such as certain antibacterial substances (e.g., the gramicidins from Bacillus brevis) and oxytocin (see pp. 40 and 367).

The Structure of Insulin:

The use of modern technics to study the structure of a protein is exemplified by the recent work on insulin. Sanger and his co-workers have utilized the DNP method and other technics to elucidate the structure of the insulin molecule. He has concluded that this protein contains four peptide chains per unit weight of 12,000. Two have glycine and two have phenylalanine at the free amino end of the chains. The chains themselves are very probably connected by S-S linkages. Oxidation with performic acid breaks these linkages. By such a reaction, Sanger obtained two major polypeptide chains from the insulin molecule which he termed A and B. The A chain has a molecular weight of approximately 2750 and the terminal amino acid is glycine; the B chain, with a molecular weight of approximately 3700, has phenylalanine as the terminal amino acid. Sanger, Thompson, and Tuppy[7] have succeeded in determining the complete sequence of all of the amino acids in both A and B chains of insulin (see next page).

*DNP derivatives of amino acids are also used to improve the separation of amino acids by paper chromatography.

Structure of the A Chain

Gly.	Ileu.	Val.	Glu.	Glu.	$CySO_3H$.	$CySO_3H$.	Ala.	Ser.	Val.	$CySO_3H$.	Ser.	Leu.	Tyr.	Glu.	Leu.	Glu.	Asp.	Tyr.	$CySO_3H$.	Asp. NH_2
1	2	3	4	5	6	7	8	9	10	11	12	13	14	15	16	17	18	19	20	21

Structure of the B Chain

Phe.	Val.	Asp.	Glu.	His.	Leu.	$CySO_3H$.	Gly.	Ser.	His.	Leu.	Val.	Glu.	Ala.	Leu.	Tyr.	Leu.	Val.	$CySO_3H$.	Gly.	Glu.	Arg.	Gly.	Phe.	Phe.	Tyr.	Thr.	Pro.	Lys.	Ala.
1	2	3	4	5	6	7	8	9	10	11	12	13	14	15	16	17	18	19	20	21	22	23	24	25	26	27	28	29	30

Amino Acid Sequences of the A and B Chains of the Insulin Molecule

The abbreviations above are those suggested by Brand for designation of the amino acids in a peptide chain. The acid at the left is the amino acid with a free amino group. $CySO_3H$ is cysteic acid (see p. 250), the oxidized form of cysteine, which would be obtained after performic acid oxidation to break the S-S linkages which connect the chains in the intact molecule of insulin. Asp. NH_2 is asparagine, the mono amide of aspartic acid.

The positions at which the chains are interconnected in an insulin residue of molecular weight 6,000 (derived from beef pancreas) were also reported by Sanger. In the A chain, the cysteine residues at 6 and 11 (counting from the N-terminal acid, glycine) are connected by an S-S linkage. The A and B chains are interconnected by S-S linkages between 7 and 7 and between 20 and 19 respectively. This may be represented diagrammatically as follows:

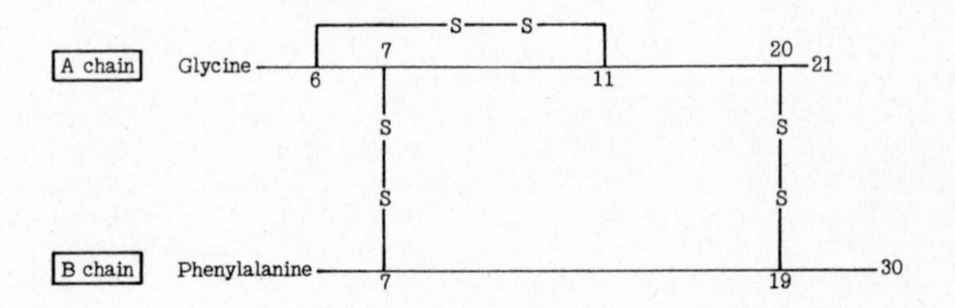

The insulins of various mammalian species (pig, sheep, horse, whale) exhibit differences in the amino acid sequences at positions 8-10 of the A chain.

Glucagon (see p. 342) is a polypeptide (minimal molecular weight 3485) containing 29 amino acid residues. It differs from insulin in several ways; only a few dipeptide sequences are similar, and there are no disulfide bridges in glucagon.

Structure of Oxytocin and of Vasopressin:

Du Vigneaud and his colleagues have established the structure of **oxytocin** and of **vasopressin**, the two hormones of the posterior pituitary gland. These hormones are polypeptides with a molecular weight of about 1000 and consist of eight different amino acids arranged in a cyclic structure through S-S linkages. The exact structures of these hormones are shown on p. 367.

Structure of Corticotropin (ACTH):

The adrenocorticotropic principle of the anterior pituitary (corticotropin; see p. 364) has been fractionated into eight equally active components which constitute about 2% of the weight of the corticotropin now used clinically. That which is most abundant and which in itself can perform all of the known functions of corticotropin is known as β-ACTH. It is a polypeptide containing 39 amino acids and has a molecular weight of 4566. The chemical structure of this molecule has been determined, and β-ACTH is thus the largest single protein chain whose structure is known.

Molecular Weights:

Molecular weights of proteins have been most successfully investigated by physical methods such as osmotic pressure measurements by freezing point (cryoscopic) determinations. In general, the results lack precision because of variables due to pH, electrolytes, and the degree of hydration of the protein molecule. The method developed by Svedberg is the best yet advanced for determination of the molecular weight of proteins. This method depends upon measurement of sedimentation rates as determined in the ultracentrifuge.

Many proteins are of extremely high molecular weight (e.g., psittacosis virus: 8,500,000,000). Such high molecular weight proteins might be considered to contain units of about 400,000. These units are thought to combine (polymerize) in a reversible manner in accordance with concentration, pH, temperature, etc.

Molecular weights of representative proteins (ultracentrifuge measurements) are as follows:

Protein	Molecular weight
Egg albumin	44,000
Insulin	12,000
Serum albumin	69,000
Hemoglobin (horse)	68,000
Serum globulin	180,000
Fibrinogen	450,000
Thyroglobulin	630,000

GENERAL REACTIONS OF PROTEINS

Precipitation Reactions:

A. Concentrated Mineral Acids: Heavy precipitates occur with small amounts of the acid, but further addition of acid redissolves the protein and, later, hydrolysis may occur.

B. Alkalis do not precipitate, but hydrolysis and oxidative decomposition occur.

C. Heavy metals act as protein precipitants, depending upon the hydrogen ion concentration, temperature, and the presence of other electrolytes. Mercuric chloride and silver nitrate produce a heavy precipitate which cannot be redissolved, whereas copper sulfate and ferric chloride cause precipitation with resolution in an excess of reagent.

D. So-called "alkaloidal" reagents (trichloroacetic acid, tannic acid, phosphotungstic acid, phosphomolybdic acid) act as protein precipitants when the pH of the solution is on the acid side of the iso-electric point of the protein.

E. Alcohol is a protein precipitant. It is most effective when the protein is at its iso-electric point.

F. Heat will coagulate many proteins, although the effective temperature may range from 38 to 75° C. Various factors influence coagulation, but the protein is most easily coagulated at its iso-electric point. The resulting coagulum is insoluble unless the solvent causes hydrolysis or other decomposition.

Denaturation:

Many of the agents listed above, as well as x-ray and ultraviolet irradiation, cause denaturation, a change in the physical and physiologic properties of the protein, as well as other changes which are not well understood. Denaturation results in an unfolding of the protein molecule. The hydrogen bond (see p. 35) is the main linkage destroyed in the course of denaturation. In some cases denaturation without coagulation may be accomplished by heating, as occurs when the protein is heated in an acid or alkaline solution. The protein is transformed into a metaprotein which is insoluble at its iso-electric point. It will flocculate when returned to the iso-electric point but will redissolve in either acid or base.

Other changes resulting from denaturation include alteration in surface tension, loss of enzyme activity, and loss or alteration of antigenicity.

Color Reactions:

A. Millon Reaction: A test for tyrosine in the molecule. Add three to four drops of Millon reagent (mercurous nitrate in nitrous acid) to 5 ml. of solution. Mix, and heat gradually. A white precipitate is formed which gradually turns red. The reaction does not occur if the protein is not precipitable by strong acid.

B. Biuret Reaction: Add alkali and two or three drops of a weak copper sulfate solution (about 0.02%). A bluish to pink color occurs, depending upon the type of protein present. The test is specific for the peptide linkage.

C. Xanthoproteic Reaction: Nitric acid added to a protein causes a yellow color which turns orange on addition of alkali. The reaction is due to the presence of an aromatic nucleus in the protein (the amino acids tryptophan, tyrosine, and phenylalanine).

D. Glyoxylic Acid Test (Hopkins-Cole Test): Specific for tryptophan. The tryptophan-containing material is treated with a small amount of a solution containing glyoxylic acid. This is then stratified above concentrated sulfuric acid. A violet ring occurs at the zone of contact.

E. Ninhydrin Reaction: Previously discussed under amino acid reactions (see p. 35). This reaction is given by all proteins and by all compounds containing at least one free amino group and one free carboxyl group.

• • •

References:

1. The official rules for the nomenclature of the amino acids appear in Chem. and Eng. News **30**:4522, 1952.
2. Hamilton, P.B., and Van Slyke, D.D.: J. Biol. Chem. **150**:231, 1943.
3. a. Moore, S., and Stein, W.H.: J. Biol. Chem. **192**:663, 1951.
   b. Moore, S., and Stein, W.H.: Ibid. **211**:893, 1954.
4. Moore, S., Spackman, D.H., and Stein, W.H.: Anal. Chem. **30**:1185, 1958.
5. Lewis, J.C., in "Methods in Medical Research," pg. 224, Year Book, 1950.
6. a. Craig, L.C., et al.: Cold Spring Harbor Symposia Quart. Biol. **14**:24, 1949.
   b. Von Tavel, P., and Signer, R.: Adv. in Protein Chem. **11**:237, 1956.
7. a. Sanger, F., and Tuppy, H.: Biochem. J. **49**:463, 1951.
   b. Sanger, F., and Thompson, E.O.P.: Ibid. **53**:353, 1953.

Bibliography:

Block, R.J., and Bolling, D.: The Amino Acid Composition of Proteins and Foods. Thomas, 1951.
Block, R.J., and Bolling, D.: Determination of the Amino Acids. Burgess, 1938.
Block, R.J., Durrum, E.L., and Zweig, G.: Paper Chromatography and Paper Electrophoresis. Academic, 1955.
Cohn, E.J., and Edsall, J.T.: Proteins, Amino Acids and Peptides. Reinhold, 1943.
Greenberg, D.M., Ed.: Amino Acids and Proteins. Thomas, 1951.
Haurowitz, F.: Chemistry and Biology of Proteins. Academic, 1950.
Neurath, H., and Bailey, K., Eds.: The Proteins. 2 Vols. Academic, 1953-54.

4...
Nucleoproteins and Nucleic Acids

Nucleoproteins are one of the groups of conjugated proteins. They are characterized by the presence of a nonprotein prosthetic group (nucleic acid) which is attached to one or more molecules of a simple protein. The simple protein is usually a basic protein such as a protamine or histone. These conjugated proteins are found in all animal and plant tissues, but are most easily isolated from yeast or from tissues with large nuclei where the cells are densely packed, such as in the thymus gland.

The nucleoproteins were so named because they constitute a large part of the nuclear material of the cell. However, these proteins are also found in the cytoplasm associated particularly with the mitochondria.

All living cells contain nucleoprotein, and some of the simplest systems, such as the viruses, seem to be entirely pure nucleoprotein. Furthermore, such an important cellular constituent as chromatin is largely composed of nucleoproteins, which indicates that these compounds are involved in cell division and the transmission of hereditary factors. For this reason any abnormality in the mechanism of nucleoprotein formation is followed by alterations in cell growth and reproduction. This may be exemplified by the effects of deficiencies of folic acid and vitamin B_{12}, radiations which induce mutations, or radiomimetic chemical agents such as the nitrogen and sulfur mustards, which not only affect chromosome structure and gene activity but also suppress mitosis.

Nucleoproteins have been extracted from a variety of plant and animal tissues; extracting agents used include water, dilute alkali, sodium chloride solutions, and buffers ranging in pH from 4 to 11. In each case extraction is followed by precipitation with acid, saturated ammonium sulfate, or dilute calcium chloride. When the purified nucleoprotein is hydrolyzed with acid or by the use of enzymes, it is broken down into various components as follows:

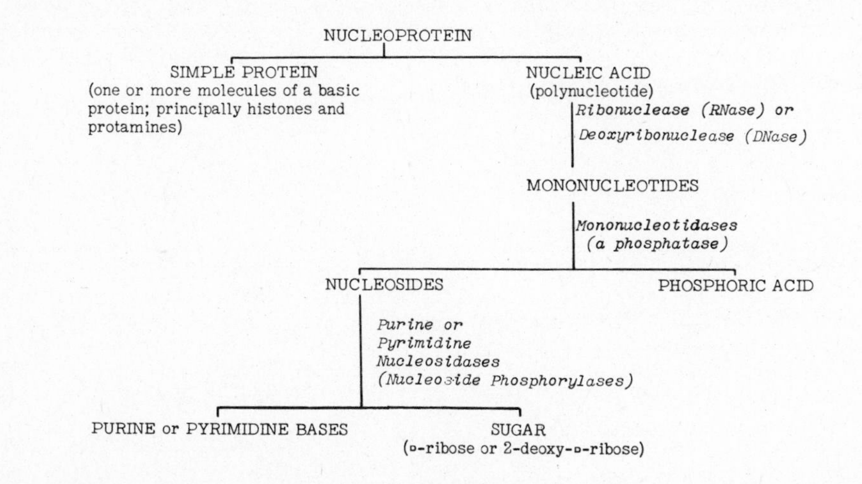

By the use of careful acid hydrolysis the purine or pyrimidine base may be removed from the nucleotide leaving the sugar attached to the phosphate:

$$\text{Mononucleotide} \xrightarrow{\text{Acid hydrolysis}} \text{Sugar-phosphate} + \text{Pyrimidine or Purine base}$$

THE PYRIMIDINE AND PURINE BASES

The various purine and pyrimidine bases which occur in the nucleotides of nucleic acids are derived by appropriate substitution on the ring structures of the parent substances, purine or pyrimidine. The structures of these parent nitrogenous bases are as follows*: (The positions on the rings are numbered according to the original Fischer system. The newer international system proposes to number the pyrimidine ring as shown in the brackets. The purine numbering system is not changed.)

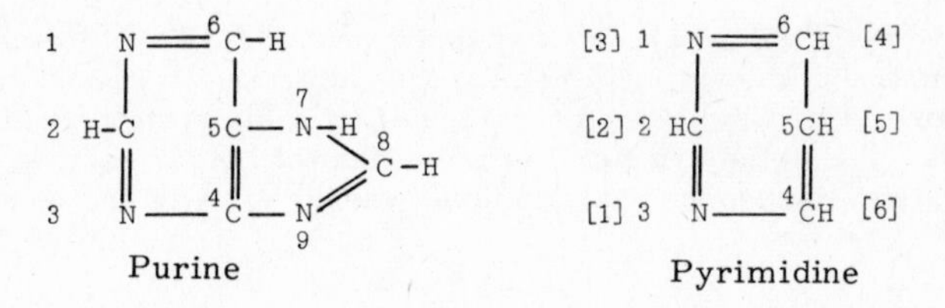

The Pyrimidine Bases:

Five pyrimidine bases have been isolated from nucleic acids. **Cytosine** (2-oxy-6-amino-pyrimidine) is found in all nucleic acids. A **5-methyl cytosine** has recently been found in comparatively small amounts in certain plant and animal (mammalian) nucleic acids. Cytosine is absent in the deoxyribonucleic acid (DNA) of certain bacterial viruses, viz., coli bacteriophage of the T-even series (T_2, T_4, T_6, etc.); in its place there is 5-hydroxymethylcytosine (HMC)[1].

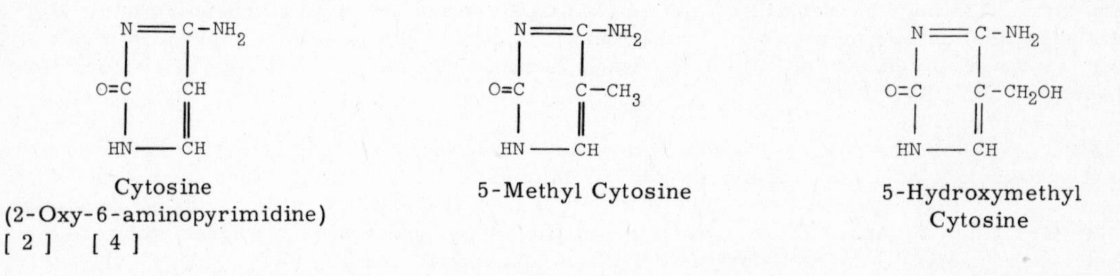

Thymine (2, 6-dioxy-5-methyl pyrimidine) occurs in nucleic acids which contain deoxyribose as the characteristic carbohydrate, the so-called deoxyribonucleic acids (DNA) (see below). **Uracil** (2, 6-dioxypyrimidine), on the other hand, is confined to the ribonucleic acids (RNA) which contain ribose rather than the deoxy sugar.

*The structures of the purine and pyrimidine bases may also be shown as follows:

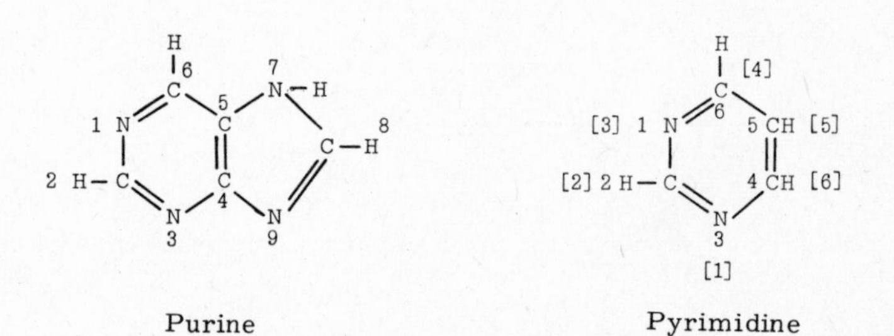

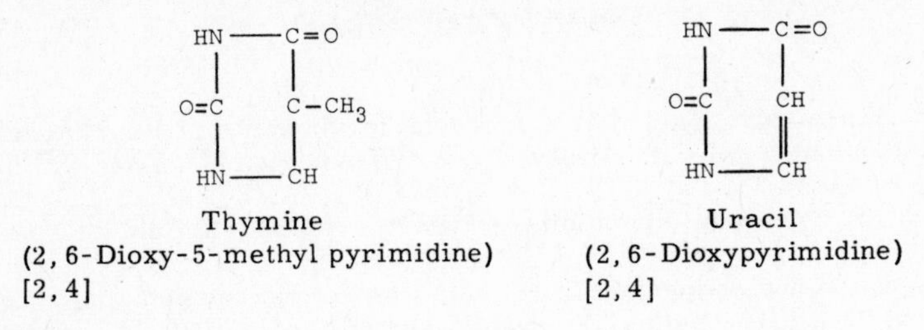

Thymine
(2, 6-Dioxy-5-methyl pyrimidine)
[2, 4]

Uracil
(2, 6-Dioxypyrimidine)
[2, 4]

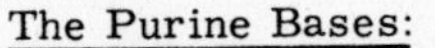

The Purine Bases:

Adenine and guanine are the two purines found in all nucleic acids.

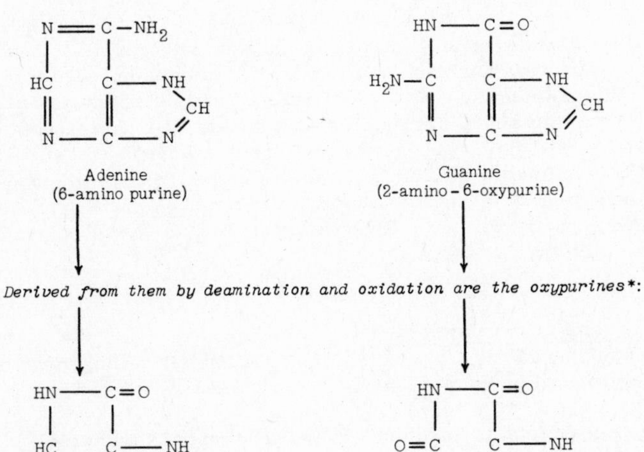

Adenine
(6-amino purine)

Guanine
(2-amino - 6-oxypurine)

Derived from them by deamination and oxidation are the oxypurines:*

Hypoxanthine
(6-oxypurine)

Xanthine†
(2, 6-dioxypurine)

Uric Acid
(2, 6, 8-trioxypurine)

---

*The oxypurines or oxypyrimidines may form enol derivatives by migration of hydrogen to the oxygen substituents. This is illustrated by the so-called lactam (hydroxy) structure of uric acid, which is formed by enolization of the lactim (oxy) form shown above.

The existence of these two forms is suggested by the fact that the oxypurines or oxypyrimidines form salts with alkali. The purines and pyrimidines also form salts with acids because of the nitrogen atoms which are weakly basic.

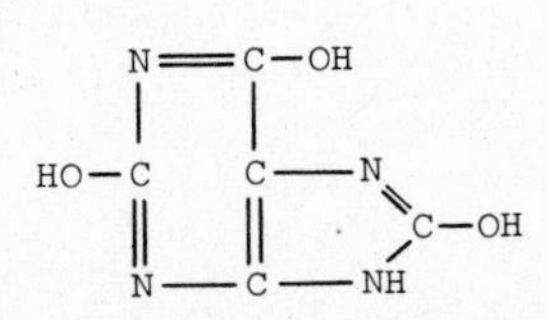

Uric Acid (Lactam Form)
(2, 6, 8-Trihydroxypurine)

†Three xanthine derivatives are important constituents of coffee, tea, and cocoa.

Coffee:	Caffeine,	1, 3, 7-trimethylxanthine.
Tea:	Theophylline,	1, 3-dimethylxanthine.
Cocoa:	Theobromine,	3, 7-dimethylxanthine.

THE NUCLEOTIDES

A nucleotide, the structural unit of nucleic acids, is composed of a purine or a pyrimidine base attached to a sugar by a glucosidic linkage; the sugar is then combined with phosphoric acid.

Purine or Pyrimidine — Sugar — Phosphoric Acid

Two general types of nucleotides are found in nucleic acids; one contains D-ribose (furanose form; see p. 6), and the other contains 2-deoxy-D-ribofuranose (see p. 11). It is therefore customary to refer to these nucleic acids as ribonucleic acid (RNA)* or deoxyribonucleic acid (DNA).

Types and Sources of Nucleotides:

The nucleotides from yeast nucleic acid contain ribose, whereas those from thymus contain considerable deoxyribose. The earliest studies of nucleic acid chemistry were made on nucleoproteins derived from these two sources, and it was therefore concluded that ribose was characteristic of plant nucleic acids and deoxyribose of animal nucleic acids. Hence the terms plant nucleic acid or animal nucleic acid were applied to what are now designated RNA or DNA, respectively. It is now realized that both types of nucleotides are found in most cells. DNA is the principal nucleic acid of the nucleus (RNA is found in the nucleolus), whereas that of the cytoplasm is largely RNA (although small amounts of DNA which cannot be histologically demonstrated are probably also present). The cytoplasmic RNA exists mainly in the mitochondria and in other very small particles within this area of the cell.

Nomenclature and Structure of the Nucleotides:

The name of each nucleotide is derived from its constituent nitrogenous base. For example, the nucleotides obtained from ribonucleic acid of yeast are:

Adenylic acid (adenine + ribose + phosphoric acid)
Guanylic acid (guanine + ribose + phosphoric acid)
Cytidylic acid (cytosine + ribose + phosphoric acid)
Uridylic acid (uracil + ribose + phosphoric acid)

Deoxyribonucleic acid contains thymidylic acid (thymine + deoxyribose + phosphoric acid) rather than uridylic acid.

In the nucleotides derived from ribonucleic acid the sugar is attached (in the furanose form) to the purine base at position 9 or to the pyrimidine at position 3, and the attachment of the phosphoric acid is to carbon number 3 of the sugar. This is illustrated in the following formulas.

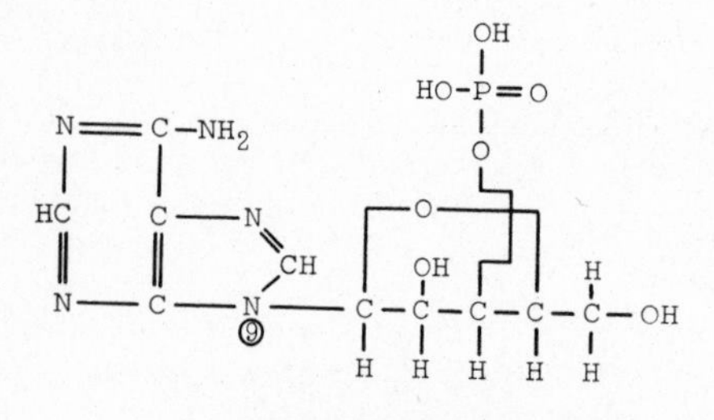

Yeast Adenylic Acid
(Adenine Ribonucleotide)

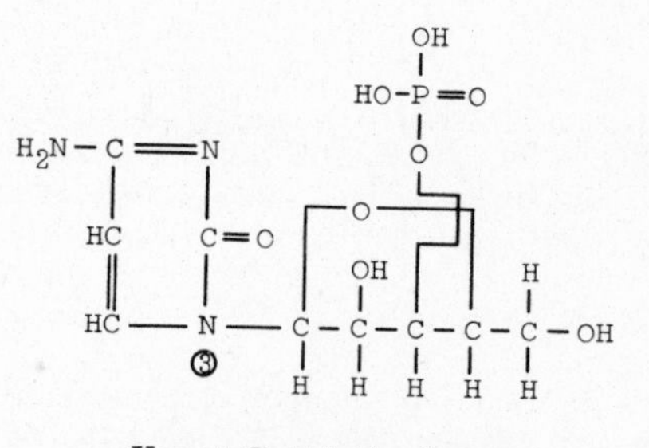

Yeast Cytidylic Acid
(Cytosine Ribonucleotide)

*To allow for the possible presence of pentoses other than ribose, some authors prefer the term pentose nucleic acid (PNA) to RNA.

The structure of the nucleotides of deoxyribonucleic acid is similar, although the phosphoric acid may be attached either to carbon number 3 of the sugar (2-deoxyribofuranose), as in the ribonucleotides shown above, or to the terminal carbon (number 5), as shown in the formulas below.

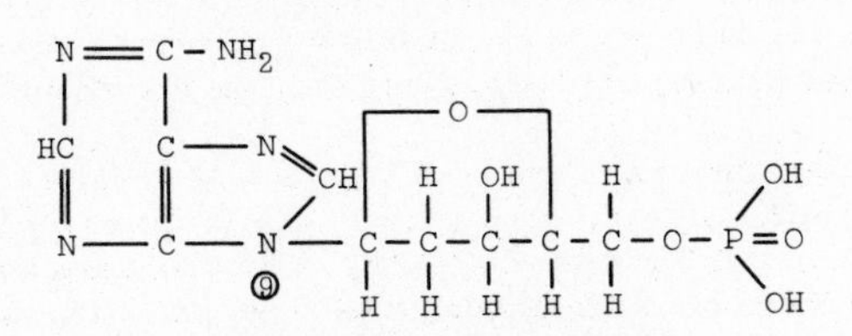

Thymus Adenylic Acid
(Adenine Deoxyribonucleotide)

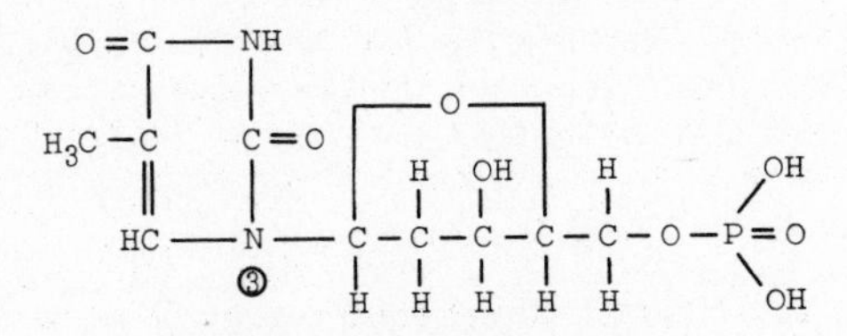

Thymus Thymidylic Acid
(Thymine Deoxyribonucleotide)

Biologically Important Nucleotides:

Nucleotides which are not combined in nucleic acids are also found in the tissues. They have important special functions. Some of these compounds will be listed below.

Adenine derivatives. Adenylic acid (AA), also designated adenosine monophosphate (AMP), found in muscle, is a combination of adenine, ribose, and phosphoric acid.

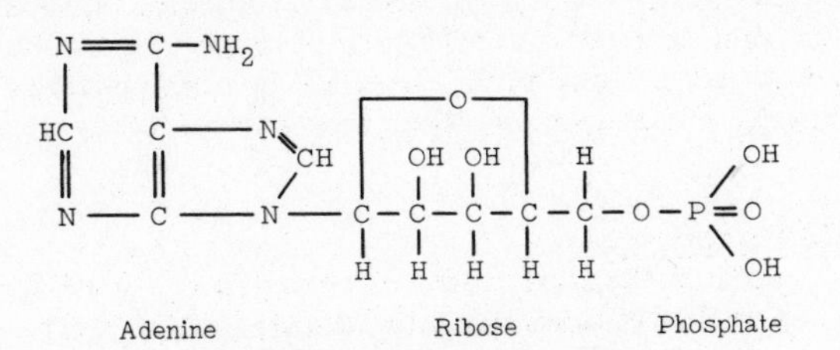

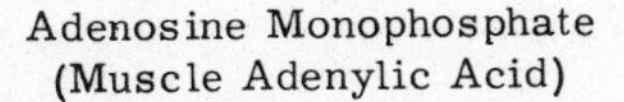

Adenosine Monophosphate
(Muscle Adenylic Acid)

Adenylic acid may also exist as the diphosphate (adenosine diphosphate, ADP) or as adenosine triphosphate (ATP). These latter compounds are very important in oxidative phosphorylation and as sources of high-energy phosphate (see pp. 119 to 122).

The incorporation of sulfate into ester linkages in such compounds as the sulfated mucopolysaccharides (e.g., chondroitin sulfuric acid and the similar mucoitin-sulfuric acid in mucosal tissues) requires the preliminary "activation" of the sulfate moiety. This is accomplished by the formation of "active sulfate" with ATP. The reaction may be depicted as follows:

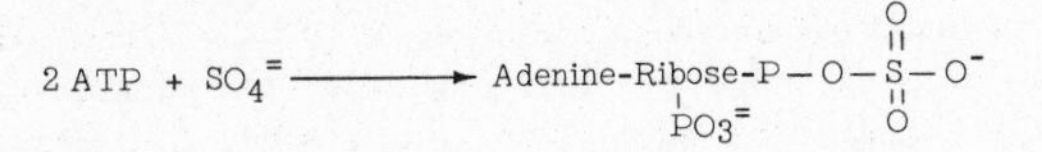

Activating system ⟶ "Active" sulfate (adenosine-3'-phosphate-5'-phosphosulfate)

"Active" sulfate is also required in conjugation reactions involving sulfate (see p. 167).

Hypoxanthine derivatives. Deamination of adenosine monophosphate produces a hypoxanthine nucleotide, usually designated inosinic acid; removal of the phosphate group of inosinic acid forms the nucleoside, inosine (hypoxanthine riboside).

Analogues of ADP and of ATP in which the purine derivative is hypoxanthine rather than adenine have also been found to participate in phosphorylation reactions. These compounds are inosine diphosphate (IDP) and inosine triphosphate (ITP; see pp. 89 and 187).

Guanine derivatives. Guanine analogues of ATP are also involved in metabolism. The oxidation of ketoglutaric acid to succinyl-Co A (see p. 88) involves oxidative phosphorylation with transfer of phosphate to guanosine diphosphate (GDP) to form guanosine triphosphate (GTP). This phosphorylation reaction is identical with similar reactions involving ADP and ATP (see p. 119).

GDP and GTP are also active in protein synthesis involving ribonucleic acid (see p. 223).

Uracil derivatives. Uridine (uracil-ribose) derivatives are important coenzymes in reactions involving epimerisation of galactose and glucose (uridine diphosphate glucose, UDPG; and uridine diphosphate galactose, UDPgal; see p. 195). A uridine coenzyme, uridine diphosphate glucuronic acid (UDPgluc; see pp. 168 and 276), serves as a source of "active" glucuronide in conjugation reactions requiring glucuronic acid, e.g., formation of menthol glucuronide[4] or bilirubin glucuronide.

Uracil also participates in the formation of high-energy phosphate compounds analogous to ATP, ITP, or GTP, mentioned above. Uridine triphosphate (UTP) is formed, for example, in the reactions involving conversion of galactose to glucose (see p. 195). It may also be produced by the transfer of phosphate from ATP to UDP. Berg and Joklik[3] described the synthesis of UTP (as well as ITP) by phosphorylation of the diphosphate with ATP as phosphate donor. The enzyme which catalyzes this phosphate transfer is called a **nucleoside diphosphokinase.** It has been partially purified from dried (brewer's) yeast and from rabbit muscle. Krebs has also detected it in pigeon breast muscle and in the intestinal mucosa of the rat.

Cytosine derivatives. Cytidine (cytosine-ribose) may form the high-energy phosphate compound, **cytidine triphosphate (CTP).** CTP is the only nucleotide which was found to be effective in an in vitro mitochondrial system for the biosynthesis of lecithin. CTP reacts with phosphoryl choline to form cytidine diphosphate choline (CDP-choline), which in turn combines with a diglyceride to form the phospholipid (see p. 208). CTP is also required in a similar series of reactions involving phosphoryl ethanolamine which leads to the biosynthesis of the cephalins.

Nucleotide Structures in B Vitamins:

In their active form, several vitamins of the B complex are combined in nucleotide linkages. These include riboflavin, as in flavin adenine dinucleotide (FAD; see p. 79) or as riboflavin phosphate; niacin, in Coenzymes I and II. [i.e., diphosphopyridine nucleotide (DPN) and triphosphopyridine nucleotide (TPN) respectively; see p. 81]; and pantothenic acid in Coenzyme A (co-acetylase; see p. 85).

Thiamine (cocarboxylase) and pyridoxal (cotransaminase) are both phosphorylated when functioning as prosthetic groups of enzymes involved in intermediary metabolism (see pp. 77 and 82). A nucleotide is also a component of the structure of vitamin B_{12} (see p. 96).

NUCLEOSIDES

When a nucleotide is treated with an appropriate enzyme (a nucleotidase, which is actually a phosphatase), the phosphoric acid is split off and a nucleoside results. These are combinations of purine or pyrimidine bases with the pentose sugars. For example, the yeast nucleotides listed on p. 46 would yield four nucleosides designated as adenosine, guanosine, cytidine, and uridine. From a DNA such as thymus nucleic acid, thymidine (thymine nucleoside) would be obtained instead of uridine.

Biologically Important Nucleosides:

The importance of the amino acid methionine as a source of labile methyl groups is discussed on pp. 205 and 247. The participation of methionine in transmethylation reactions requires first the formation of ''active'' methionine. This compound has been found to be an adenine nucleoside, S-adenosyl methionine[5].

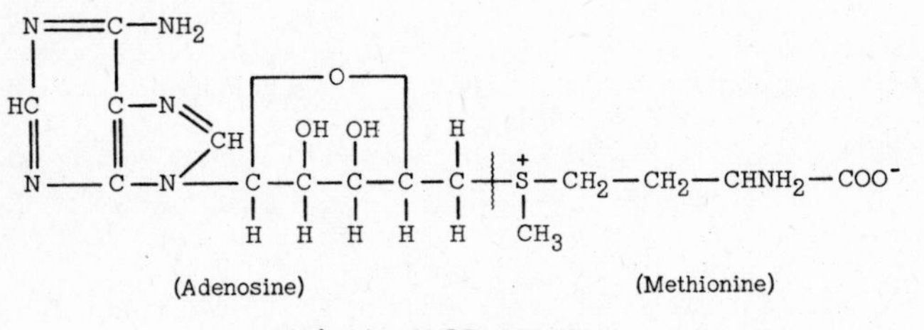

''Active'' Methionine

A free nucleoside containing a thiopentose has been isolated from yeast.

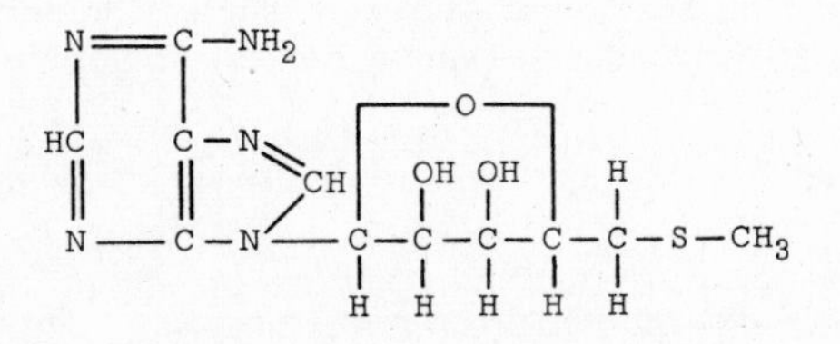

Adenine-5′-thiomethylriboside

NUCLEIC ACIDS

Structure of Nucleic Acids:

Early analyses of yeast or thymus nucleic acids indicated that the four nitrogenous bases were present in approximately equal amounts. This suggested that nucleic acid was composed of four nucleotides, each containing one purine or pyrimidine base. The so-called ''tetranucleotide'' structure was thus advanced as a hypothetical formula for the nucleic acid prosthetic groups in the nucleoproteins. More recent studies show that the relative content of each nitrogenous base is not equal but that these ratios vary according to the source of the nucleic acid. The tetranucleotide hypothesis has now been abandoned.

The current concept of the linkage of nucleotides in nucleic acid is shown on p. 50. It will be noted that the nitrogenous purine or pyrimidine bases are linked to a terminal carbon (number 1) of the sugar, and the individual nucleotides, through phosphodiester groups connected at carbon atom number 3 of one sugar molecule to a terminal carbon (number 5) of the next. This (3, 5) linkage seems characteristic of DNA preparations, a finding which is to be expected since the only sugar hydroxyl groups available in deoxyribose for the formation of a phosphoric acid ester are those in the 3 or 5 position. In RNA preparations 3, 5 linkages also predominate, although 2-3 linkages may also be present. 2-5 linkages seem to be excluded by recent studies.

X-ray diffraction patterns of fibrous DNA suggest that its form is that of a double helix. Cytochemical studies of DNA obtained from a variety of cell types indicate that the amount of adenine always equals the amount of thymine and the amount of guanine always equals the amount of cytosine. This ''pairing'' of the nitrogenous bases together with the x-ray diffraction data are the bases for the Watson and Crick[6] working model of DNA. These authors have proposed a double helical structure in which the two polynucleotides are coiled in such a manner that an adenine of one chain is bonded by hydrogen bonds to a thymine of the other, and a guanine of one chain is similarly bonded to a cytosine of the other. The hydrogen bonding is believed to involve the keto and amino groups on position 6 of the nitrogenous bases.

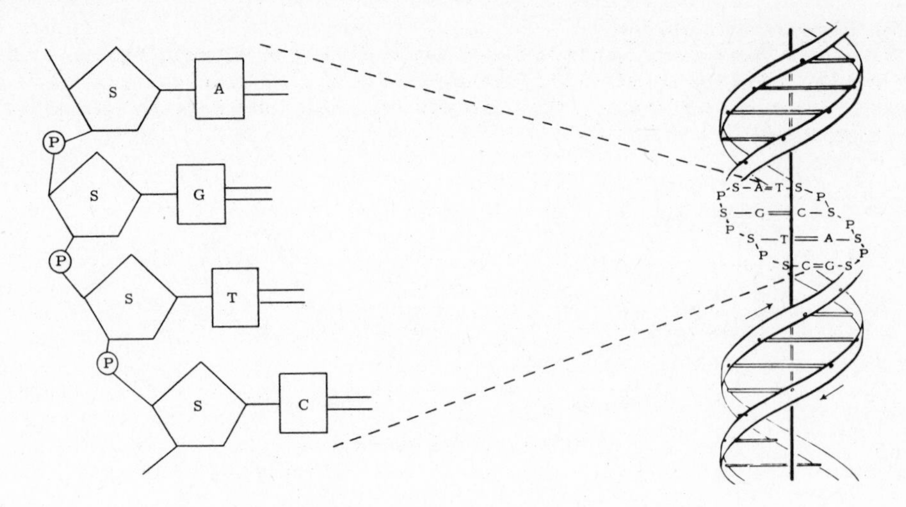

A Tetranucleotide Portion of Deoxyribonucleotide and the Watson and Crick Model of the Double Helical Structure of DNA

Deoxyribonucleic acids are extremely fibrous or elongated. This accounts for the high viscosity which is so characteristic of DNA solutions. RNA solutions, on the other hand, do not exhibit considerable viscosity, and these molecules appear to be spherical in shape with a tendency to form side branches.

Molecular weights of 1 to 5×10^6 have been found for DNA. Somewhat lower weights are reported for yeast RNA and for that of tobacco mosaic virus. The molecular weights for a given preparation are likely to vary considerably, however, depending on the method used to obtain it. This suggests that the molecules are highly polymerized and tend to depolymerize in the course of extraction and purification.

Action of Reagents on DNA:

Acids and alkalies cause irreversible reductions in viscosity, and certain chemical groups which were not available prior to the acid or alkaline treatment now become titratable. It is suggested that this is caused by rupturing of the hydrogen bonds between amino and hydroxyl groups of adjacent bases, and these groups are thus freed to react in the titrations. Guanidine, urea, and phenol in relatively high concentrations also irreversibly decrease the viscosity of DNA solutions. These observations all point to the loss of viscosity as due mainly to a destruction of the rigidity of the molecule, which is normally maintained by the inter- and intramolecular hydrogen bonds.

The nitrogen and the sulfur mustards are known to cause abnormalities in mitosis as well as a high incidence of mutations. In higher concentrations they may inhibit growth.* These compounds, even in low concentrations, also reduce the viscosity of DNA solutions. It is believed that one chemical explanation for the biologic activity of the mustards is their ability to combine with (alkylate) the free amino groups of the purine or pyrimidine bases or of the proteins. In so doing, in the case of the nucleic acids, the bonding described above as maintaining the stiffness of the molecule is destroyed. This is evidenced by the reduction in viscosity.

*It is noteworthy that those radiations in the ultraviolet which induce mutations most readily are of the same wavelength as those which characterize the absorption spectra of nucleic acids. This further supports the idea that alterations in nucleic acid structure are involved in the causation of mutations.

Biologic Significance of Nucleic Acids:

Viruses are notably rich in nucleoprotein, but their nucleic acid content varies both in amount and in composition. Influenza and tobacco mosaic viruses, for example, each contain about 6% as contrasted with as much as 40% in certain Escherichia coli bacteriophages. In the plant viruses only RNA is found, and in many animal and bacterial viruses such as the coli phages only DNA is found. Other animal viruses have both DNA and RNA. It seems to be definitely established that no living cell can exist independently unless it possesses both types of nucleic acid. Thus viruses, which lack one or the other type, must associate with other living cells to provide for their reproduction.

The function of DNA in connection with the genetic function of the nucleus of the cell is indicated by several pieces of evidence. DNA is found in all nuclei and it is confined to the chromosomes. In studies of the DNA content of cell nuclei, the relationship between DNA molecules and the estimated number of genes present is such as to suggest an equality of nucleic acid molecules to genes, i.e., that each gene is composed of one nucleic acid molecule. The amount of DNA in each somatic cell is constant for a given species, but the amount in the germ cell, which has only half the number of chromosomes, is half that in the somatic cell. In cells containing multiple sets of chromosomes (polyploidy), the amount of DNA is correspondingly increased. It is also reported that there is virtually no variation in the composition of DNA found in the sperm cell and that of other cells of the same organism, although the cell proteins are quite different. This would be expected if DNA is an integral part of the chromosomes, which are reduplicated in every other cell from the parent cells.

The role of DNA in transmission of heritable variations is further supported by the results of experiments (Avery) in which a DNA preparation from type III pneumococci was able to promote the transformation of the uncapsulated (rough) strain type II pneumococci to the capsulated (smooth) form of the type III organisms. The transformation resulted in the production of a virulent organism from an avirulent one, and the transformed cells continued to produce cells of similar type and virulence. Using specific DNA preparations, this phenomenon of "transformation" has also been accomplished with other bacteria as well as with bacteriophage.

Action of Ribonucleases and Deoxyribonucleases on Nucleic Acids:

The highly polymerized nucleic acids or polynucleotides as they exist in nature are broken down into smaller components (oligonucleotides) by the action of specific enzymes. Examples are the enzymes ribonuclease (RNase) and deoxyribonuclease (DNase), which split RNA or DNA, respectively. **Streptodornase** is a deoxyribonuclease-containing enzyme found in certain streptococci. It is used clinically, together with **streptokinase***, which is also derived from certain hemolytic streptococci. The mixture attacks fibrin blood clots and digests and liquefies viscous accumulated pyogenic material, e.g., in the chest cavity. The liquefied material may then be more easily removed by aspiration.

Streptodornase is actually a mixture of "nucleolytic" enzymes rather than a single one, so that it carries the breakdown of DNA to completion. A highly purified DNase has been isolated and crystallized. It acts only on the initial depolymerization of DNA; further breakdown must be brought about by nucleotidases or nucleosidases as shown on p. 43.

Identification of Nucleic Acids in Cells:

Ribonucleic acids have been identified histologically by the orcinol color reactions for pentoses (see p. 14); and DNA by the color reaction of Feulgen, which is generally believed to be specific for deoxyribonucleic acid.

Thymine, 5-methylcytosine, thymidine, thymidylic acid, and DNA can be determined fluorometrically by reacting acetol (an alkaline hydrolysis product of the 5-methyl pyrimidine ring common to all of these compounds) with *o*-amino benzaldehyde to form strongly fluorescent 3-hydroxy quinaldine. The new, sensitive method which is based on the thymine or 5-methyl linkage in DNA[7] has been used for the determination of the DNA content of various rabbit tissues. The thymine in as little as 50 mμg. of DNA in tissue homogenates can be measured.

*Streptokinase serves as an activator of plasminogen (profibrinolysin, a fibrinolysokinase) which exists in normal blood serum. The active product, plasmin (fibrinolysin), is the proteolytic enzyme which dissolves a fibrin clot (see p. 127).

Another very sensitive fluorometric method for measurement of DNA is based on the formation of a fluorescent quinaldine by reacting deoxyribose with 3, 5-diaminobenzoic acid[8]. As little as 2.4 mμg. of DNA can be measured by this method.

By the use of the quartz microscope, quantitative spectrophotometry can be performed on various cellular constituents. This instrument is equipped with quartz optics and thus transmits ultraviolet light. It possesses about twice the resolving power of the ordinary microscope and can detect substances in concentrations as low as 10^{-6} to 10^{-9} milligrams. The nucleic acids and the proteins are most easily studied by these microtechnics involving differential absorption in the ultraviolet. The nucleic acids absorb ultraviolet light strongly at a wavelength of 2600 Å. This is due to the presence of the double bonds in the pyrimidine bases. With equal amounts of nucleic acids and protein in solution, the nucleic acid absorption band completely dominates the protein band at 2600 Å. Thus the pyrimidines can be easily localized in the cell. The cell proteins, on the other hand, have a maximum absorption nearer to 2800 Å. This is due to the presence of the amino acids tryptophan, tyrosine, and phenylalanine, which contain an aromatic nucleus. By the use of these absorption technics many studies are being carried out on the relationship of various aspects of cell growth and reproduction to nucleoprotein metabolism.

• • •

References:

1. Wyatt, G.R., and Cohen, S.S.: Nature **170**:1072, 1952.
2. Lipman, F.: Science **128**:575, 1958.
3. Berg, P., and Joklik, W.K.: J. Biol. Chem. **210**:657, 1954.
4. Dutton, G.J., and Storey, I.D.E.: Biochem. J. **53**:XXXVII, 1953.
5. Cantoni, G.L.: J. Biol. Chem. **204**:403, 1953.
6. Crick, F.H.C., and Watson, J.D.: Nature **171**:737, 1953; Proc. Royal Soc., **223 A**:80, 1954.
7. Roberts, D., and Friedkin, M.: J. Biol. Chem. **233**:483, 1958.
8. Kissane, J., and Robins, E.: J. Biol. Chem. **233**:184, 1958.

Bibliography:

Chargaff, E., and Davidson, J.N., Eds.: The Nucleic Acids. 2 Vols. Academic, 1955.

5...
Porphyrins and Bile Pigments

Porphyrins are cyclic compounds formed by the linkage of four pyrrole rings through methylene bridges. A characteristic property of the porphyrins is the formation of complexes with metal ions. The metal ion is bound to the nitrogen atom of the pyrrole rings. Examples are the iron porphyrins such as **heme** of hemoglobin and the magnesium-containing porphyrin **chlorophyll**, the photosynthetic pigment of plants.

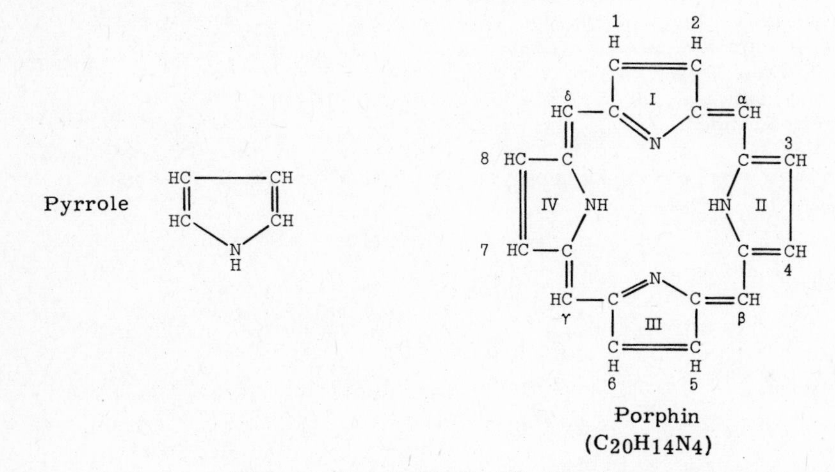

Porphin
($C_{20}H_{14}N_4$)

Rings are labeled I, II, III, IV. Substituent positions on rings are labeled 1, 2, 3, 4, 5, 6, 7, 8. Methylene bridges are labeled α, β, γ, δ.

In nature, the metalloporphyrins are conjugated to proteins to form a number of compounds of importance in biological processes. These include:

A. Hemoglobins: Iron porphyrins attached to the basic protein, globin. These conjugated proteins possess the ability to combine reversibly with oxygen. They serve as the transport mechanism for oxygen within the blood. Hemoglobin has a molecular weight of about 67,000; it contains four atoms of iron per mol in the ferrous (Fe^{++}) state.

B. Erythrocruorins: Iron porphyrinoproteins which occur in the blood and tissue fluids of some invertebrates. They correspond in function to hemoglobin.

C. Myoglobins: Respiratory pigments which occur in the muscle cells of vertebrates and invertebrates. An example is the myoglobin obtained from the heart muscle of the horse and crystallized by Theorell (1934). The purified porphyrinoprotein has a molecular weight of about 17,000 and contains only one atom of iron per mol.

D. Cytochromes: Compounds which act as electron transfer agents in oxidation-reduction reactions (see p. 114). The most important cytochrome is **cytochrome c**, which has a molecular weight of about 13,000 and contains one atom of iron per mol.

E. Catalases (see p. 114): Iron porphyrin enzymes, several of which have been obtained in crystalline form. They are assumed to have a molecular weight of about 225,000 and to contain four atoms of iron per mol. In plants, catalase activity is minimal, but the iron porphyrin enzyme, peroxidase (see p. 114), performs similar functions. A peroxidase from horse-radish has been crystallized; it has a molecular weight of 44,000 and contains one atom of iron per mol.

Structure of Porphyrins:

The porphyrins found in nature are compounds in which various side chains are substituted for the eight hydrogen atoms numbered in the porphin nucleus as shown on p. 53. As a simple means of showing these substitutions, Fischer proposed a shorthand formula in which the methylene bridges are omitted and each pyrrole ring is shown as a bracket with the eight substituent positions numbered as shown below. Uroporphyrin, whose detailed structure is shown on p., 58 would be represented as shown below (A = $-CH_2.COOH$) (P = $-CH_2.CH_2.COOH$):

It will be noted that the arrangement of the A and P substituents in the uroporphyrin shown is asymmetrical (in ring IV, the expected order of the acetate and propionate substituents is reversed). This type of asymmetrical substitution is classified as a Type III porphyrin. A porphyrin with a completely symmetrical arrangement of the substituents is classified as a Type I porphyrin. Only Types I and III are found in nature, and the Type III series is by far the more abundant.

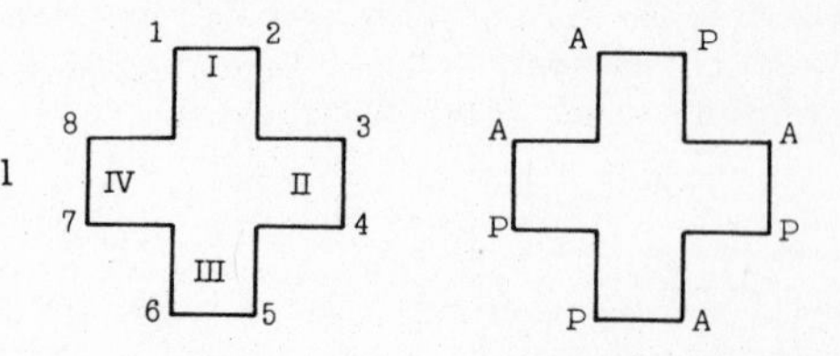

Uroporphyrin III

The formation and occurrence of other porphyrin derivatives may be depicted as follows:

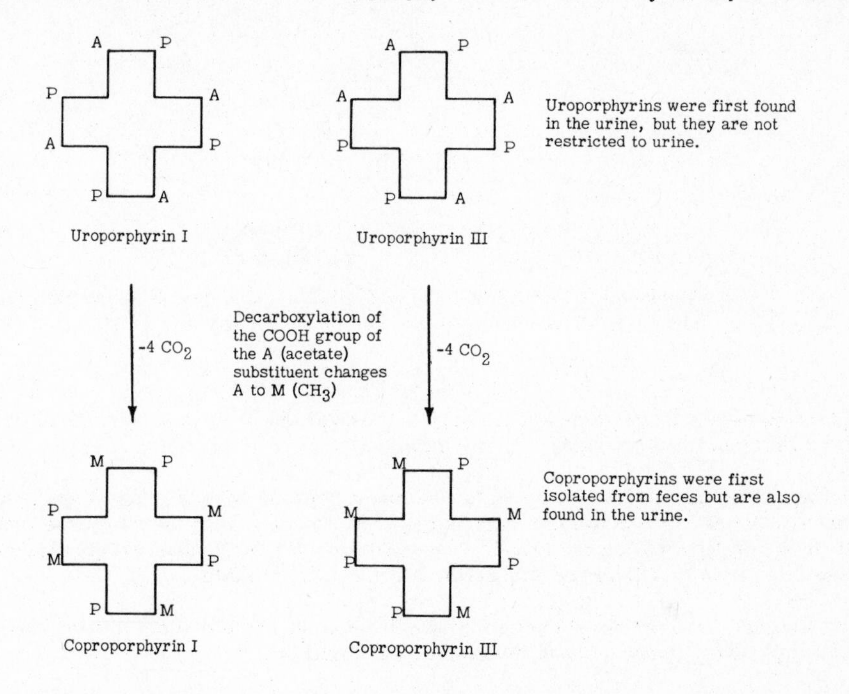

The following compounds are all Type III porphyrins (i.e., the methyl groups are in the same substituent position as in Type III coproporphyrin). However, they are sometimes identified as belonging to series 9 because they were designated ninth in a series of isomers isolated by Hans Fischer, the pioneer worker in the field of porphyrin chemistry. These derivatives have three types of substituents: ethyl ($-CH_2CH_3$), E; hydroxyethyl ($-CH_2CH_2OH$), EOH; or vinyl ($-CH{=}CH_2$), V.

BIOSYNTHESIS OF PORPHOBILINOGEN

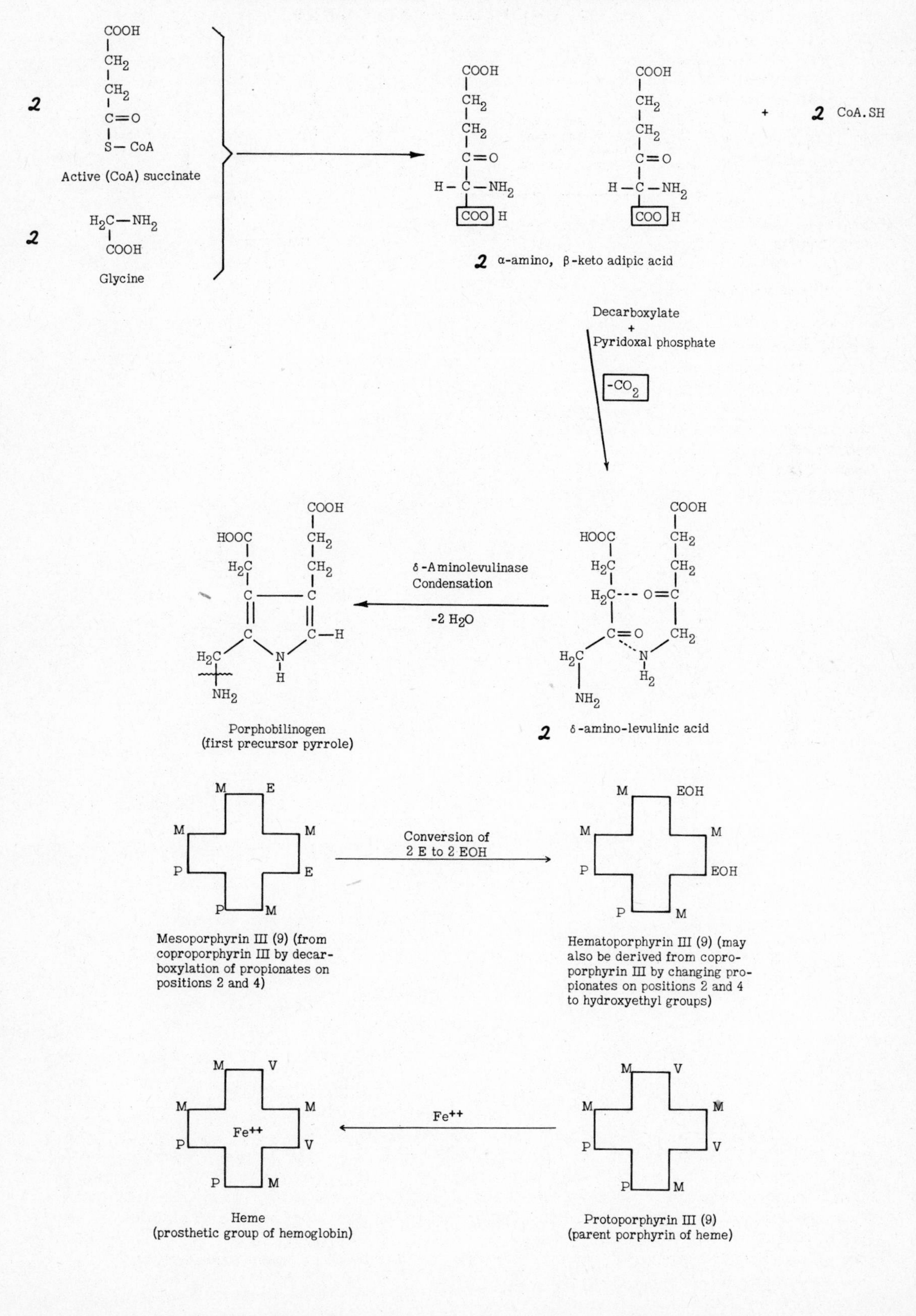

CONVERSION OF PORPHOBILINOGEN TO UROPORPHYRINOGENS

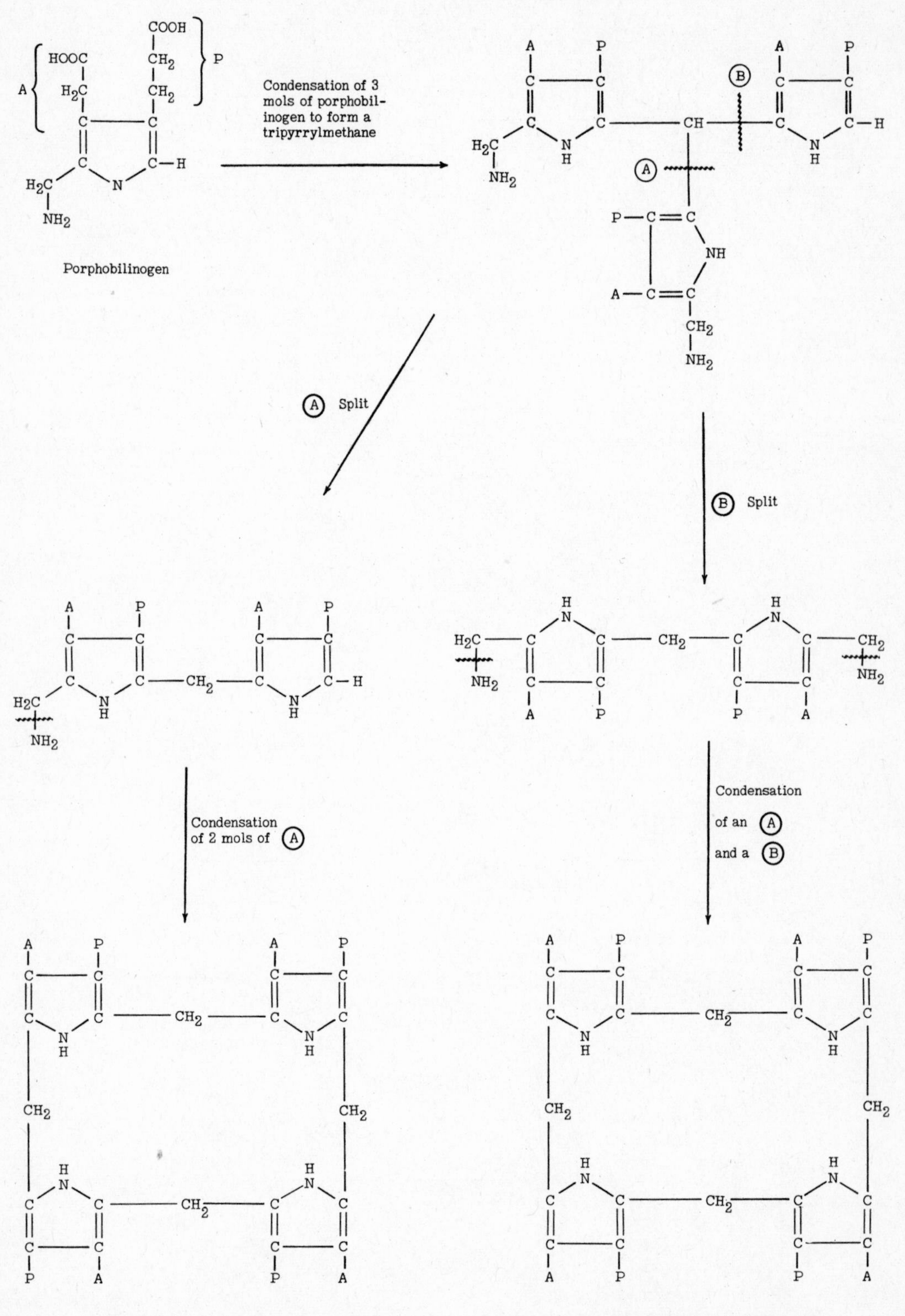

Deuteroporphyrins and mesoporphyrins may be formed in the feces by bacterial activity on protoporphyrin III.

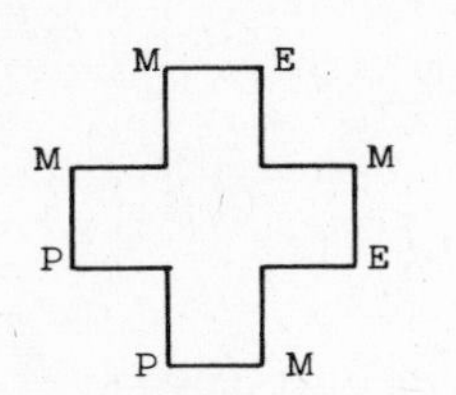

Mesoporphyrin (Type III)

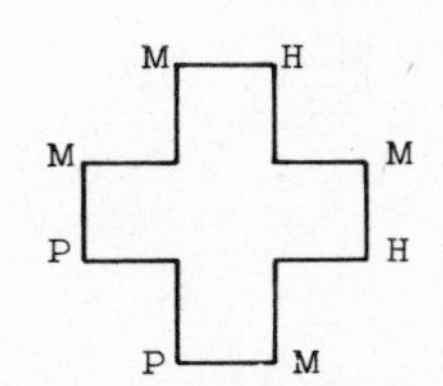

Deuteroporphyrin (Type III)

Biosynthesis of Porphyrins:

Both chlorophyll, the photosynthetic pigment of plants, and heme, the iron protoporphyrin of hemoglobin in animals, are synthesized in living cells by a common pathway. The two starting materials are "active succinate," the Coenzyme A derivative of succinic acid (see p. 88), and the amino acid, glycine. These two compounds combine to form α-amino-β-keto adipic acid, which undergoes decarboxylation to produce δ-amino levulinic acid. Two δ-amino levulinic acid residues then condense to form the first precursor pyrrole, porphobilinogen. These reactions are shown on p. 55.

Schulman and Richert[1] studied the incorporation of glycine and succinate into heme by avian red blood cells. It was found that blood samples from vitamin B_6 and from pantothenic acid-deficient ducklings utilized glycine and succinate for heme synthesis at a reduced rate, although δ-amino levulinic acid incorporation into heme was essentially normal. The addition in vitro of pyridoxal-5-phosphate restored glycine and succinate incorporation without affecting amino levulinic acid incorporation. Added Coenzyme A was without effect in vitro, but the injection of calcium pantothenate one hour before the blood specimens were drawn restored to normal the rate of glycine incorporation into heme. These experiments suggest that the block in heme synthesis in pantothenic acid or in vitamin B_6 deficiency occurs at a very early step in heme synthesis, presumably in the formation of Coenzyme A-succinate and in the pyridoxal-dependent decarboxylation of α-amino-β-keto adipic acid. The anemia which has been found to accompany vitamin B_6 or pantothenic acid deficiency in several species of experimental animals may be explained on a biochemical basis by these observations.

The pathway for the synthesis of porphobilinogen as described above has received support by the finding of both δ-amino levulinic acid and porphobilinogen in the urine of normal subjects[2] as well as in the urine of patients with acute porphyria[3]. In this group of patients these compounds, which are involved in the synthesis of porphyrins, occur in greatly increased amounts. The conversion of δ-amino levulinic acid to porphobilinogen has been shown to occur by in vitro experiments using enzyme preparation from aqueous extracts of hemolysates of chick red blood cells[4] as well as from liver tissue[5].

The formation of a tetrapyrrole, i.e., a porphyrin, occurs by the condensation of four monopyrroles derived from porphobilinogen. In each instance, the amino carbon (originally derived from the α carbon of glycine) serves as the source of the methylene (α, β, γ, δ) carbons which connect each pyrrole in the tetrapyrrole structure. The conversion of porphobilinogen to a porphyrin can be accomplished simply by heating under acid conditions. In the tissues, this reaction is catalyzed by a specific enzyme such as that which has been isolated from avian red blood cells. Lockwood and Rimington have purified such an enzyme which transforms porphobilinogen to uroporphyrin, the first tetrapyrrole. They have designated this enzyme **porphobilinogenase.**

It has been pointed out (see p. 54) that only Types I and III porphyrins occur in nature, and it may be assumed that the Type III isomers are the more abundant since the biologically important porphyrins such as heme and the cytochromes are Type III isomers. Shemin, Russell, and Abramsky[6] have suggested that both Type I and Type III porphyrinogens may be formed as diagrammed on p. 56. In this scheme, 3 mols of porphobilinogen condense first to form a tripyrrylmethane, which then breaks down into a dipyrrylmethane and a monopyrrole. The dipyrryl compounds are of two types, depending upon where the split occurs on the tripyrryl precursor: at the point marked (A) or at (B). The formation of the tetrapyrrole occurs by condensation of two dipyrrylmethanes. If two of the (A) components condense, a Type I porphyrin results; if one

(A) and one (B) condense, a Type III results. Because of the structure of the side chains on porphobilinogen (acetate and propionate), it is clear that uroporphyrinogens Types I and III would result as the first tetrapyrroles to be formed.

The uroporphyrinogens are converted to coproporphyrinogens by decarboxylation of all of the acetate (A) groups, which changes these to methyl (M) substituents (see p. 54). Only Type III coproporphyrinogen is then oxidatively decarboxylated to form protoporphyrinogen, which is auto-oxidized to protoporphyrin. The inclusion of ferrous iron into protoporphyrin completes the synthesis of heme.

Granick and Mauzerall[7a] have obtained three soluble enzyme fractions which are involved in porphyrin biosynthesis from the red blood cells of chickens, rabbits, and man. One fraction, δ-amino levulinase, condenses δ-amino levulinic acid to porphobilinogen (see p. 55). The steps involving condensation of porphobilinogen to uroporphyrinogens (see p. 56) seem to require two kinds of enzymic reactions. By the action of an **isomerase**, uroporphyrinogen III is produced. The absence of this enzyme leads to the formation of uroporphyrinogen I. This observation may be significant as an explanation for a biochemical error in porphyria (see p. 61). In the third enzymatic step, both uroporphyrinogens I and III are acted upon by a **decarboxylase** removing carboxyl groups from the acetic acid side chains to produce coproporphyrinogens Types I and III. Finally, as noted above, only coproporphyrinogen III is further metabolized to protoporphyrin III and to the iron protoporphyrin, heme.

The porphyrinogens which have been described above are colorless reduced porphyrins containing six extra hydrogen atoms as compared to the corresponding porphyrins (see below). They have been reported to occur in biological material by Fischer and by Watson, Schwartz, and their associates. It is now apparent that these reduced porphyrins (the porphyrinogens) and not the corresponding porphyrins are the actual intermediates in the biosynthesis of protoporphyrin and of heme. This idea is supported by the observation that the oxidized porphyrins cannot be used for heme or for chlorophyll synthesis either by intact or disrupted cells. Furthermore, the condensation of four mols of porphobilinogen (see p. 56) would directly give rise to uroporphyrinogens rather than to uroporphyrins.

The porphyrinogens are readily auto-oxidized to the respective porphyrins (see below). These oxidations are catalyzed in the presence of light and by the porphyrins that are formed. The amounts of these porphyrin by-products that are produced depends not only on the activities of the various enzymes involved but also on the presence of catalysts (light) or inhibitors (reduced glutathione) of their auto-oxidation.

A summary of the steps in the biosynthesis of the porphyrin derivatives from porphobilinogen is given on p. 59.

Oxidation of Uroporphyrinogen to Uroporphyrin

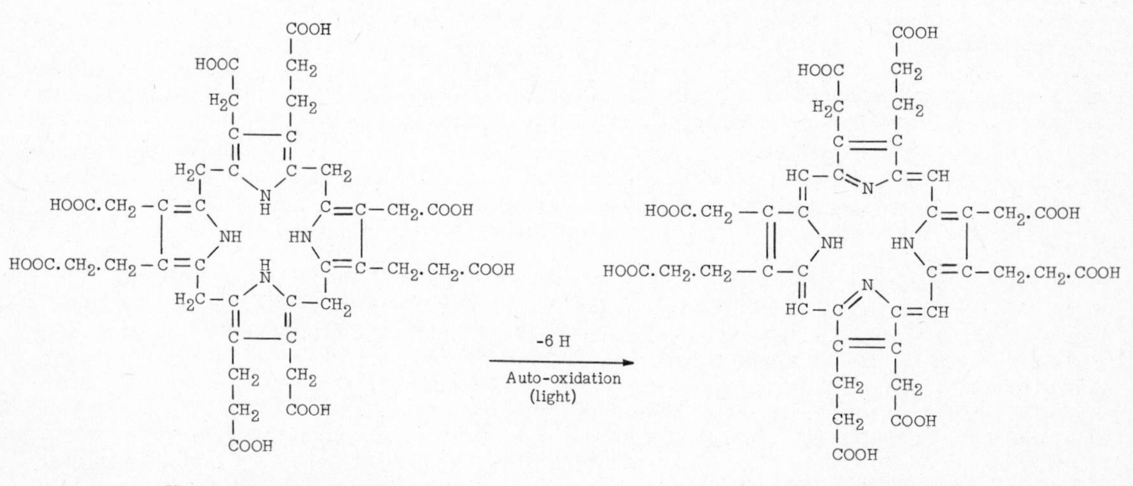

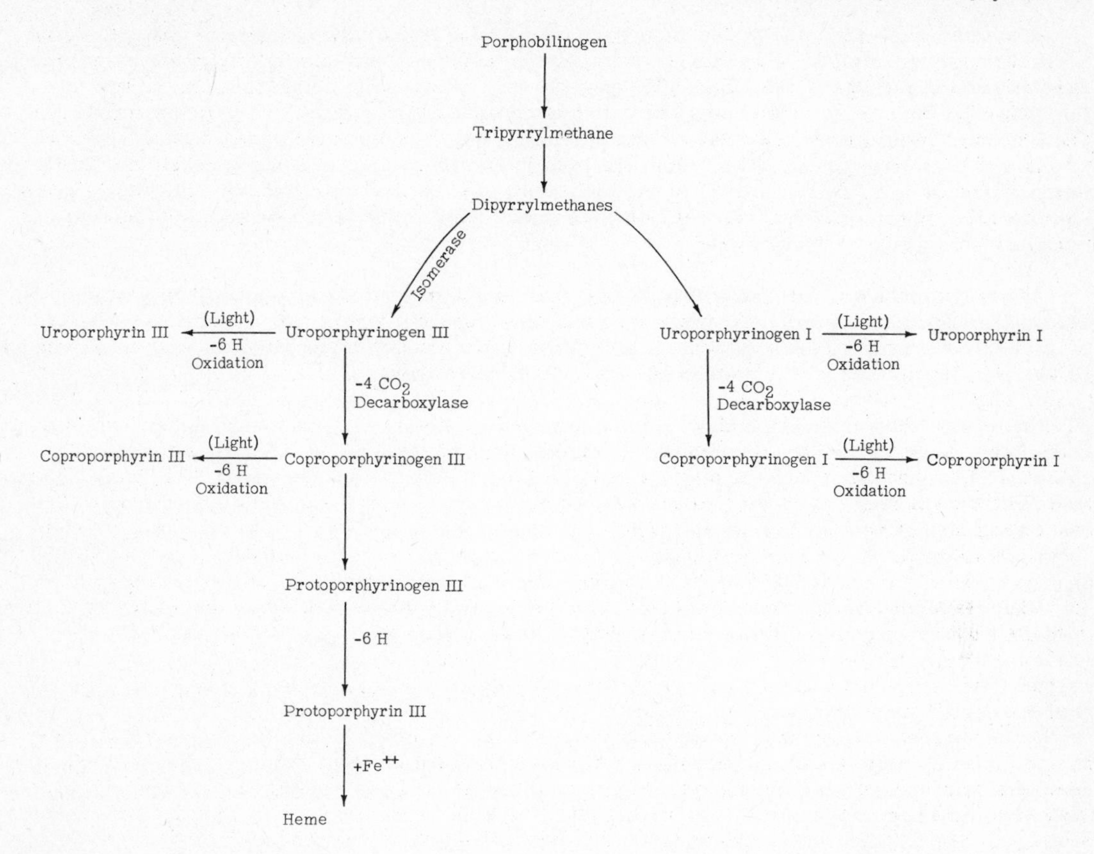

Chemistry of Porphyrins:

Because of the presence of tertiary nitrogens in the two pyrrolene rings contained in each porphyrin, these compounds act as weak bases. Those which possess a carboxyl group on the side chain act also as acids. As a result, such compounds are weakly amphoteric (see p. 31). Their isoelectric points range from pH 3.0 to 4.5, and within this pH range the porphyrins may easily be precipitated from an aqueous solution.

The various porphyrinogens are colorless, whereas the various porphyrins are all colored. In the study of porphyrins or porphyrin derivatives, the characteristic absorption spectrum which each exhibits, both in the visible and the ultraviolet regions of the spectrum, is of great value. An example is the absorption curve for a solution of hematoporphyrin in 5% hydrochloric acid (see below). Note the sharp absorption band near 400 mμ. This is a distinguishing feature of the porphin ring and is characteristic of all porphyrins regardless of the side chains present. This band is termed the **Soret band**, after its discoverer. Hematoporphyrin in acid solution, in addition to the Soret band, has two weaker absorption bands with maxima at 550 and 592 mμ.

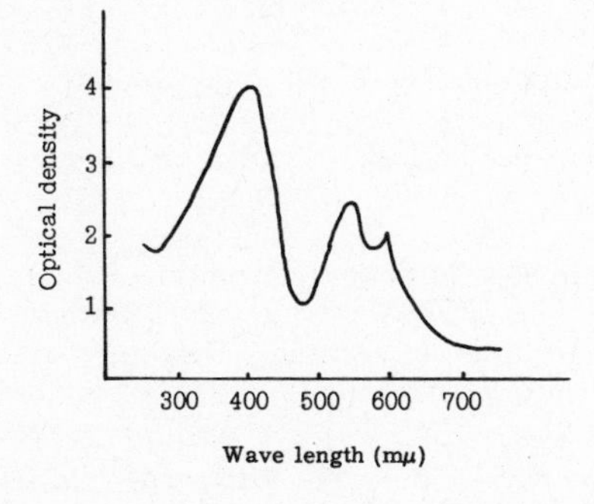

Absorption Spectrum of Hematoporphyrin (0.01% solution in 5% hydrochloric acid)

In organic solvents, porphyrins have four main bands in the visible spectrum as well as the Soret band. For example, a solution of protoporphyrin in an ether–acetic acid mixture exhibits absorption bands at 632.5, 576, 537, 502, and 395 mμ. When porphyrins dissolved in strong mineral acids or in organic solvents are illuminated by ultraviolet light, they emit a strong red fluorescence. This fluorescence is so characteristic that it is frequently used to detect small amounts of free porphyrins. The double bonds in the porphyrins are responsible for the characteristic absorption and fluorescence of these compounds, and as previously noted, the reduction (by addition of hydrogen) of the methene (C=H) bridges to CH_2 leads to the formation of colorless compounds termed **porphyrinogens.**

When a porphyrin combines with a metal, its absorption in the visible spectrum becomes changed. This is exemplified by protoporphyrin, the iron-free precursor of heme. In alkaline solution, protoporphyrin shows several sharp absorption bands (at 645, 591, and 540 mμ), whereas heme has a broad band with a plateau extending from 540 to 580 mμ.

Heme and other ferrous porphyrin complexes react readily with basic substances such as hydrazines, primary amines, pyridines, ammonia, or an imidazole such as the amino acid, histidine. The resulting compound is called a hemochromogen (hemochrome). The hemochromogens still show a Soret band but also exhibit two absorption bands in the visible spectrum. The band at the longer wave length is called the alpha band; that at the shorter wave length, the beta band. An example is the hemochromogen formed with pyridine. This has its alpha band at 559 mμ and its beta band at 527.5 mμ. In the formation of a hemochromogen, it is believed that two molecules of the basic substance replace two molecules of water which were loosely bound to iron in the ferrous porphyrin. This is illustrated by the structure of hemoglobin (described below).

Structures of Hemoglobin and Cytochrome c:

In hemoglobin, the iron in the protoporphyrin III (9) structure (heme) is conjugated to imidazole nitrogens contained in two histidine residues within the protein, globin. When hemoglobin combines with oxygen, the iron is displaced from one imidazole group and one molecule of oxygen is bound by one atom of iron. At the same time, there is an increase in the acidity of the compound, i.e., oxyhemoglobin is a stronger acid than is reduced hemoglobin.

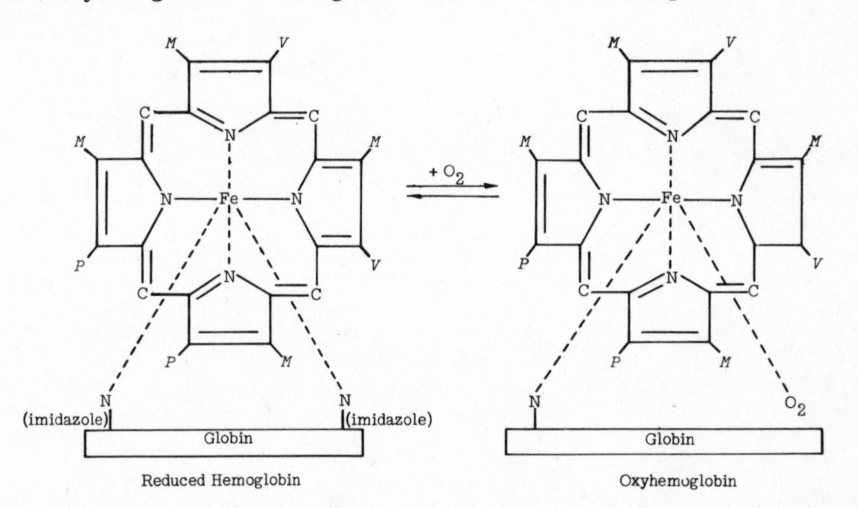

Cytochrome c is also an iron porphyrin; it contains about 0.43% iron. The iron porphyrin is derived from protoporphyrin III (9), but the two vinyl side chains are reduced and linked by a thioether bond to two cysteine residues of the protein; in addition, the iron is linked to a histidine residue in the protein, as in the case of hemoglobin.

The amino acid sequence of the hemin-containing peptide fragment obtained from the proteolytic breakdown of cytochrome c has recently been reported[7b]. In order, the 12 amino acids which it contains are valine, glutamine, lysine, cysteine, alanine, glutamine, cysteine, histidine, threonine, valine, glutamine, and lysine. In the structure shown on p. 61, the two cysteine residues of the peptide are shown attached by sulfur bridges to the two saturated vinyl side chains of hemin; the linkage of iron is depicted through the imidazole nitrogen of the histidine in the peptide.

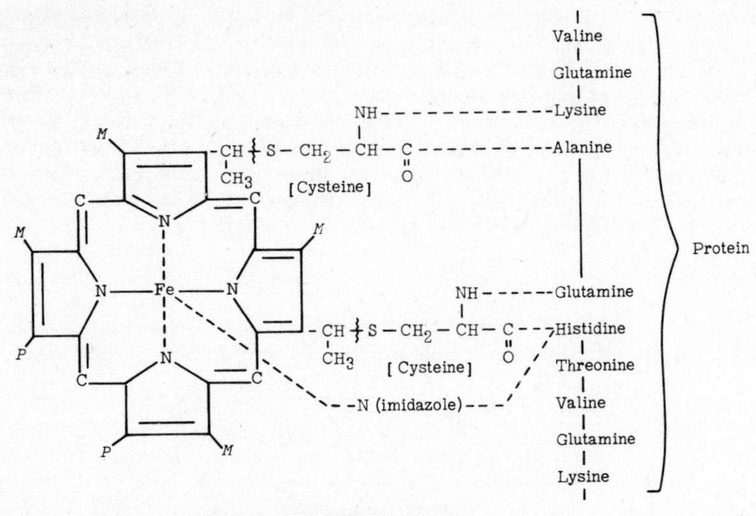

Structure of Cytochrome c

Tests for Porphyrins:

The presence of coproporphyrins or of uroporphyrins is of clinical interest since these two types of compounds are excreted in increased amounts in the porphyrias. Coproporphyrins I and III are soluble in glacial acetic acid–ether mixtures, from which they may then be extracted by hydrochloric acid. Uroporphyrins, on the other hand, are not soluble in acetic acid–ether mixtures but are partially soluble in ethyl acetate, from which they may be extracted by hydrochloric acid. In the HCl solution ultraviolet illumination gives a characteristic red fluorescence. A spectrophotometer may then be used to demonstrate the characteristic absorption bands. The melting point of the methyl esters of the various porphyrins may also be used to differentiate them[8, 9]. Paper chromatography (p. 36) has recently been employed as a means to separate and identify the porphyrins.

Porphyrinuria:

The excretion of coproporphyrins may be increased under many circumstances, e.g., acute febrile states; after the ingestion of certain poisons, particularly heavy metals such as lead or arsenic; blood dyscrasias, hemolytic anemia, or pernicious anemia; and sprue, cirrhosis of the liver, acute pancreatitis, and malignancies (e.g., Hodgkin's disease). In these conditions, uroporphyrin is not present in the urine.

The Porphyrias:

When the excretion of both coproporphyrin and uroporphyrin is increased because of their presence in the blood, the condition is referred to as **porphyria.** Under this term a number of syndromes are included; some are hereditary and familial and some are acquired, but all are characterized by increased excretion of uroporphyrin and coproporphyrin in the urine or feces or both. Reduced catalase activity of the liver has also been reported in these cases. The origin of the metabolic disturbance which produces the porphyrias is not known.

A summary of these observations on the excretion of porphyrins is given in the table on p. 62.

Acute porphyria. This is the most common type of porphyria. It is believed to be primarily related to a hepatic defect (hepatogenic porphyria), but is often described as the gastrointestinal-nervous type or as acute relapsing, toxic, or acquired porphyria. It is probably inherited as a Mendelian-dominant trait, although it may not manifest itself until the third decade. The presenting symptoms occur most often in the gastrointestinal tract and nervous system. In the urine, which is characteristically pigmented and darkens on standing, there is an increased excretion of both Type III coproporphyrin and uroporphyrin. δ-Aminolevulinic acid and porphobilinogen have also been isolated from the urine in acute porphyria[3].

Congenital porphyria. A second form of porphyria, sometimes referred to as congenital or erythrogenic porphyria, is the light-sensitive type. It is different from the acute porphyria described above and, since it is transmitted as a Mendelian-recessive trait, is rare. It occurs more often in males than in females and usually manifests itself early in life. The signs include sensitivity to light (as a result of accumulation of porphyrins under the skin), anemia, and enlargement of the liver and spleen. The bone marrow appears to be the site of the metabolic error. The areas of the body which are exposed to light become necrotic, scarred, and deformed, and the patient usually dies from the severe anemia. Large amounts of Type I coproporphyrin and uroporphyrin appear in the feces and in the urine, which is characteristically red.

Excretion of Porphyrins

Type	Porphyrins Excreted	Remarks
Normal	Urine: Copro. I, Copro. III } 60-280 (avg. 160) γ/day; about 30% is Type III.	The excretion of uroporphyrins is negligible in normal individuals, averaging 15-30 γ/day, mostly Type I.
	Feces: Copro. I, Copro. III } 300-1100 γ/day; 70-90% is Type I.	
HEREDITARY		
Acute (hepatogenic) porphyria (increased porphyrins in the liver)	Mainly Type III porphyrins. Copro. III: 144-2582 γ/day. Copro. I: Small amounts. Uro. I and III*: 61-147,000 γ/day. Porphobilinogen, δ-aminolevulinic acid } in urine.	Hereditary as a Mendelian dominant. Relatively common; metabolic defect is in the liver. Catalase activity of liver is markedly reduced. Patients are not light-sensitive.
Congenital (erythrogenic) porphyria (increased porphyrins in the marrow)	Mainly Type I porphyrins. Type III coproporphyrin was reported in 2 cases. The fecal content of porphyrin is high.	Hereditary as a Mendelian recessive. Rare. Patient shows sensitivity to light. Marrow is site of metabolic error.
Chronic porphyria (mixed)	Varies. There may be increased amounts of Copro. I and III and no uroporphyrins; in other cases, Uro. I and III are increased; still others have mixtures of copro- and uroporphyrins. Porphobilinogen may also be found in the urine.	Hereditary or acquired? Patient may be light-sensitive. Frequently associated with enlargement of the liver.
ACQUIRED		
Toxic agents	Copro. III.	E.g., heavy metals, chemicals, acute alcoholism, and cirrhosis in alcoholics.
Liver disease	Copro. I.	E.g., infectious hepatitis, cirrhosis not accompanied by alcoholism, obstructive jaundice.
Blood dyscrasias	Copro. I.	E.g., leukemia, pernicious anemia, hemolytic anemias.
Miscellaneous	Copro. III.	E.g., poliomyelitis, aplastic anemias, Hodgkin's disease.

*According to Watson[10], the uroporphyrins excreted in acute porphyria are a complex mixture of Uroporphyrin I and a peculiar Uroporphyrin III which has seven rather than eight carboxyl groups. The mixture is referred to as Waldenström porphyrin. It is extractable from urine by ethyl acetate but not by ether.

STRUCTURE OF SOME BILE PIGMENTS

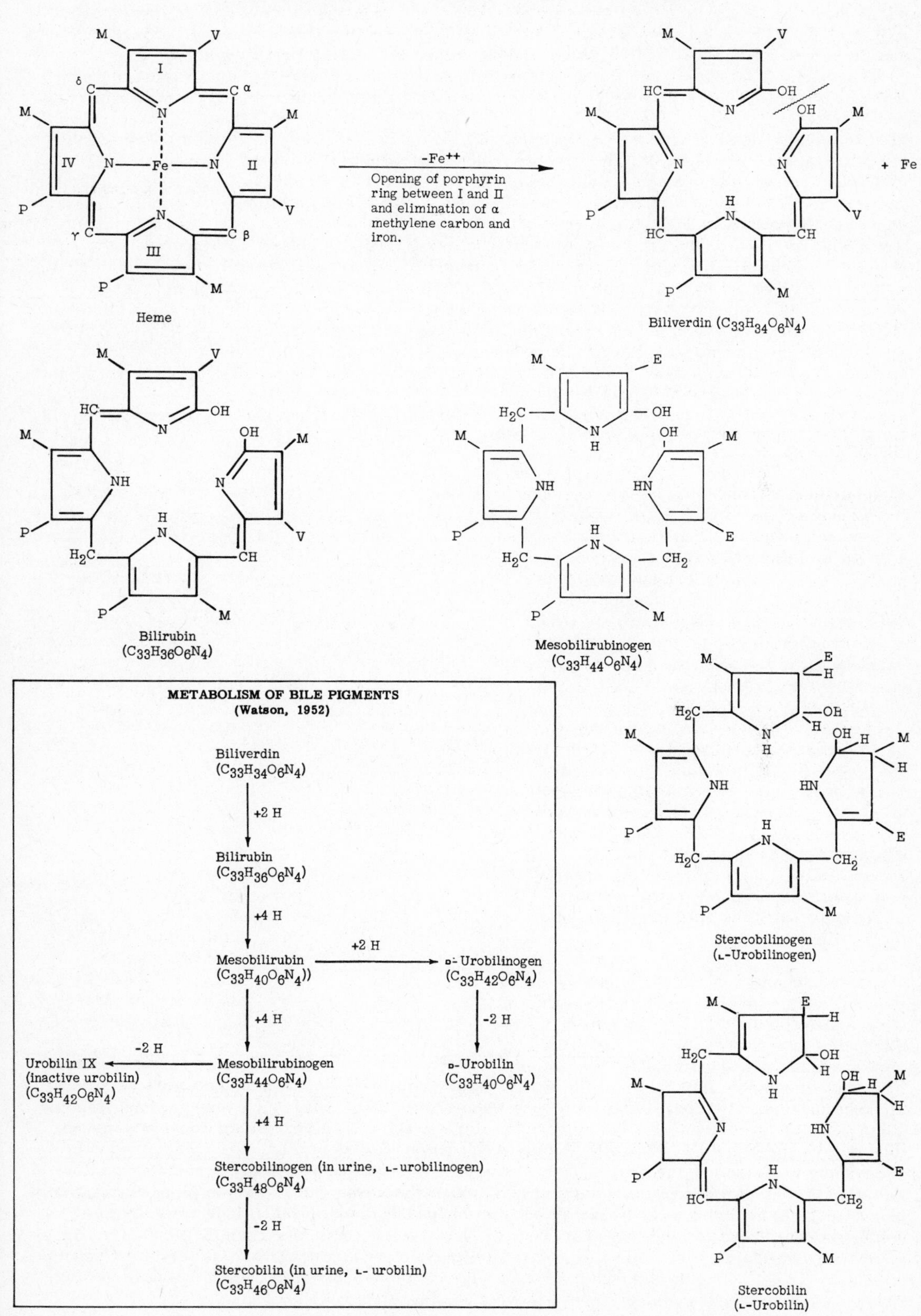

Other porphyrias. A chronic porphyria has also been described which is characterized in some cases by the excretion of increased amounts of coproporphyrins I and III without uroporphyrin; in others, by the opposite situation; or, in still others, by increases in both coproporphyrins and uroporphyrins. Finally, there also occurs within certain families a type of chronic coproporphyrinuria ("idiopathic porphyria") in which the only abnormality is the finding of increased amounts of coproporphyrin in the urine without any apparent accompanying symptoms of disease.

Porphyria has been produced experimentally in rabbit, rat, and chick embryos by the injection of the sedative drug, Sedormid® (allylisopropylacetylcarbamide). Labbe, Talman, and Aldrich[11] noted that the excretion of uric acid in the porphyric embryos was markedly decreased and the reduction was inversely proportionate to the increase in porphyrin excretion. This suggested that some manifestations of porphyria may result from inhibited synthesis of purines rather than inhibited synthesis of porphyrins, as was formerly believed. The hypothesis is supported by a consideration of the succinate-glycine cycle as proposed by Shemin and Russell[12] (see p. 236). It will be noted that according to this scheme, δ-aminolevulinic acid is a common precursor of purines as well as porphyrins. Inhibition of the oxidative deamination of δ-aminolevulinic acid might block the pathway to purine synthesis and thus shunt δ-aminolevulinic acid entirely in the direction of porphyrin synthesis with consequent overproduction and increased excretion of these compounds. This hypothesis is of considerable interest in view of the well-known clinical observation that overdosage with barbiturate and other sedative drugs, particularly those containing an allyl side chain, may induce porphyria, and that the administration of barbiturates to porphyric patients enhances the excretion of porphyrins in such patients.

Catabolism of Heme; Formation of the Bile Pigments:

When hemoglobin is destroyed in the body the protein portion, globin, may be re-utilized, either as such or in the form of its constituent amino acids, and the iron enters the iron "pool" - also for re-use. However, the porphyrin portion, heme, is broken down in all likelihood mainly in the reticuloendothelial cells of the liver, spleen, and bone marrow. The initial step in the metabolic degradation of heme involves opening of the porphyrin ring between I and II and elimination of the alpha methylene carbon (see p. 63). It is possible that the porphyrin ring may be opened and that the iron is still present before the protoporphyrin is released from the globin. Such a green, conjugated protein has been produced by oxidation of hemoglobin by oxygen in the presence of ascorbic acid. The prosthetic group of the protein is the iron complex of a bile pigment resembling biliverdin. It has been named **choleglobin** by Lemberg.

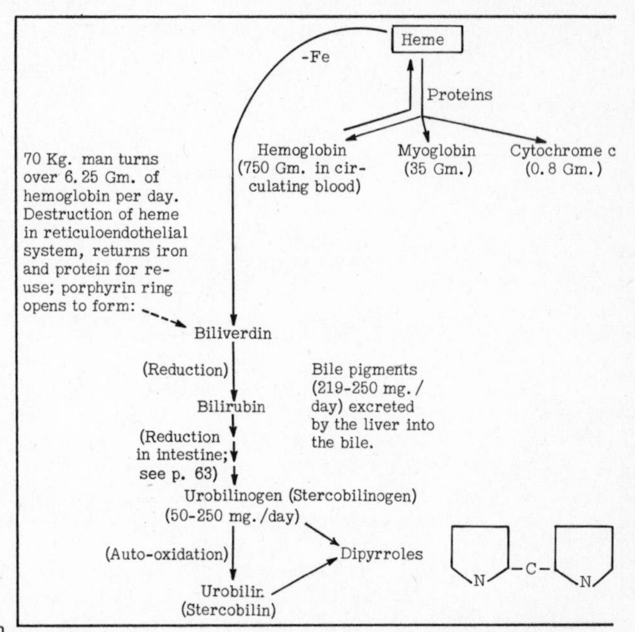

After removal of iron and cleavage of the porphyrin ring of heme, as described above, **biliverdin,** the first of the bile pigments, is formed (see p. 63). Biliverdin is easily reduced to bilirubin, which is the major pigment in human bile; there are only slight traces of biliverdin, which is the chief pigment of the bile in birds. Bilirubin is transported in the plasma to the liver, where it is conjugated with glucuronic acid to form a bilirubin diglucuronide[13] (see p. 274) and excreted by way of the bile into the intestine. The bile pigments constitute 15-20% of the dry weight of human bile.

In the lower portions of the intestinal tract, especially the colon, the bile pigments are subjected to the reductive action of enzyme systems present in the intestinal bacteria (mainly the coliform organisms). If the intestinal flora are modified or diminished, as by the administration of oral antibiotic agents which are capable of producing partial sterilization of the intestinal tract, bilirubin may not be further reduced and may later be auto-oxidized in contact with air to biliverdin. Thus the feces acquire a green tinge under these circumstances.

Under normal circumstances, the first product of the reduction of bilirubin within the intestine is mesobilirubin, a compound which contains an additional four hydrogen atoms acquired by reduction of the two vinyl groups of bilirubin to ethyl (E) groups. When four more hydrogen atoms are added (two at the β and δ methylene carbons, and two at the pyrrole nitrogens of rings I and II), mesobilirubinogen is formed. Further reduction produces stercobilinogen (also designated L-urobilinogen); and auto-oxidation in the presence of air of stercobilinogen, which is colorless, produces stercobilin (L-urobilin), a brown-red pigment which gives the normal color to the feces. Stercobilin is strongly levorotatory ($[\alpha]_D = -3600°$).

The various reduction products of bilirubin may in part be absorbed from the intestine and returned to the liver for re-excretion, or may be excreted in the urine, which normally contains traces of urobilinogen and urobilin as well as mesobilirubinogen. The great majority of the bile pigments are, however, excreted with the feces.

Urobilin IX α, or inactive (i) urobilin, is an optically inactive bile pigment which has been identified in the feces. It is less stable than stercobilin, becoming oxidized in air to form violet and red pigments. In the feces of patients whose intestinal flora has been changed by oral administration of oxytetracycline or chlortetracycline, a dextrorotatory bile pigment, D-urobilin ($[\alpha]_D = +5000°$) has been found. It is believed to be derived from mesobilirubin by way of a D-urobilinogen[14].

The metabolism of bile pigments is summarized and the structures of the more important compounds are shown on p. 63.

The presence of dipyrroles in association with other bile pigments has recently been reported. Such compounds have been designated **mesobilifuscins**, and are said to be identical with the copronigrin isolated from feces by Watson. It is not known whether the dipyrroles are precursors or breakdown products of the tetrapyrroles, or both. It will be recalled that dipyrroles may be intermediates in the formation of tetrapyrroles (see p. 57).

Summary of Porphyrin Metabolism:

Types I and III uroporphyrinogens are synthesized from glycine and succinate precursors. By decarboxylation, these are converted to coproporphyrinogens. Both uroporphyrinogens and coproporphyrinogens are readily auto-oxidized to the corresponding uroporphyrins or coproporphyrins. The Type I coproporphyrin is apparently a by-product of the synthesis of heme protoporphyrin because the amount produced is proportional to the production of hemoglobin. However, this pigment is useless and is excreted in the urine in amounts of 40 to 190 γ per day except under conditions of rapid hemopoiesis, as in hemolytic disease, where the daily urinary excretion of Coproporphyrin I may exceed 200 to 400 γ per day. A somewhat larger amount of coproporphyrin is normally excreted in the feces: 300 to 1100 γ per day, 70-90% of which is Coproporphyrin I.

Type III coproporphyrinogen is largely converted to protoporphyrin and thence to heme, which conjugates with protein to form the hemoproteins: hemoglobin, myoglobin, cytochrome, etc. A small amount (20 to 90 γ per day) of Type III coproporphyrin is normally excreted as such in the urine.

The excretion of uroporphyrins by normal subjects is negligible.

The breakdown of heme leads to the production of the bile pigments, so that the amount of bile pigment formed each day is closely related to the amount of hemoglobin created and destroyed. In hemoglobin, the porphyrin portion, exclusive of the iron, makes up 3.5% by weight of the hemoglobin molecule. Thus, 35 mg. of bilirubin could be expected to appear for each gram of hemoglobin destroyed. It is estimated that in a 70 Kg. man, about 6.25 grams of hemoglobin are produced and destroyed each day (normal is 90 mg./Kg./day). This means that about 219 mg. of bilirubin (6.25 x 35) should be produced per day in this same individual. The bilirubin is excreted by the liver into the intestine by way of the bile and may be measured as fecal urobilinogen (stercobilinogen), to which it is converted by the action of the intestinal bacteria. However, 10 to 20% more than this estimated quantity actually appears each day as urobilinogen. This is probably due to some porphyrin which is synthesized probably in the liver but never actually incorporated into red cells (early urobilinogen), as well as porphyrin derived from the catabolism of hemoproteins other than hemoglobin. Total daily bile pigment production is therefore close to 250 mg., but not all of this is recovered as urobilinogen because some of the urobilinogen is broken down by bacterial action to dipyrroles (see p. 56), which do not yield a color with the Ehrlich reagent (see p. 278) used to measure urobilinogen. The diagram on p. 64 summarizes the catabolism of heme and the formation of bile pigments.

• • •

References:

1. Schulman, M. P., and Richert, D. A.: J. Biol. Chem. **226**:181, 1957.
2. Mauzerall, D., and Granick, S.: J. Biol. Chem. **219**:435, 1956.
3. Granick, S., and Vanden Schrieck, H. G.: Proc. Soc. Exper. Biol. and Med. **88**:270, 1955.
4. Granick, S.: Science **120**:1105, 1954.
5. Gibson, K. D., Neuberger, A., and Scott, J. J.: Biochem. J. **58**:XLI, 1954.
6. Shemin, D., Russell, C. S., and Abramsky, T.: J. Biol. Chem. **215**:613, 1955.
7. a. Granick, S., and Mauzerall, D.: J. Biol. Chem. **232**:1119, 1958.
   b. Theorell, H.: Science **124**:467, 1956.
8. Dobriner, K., and Rhoads, C. P.: Physiol. Rev. **20**:416, 1940.
9. Watson, C. J., and Larson, E. A.: Physiol. Rev. **27**:478, 1947.
10. Watson, C. J., Schwartz, S., and Hawkinson, V.: J. Biol. Chem. **157**:345, 1945.
11. Labbe, R. F., Talman, E. L., and Aldrich, R. A.: Biochim. Biophys. Acta **15**:590, 1954.
12. Shemin, D., and Russell, C. S.: J. Am. Chem. Soc. **74**:4873, 1953.
13. Schmid, R.: Science **124**:76, 1956.
14. Watson, C. J., and Lowry, P. T.: J. Biol. Chem. **218**:633, 641, 1956.

Bibliography:

Lemberg, R., and Legge, J. W.: Hematin Compounds and Bile Pigments. Interscience, 1949.

Symposium, Porphyrin Biosynthesis and Metabolism. Ciba Foundation. Little, Brown and Co., 1954.

Rimington, C.: Haem Pigments and Porphyrins. Ann. Rev. Biochem. **26**:561. Annual Reviews, 1957.

6...
Vitamins

When animals are maintained on a chemically defined diet containing only purified proteins, carbohydrates, and fats, and the necessary minerals, it is not possible to sustain life. Additional factors present in natural foods are required, although often only minute amounts are necessary. These ''accessory food factors'' are called vitamins. The vitamins have no chemical resemblance to each other, but because of a similar general function in metabolism they are considered together.

Early studies of the vitamins emphasized the more obvious anatomic changes which occurred when animals were maintained on vitamin-deficient diets. Increased knowledge of the physiologic role of each vitamin has enabled us to concentrate attention on the metabolic defects which occur when these substances are lacking, and we may therefore refer to the biochemical changes as well as the anatomic lesions which are characteristic of the various vitamin deficiency states.

Before the chemical structures of the vitamins were known it was customary to identify these substances by letters of the alphabet. This system is gradually being replaced by a nomenclature based on the chemical nature of the compound or a description of its source or function.

The vitamins are generally divided into two major groups: fat-soluble and water-soluble. The fat-soluble vitamins, which are usually found associated with the lipids of natural foods, include vitamins A, D, E, and K. The vitamins of the B complex and vitamin C comprise the water-soluble group.

FAT-SOLUBLE VITAMINS

VITAMIN A

Chemistry:
Vitamin A is an alcohol with a high molecular weight. On the following page is shown the derivation of vitamin A_1 from β-carotene by a hydrolytic cleavage at the mid-point in the polyene chain connecting the two β-ionone rings.

Vitamin A_2 has also been described. Its potency is one hundred times less than that of vitamin A_1, and it also differs from A_1 in the structure of the polyene side chain.

Vitamin A alcohol is found only in the animal kingdom. It occurs, mainly as an ester with higher fatty acids, in the liver, kidney, lung, and fat depots. The sources of all of the vitamin A in animals are probably certain plant pigments known as carotenes or carotenoid pigments, the provitamins A, which are synthesized by all plants except parasites and saprophytes. These provitamins A are transformed into vitamin A in the animal body. β-Carotene yields two molecules of vitamin A, whereas the α- and γ-carotenes yield only one molecule since they are not symmetrical. The conversion of the carotenes to vitamin A occurs in the intestinal wall in rats, pigs, goats, rabbits, sheep, and chickens; but in man the liver is believed to be the only organ which is capable of accomplishing this transformation.

These provitamins of the diet are less well absorbed from the intestine than vitamin A alcohol. Many factors affect the efficiency of absorption and utilization of preformed vitamin A and carotene. Small amounts of mineral oil added to the diet of experimental animals were found to interfere with the utilization of both vitamin A and carotene, although the inhibition of carotene utilization was much greater than that of vitamin A.

In general it is assumed that carotene has one-half or less the value of an equal quantity of vitamin A in the diet of man. This is due to the differences in intestinal absorption and the necessity for conversion to vitamin A by the liver.

β-Carotene

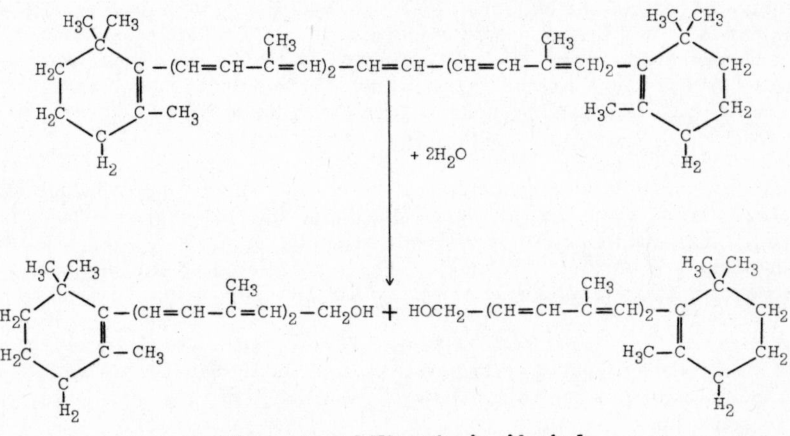

2 Molecules of Vitamin A_1 Alcohol

Role in Physiology:

The maintenance of the integrity of epithelial tissue is an important function of vitamin A. In its absence normal secretory epithelium is replaced by a dry, keratinized epithelium which is more susceptible to invasion by infectious organisms. Vitamin A is also a constituent of visual purple, the pigment involved in rod vision in the retina. Night blindness (nyctalopia), which is caused by a disturbance of rod vision, is therefore a manifestation of vitamin A deficiency. Xerophthalmia, i. e., keratinization of ocular tissue, which may progress to blindness, is a late result of vitamin A deficiency.

In the absence of vitamin A the growth of experimental animals does not progress normally. The skeleton is affected first, then the soft tissues. Mechanical damage to the brain and cord occurs when these structures attempt to grow within the arrested limits of the bony framework of the cranium and vertebral column.

Sources and Daily Allowance:

All yellow vegetables and fruits (e. g., sweet potatoes, apricots, and yellow peaches) and the leafy green vegetables supply provitamin A in the diet. Preformed vitamin A is supplied by milk fat, liver, and, to a lesser extent, by kidney and the fat of muscle meats.

The recommended daily allowance is 5000 I. U. per day for an adult (increase to 6000 to 8000 units during pregnancy and lactation). An international unit is equivalent to the activity of 0.6 micrograms (0. 0006 milligrams) of pure β-carotene; 0. 3 micrograms of vitamin A alcohol; 0. 344 micrograms of vitamin A acetate.

A number of reports[1] have emphasized the possibility of toxic effects as a result of the ingestion of excess amounts of vitamin A. Hypervitaminosis A may occur as a consequence of the administration of large doses (in the form of vitamin A concentrates) to infants and small children. The principal symptoms are painful joints, periosteal thickening of long bones, and loss of hair.

Vitamin A deficiency can occur not only from inadequate intake but also because of poor intestinal absorption or inadequate conversion of provitamin A, as occurs in diseases of the liver. In such cases, a high plasma carotene content may coincide with a low vitamin A level.

It is important to point out that the fat-soluble group of vitamins cannot be absorbed from the intestine in the absence of bile. For this reason any defect in fat absorption is likely to foster deficiencies of fat-soluble vitamins as well.

Vitamin A was formerly assayed by means of tedious and expensive technics, using rats as test animals. Chemical and physical methods of determination based on spectrophotometric measurements are now available. Vitamin A_1 absorbs maximally at 610 to 620 mμ and A_2 at 692 to 696 mμ. A colorimetric determination of vitamin A is based on the Carr-Price reaction, in which a blue color is obtained when a solution of antimony trichloride in chloroform is added to the vitamin-containing mixture. This reaction may be used to determine the vitamin A content of blood plasma.

THE VITAMINS D

The vitamins D are actually a group of compounds. All are sterols which occur in nature, chiefly in the animal organism. Certain of these sterols (known as provitamins D), when subjected to long wave ultraviolet light (about 265 mμ), acquire the physiologic property of curing or preventing rickets, a disease characterized by skeletal abnormalities, including failure of calcification.

Although all of the vitamins D possess antirachitic properties, there is a considerable difference in their potency when tested in various species. For example, irradiated ergosterol (vitamin D_2) is a powerful antirachitic vitamin for man and for the rat but not for the chicken. Vitamin D_3, on the other hand, is much more potent for the chicken than for the rat or the human organism.

Chemistry:

The two most important vitamins D for nutritional purposes are D_2 (activated ergosterol; also known as calciferol or viosterol) and D_3 (activated 7-dehydrocholesterol), the form which occurs in nature in the fish liver oils. Provitamin D_2 (ergosterol) occurs in the plant kingdom (e.g., in ergot and in yeast).

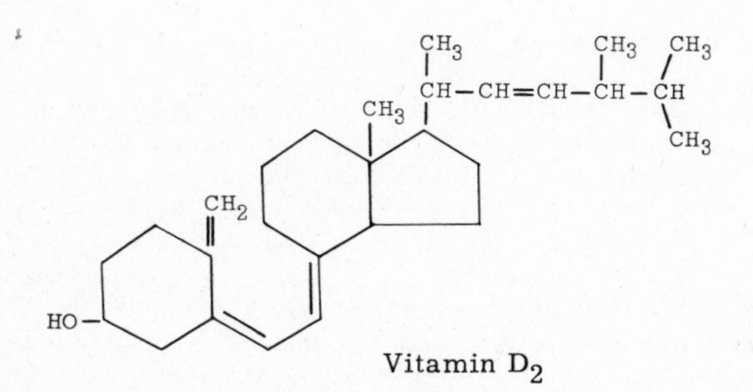

Vitamin D_2

The structure of vitamin D_3 is the same as that of D_2 given above except that the side chain on position 17 is that of cholesterol:

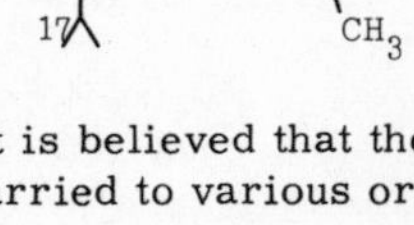

CH_3 CH_3
CH– $(CH_2)_3$ – CH
17 CH_3

Man and other mammals can synthesize provitamin D_3 in the body. It is believed that the vitamin is then activated in the skin by exposure to ultraviolet rays and carried to various organs in the body for utilization or storage (in liver).

Role in Physiology:

The principal action of vitamin D is to increase the absorption of calcium and phosphorus from the intestine. The vitamin also has a direct effect on the calcification process. Evidence for this has been obtained by isotopic tracer studies which indicate that the administration of vitamin D to animals deficient in this vitamin increases the rate of accretion and resorption of minerals in bone.

An effect of vitamin D on citrate metabolism was reported by Steenbock and Bellin[2]. In physiologic doses, vitamin D caused an increase in the citrate content of blood, bone, kidney, heart, and small intestine of rats maintained on a normal or on a rachitogenic ration low in phosphorus. There was no effect on the citrate content of the liver. The excretion of citrate in the urine of these animals was also elevated, presumably as a result of the accumulation of citrate within the tissues.

Vitamin D also influences the handling of phosphate by the kidney. In animals deficient in vitamin D, the excretion of phosphate and its renal clearance are decreased; in parathyroidectomized animals, vitamin D increases the clearance of phosphate and promotes lowering of the serum phosphate concentration[3].

The mode of action of vitamin D is not known. A function in connection with the action of the alkaline phosphatases has been suggested by Zetterström[4]. The alkaline phosphatases of the serum inevitably increase in rickets; in fact, this increase is virtually diagnostic of the disease Very similar if not identical phosphatases occur in bone, the intestinal mucosa, the kidney, and the liver. The phosphatases catalyze the removal of combined phosphate from such organic phosphates as hexosephosphate or glycerophosphate, liberating inorganic phosphate. In bone, this inorganic phosphate influences the deposition of calcium phosphate in the ossification process. A similar mechanism may operate in intestinal absorption or renal reabsorption of phosphate. Zetterström and co-workers prepared a water-soluble phosphorylated vitamin D_2 and demonstrated that the vitamin D_2-phosphate activated in vitro the alkaline phosphatases of bone, kidney, and intestinal mucosa. The effect was greatest with the enzymes from bone.

These observations suggest that a deficiency of vitamin D may be characterized by a lack of activation of alkaline phosphatases with resultant effects on phosphate metabolism in the intestine, kidney, and bone. The great importance of phosphorylation processes in other aspects of metabolism, particularly of carbohydrates, indicates that vitamin D may play a more general role in metabolism than hitherto suspected.

Sources and Daily Allowance:

In its active form vitamin D is not well distributed in nature, the only rich sources being the liver and viscera of fish and the liver of animals which feed on fish. However, certain foods, such as milk, may have their vitamin D content increased by irradiation with ultraviolet light.

For children a requirement of 400 I. U. per day has been suggested. Although the adult requirement is not known, 400 I. U. have been proposed for women during pregnancy and lactation. The ingestion of large amounts of vitamin D has been shown to cause toxic reactions and widespread calcification of the soft tissues, including lungs and kidney. The quantities of the vitamin required to induce a state of hypervitaminosis are not obtainable from natural sources, and so this fact is of importance only when massive doses of vitamin D are given for therapeutic purposes.

One unit of vitamin D (one U. S. P. unit or I. U.) is defined as the biologic activity of 0.025 micrograms of calciferol. Substances intended for human nutrition are biologically assayed for vitamin D_2 by the "line test." Twenty-eight day-old rats are put on a rachitogenic diet, characterized by a high cereal content and a high (4:1) ratio of calcium to phosphorus, until depleted of vitamin D (18 to 25 days). For the following eight days one group, the **reference group**, is given daily test doses of cod liver oil of known potency (the U. S. P. Reference Oil) in a quantity sufficient to produce a narrow continuous line of calcification across the metaphysis of the tibia. This degree of healing is designated in the experimental protocol as "unit" or two-plus healing. Other groups of rachitic test animals are given the test substance in varying amounts for a similar eight-day test period. All animals are sacrificed ten days after the reference or test samples were first administered and the tibial bones examined for the degree of healing. The potency of the test material is calculated from the quantity required to produce healing equivalent to that of the reference sample. Substances intended for the chick, which responds better to vitamin D_3, are assayed by a method which determines the amount of ash in the bones of growing chicks which have been fed with the test materials.

Physical (spectrophotometric) methods are also used in assay of the vitamins D.

THE VITAMINS E

Chemistry:

Compounds possessing vitamin E activity are known chemically as tocopherols. There are three such substances, designated as α-, β-, and γ-tocopherols.

The structure of α-tocopherol is shown below:

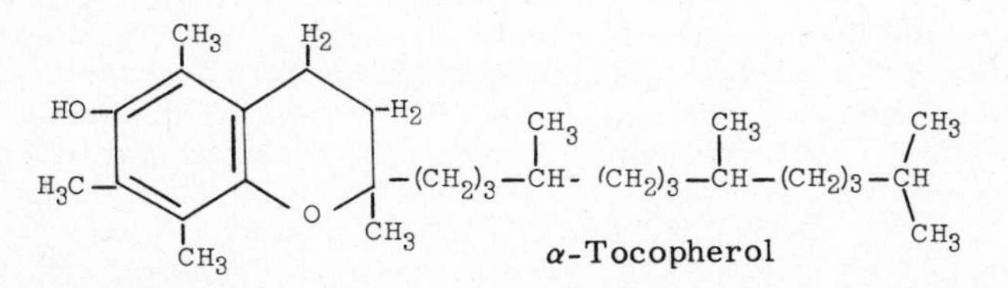

α-Tocopherol

A deficiency in rats and some other animals causes resorption of the fetus in the female and, in the male, atrophy of spermatogenic tissue and permanent sterility. No satisfactory evidence of a beneficial effect of vitamin E in human beings has yet been advanced.

The most striking chemical characteristic of the vitamins E is their antioxidant property. In fact, these compounds are used in foods as antioxidants in order to protect other vitamins (e.g., vitamin A) from oxidative destruction. It is also of interest in this connection that the body fat from the carcass of a vitamin E–deficient rat is abnormally susceptible to oxidation and can be stabilized by tocopherol. The relationship between fat and vitamin E is further exemplified by the observation that the amount and quality of fat in the diet affect the production of vitamin E deficiency.

Role in Physiology:

A specific role for vitamin E in metabolism has not yet been discovered. However, there is evidence that it functions as a co-factor in the electron transfer system operating between cytochromes b and c[5] (see p. 116).

In some animal species a lack of vitamin E produces muscular dystrophy. Such dystrophic muscles exhibit increased respiration (oxygen uptake). Treatment with tocopherol reduces the oxygen uptake of such tissue. However, vitamin E has not been shown to benefit any type of muscular dystrophy seen in man.

The skeletal muscle from vitamin E–deficient rabbits contains more ribonucleic and desoxyribonucleic acid than muscle from normal animals, which suggests that the turnover rate of nucleic acids is accelerated by vitamin E deficiency.

The placental transfer of vitamin E is limited; mammary transfer is much more extensive. Thus the serum alpha-tocopherol level of breast-fed infants is increased more rapidly than that of bottle-fed children.

The level of tocopherol in the plasma after oral administration of DL-α-tocopherol acetate has been measured. Single doses of 200, 400, 500, and occasionally of 100 mg. were effective in increasing the free tocopherol levels of the plasma to a significant degree after six hours. Repeated daily oral doses produced somewhat greater maximum increases than did single doses. However, the parenteral administration of vitamin E failed to increase the free tocopherol level of the plasma regardless of the type of compound (free alcohol or monosodium phosphate) and the type of vehicle (oil or water) used.

Certain diets low in protein and especially in the sulfur-containing amino acids (particularly cystine) were found to produce an acute massive hepatic necrosis in experimental animals[6].

A vitamin E deficiency enhances the effects of such diets, whereas added vitamin E exerts a preventive action upon the necrosis. Rats which have been kept on the deficient ration develop the fatal hepatic lesion suddenly (within a few hours or days) after a symptom-free latent period which averages 45 days. However, the occurrence of a metabolic defect in the livers of these animals can be demonstrated several weeks before the development of the necrotic lesion itself. Liver slices from these animals which are still histologically normal are able to respire in the Warburg apparatus for only 30 to 60 minutes; subsequently, oxygen consumption declines as incubation continues. The three major metabolic pathways for the utilization of acetate by the liver, viz., ketogenesis, lipogenesis, and oxidation to CO_2, are also deficient in the prenecrotic liver slice, probably as a result of the respiratory defect. If the diet is supplemented with cystine, vitamin E, or preparations of "Factor 3"[7], both the metabolic and the histologic lesions are prevented. A reversal of the respiratory decline in the necrotic liver can also be prevented by direct infusion of tocopherols into the portal vein.

Factor 3 has recently been identified as a selenium compound[8], and sodium selenite is entirely protective against dietary liver necrosis in rats. The apparent interchangeability of selenium, vitamin E, and cystine, with the first-named being the most potent, in protection against dietary liver necrosis suggests that selenium is part of an enzyme essential for an oxidation-reduction reaction involving cystine, and that vitamin E is also involved as an antioxidant.

Sources:

Good sources of vitamin E include milk, eggs, muscle meats, fish, cereals, and leafy vegetables. Concentrates of natural tocopherol mixtures are prepared by molecular distillation of wheat germ oil, which is particularly rich in vitamin E.

The vitamin is usually measured by a biologic assay based on the ability of the material under test to support gestation when the pregnant female rat is maintained on a vitamin E–deficient diet. A chemical method which permits the estimation of 2 to 5 micrograms of vitamin E has also been described[9]. The vitamin E content of human blood as measured by this method was found to range from 361 to 412 micrograms per 100 ml.

THE VITAMINS K

Chemistry:

A large number of chemical compounds which are related to 2-methyl-1, 4-naphthoquinone possess some degree of vitamin K activity.

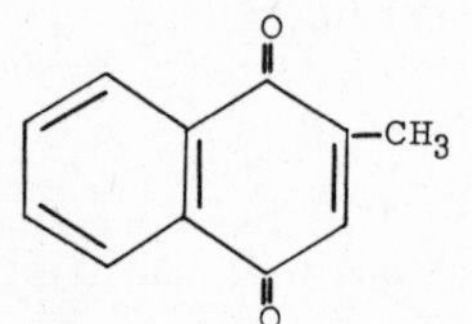

Vitamin K_3; Menadione
(2-Methyl-1, 4-naphthoquinone)

The naturally occurring vitamins K possess a phytyl radical on position 3 (vitamin K_1) or a difarnesyl radical (K_2).

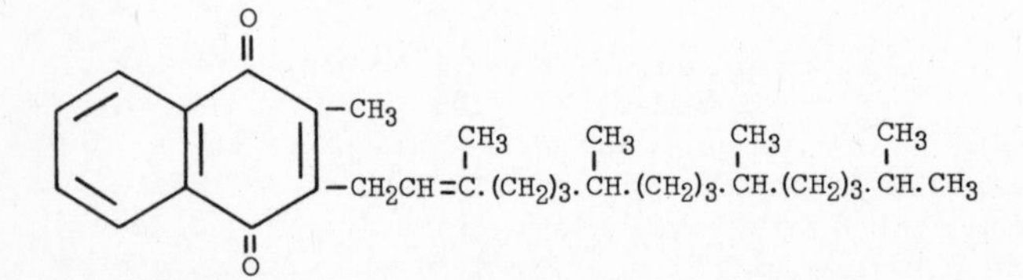

Vitamin K_1; Phytonadione (Mephyton®)
(2-Methyl-3-phytyl-naphthoquinone)

The natural vitamins (K_1 and K_2) are of about the same potency in the human. Vitamin K_3, which was the first vitamin to be produced synthetically, is about three times as potent as the natural vitamin on a weight basis when assayed in chicks. The potency of the parent naphthoquinone suggests that this portion of the molecule is essential for the formation of a second substance which actually exerts the biologic effects of the vitamin.

Vitamin K_1 has now also been produced synthetically.

Role in Physiology:

The best known function of vitamin K is to catalyze the synthesis of prothrombin by the liver. In the absence of vitamin K a hypoprothrombinenia occurs in which blood clotting time may be greatly prolonged. It must be emphasized that the effect of vitamin K in alleviation of hypoprothrombinemia is dependent upon the ability of the hepatic parenchyma to produce prothrombin. Advanced hepatic damage, as in carcinoma or cirrhosis, may be accompanied by a prothrombin deficiency which cannot be relieved by vitamin K.

Recent work indicates that the activities of ''stable factors'' (II, B, 4, p. 126) and of PTC (I, A, 2, p. 125) are also reduced in states of vitamin K deficiency or after administration of vitamin K antagonists such as bishydroxycoumarin (Dicumarol®). The cause of delayed clotting in vitamin K deficiency states is therefore not confined to a prothrombin deficiency, although this is perhaps the most important factor.

The vitamins K, especially vitamin K_3 (menadione), have been postulated to be components of electron-transfer systems in citric acid cycle oxidations which are coupled to phosphorylation (see p. 121). The coenzyme which is probably formed from vitamin K has been designated Coenzyme Q (quinone). Its structure, which is shown below, is that of a 2, 3-dimethoxy-5-methylbenzoquinone with a polyisoprenoid side chain at carbon 6. The number of isoprenoid units (formula: ($-CH_2-CH{=}C.CH_3-CH_2-$) varies with the source; for example, Coenzyme Q from beef heart has 10 units, whereas coenzymes with 6-9 units have been isolated from microorganisms.

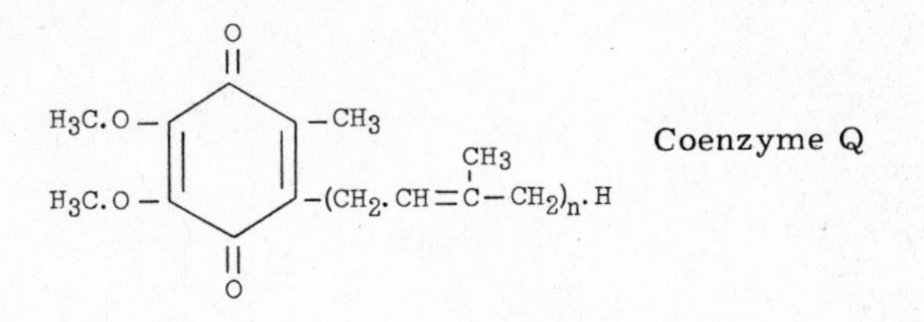

Coenzyme Q

Further discussion of the probable functions of Coenzyme Q in respiration and oxidative phosphorylation will be found on p. 121.

A dietary deficiency of vitamin K is unlikely since the vitamin is fairly well distributed in foods. Moreover, the intestinal microorganisms synthesize considerable vitamin K in the intestine. However, as has already been noted for the other fat-soluble vitamins, the absorption of vitamin K from the intestine depends on the presence of bile. A deficiency state will therefore result whenever there is biliary tract obstruction or a defect in fat absorption such as in sprue and celiac disease. Short-circuiting of the bowel as a result of surgery also may foster a deficiency which will not respond even to large oral doses of vitamin K. However, water-soluble forms of vitamin K are available. Two of these are shown below. They may be absorbed even in the absence of bile.

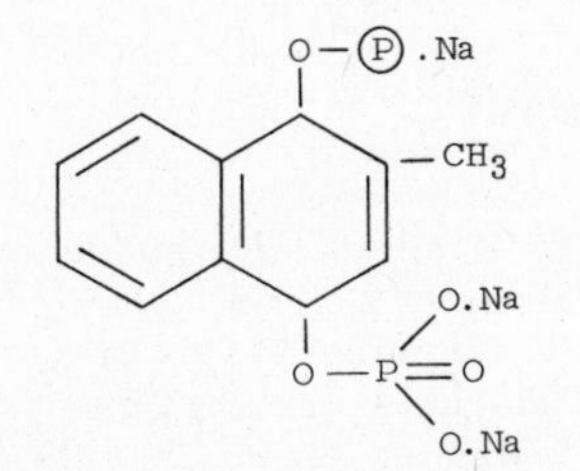

Sodium Menadiol Diphosphate

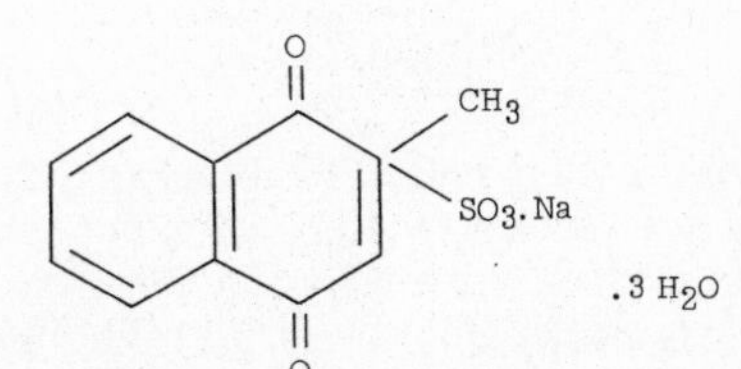

Menadione Sodium Bisulfite

Immediately after birth the intestinal flora produces insufficient vitamin K for the needs of the organism since there is a delay in establishing the normal flora in the intestine, which is, of course, sterile at birth. The quantity of the vitamin supplied by the mother during gestation is apparently not large. Thus during the first few days of life a hypoprothrombinemia may appear which will persist until the intestinal flora becomes active in the manufacture of the vitamin. This can be prevented by administering vitamin K to the mother before parturition or by giving the infant a small dose of the vitamin.

The parenteral administration to infants of too large doses of vitamin K (e.g., 30 mg. per day for three days) has been shown to produce hyperbilirubinemia in some cases. Three mg. of sodium menadiol diphosphate, which is equivalent to 1 mg. of vitamin K_1, is adequate to prevent hypoprothrombinemia in the newborn, and there is no danger of provoking jaundice with this dosage. The oral administration of vitamin K has not been found to produce jaundice.

Uncontrollable hemorrhage is a symptom of vitamin K deficiency. The newborn child may bleed into the adrenal and the brain, from the umbilical cord and into the gastrointestinal tract. In the adult hemorrhage may also occur, most commonly after an operation on the biliary tract.

An important therapeutic use for vitamin K is as an antidote to the anticoagulant drugs such as bishydroxycoumarin (Dicumarol®; see p. 127). For this purpose large doses of vitamin K_1 may be used, either orally or, as an emulsion, intravenously. The prothrombin time, which is lengthened by the use of the anticoagulant drug, will usually return to normal in 12 to 36 hours after the administration of the vitamin provided liver function is adequate to manufacture prothrombin.

Sources:

The green leafy tissues of plants are good sources of the vitamin. Fruits and cereals are poor sources. Molds, yeasts, and fungi contain very little; but since it occurs in many bacteria, most putrefied animal and plant materials contain considerable quantities of vitamin K.

THE WATER-SOLUBLE VITAMINS

VITAMIN C

Chemistry:

The chemical structure of vitamin C (ascorbic acid, cevitamic acid) resembles that of a monosaccharide.

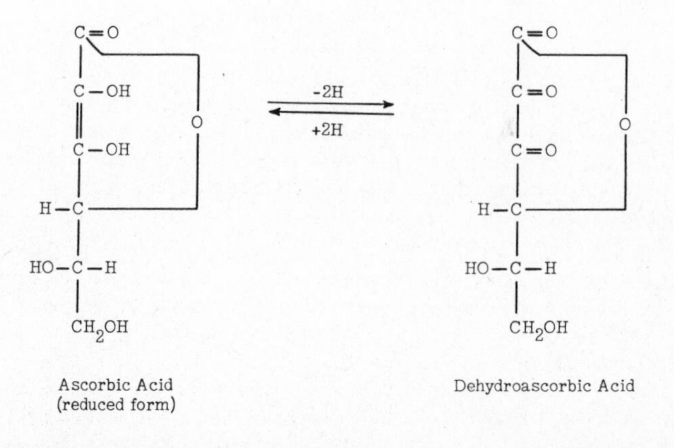

Ascorbic Acid (reduced form) Dehydroascorbic Acid

Vitamin C is readily oxidized to the dehydro form. Both forms are physiologically active, and both are found in the body fluids. The enediol group of ascorbic acid (from which removal of hydrogen occurs to produce the dehydro form, as shown in the formulas above) may be involved in the physiologic function of this vitamin. It is conceivable that this chemical grouping functions in a hydrogen transfer system; a role of the vitamin in such a system, i.e., the oxidation of tyrosine, is described below.

The reducing action of ascorbic acid is the basis of the chemical determination of the compound. In most plant and animal tissues this is the only substance which exhibits this reducing action in acid solution. One of the most widely used analytic reactions for vitamin C is the quantitative reduction of the dye, 2, 6-dichlorophenolindophenol, to the colorless leuco base by the reduced form of ascorbic acid. The method has been adapted to the microdetermination of blood ascorbic acid, so that only 0.01 ml. of serum is needed for assay. Dehydroascorbic acid can be determined colorimetrically by the formation of a hydrazone with 2, 4-dinitrophenylhydrazine. This method may also be used for the assay of total vitamin C.

Role in Physiology:

Although ascorbic acid is undoubtedly widely required in the metabolism of animals, most mammals do not require an extrinsic source of the vitamin since they have acquired the ability to synthesize it in their own tissues. In the rat, for example, ascorbic acid is synthesized from glucose and glucuronic acid (see p. 193).

Severe ascorbic acid deficiency produces scurvy. The pathologic signs of this deficiency are almost entirely confined to supporting tissues of mesenchymal origin (bone, dentine, cartilage, and connective tissue). Scurvy is characterized by failure in the formation and maintenance of intercellular materials, which in turn causes typical symptoms, such as hemorrhages, loosening of the teeth, poor wound healing, and the easy fracturability of the bones.

An increase in the excretion of certain amino acids (threonine, serine, glycine, histidine, tyrosine, lysine, and β-amino isobutyric acid) was reported in two cases of scurvy in infants by Jonxis and Huisman[10]. The blood levels of the amino acids were normal, suggesting that the amino-aciduria was of renal origin (see also p. 291). After three or more weeks of treatment with ascorbic acid, the amino-aciduria disappeared.

The biochemical function of ascorbic acid is still unknown. Sealock and Goodland[11] have found that ascorbic acid functions as a coenzyme for the metabolic oxidation of tyrosine (see p. 253). The mechanism of its action is not known, but it was suggested by these authors that the deaminated amino acid (*p*-hydroxyphenylpyruvic acid; tyrosine keto acid) is oxidized by transfer of hydrogen to dehydroascorbic acid, which acts as a hydrogen acceptor and is thus reduced to ascorbic acid. Subsequent transfer of the hydrogen to oxygen regenerates the dehydroascorbic acid.

A relation between ascorbic acid and tyrosine metabolism had previously been suggested by the observation that guinea pigs deficient in vitamin C excrete considerable quantities of homogentisic acid after administration of tyrosine (see p. 251). Premature infants may also occasionally exhibit an alcaptonuria which is abolished by the administration of vitamin C. A requirement for ascorbic acid in connection with homogentisic acid oxidation (see p. 256) has also been postulated.

The conversion of the active tetrahydro formyl derivatives (see p. 92) by liver slices is enhanced by the addition of vitamin C; ascorbic acid was also found to increase the urinary excretion of folinic acid after folic acid was administered.

The adrenal cortex contains a large quantity of vitamin C, and this is rapidly depleted when the gland is stimulated by adrenocorticotropic hormone. A similar depletion of adrenocortical vitamin C is noted when experimental animals (guinea pigs) are injected with large quantities of diphtheria toxin. Increased losses of the vitamin accompany infection and fever. These losses are particularly notable when bacterial toxins are present. All of these observations suggest that the vitamin may play an important role in the reaction of the body to stress.

Sources and Daily Allowances:

The best food sources of vitamin C are citrus fruits, berries, melons, tomatoes, green peppers, raw cabbage, and leafy green vegetables, particularly salad greens. Potatoes, while only a fair source of vitamin C on a per gram basis, constitute an excellent source in the average diet because of the quantities which are commonly consumed.

The vitamin is easily destroyed by cooking, since it is readily oxidized. There may also be a considerable loss in mincing of fresh vegetables such as cabbage, or in the mashing of potatoes. Losses of vitamin C during storage and processing of foods are also extensive, particularly where heat is involved. Traces of copper and other metals accelerate this destruction.

The tissues and body fluids contain varying amounts of vitamin C. As a rule the tissues of the highest metabolic activity (excepting muscle) have the highest concentration. Fasting individuals who receive liberal intakes of vitamin C (75 to 100 mg. per day) have serum ascorbic acid levels of 1 to 1.4 mg./100 ml. Those on diets which provide only 15 to 25 mg. per day will have serum levels correspondingly lower: 0.1 to 0.3 mg./100 ml. When the blood levels of ascorbic acid exceed 1 to 1.2 mg./100 ml., excretion of the vitamin occurs readily. For this reason the intravenous administration of vitamin C is usually attended by a considerable urinary loss.

The recommended intake is 70 to 75 mg. per day for all adults except during pregnancy and lactation, when the intake should be increased to 100 to 150 mg. per day. Children up to ten years of age require 30 to 60 mg. per day. Children from ten to 12 years old require 75 to 100 mg. per day.

THE VITAMINS OF THE B COMPLEX

The term "vitamine" was originally coined by Funk to characterize the substance which was lacking in the nutritional deficiency disease, beriberi. Other workers soon discovered in the same foodstuffs other water-soluble factors which exhibited various activities attributed to the anti-beriberi factor. Eventually the term "water-soluble vitamin B" came to be applied to Funk's original substance; and as additional water-soluble vitamins were characterized and new vitamins were added to the list, a considerable amount of confusion arose in the nomenclature. For example, in the American literature the anti-pellagra vitamin was designated vitamin G, whereas recognition of the close association of these vitamin B factors in natural foodstuffs led British workers to propose that all of this closely related group of water-soluble vitamins be designated as vitamins of the B complex, using subscript numbers to distinguish individual members. Under this system vitamin G was redesignated vitamin B_2. Vitamin B_2 now refers to riboflavin, and the original pellagra-preventive vitamin of Goldberger is designated by its chemical name, nicotinic acid.

Many of the individual B factors which were subsequently reported from different laboratories proved to be identical with one another. The elucidation of the chemical nature of each factor, and its synthesis, have served to clarify the problem. It is now customary to assign to each vitamin a name descriptive of its origin, function, or chemical structure rather than a lettered designation.

There is still not complete agreement as to what substances should be considered as B vitamins, but those substances listed below are generally accepted as members of the B complex. It is possible that the requirements of different animal species vary. In recent years it has become apparent that the activity of the intestinal flora in synthesizing vitamins of the B group explains species differences as well as the wide variation in requirements noted in human experiments. This is supported by the observation that previously unrecognized deficiencies appear in animals when the intestinal bacteria are destroyed by the use of sulfonamide drugs or antibiotics; and that careful balance experiments in humans indicate a higher fecal or urine content of certain B vitamins than can be accounted for by the dietary intake.

A requirement for many of the B vitamins seems universal in the plant and animal world. In most cases these vitamins function as constituents of cellular enzyme systems. The demands of many microorganisms for a preformed source of most of the B factors has been used as a valuable tool in the discovery and assay of these vitamins. In fact, recently discovered members of the group, notably folic and folinic acids and vitamin B_{12}, were first shown to be required nutrients for microorganisms and later found to be necessary in higher animals as well.

Vitamins of the B Complex:

1. Thiamine - Vitamin B_1, anti-beriberi substance, anti-neuritic vitamin, aneurin.
2. Riboflavin - Vitamin B_2, lactoflavin.
3. Niacin - P-P factor of Goldberger, nicotinic acid.
4. Pyridoxine - Vitamin B_6, rat antidermatitis factor.
5. Pantothenic acid - Filtrate factor, chick antidermatitis factor.
6. Lipoic acid - Thioctic acid, protogen, acetate-replacement factor.
7. Biotin - Vitamin H, anti-egg-white-injury factor.

8. Folic acid group - Liver Lactobacillus casei factor, vitamin M, Streptococcus lactis R (SLR) factor, vitamin B_c, fermentation residue factor.
9. Inositol - Mouse anti-alopecia factor.
10. *p*-Aminobenzoic acid (PABA).
11. Vitamin B_{12} - Cyanocobalamin, anti–pernicious anemia factor, extrinsic factor of Castle.

THIAMINE

Chemistry:

The crystalline vitamin, thiamine hydrochloride ($C_{12}H_{17}ClN_4OS.HCl$), is a 2, 5-dimethyl-6-amino pyrimidine combined with 4-methyl-5-hydroxyethyl thiazole. It is shown below as the chloride hydrochloride.

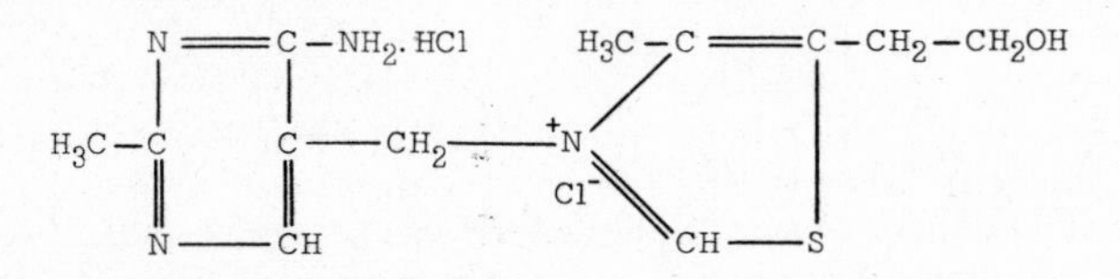

Role in Physiology:

Thiamine, in the form of thiamine diphosphate (thiamine pyrophosphate), is the coenzyme for the decarboxylation of α-keto acids such as pyruvic acid or α-ketoglutaric acid. As such it is often referred to as cocarboxylase. The decarboxylation reaction with pyruvic acid is as follows:

$$\underset{\text{[Pyruvic acid]}}{CH_3-\overset{\overset{O}{\|}}{C}-COOH} \xrightarrow[\text{Thiamine diphosphate (cocarboxylase)}]{\boxed{\text{Carboxylase}}} \underset{\text{[Acetaldehyde]}}{CH_3-\overset{\overset{O}{\|}}{C}-H + CO_2}$$

This reaction as it occurs in yeasts is one of ''straight'' decarboxylation or simple removal of carbon dioxide. It results in the production of acetaldehyde, which is reduced subsequently to ethyl alcohol. In animal tissues the decarboxylation of pyruvic acid results in the formation of acetic acid, which is an oxidation product of acetaldehyde. This reaction is therefore referred to as ''oxidative'' decarboxylation. It involves not only thiamine but also participation of other coenzymes, specifically lipoic acid, Coenzyme A, and diphosphopyridine nucleotide (DPN). These reactions will be shown later (see p. 88).

Thiamine diphosphate is also a coenzyme in the reactions of transketolation which occur in the direct oxidative pathway for glucose metabolism (see p. 187). The operation of this pathway in erythrocytes from thiamine-deficient rats is markedly retarded at the transketolase step so that pentose sugars accumulate to levels three times normal[12]. The biochemical defect appears before growth ceases in the thiamine-deficient animal, and the defect can be significantly alleviated by addition of thiamine or of co-carboxylase to the cells in vitro or by the intraperitoneal injection of thiamine in vivo.

Thiamine deficiency affects predominantly the peripheral nervous system, the gastrointestinal tract, and the cardiovascular system. Thiamine has been shown to be of value in the treatment of beriberi, alcoholic neuritis, and the neuritis of pregnancy or of pellagra. The vitamin is necessary for the optimal growth of infants and children.

In certain fish there is a heat-labile enzyme which destroys thiamine. Attention was drawn to this ''thiaminase'' by the appearance of ''Chastek's paralysis'' in foxes fed a diet containing 10% or more of uncooked fish. The disease is characterized by anorexia, weakness, progressive ataxia, spastic paraplegia, and hyperesthesia. The similarity between the focal lesions of the nervous system in this paralysis in the fox and the lesions seen in Wernicke's syndrome in man have lent support to the concept that the latter is in part attributable to thiamine deficiency.

Chemical or microbiologic methods are used to determine thiamine in foods or in body fluids. The chemical procedures are based on conversion of thiamine to a compound which fluoresces under ultraviolet illumination. Quantitative measurement of the vitamin may then be accomplished with a photofluorometer. The older methods were based on the conversion of thiamine by oxidation to the fluorescent compound, thiochrome. In a new procedure suggested by Teeri[13] the vitamin reacts with cyanogen bromide to form a highly fluorescent compound.

In detection of thiamine deficiency in man, a determination of the amount of thiamine excreted in four hours may be used. This is sometimes modified to include the prior administration of a test dose of thiamine, and the percentage of the test dose which is excreted is observed (thiamine load test). Such studies may differentiate between persons with very high or moderate to low thiamine intakes, but the principal value of such tests in individual cases is to rule out thiamine deficiency. Another diagnostic test for thiamine deficiency is based on the measurement of the ratio of lactic to pyruvic acids in the blood after administration of glucose. Blood and urinary pyruvic acid levels are characteristically elevated in thiamine deficiency, as would be expected from the role of thiamine in pyruvic acid metabolism; but abnormal blood lactic acid–pyruvic acid ratios are said to be more specific indicators of vitamin B_1 deficiency than the levels of pyruvic acid alone.

Sources and Daily Allowances:

Thiamine is present in practically all of the plant and animal tissues commonly used as food, but the content is small. Among the more abundant sources are unrefined cereal grains, liver, heart, kidney, and lean cuts of pork. With improper cooking the thiamine contained in these foods may be destroyed. Since the vitamin is water-soluble and somewhat heat-labile, particularly in alkaline solutions, it may be lost in the cooking water. The enrichment of flour, bread, corn, and macaroni products with thiamine has increased considerably the availability of this vitamin in the diet. On the basis of the average per capita consumption of flour and bread in the United States, as much as 40% of the daily thiamine requirement is now supplied by these foods.

It is difficult to fix a single requirement for vitamin B_1. The requirement is increased when metabolism is heightened, as in fever, hyperthyroidism, increased muscular activity, pregnancy, and lactation. There is also a relationship to the composition of the diet. Fat and protein reduce, while carbohydrate increases the quantity of the vitamin required in the daily diet. It is also possible that some of the thiamine synthesized by the bacteria in the intestine may be available to the organism. Deficiencies of thiamine are likely not only in persons with poor dietary habits, or in the indigent, but also in many patients suffering from organic disease.

The Committee on Foods and Nutrition of the National Research Council has recommended a thiamine intake of about 0.5 mg. per 1000 Calories. In a population subsisting on a high-carbohydrate (rice) diet low in fat and protein, an average daily intake of about 0.2 mg. per 1000 Calories is associated with widespread beriberi. Mild polyneuritis was produced in two women maintained on a daily thiamine intake of 0.175 mg. per 1000 Calories for a period of about four months.

RIBOFLAVIN

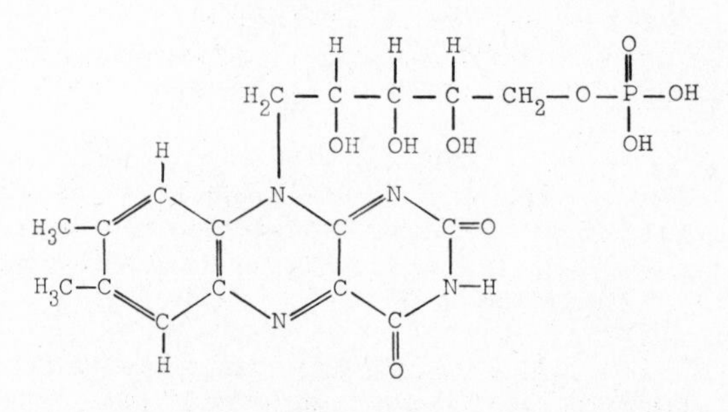

6,7-Dimethyl-9(D-ribityl-5-phosphate)-isoalloxazine

Riboflavin Phosphate
(Riboflavin Mononucleotide)

The existence of a water-soluble, yellow-green, fluorescent pigment in milk whey was noted as early as 1879; but this substance, riboflavin, was not isolated in pure form until 1932. At that time it was shown to be a constituent of oxidative tissue-enzyme systems and an essential growth factor for laboratory animals. *

Riboflavin is relatively heat-stable but sensitive to light. On irradiation with ultraviolet rays or visible light it undergoes irreversible decomposition.

Role in Physiology:

Riboflavin is a constituent of several enzyme systems which are involved in intermediary metabolism. These enzymes are called flavoproteins (see p. 117). Riboflavin acts as a coenzyme for hydrogen transfer in the reactions catalyzed by these enzymes as described on p. 116. In its active form riboflavin is combined with phosphate. This phosphorylation of riboflavin occurs in the intestinal mucosa as a condition for its absorption.

Two forms of riboflavin are known to exist in various enzyme systems. The first, riboflavin phosphate (riboflavin mononucleotide), is a constituent of the Warburg yellow enzyme (see p. 117), cytochrome c reductase (see p. 116), and the amino acid oxidase for the naturally-occurring L-amino acids (see p. 228). The other form, whose structure is shown below, is called flavin adenine dinucleotide (FAD) because it contains two phosphate groups and adenine as well as ribose. FAD is the prosthetic group of diaphorase (see p. 116), the D-amino acid oxidases (see p. 228), glycine oxidase (see p. 233), and xanthine oxidase, which contains also iron and molybdenum (see p. 269). It is also an integral part of the prosthetic group of butyryl-coenzyme A-dehydrogenase, the enzyme which mediates the first oxidative step in the oxidation of the lower fatty acids, such as butyric acid.

Characteristic lesions of the lips, fissures at the angles of the mouth (cheilosis), localized seborrheic dermatitis of the face, a particular type of glossitis (magenta tongue), and certain functional and organic disorders of the eyes may result from riboflavin deficiency. However, these are not due to riboflavin deficiency alone and may result from various other conditions.

It has recently been suggested that a determination of the riboflavin content of the serum is of value in the diagnosis of riboflavin deficiencies. Methods for such determinations are given by Suvarnakich, et al.[14] These authors report that the normal concentration of riboflavin in the serum is 3.16 micrograms per 100 ml. Most of this is present as flavin adenine dinucleotide (FAD) (2.32 micrograms per 100 ml.); the remainder exists as free riboflavin (0.84 mcg. per 100 ml.).

Riboflavin deficiency, according to some clinicians, is the most prevalent avitaminosis in the United States, although it is always to be remembered that persons with one deficiency syndrome are likely to have other associated vitamin deficiencies, particularly those of thiamine and niacin.

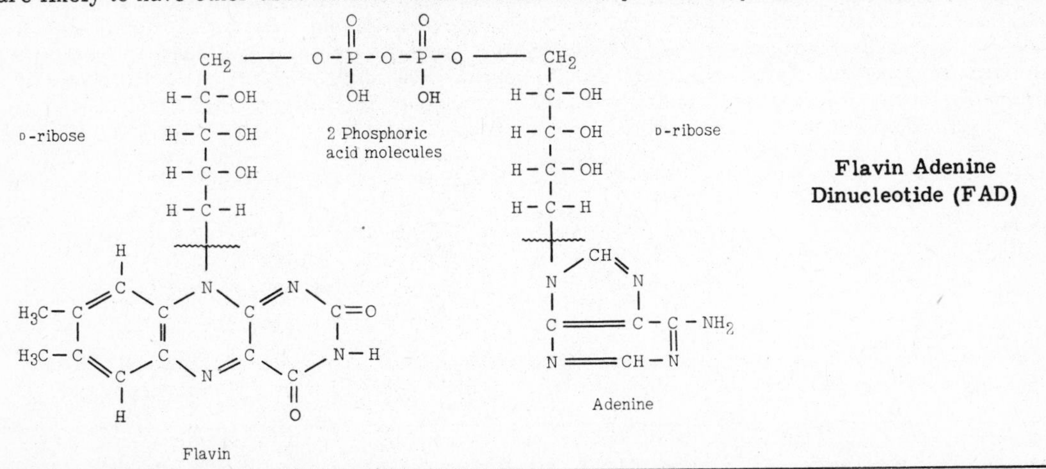

Flavin Adenine Dinucleotide (FAD)

*A lyxoflavin has been isolated from heart tissue. Its structure is identical with that of riboflavin except that the sugar attached to the flavin nucleus is lyxose rather than ribose. When lyxoflavin was added to the diet of rats or baby pigs, weight gains were reported. However, this substance does not replace riboflavin in biologic tests with microorganisms (Lactobacillus casei) or rats.

Sources and Daily Allowances:

Riboflavin is widely distributed throughout the plant and animal kingdoms, with very rich sources in anaerobic fermenting bacteria. Milk, liver, kidney, and heart are excellent sources. Many vegetables are also good sources, but the cereals are rather low. The riboflavin concentration in oats, wheat, barley, and corn is increased strikingly during germination.

Ordinary cooking procedures do not affect the riboflavin content of foods. Roasted, braised, or boiled meats retain 70 to 85% of the vitamin; an additional 15% is recovered in the drippings.

Unless proper precautions are taken, extensive losses of riboflavin in milk may occur during pasteurization, exposure to light in the course of bottling, or as a result of the irradiation of milk to increase its vitamin D content. Flour and bread, as a result of enrichment with crystalline riboflavin, may provide as much as 16% of the daily per capita requirement for this vitamin in the United States.

The riboflavin requirement for adults is 1.5 to 1.8 mg. per day, with an increase to 2 to 2.5 mg. during pregnancy and lactation. For adolescents, 2 to 2.5 mg. per day may be required.

NIACIN AND NIACINAMIDE

Niacin and niacinamide are specific for the treatment of acute pellagra. It is important to remember that vitamin deficiencies seldom occur singly, as is well illustrated by patients with pellagra. Very often these patients exhibit symptoms caused by a lack of vitamins other than niacin, particularly a polyneuritis amenable to thiamine administration. Nevertheless, the dermatitis, diarrhea, dementia, stomatitis, and glossitis observed respond, often spectacularly, to niacin. Niacin is the P-P (pellagra-preventive) factor originally named by Goldberger.

Pellagra remains one of the most serious deficiency diseases in the United States. This is due for the most part to the individual's failure to consume an adequate diet, although increased requirement, diminished absorption, or diminished utilization may be contributing factors.

It has been shown that the amino acid, tryptophan, normally contributes to the niacin supply of the body (see p. 256). Many of the diets causing pellagra are low in good quality protein as well as in vitamins. For this reason, pellagra is usually due to a combined deficiency of tryptophan and niacin.

Niacin is not excreted to any extent as the free nicotinic acid. A small amount may occur in the urine as niacinamide or as nicotinuric acid, the glycine conjugate. By far the largest portion is excreted as methyl derivatives, viz., N-methyl nicotinamide and the 6-pyridone of N-methyl nicotinamide, and N-methyl nicotinic acid and the glycine conjugates of these methyl derivatives. This methylation is accomplished in the liver at the expense of the labile methyl supply of the body. Methionine is the principal source of these methyl groups.

Chemistry:

See p. 81.

Role in Physiology:

Niacinamide functions as a constituent of two coenzymes: Coenzyme I, diphosphopyridinenucleotide (DPN); and Coenzyme II, triphosphopyridinenucleotide (TPN)*. These coenzymes,

---

*Another pyridine nucleotide, which has been named Coenzyme III[15], is found in bacterial as well as animal tissue extracts. It functions in the oxidation of cysteine sulfinic acid to cysteic acid (see p. 250).

Chemistry:

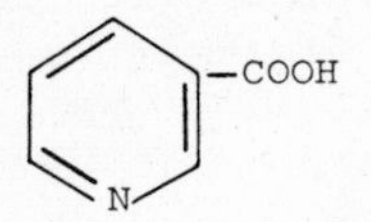

Niacin (Nicotinic Acid), $C_6H_5O_2N$

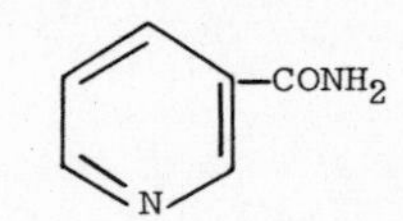

Niacinamide (Nicotinic Acid Amide), $C_6H_6ON_2$

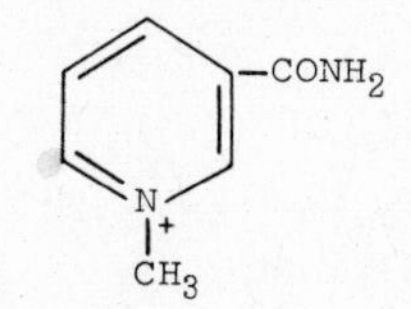

N-Methylnicotinamide

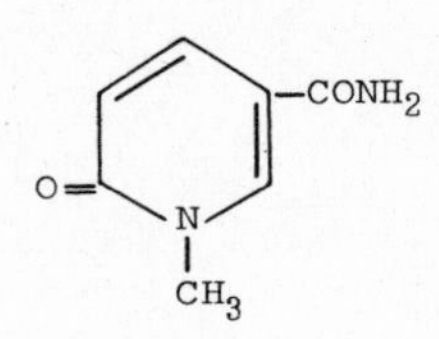

6-Pyridone-N-methylnicotinamide

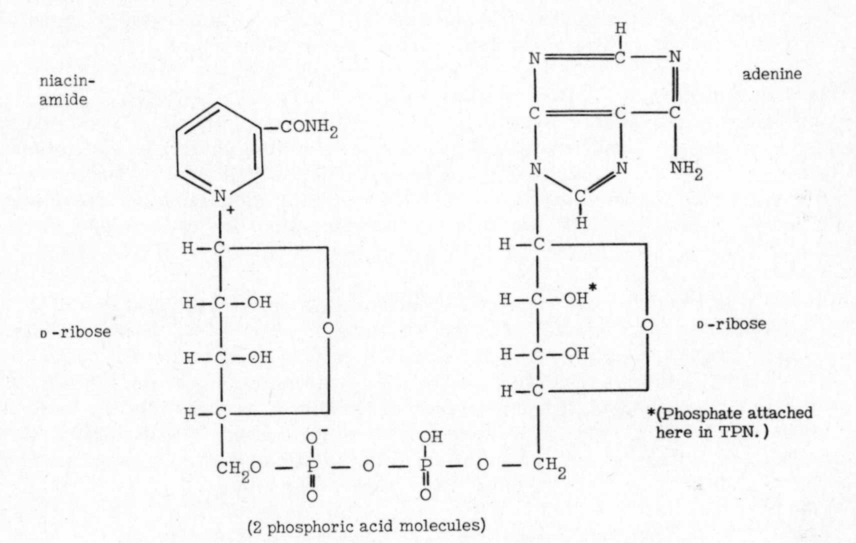

Diphosphopyridine Nucleotide (DPN)
(Oxidized Form)

which operate as hydrogen and electron transfer agents by virtue of reversible oxidation and reduction (see p. 115), play a vital role in metabolism. The function of niacin in metabolism explains its great importance in human nutrition and its requirement by many other organisms, including bacteria and yeasts.

The structure of DPN is known to be a combination of niacinamide with two molecules of the pentose sugar, D-ribose; two molecules of phosphoric acid; and a molecule of the purine base, adenine. It is shown in the oxidized form (see above). In changing to the reduced form it accepts hydrogen and electrons.

The mechanism of the transfer of hydrogen from a metabolite to oxidized DPN, thus completing the oxidation of the metabolite and the formation of reduced DPN, is shown in the abbreviated formula below. These reactions have been studied by observing the transfer of deuterium (heavy hydrogen) from labeled ethanol, CH_3CD_2OH, as catalyzed by alcohol dehydrogenase[16].

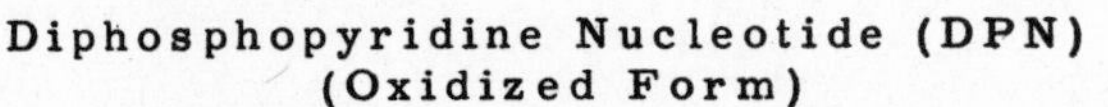

The reduced DPN was found to have but one atom of deuterium per molecule. Two atoms of deuterium are transferred, but the second one loses an electron, enters the medium as D^+, and equilibrates with the normal H^+ of the water. It is therefore not present in the reduced DPN. When the labeled DPN*D* is incubated with acetaldehyde to bring about a reversal of the above reaction, the alcohol formed has only one atom of deuterium per molecule.

According to Pullman, et al.[17], reduction of DPN occurs in the **para-** position rather than in the **ortho-** position as previously assumed.

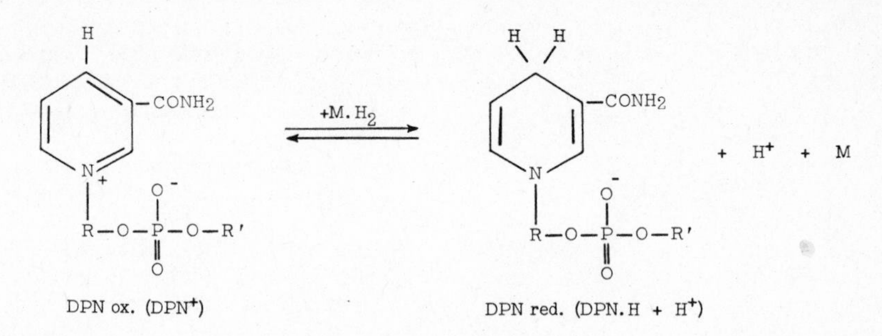

DPN ox. (DPN+) DPN red. (DPN.H + H+)

TPN differs only in the presence of one more phosphate moiety, esterified to the OH group on the second carbon of the ribose attached to the adenine. Its function is similar to that of DPN in hydrogen and electron transport. The two coenzymes are interconvertible.

Sources and Daily Allowances:

Niacin is found most abundantly in yeast. Lean meats, liver, and poultry are good sources. Milk, tomatoes, canned salmon, and several leafy green vegetables contribute sufficient amounts of the vitamin to prevent disease, although they are not in themselves excellent sources. On the basis of the average per capita consumption, enriched bread and other enriched flour products may provide as much as 32% of the daily niacin requirement. Most fruits and vegetables are poor sources of niacin.

The recommended daily allowance for niacin in adults is from 17 mg. daily on an 1800 Calorie intake to 21 mg. on a 3200 Calorie intake. Adolescents may require 17 to 25 mg. niacin per day. An increase over normal adult requirements of 2-3 mg. per day is recommended during pregnancy and lactation. These requirements are influenced by the protein content of the diet because of the ability of the amino acid, tryptophan, to supply much of the niacin required by the body (see p. 257). Sixty mg. tryptophan equals 1 mg. niacin. There is also evidence that niacin may be synthesized by bacterial activity in the intestine and that some of this may be absorbed and utilized by the tissues.

PYRIDOXINE

This substance was first discovered as essential for rats and named the rat antidermatitis factor, or the rat antipellagra factor. Later work has shown that the rat and man convert pyridoxine to other substances which far surpass pyridoxine in potency when tested with the lactobacilli or yeasts which are now used to assay foodstuffs for this vitamin. This suggests that pyridoxine may not be the most active form of the vitamin in nature but that it is convertible to other derivatives which function as described below. These more active derivatives are pyridoxal and pyridoxamine phosphates. Vitamin B_6 as it occurs in nature is probably a mixture of all three.

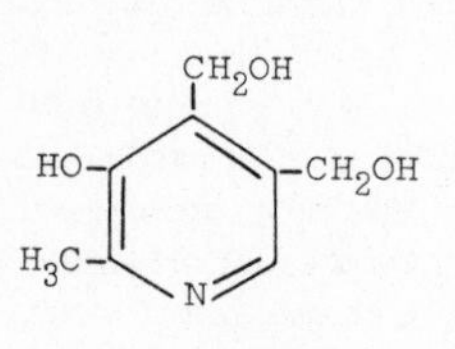

Pyridoxine

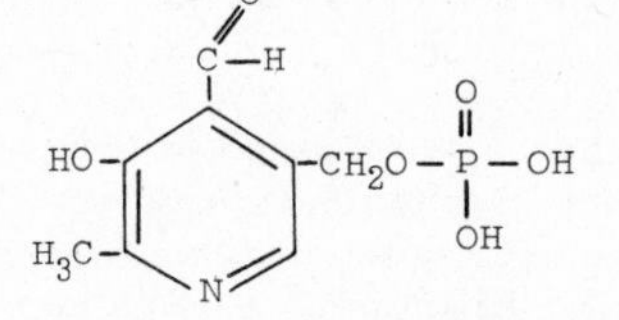

Pyridoxal Phosphate

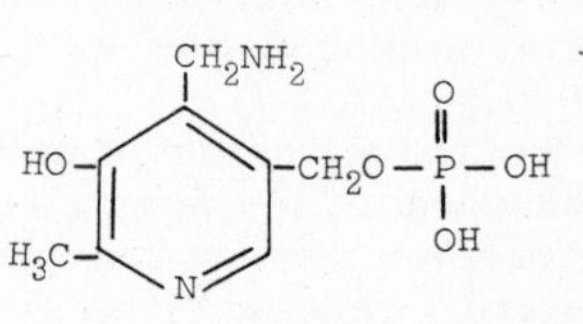

Pyridoxamine Phosphate

The predominant metabolite of vitamin B_6, which is excreted in the urine either from dietary B_6 or after ingestion of any of the three B_6 derivatives, is 4-pyridoxic acid (2-methyl-3-hydroxy-4-carboxy-5-hydroxymethyl pyridine). This metabolite can be measured by a fluorometric method described by Reddy at al.[18]

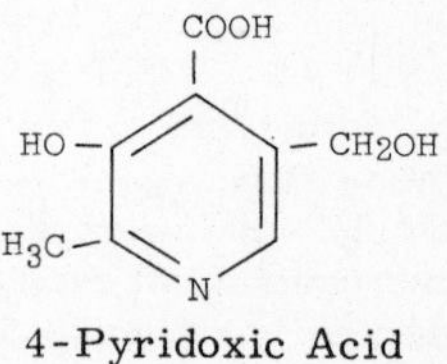

4-Pyridoxic Acid

Role in Physiology:

Pyridoxal phosphate is the prosthetic group of enzymes which decarboxylate tyrosine, arginine, glutamic acid, and certain other amino acids. In this way it functions as a co-decarboxylase. The deaminases (dehydrases) for serine and threonine are also catalyzed by pyridoxal phosphate acting as coenzyme. A third and very important function of the vitamin is as a coenzyme for enzymes involved in transamination (see p. 226), i.e., a cotransaminase. This function of pyridoxal phosphate probably is carried out by conversion to pyridoxamine phosphate. The reaction is reversible, so that the vitamin is actually functioning in an amino transfer system analogous to the hydrogen transfer systems described above in connection with niacinamide and riboflavin.

There is a specific relationship between vitamin B_6 and the metabolism of tryptophan because of the requirement for pyridoxal phosphate as a coenzyme for kynureninase (see p. 257). Failure to convert kynurenine to anthranilic acid results in the production of xanthurenic acid from kynurenine. In pyridoxine-deficient rats, dogs, swine, monkeys, and man, xanthurenic acid is found in the urine. When the vitamin is administered to the vitamin-deficient animals, xanthurenic acid disappears from the urine, and none can be found in the urine of normal animals. The examination of the urine for this metabolite after the feeding of a test dose of tryptophan has been used to diagnose vitamin B_6 deficiency.

The metabolism of cysteine is described on p. 250. In these reactions, vitamin B_6 is concerned with the transfer of sulfur from methionine to serine to form cysteine. This relates the vitamin to **transulfuration** as well as to transamination described above. The removal of sulfur from cysteine or homocysteine is catalyzed by desulfhydrases. These enzymes also require pyridoxal phosphate as coenzymes.

In studies on the factors which affect the transport of amino acids into the cells, Christensen[19] has concluded that pyridoxal participates in the mechanisms which influence intracellular accumulation of these metabolites.

It is thus apparent that vitamin B_6 is essential to amino acid metabolism in several roles: as a coenzyme for decarboxylation, deamination of serine and threonine, transamination, transulfuration, desulfuration of cysteine and homocysteine, the activity of kynureninase, and the transfer of amino acids into cells.

Vitamin B_6 is required by all animals investigated so far. Impaired growth results when immature animals are maintained on a vitamin B_6-free diet. Specific defects include acrodynia, edema of the connective tissue layer of the skin, convulsive seizures and muscular weakness in rats, and severe microcytic hypochromic anemia in dogs, swine, and monkeys accompanied by a two- to four-fold increase in the level of plasma iron and by hemosiderosis in the liver, spleen, and bone marrow. The anemia is not hemolytic in character, since there is no rise in icterus index or serum bilirubin. It will be recalled that pyridoxal is a coenzyme in the reaction by which α-amino-β-keto adipic acid is decarboxylated to δ-amino levulinic acid (see p. 57). The anemia of pyridoxine-deficient animals may be attributed to a defect at this point in the synthesis of heme.

No counterpart of the anemias produced in animals by pyridoxine deficiency has been recognized in man. There is, however, no question that vitamin B_6 is required in the diet of humans although this vitamin is adequately supplied in the usual diets of adults, children, and all but very young infants. However, deficiency states in infants and in pregnant women have been described. In the first instance, epileptiform seizures were reported in a small percentage (three to five per thousand) of very young infants maintained on an unsupplemented diet of a liquid infant food preparation which had been autoclaved at a very high temperature. Presumably the method of preparation destroyed most of the vitamin B_6 content of the product. Supplementation of this material with pyridoxine promptly alleviated the symptoms[20]. In the second instance, pregnant women given 10 grams of DL-tryptophan excreted various intermediary metabolites of tryptophan breakdown, including xanthurenic acid. The administration of vitamin B_6 to these women suppressed to a large degree the excretion of these metabolites. This suggested that in pregnancy there may exist a B_6 deficiency which is brought about by the increased demand of the fetus for this essential vitamin[20].

There is increasing evidence that vitamin B_6 is intimately concerned with the metabolism of the central nervous system. Swine, after nine to ten weeks on a vitamin B_6-free diet, exhibit demyelinization of the peripheral nerves and degeneration of the axon. In humans, the effects of pyridoxine deficiency have been best demonstrated in infants and children. The epileptiform seizures in infants which were described above are examples. The abnormal central nervous system activity that accompanies low vitamin B_6 intake during infancy is characterized by a syndrome of increasing hyperirritability, gastrointestinal distress, and increased startle responses as well as convulsive seizures. During the actual periods of seizure electroencephalographic changes may be noted. The clinical and encephalographic changes both respond quickly to pyridoxine therapy.

A syndrome resembling vitamin B_6 deficiency as observed in animals has also been noted in man during the treatment of tuberculosis with high doses of the tuberculostatic drug, isoniazid (isonicotinic acid hydrazide, INH). From 2 to 3% of patients receiving conventional doses of INH (2 to 3 mg. /Kg.) developed neuritis; 40% of patients receiving 20 mg. /Kg. developed neuropathy. Tryptophan metabolism (as indicated by xanthurenic acid excretion) was also altered. The signs and symptoms were alleviated by the administration of pyridoxine. Fifty mg. of pyridoxine per day completely prevented the development of the neuritis. It is believed that INH forms a hydrazone complex with pyridoxal, resulting in incomplete activation of the vitamin.

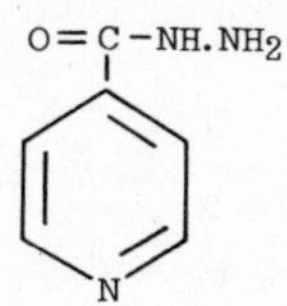

Isonicotinic Acid Hydrazide
(Isoniazid, INH)

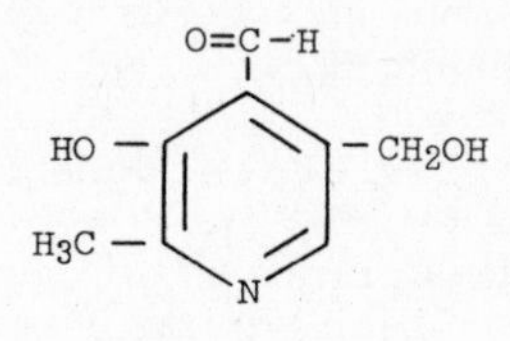

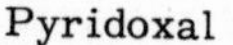

Pyridoxal

The role of pyridoxal in the metabolism of brain has recently been elucidated[22]. In connection with the function of this vitamin as a co-decarboxylase for amino acids, there is one pyridoxal-dependent reaction which is specific to the central nervous system. This reaction is the decarboxylation of glutamic acid to γ-amino butyric acid, which is further metabolized to succinic acid by way of a DPN-dependent soluble dehydrogenase in brain[23]. The glutamic decarboxylase and the product of the decarboxylation, γ-amino butyric acid, are found only in the central nervous system, principally in the gray matter.

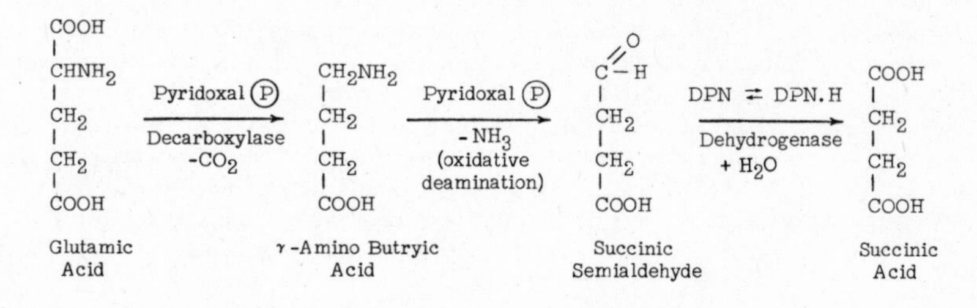

The effects of γ- amino butyric acid on peripheral as well as central synaptic activity suggest that this compound may function as a regulator of neuronal activity. It is now believed that the epileptiform seizures in animals produced by a deficiency of B_6, the action of INH, or the administration of pyridoxine antimetabolites, e.g., deoxypyridoxine (see p. 107), may be related to a decrease in the activity of the glutamic acid decarboxylase with a resultant decrease in the amounts of γ-amino butyric acid necessary to regulate neuronal activity in a normal manner. This idea is supported by the fact that the seizures can be controlled not only by the administration of vitamin B_6 but also by the administration of γ-amino butyric acid.

Sources and Daily Allowances:

It has been difficult to establish definitely the human requirement for vitamin B_6, probably because the quantity needed is not large and because bacterial synthesis in the intestine provides a portion of that requirement. There is some evidence that the requirement for vitamin B_6 is related to the dietary protein intake.

For an adult, 2 mg. per day has been recommended. In a significant number of infants, intakes of less than 0. 1 mg. per day were associated with clinical manifestations of deficiency, as described above; with 0. 3 mg. per day, no symptoms developed, and in most cases there was no increase in the excretion of xanthurenic acid after a tryptophan load. However, in the case of some infants studied, as much as 2 mg. per day were required to prevent xanthurenic acid excretion after a tryptophan load, suggesting the existence of an abnormality in the metabolism of vitamin B_6 in these cases[24].

Good sources of the vitamin include yeast and certain seeds, such as wheat and corn, liver, and, to a limited extent, milk, eggs, and leafy green vegetables.

PANTOTHENIC ACID

Pantothenic acid is essential to the nutrition of many species of animals, plants, bacteria, and yeasts as well as for man. In experimental animals, symptoms due to pantothenic acid deficiency occur in such a wide variety of tissues that the basic function of this vitamin in cellular metabolism is amply confirmed. Gastrointestinal symptoms (gastritis and enteritis with diarrhea) are common to several species when a deficiency of this vitamin occurs. Skin symptoms, including cornification, depigmentation, desquamation, and alopecia also occur frequently. Lack of this vitamin also affects the adrenals. Animals deficient in pantothenic acid exhibit hemorrhage and necrosis of the adrenal cortex and an increased appetite for salt. If this condition persists the gland becomes exhausted, as shown by disappearance of lipoid material from the cortex and an acute state of adrenal cortical insufficiency, with sudden prostration and terminal dehydration.

Chemistry:

In its active form, pantothenic acid is a constituent of Coenzyme A, also known as coacetylase, the coenzyme for acetylation reactions. The coenzyme has a nucleotide structure, as shown below. Synthesis of the coenzyme has been elucidated by Hoagland and Novelli[25]. The biosynthesis in mammalian tissues appears to proceed as follows:

1. Pantothenic acid + cysteine + ATP ⟶ Pantothenyl cysteine
2. Pantothenyl cysteine $\xrightarrow{-CO_2}$ Pantotheine (*Lactobacillus bulgaricus factor; LBF*)
3. Pantotheine + ATP $\xrightarrow{\text{Pantotheine kinase}}$ 4′-phosphopantotheine
4. 4′-phosphopantotheine + ATP ⟶ Dephospho-Coenzyme A + 2 P
5. Dephospho-CoA + ATP ⟶ Coenzyme A + ADP

Magnesium ion is necessary for these reactions, and the substrates must be in their reduced (SH) forms.

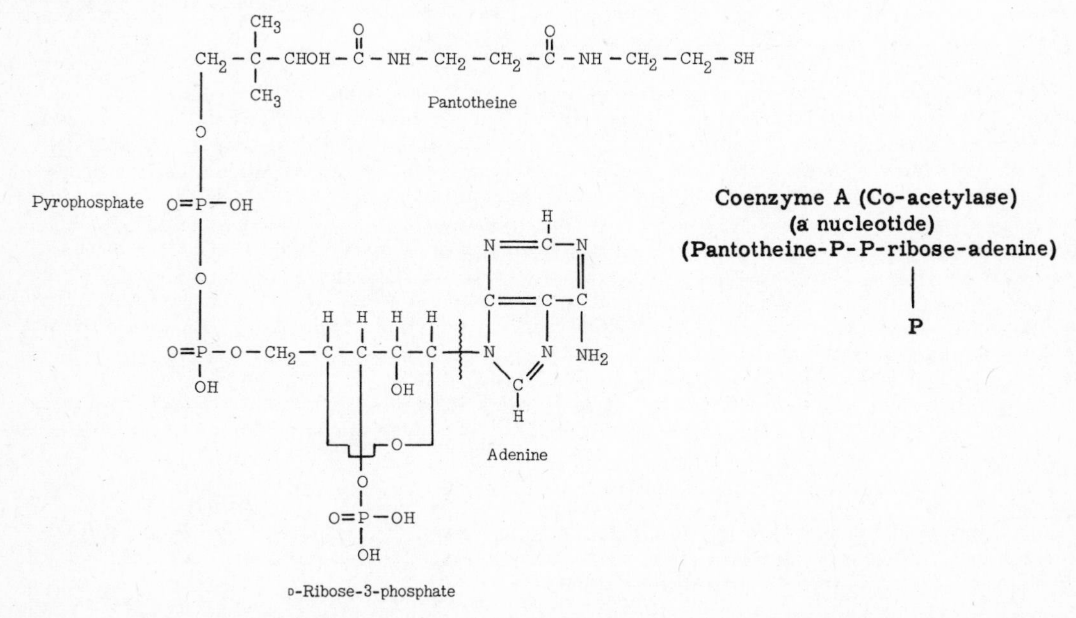

In reactions involving Coenzyme A, combination of the metabolite activated by the coenzyme occurs at the sulfhydryl (SH) group of the pantotheine moiety, through a high-energy sulfur bond. It is therefore customary to abbreviate the structure of the free (reduced) coenzyme as CoA. SH, in which only the reactive SH group or the coenzyme is indicated.

Role in Physiology:

As a constituent of Coenzyme A pantothenic acid is essential to several fundamental reactions in metabolism. Most important is the combination of Coenzyme A with acetate to form "active acetate." In the form of acetyl-Coenzyme A (active acetate), acetic acid participates in a number of important metabolic processes. For example, it is utilized directly by combination with oxaloacetic acid to form citric acid, which initiates the tricarboxylic acid cycle (see p. 183). Thus acetic acid derived from carbohydrates, fats, or many of the amino acids undergoes further metabolic breakdown via this "final common pathway" in metabolism. In the form of active acetate, acetic acid also combines with choline to form acetylcholine, or with the sulfa drugs which are acetylated prior to excretion.

The product of decarboxylation of ketoglutaric acid in the Krebs cycle is a Coenzyme A derivative called "active" succinate (see p. 88). Active succinate and glycine are involved in the first step leading to the biosynthesis of heme (see p. 57). Anemia frequently occurs in animals deficient in pantothenic acid. It may be assumed that this is referable to difficulty in formation of CoA-succinate.

Coenzyme A has also an essential function in lipid metabolism. The first step in the oxidation of fatty acids catalyzed by thiokinases involves the "activation" of the acid by formation of the Coenzyme A derivative, and the removal of a two-carbon fragment in beta oxidation is accomplished by a "thiolytic" reaction, utilizing another mol of Coenzyme A (see p. 209). The two-carbon fragments thus produced are actually in the form of active acetate which may directly enter the citric acid cycle for degradation to carbon dioxide and water or combine to form ketone bodies (see p. 211).

Activation of some amino acids may also involve Co A (see p. 239).

Acetic acid is a precursor of cholesterol and thus of the steroid hormones. The utilization of acetic acid for this purpose is also catalyzed by pantothenic acid as Coenzyme A. As was noted above, a pantothenic acid deficiency inevitably produces profound effects on the adrenal gland. The anatomic changes are accompanied by evidence of functional insufficiency as well. This is due to poor synthesis of cholesterol by the pantothenic acid–deficient gland.

All of these facts point to an extremely important function for this vitamin in metabolism, involving as it does the utilization of carbohydrate, fat, and protein and the synthesis of cholesterol and steroid hormones as well as various acetylation reactions.

The combination of acetic acid or of fatty acids with Co A occurs at the terminal sulfhydryl group (SH) of the pantotheine residue. When acetate is transferred, the SH group is liberated to participate in the activation of another mol of acetate. Acetate may also be converted to acetyl phosphate, utilizing ATP as a phosphate donor. This high-energy compound can then be transferred directly to the SH site on Co A. The sulfur bond of acetyl-Co A (Co A. SH) is a high-energy bond equivalent to that of the high-energy phosphate bonds of ATP and other high-energy phosphorylated compounds (see p. 119). A similar high-energy sulfur bond is found in the derivatives of lipoic acid (thioctic acid; see p. 88). The formation of these high-energy bonds requires, therefore, a source of energy, either from a coupled exothermic reaction which yields the energy for incorporation into the bond, or from the transfer of energy from a high-energy phosphate bond or from another high-energy sulfur bond.

Examples of the function of Coenzyme A in the formation of active acetate from pyruvate and in the formation of active succinate from ketoglutarate will be shown in the next section (see p. 88).

Sources:

Excellent food sources (100 to 200 micrograms per gram of dry material) include egg yolk, kidney, liver, and yeast. Broccoli, lean beef, skimmed milk, sweet potatoes, and molasses are fair sources (35 to 100 micrograms per gram).

A 57% loss of pantothenic acid in wheat may occur during the manufacture of patent flour, and up to 33% is lost during the cooking of meat. Only a slight loss occurs in the preparation of vegetables.

LIPOIC ACID

In connection with the oxidative decarboxylation of pyruvate (see p. 77) to acetate it has been noted that in addition to thiamine, certain other coenzymes are required. Closely associated with thiamine in the initial decarboxylation of α-keto acids is **lipoic acid.** This factor was first detected in studies of the nutrition of lactic acid bacteria, where it was shown to replace the growth-stimulating effect of acetate. For this reason, the designation "acetate replacement factor" was assigned to it. A vitamin required for the nutrition of the protozoon, Tetrahymena geleii, to which the term "protogen" was applied, and a factor from yeast (pyruvate oxidation factor) necessary for the oxidation of pyruvate to acetate by Streptococcus fecalis, which were studied at about the same time, were both later shown to be identical with the original acetate replacement factor of Guirard, Snell, and Williams. In attempts to isolate the active compound from liver the factor was found to be fat-soluble, and for this reason it was renamed lipoic acid. It has now been extracted from natural materials and its structure shown to be a sulfur-containing fatty acid, **6, 8-dithio-octanoic acid.**

Chemistry:

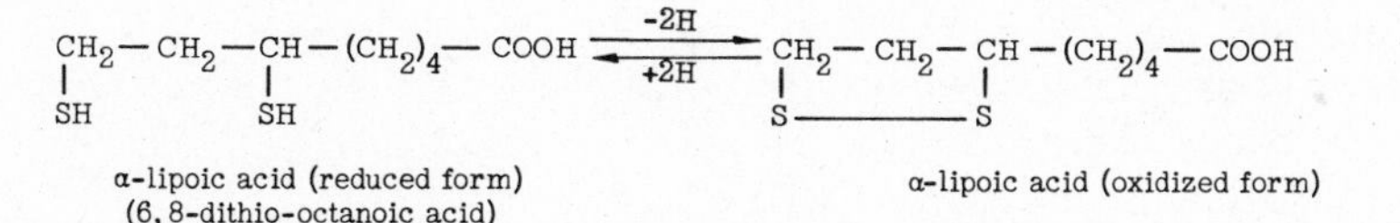

In its active state in the tissues, lipoic acid is closely associated with thiamine pyrophosphate (diphosphate). This "active form" of lipoic acid is referred to as lipothiamide pyrophosphate (LTPP). It is believed to have the following structure:

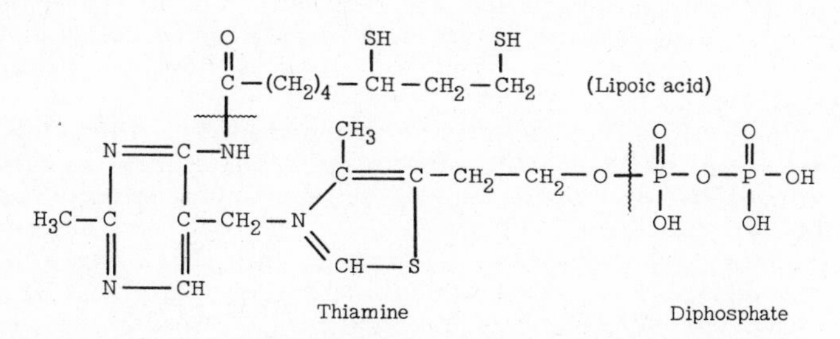

Lipothiamide Pyrophosphate (LTPP; Reduced Form)

Role in Physiology:

Lipoic acid occurs in a wide variety of natural materials. It is recognized as an essential component in metabolism, although it is active in extremely minute amounts. It has not yet been demonstrated to be required in the diet of higher animals, and attempts to induce a lipoic acid deficiency in animals have so far been unsuccessful.

The oxidative decarboxylation of pyruvic acid and of α-ketoglutaric acid (see pp. 77 and 181) involves both thiamine and lipoic acid as lipothiamide pyrophosphate. The series of reactions by which pyruvate is converted to acetate (reaction No. 1 of the citric acid cycle; see p. 181) is shown below. It will be noted that pantothenic acid and nicotinic acid derivatives, as Coenzyme A and DPN respectively, are also required.

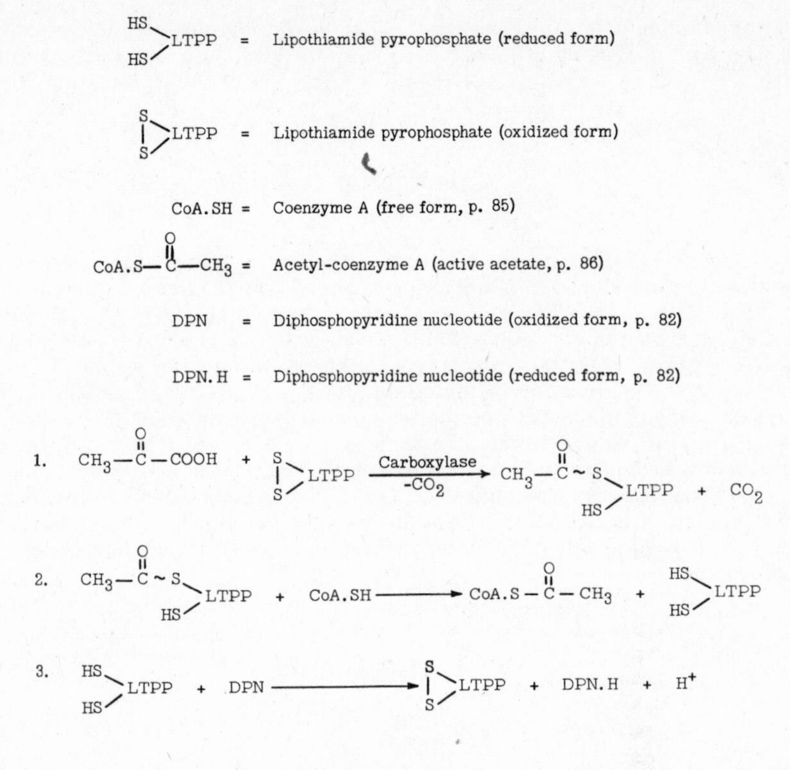

In reaction No. 1, the energy of the decarboxylation step is utilized to form a high-energy bond at the linkage between the acetyl group and a sulfur of the lipothiamide moiety. In reaction No. 2, the high-energy bond is maintained when the acetyl group is transferred to Co A in the formation of acetyl-Co A ("active acetate"). The reoxidation of LTPP is accomplished in reaction No. 3 with the aid of DPN, so that lipothiamide is again ready to function as in reaction No. 1.

A similar series of reactions accomplishes the decarboxylation of α-ketoglutaric acid to form succinyl Co A ("active succinate") (reaction No. 7 of the citric acid cycle; see p. 181). The high-energy sulfur bond is formed at the expense of the decarboxylation reaction.

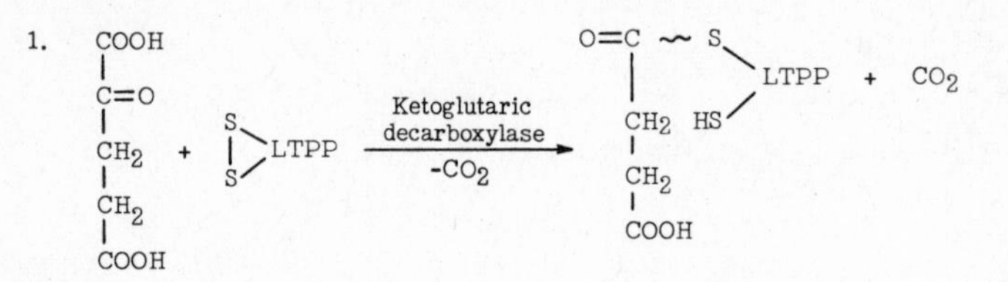

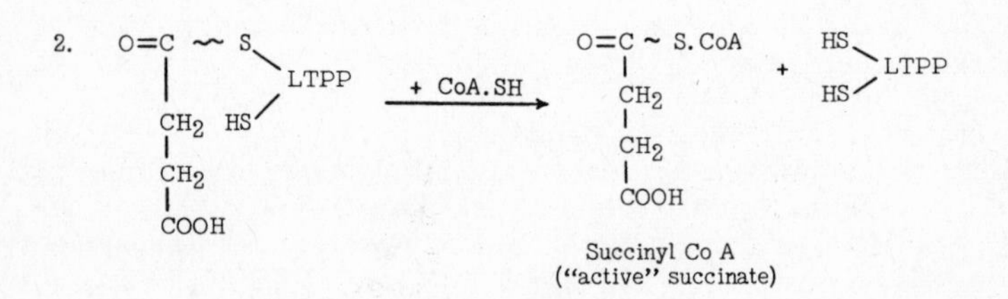

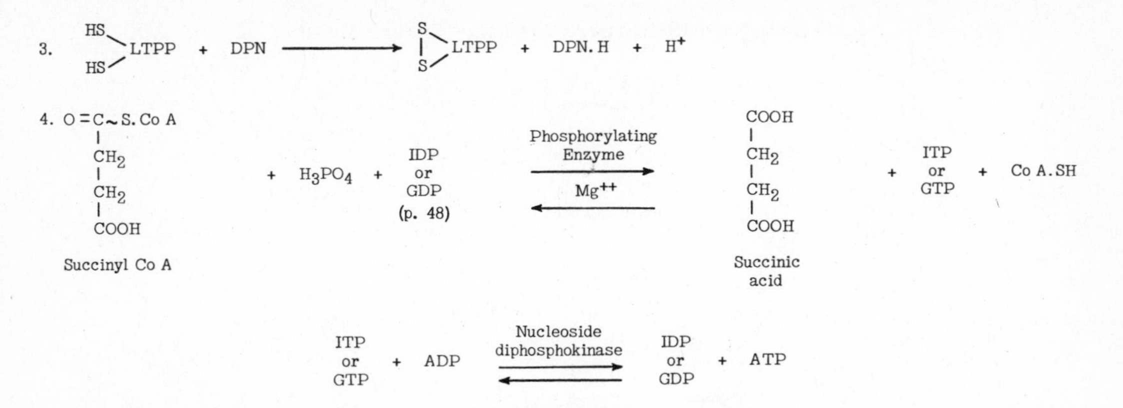

Succinyl Co A ("active succinate") is a precursor of porphobilinogen, the first pyrrole to be synthesized in the production of the porphin nucleus (see p. 57).

In reaction No. 4 above, the conversion of succinyl Co A to succinic acid is shown. In the process, a high-energy phosphate bond is created by transfer of the energy of the sulfur bond to inosine diphosphate (IDP) or to guanosine diphosphate (GDP) (see p. 48), catalyzed by a magnesium-activated phosphorylating enzyme and resulting in the formation of inosine triphosphate (ITP) or guanosine triphosphate (GTP) as well as succinic acid and free Co A. Subsequently, ITP or GTP may transfer a high-energy phosphate to ADP to form ATP. The formation of high-energy phosphate from succinyl Co A is an example of "phosphorylation at the substrate level" (see p. 120).

BIOTIN

Biotin was first shown to be an extremely potent growth factor for microorganisms. As little as 0.005 micrograms permits the growth of test bacteria. In experiments with rats, dermatitis, retarded growth, loss of hair, and loss of muscular control occurred when egg white was fed as the sole source of protein. Certain foods were then found to contain a protective factor against these injurious effects of egg white protein, and this protective factor was named biotin, or the anti-egg-white-injury factor. The antagonistic substance in raw egg white is a protein (avidin) which combines with biotin, even in vitro, to prevent its absorption from the intestine.

It is difficult to obtain a quantitative requirement for biotin. A large proportion of the biotin requirement probably is supplied by the action of the intestinal bacteria. It has been found that a biotin deficiency can be induced more readily in animals after the feeding of those sulfonamide drugs which reduce intestinal bacteria to a minimum. Careful balance studies in man showed that in many instances urinary excretion of biotin exceeded the dietary intake, and that in all cases fecal excretion was as much as three to six times greater than the dietary intake. It is therefore difficult to conceive of a dietary biotin deficiency under the usual circumstances. Highly purified diets have produced the deficiency in chickens and monkeys, but it is possible that the observed symptoms were really due to the effect of the diets on the intestinal flora.

Chemistry:

$C_{10}H_{16}O_3N_2S$, (Hexahydro-2-oxo-1-thieno-3, 4-imidazole-4-valeric acid).

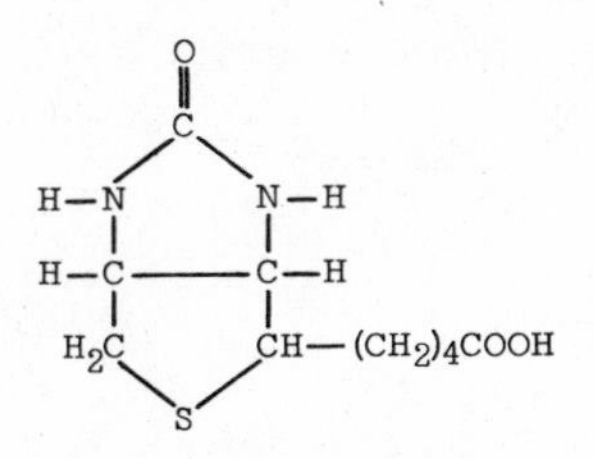

A combined form of biotin has been identified in natural materials. The crystalline compound, which is termed "biocytin," is a combination of biotin through the terminal (epsilon) nitrogen of

Sources and Utilization of the One Carbon Moiety

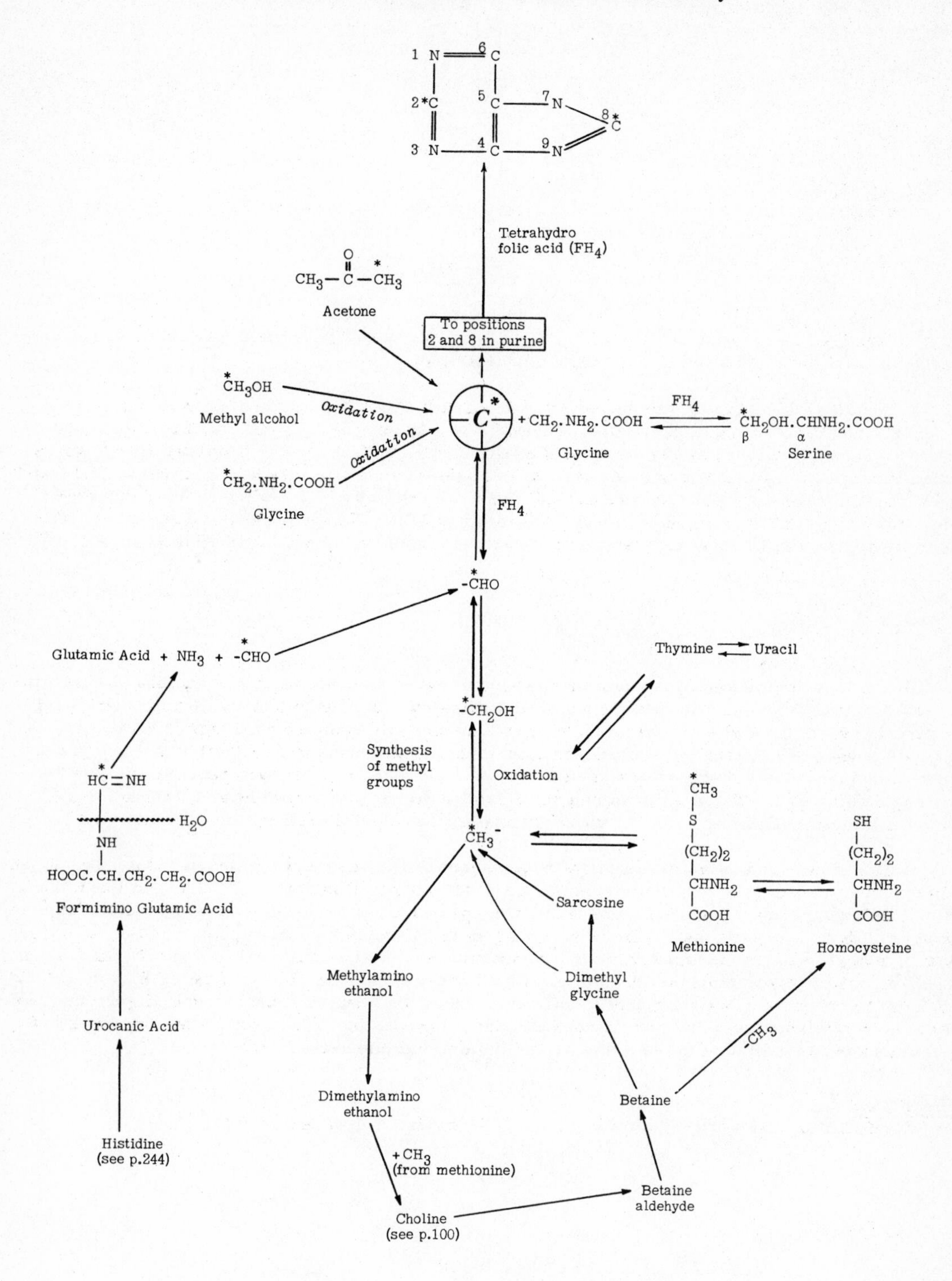

the amino acid, lysine (i.e., ε-N-biotinal lysine). This bound form of biotin is very similar to the vitamin compound as it exists in the cells. It is possible that the natural compound is a coenzyme for certain reactions as described below. When biocytin was administered to human subjects, only free biotin appeared in the blood and urine, indicating that the tissues possess an enzyme, **biocytinase**, capable of releasing the vitamin from its bound form.

Role in Physiology:

Human subjects placed on a refined diet containing a large amount of dehydrated egg white developed symptoms resembling those caused by biotin deficiency in animals. Urinary biotin decreased from a normal level of 30 to 60 micrograms to 4 to 7 micrograms. Prompt relief of symptoms occurred when a biotin concentrate was administered.

Biotin is thought to function in the "fixation" of carbon dioxide. In bacterial metabolism the reaction is illustrated by the conversion of pyruvic acid to oxaloacetic acid (Wood-Werkman reaction). This serves as a method for the production of dicarboxylic acids to maintain the tricarboxylic acid cycle (see p. 184) and also for the synthesis of aspartic acid as shown in the reactions below. These functions of biotin are supported by the observation that both aspartic acid and oxaloacetic acid, when added to the nutrient solution medium, reduce the requirement of the bacteria for the vitamin. The "sparing" action of the two compounds is presumably achieved by reducing the need for the vitamin in the synthesis of aspartic and oxaloacetic acids. Fixation of CO_2 in formation of carbon No. 6 in purine synthesis (see p. 264) is impaired in biotin-deficient yeast[26], which suggests that the vitamin plays a role in purine synthesis in connection with this step in formation of the purine structure.

A number of other enzyme systems[35, 36] are reportedly influenced by biotin. These include succinic acid dehydrogenase and decarboxylase, and the deaminases of the amino acids, aspartic acid, serine, and threonine. The synthesis of citrulline (see p. 230) is also related to the biotin content of the medium[27].

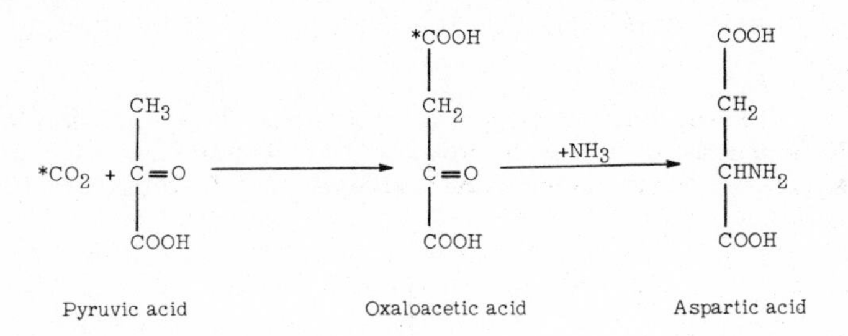

Sources:

It seems doubtful whether any but the most severely deficient diet would result in a biotin deficiency in man. The vitamin is widely distributed in natural foods. Egg yolk, kidney, liver, tomatoes, and yeast are excellent sources.

THE FOLIC ACID GROUP

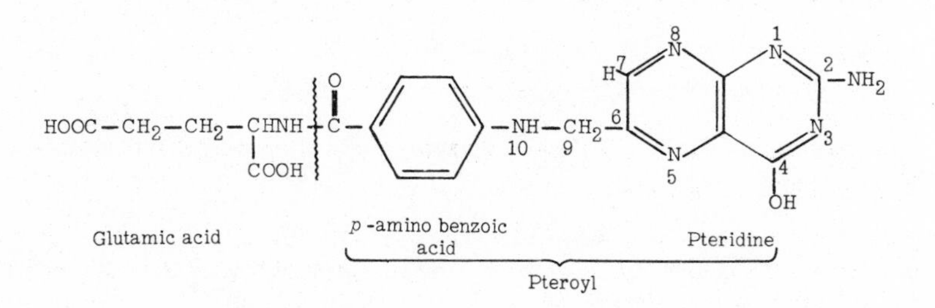

Folic acid (folacin, pteroyl glutamic acid, PGA) is a combination of the pteridine nucleus, *p*-aminobenzoic acid, and glutamic acid.

There are at least three chemically related compounds of nutritional importance which occur in natural products. All may be termed pteroyl glutamates. These three compounds differ only in the number of glutamic acid residues attached to the pteridine-PAB complex. The formula for folic acid shown above is the monoglutamate. This is synonymous with vitamin B_c. The substance once designated as the fermentation factor is a triglutamate, and vitamin B_c conjugate of yeast is a heptaglutamate. The deficiency syndrome which is now thought to be due to a lack of these substances has been recognized in the monkey since 1919. Since many investigators discovered the same deficiency state by varying technics, several names for these factors, in addition to those already mentioned, have appeared. These include vitamin M, factor U, factors R and S, norit eluate factor, S lactis R factor, and L. casei factor.

Folic acid, when added to liver slices, is converted to a formyl derivative. Ascorbic acid enhances the activity of the liver in this reaction. This form of folic acid was first discovered in liver extracts when it was found to supply an essential growth factor for a lactobacillus, Leuconostoc citrovorum. Thus it was termed the **citrovorum factor (CF)**. When its chemical structure was determined, the name **folinic acid** was applied.

The structure of folinic acid (citrovorum factor) is shown below. It is the reduced (tetrahydro) form of folic acid with a formyl group on position 5.

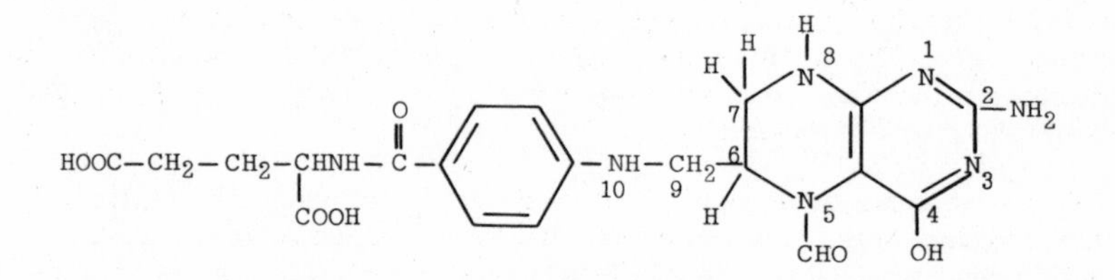

Folinic Acid, $f^5.FH_4$ (Citrovorum Factor, CF)
(5-Formyl-5, 6, 7, 8-tetrahydro-folic Acid)

A similar compound but with the formyl group on position 10 has been synthesized. It is called folinic acid-SF (synthetic factor) or leucovorin. Rhizopterin, or the Streptococcus lactis R (SLR) factor, is a naturally occurring compound which is a 10 formyl derivative of pteroic acid.

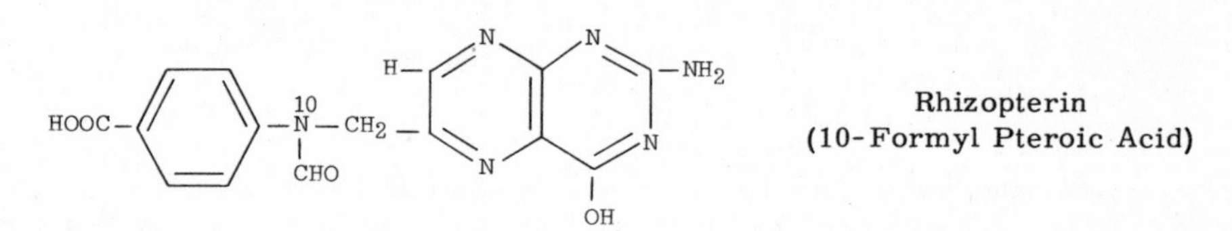

Rhizopterin
(10-Formyl Pteroic Acid)

Role in Physiology:

The folic acid coenzymes are specifically concerned with biochemical reactions involving the transfer and utilization of the single carbon (C_1) moiety (see p. 90). Before functioning as a C_1 carrier, folic acid must be reduced, first to dihydro folic acid, FH_2, and then to the tetrahydro compound, FH_4, catalyzed by a folic acid reductase and using TPN.H as hydrogen donor, as follows:

$$FH_2 + TPN.H + H^+ \longrightarrow FH_4 + TPN$$

The folic acid antagonists (see p. 94) are extremely potent competitive inhibitors of this reaction.

The "one carbon" (C_1) moiety may be either formyl (-CHO) or hydroxy methyl ($-CH_2OH$). The two are metabolically interconvertible in a reaction catalyzed by a TPN-dependent hydroxy methyl dehydrogenase.

$$-CH_2OH \underset{TPN \rightleftharpoons TPN.H}{\overset{\text{Dehydrogenase}}{\rightleftharpoons}} -CHO$$

As has been noted above, folinic acid is a 5-formyl (f^5) tetrahydro folic acid (FH_4): in abbreviated form, $f^5.FH_4$. However, except for the formylation of glutamic acid in the course of the metabolic degradation of histidine (see p. 245), the f^5 compound (folinic acid) is metabolically inert. Instead, the f^{10} ($f^{10}.FH_4$) or f^{5-10} ($f^{5-10}.FH_4$) tetrahydro derivatives, in which the single carbon is bound between positions 5 and 10 on tetrahydro folic acid, are the active forms of the folic acid coenzymes in metabolism.

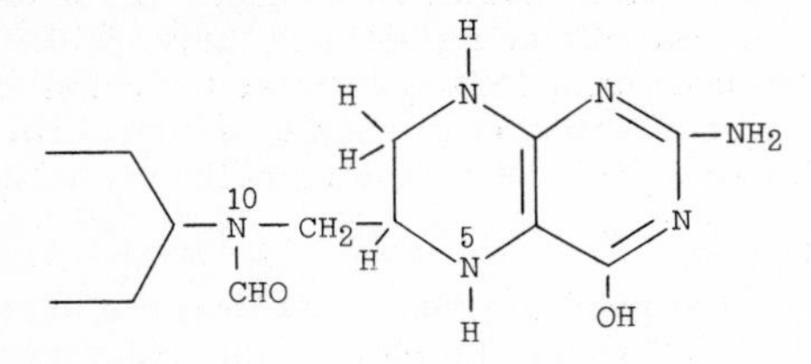

$f^{10}.FH_4$ (10 formyl tetrahydro folic acid)

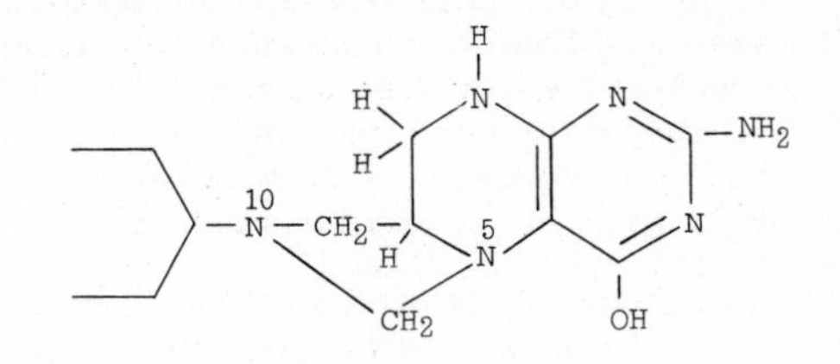

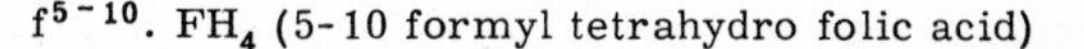

$f^{5-10}.FH_4$ (5-10 formyl tetrahydro folic acid)

The f^5 can be converted to f^{10} by the action of an enzyme system, formyl tetrahydro folic acid isomerase, as follows:

$$f^5.FH_4 + ATP \xrightarrow[\text{cyclo dehydrase}]{Mg^{++}} f^{5-10}.FH_4 + ADP + P \qquad f^{5-10}.FH_4 + H_2O \xrightarrow[\text{cyclo hydrolase}]{} f^{10}.FH_4$$

The one-carbon moiety on tetrahydro folic acid can be transferred to amino or to SH groups. An example of the first is the formimino (HC = NH) group on glutamic acid, a product of histidine breakdown (see p. 245); the second is exemplified by the formation of thiazolidine carboxylic acid with cysteine.

The formimino group (fi) on glutamic acid can serve as a source of the one-carbon moiety as follows:

$$\text{fi-Glutamic acid} + FH_4 \longrightarrow fi^5.FH_4 + \text{Glutamic acid} \qquad fi^5.FH_4 \longrightarrow f^{5-10}.FH_4 + NH_3$$

Other sources of the one-carbon moiety are the methyl groups of (a) methionine, (b) choline, by way of betaine (see p. 99), and (c) of thymine - all of which are oxidized to hydroxy methyl ($-CH_2OH$) groups and carried as such on f^{5-10} FH_4. The hydroxy methyl group (h) is then oxidized in a TPN-dependent reaction to a formyl (f) group:

$$h^{5-10}.FH_4 \xrightarrow[\text{oxidize}]{TPN \rightarrow TPN.H} f^{5-10}.FH_4$$

The beta carbon of serine as a hydroxy methyl group may also contribute to the formation of a single carbon moiety.

The single formyl carbon which is present on the tetrahydro folic acids is utilized in several important reactions (see p. 90): The first is as a source of carbons 2 and 8 in the purine nucleus as described on p. 264. A second reaction involving the formyl carbon on $f^{10}.FH_4$ is formation of the beta carbon of serine in conversion of glycine to serine (see p. 234). A third reaction is in the synthesis of methyl groups for (a) methylation of homocysteine to form methionine (see p. 248) or (b) of uracil to form thymine (see p. 267), and (c) for the synthesis of choline (see p. 100). In the methylation reactions, the formyl group is first reduced to hydroxyl methyl before methylation will occur. Vitamin B_{12} (see p. 96) may also be required in these methylation reactions.

Huennekens, Osborn, and Whiteley have recently summarized the biochemistry of the folic acid coenzymes [28].

The participation of the folic acid coenzymes in reactions leading to synthesis of purines and to thymine, the methylated pyrimidine of DNA, emphasizes the fundamental role of folic acid in growth and reproduction of cells. Because the blood cells are subject to a relatively rapid rate of synthesis and destruction, it is not surprising that interference with red blood cell formation would be an early sign of a deficiency of folic acid, or that the folic acid antagonists would readily inhibit the formation of leukocytes. But it must be remembered that the requirement for the folic acid coenzymes is undoubtedly generalized throughout the body and not confined to the hematopoietic system. This is supported by the observation that in folic acid–deficient monkeys there was a considerable decrease in the rate of synthesis of nucleoprotein, which rose to normal after administration of the vitamin. The function of the folic acid coenzymes in synthesis and utilization of methyl groups relates these vitamins also to phospholipid metabolism (choline synthesis; see p. 90) and to amino acid metabolism.

An experimental deficiency of folic acid has not been produced in man. But in sprue the administration of synthetic folic acid (5 to 15 mg. per day) has been followed by rapid and impressive remissions, both clinically and hematologically. The glossitis and diarrhea subside in a few days, and a reticulocytosis occurs which is followed by regeneration of the erythrocytes and hemoglobin. Roentgenologic evidence of improved gastrointestinal function, improved fat absorption, and a return of the glucose tolerance curve to normal are also observed. The vitamin seems therefore to correct both the hematopoietic and gastrointestinal abnormalities in sprue.

Folic acid, when originally made available in crystalline form, excited considerable interest because of its therapeutic effect in nutritional macrocytic anemia, pernicious anemia, and the related macrocytic anemias, as well as in sprue. Unfortunately, the hematologic response to the vitamin is not permanent, and the neurologic symptoms in pernicious anemia (combined system disease) remain unchanged. It is now apparent that while folic acid derivatives do have an effect on hematopoiesis, other factors are also necessary for the complete development of the blood cells. In pernicious anemia the most important factor seems to be vitamin B_{12}, since uncomplicated cases respond completely to this vitamin alone. Furthermore, this factor controls both the hematologic and the neurologic defect. It has therefore been concluded that folic acid has no place in the treatment of uncomplicated pernicious anemia and, while it is harmless when administered to patients adequately treated with vitamin B_{12}, its use is unnecessary in patients receiving a normal diet[37].

Sources:

Knowledge of the distribution of the folic acid group in natural foods is still incomplete. The data at hand indicate that fresh leafy green vegetables (folia-leaf), cauliflower, kidney, and liver are rich sources. The lability of the vitamin to the cooking process is said to be similar to that of thiamine.

Folic Acid Antagonists:

The concept of competitive inhibition or metabolic antagonism is discussed on p. 107. Antagonists to folic acid have found some clinical application in the treatment of malignant disease, and confirmation of the action of folic acid in cell growth has been obtained in studies of the effect of these antagonists on cells maintained in tissue culture.

Maximal inhibitory action is obtained when an amino group is substituted for the hydroxy group on position 4 of the pteridine nucleus. Thus **aminopterin** (4-amino folic acid) is the most potent folic acid inhibitor yet discovered. Another antagonist is α-methopterin (4-amino-10-methyl-folic acid). In animals the inhibitory effect of aminopterin cannot be reversed by folic acid but only by folinic acid. This suggests that aminopterin interferes with the formation of folinic acid from folic acid or with the utilization of the formyl group. Recent work suggests that the interference of the antimetabolites occurs in the reduction of folic acid to the tetrahydro compound. Reduction is a necessary preliminary to the carriage of the one-carbon moiety.

In tissue cultures it has been found that aminopterin blocks the synthesis of nucleic acids, presumably by preventing the reduction of folic acid to the tetrahydro derivative and thus transport of the formyl carbon into the purine ring. Such inhibited cells fail to complete their mitoses; they do not progress from metaphase to anaphase because of a failure in the synthesis of nucleoprotein, a synthesis which is essential to chromosome reduplication.

Aminopterin has been used in the treatment of leukemia, particularly in children. A remission is induced temporarily in some patients, but after a time the leukemic cells apparently acquire the power to overcome the effects of the antagonist.

INOSITOL

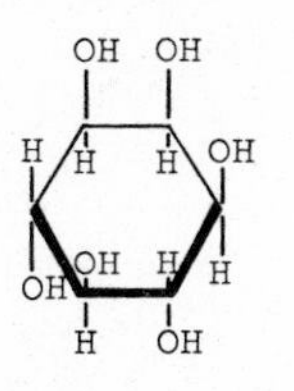

$C_6H_{12}O_6$ (Hexahydroxy Cyclohexane)

There are nine isomers of inositol. Meso-inositol, also called myo-inositol (shown above), is the most important one in nature and the only isomer which is biologically effective.

The significance of this compound in human nutrition has not been established. However, in studies on the nutrient requirements of cells in tissue culture, Eagle[29] found that 18 different human cell strains maintained on a semi-synthetic medium failed to grow without the addition of meso-inositol. None of the other isomers were effective, a finding which is in agreement with the results of similar experiments in animals.

Together with choline, inositol has a lipotropic action in experimental animals (see p. 205). This lipotropic activity may be associated with the formation of inositol-containing lipids (lipositols; see p. 22).

Deficiency symptoms in mice include so-called spectacled eye, alopecia, and failure of lactation and growth. In inositol-deficient chicks an encephalomalacia and an exudative diathesis have been reported.

Sources:
Inositol is found in fruits, meat, milk, nuts, vegetables, whole grains, and yeast.

PARA-AMINOBENZOIC ACID (PABA)

p-Aminobenzoic acid is a growth factor for certain microorganisms and an antagonist to the bacteriostatic action of sulfonamide drugs. It forms a portion of the folic acid molecule (see p. 91), and it is suggested that its actual role is to provide this component for the synthesis of folic acid by those organisms which do not require a preformed source of folic acid.

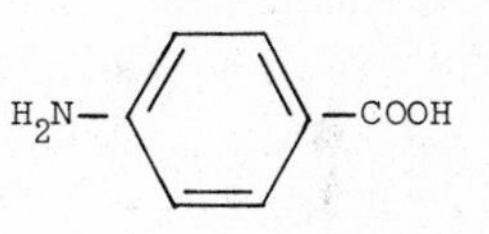

The successful use of *p*-aminobenzoic acid in the treatment of certain rickettsial diseases has been reported. It has recently been found that *p*-oxybenzoic acid is an essential metabolite for these organisms and that *p*-aminobenzoic acid acts as an antagonist to that substance.

VITAMIN B_{12}

Vitamin B_{12}, the anti-pernicious anemia factor (extrinsic factor of Castle) was first isolated in 1948 from liver as a red crystalline compound containing cobalt and phosphorus. The vitamin is now obtained as a product of fermentation by Streptomyces griseus. Its concentration, either in liver or in the fermentation liquor, is only about one part per million.

Chemistry:
The structure of vitamin B_{12} is shown below. The central portion of the molecule consists of four reduced and extensively substituted pyrrole rings surrounding a single cobalt atom. Below this is a 5, 6-dimethylbenzimidazole riboside which is connected at one end to the central cobalt atom and at the other end from the ribose moiety through phosphate and aminopropanol to a side chain on ring IV of the tetrapyrrole nucleus. A cyanide group which is coordinately bound to the cobalt atom may be removed; the resulting compound is called "cobalamin." Addition of cyanide forms "cyanocobalamin," identical with the originally isolated vitamin. Substitution of the cyanide group with a hydroxy group forms "hydroxycobalamin"; with a nitro group, "nitrocobalamin." The biologic action of these derivatives is similar to that of cyanocobalamin.

Crystalline vitamin B_{12} is stable to heating at 100° C. for long periods, and aqueous solutions at pH 4 to 7 can be autoclaved with very little loss. However, destruction is rapid when the vitamin is heated at pH 9 or above.

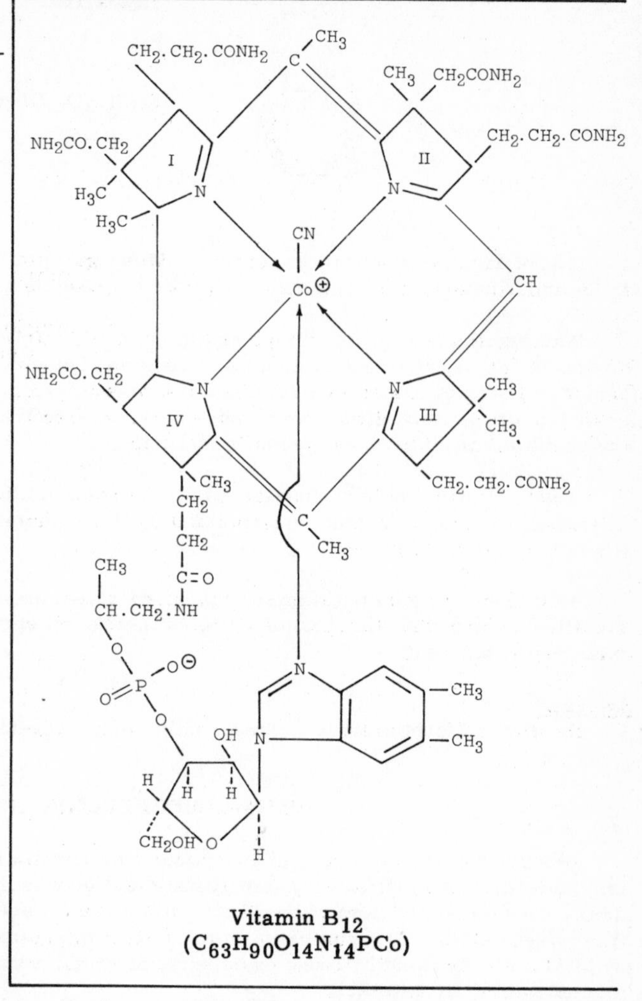

Vitamin B_{12}
($C_{63}H_{90}O_{14}N_{14}PCo$)

Role in Physiology:

The absorption of vitamin B_{12} from the gastrointestinal tract is dependent upon the presence of hydrochloric acid and a constituent of normal gastric juice which has been designated as "intrinsic factor" by Castle. The intrinsic factor is a constituent of gastric mucoprotein. It is found in the cardia and fundus of the stomach but not in the pylorus. Atrophy of the fundus and a lack of free hydrochloric acid (achlorhydria) are usually associated with pernicious anemia. Patients who have sustained total removal of the stomach will also develop a vitamin B_{12} deficiency and anemia because complete absence of intrinsic factor prevents absorption of vitamin B_{12} from their intestines, although as long as three years may elapse after the operation before anemia will be apparent. This is because the vitamin B_{12} stores disappear very slowly. In the liver the biological half-life for the vitamin is estimated to be about 400 days.

Intrinsic factor concentrates prepared from one animal species are not able in all cases to increase the intestinal absorption of vitamin B_{12} in other species of animals or in man. With intrinsic factor concentrates prepared from hog mucosa, a refractory state may eventually develop in some pernicious anemia patients given also B_{12} by mouth, which suggests that a block in the absorptive mechanism has developed. However, human gastric juice remains effective in these "refractory" patients as a means of facilitating absorption of vitamin B_{12}.

The chemical nature of the intrinsic factor is not yet known, although it is believed to be a mucoprotein. It is destroyed by heating 30 minutes at 70 to 80° C. and inactivated by prolonged digestion with pepsin or trypsin.

If very large doses of vitamin B_{12} (3000 micrograms) are given by mouth to a patient with pernicious anemia, an increase in the concentration of the vitamin in the plasma is observed. If the vitamin is given intramuscularly in a dose of 10 to 25 micrograms, the rise in its concentration in the serum is similar, and the vitamin is entirely effective without intrinsic factor. Apparently the only function of intrinsic factor is to provide for the absorption of the vitamin from the intestine, and then only when it is present in very small amounts, as it is in natural foods. Thus vitamin B_{12} itself is both the extrinsic factor and the anti–pernicious anemia factor (APA) as originally described by Castle.

The normal partial intestinal mucosal block to absorption of B_{12} is complete or almost complete in sprue and in pernicious anemia when tested by the oral administration of radioactive cobalt-labeled B_{12} followed by measurements of hepatic uptake. The defect in sprue is not corrected by the administration of intrinsic factor because it is due to a generalized defect inherent in the absorptive mechanisms in the intestinal wall. In pernicious anemia caused by a lack of intrinsic factor, the administration of a test dose of labeled B_{12} together with 75 to 100 ml. of normal human gastric juice, or with a potent source of intrinsic factor, results in a satisfactory hepatic uptake[30].

The vitamin B_{12} content of the serum can be measured by microbiologic methods (see p. 37). According to Rosenthal and Sarett[31] the content of vitamin B_{12} in human serum is between 0.008 and 0.042 micrograms per 100 ml., with an average in 24 normal individuals of 0.02 micrograms per 100 ml. In the pernicious anemia patient in relapse, the vitamin is reported to be absent or present only in very small amounts (less than 0.004 micrograms per 100 ml.).

The intramuscular injection of 10 to 100 micrograms of vitamin B_{12} produces, both in normal and in pernicious anemia patients, a prompt increase in the serum levels. This increase reaches a maximum in one hour, then falls rapidly for three hours, after which there is a slow decline over a 24-hour period. Much larger doses (1000 to 3000 micrograms) are required by mouth to produce significant rises in the serum levels.

Functions in Metabolism:

The most characteristic sign of a deficiency of vitamin B_{12} is the development of a macrocytic anemia. This indicates that when B_{12} intake is low, the demand for this vitamin in hemopoiesis exceeds that for any other clinically recognizable physiologic function. Macrocytosis is therefore the most sensitive indicator of a B_{12} deficiency. However, there is much experimental evidence for a fundamental role of this vitamin in metabolic processes which are not limited to the hematopoietic system. The most consistent evidence for the function of vitamin B_{12} in metabolism is in connection with the neogenesis of methyl groups and/or as a co-factor in transmethylation reactions, although the vitamin has not actually yet been identified as a co-factor in any known enzymatic system. Other experiments have been interpreted as suggesting a relationship of B_{12} to nucleic acid synthesis (possibly in methylation of uridine to thymidine), and as contributing to the maintenance of SH groups in the reduced state as well as to the incorporation of amino acids into protein[32].

Sources and Daily Allowances:

The exact amount of vitamin B_{12} required by a normal human subject is not known. However, pernicious anemia patients have been satisfactorily maintained by an intramuscular injection of 45 micrograms of B_{12} every six weeks. According to the most recent evidence the minimum daily requirement for the vitamin has been set at 0.6-1.2 micrograms per day, with the recommended allowance extending to 2.8 micrograms per day. As noted below, foods of animal origin are the only important dietary sources of B_{12}. The ingestion of 1 cup of milk, 4 ounces of meat, and 1 egg per day provides 2-4 micrograms of vitamin B_{12}. The use of beef liver or kidney would increase the intake to 15-20 micrograms per day.

The amounts of B_{12} in foods are very low. The dietary sources of the vitamin are predominantly foods of animal origin, the richest being liver and kidney, which may contain as much as 40 to 50 micrograms per 100 grams. Muscle meats, milk, cheese and eggs contain 1 to 5 micrograms per 100 grams. The vitamin is almost if not entirely absent from the products of higher plants. Symptoms including sore tongue, paresthesia, amenorrhea, and nonspecific "nervous symptoms" have been reported from Great Britain and the Netherlands in groups living exclusively on vegetable foods. These are the only instances in which a true dietary deficiency of the vitamin has been discovered; in all other cases of B_{12} deficiency an intestinal absorptive defect is responsible.

It is of great interest that probably the only original source of vitamin B_{12} is by microbial synthesis. There is no evidence for its synthesis by the tissues of higher plants or animals. The activity of microorganisms in synthesizing B_{12} extends to the bacteria of the intestine. This is best illustrated by the microbial flora of the rumen in ruminant animals. A vitamin B_{12} concentration of 50 micrograms per 100 grams of dried rumen contents has been reported. Presumably this accounts for the superior B_{12} content of livers from ruminant animals as compared to other animals such as the pig or rat. It is probable that the synthetic activity of the intestinal bacteria also provides B_{12} for herbivorous animals other than ruminants.

Vitamin B_{12} is excreted in the feces of human beings, and the amounts found in the feces of pernicious anemia patients may be even larger than that in normal subjects. This is due to an absorptive defect incident to a lack of intrinsic factor in these patients. It is thus apparent that B_{12} is synthesized by the bacteria present in the human intestine as well as by those in the intestine of other animals as described above. The contribution of this source of B_{12} in normal subjects is not known, but the restricted amounts available in the diet suggest that this may be an important auxiliary source of this vitamin.

CHOLINE

Chemistry:

Trimethyl-hydroxyethyl-ammonium hydroxide.

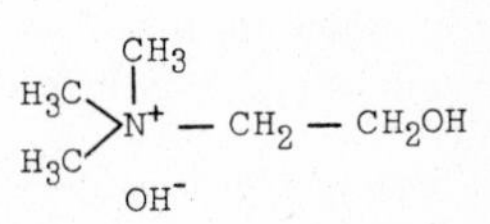

Choline is an essential metabolite, although probably it cannot be classified as a vitamin since it is synthesized by the body. Furthermore, the quantities of choline which are required by the organism are considerably larger than most substances considered as vitamins. However, in many animal species, a deficiency of choline, or of choline precursors, leads to certain well-defined symptoms which are suggestive of vitamin deficiency diseases. Disturbances in fat metabolism are most prominently evidenced by the development of fatty livers (see p. 204). In the young growing rat there is also hemorrhagic degeneration of the kidneys and hemorrhage into the eyeballs and other organs. Older rats and the young animals which survive the acute stage develop cirrhosis. In chicks and young turkeys, choline deficiency causes perosis, or slipped tendon disease, a condition in which there is a defect at the tibiotarsal joint of the bird. Many other animals, such as rabbits and dogs, are also susceptible to choline deficiency.

Role in Physiology:

Acetylcholine is well known as the chemical mediator of nerve activity. It is produced from choline and acetic acid. The reaction is preceded by the synthesis of ''active acetate,'' i.e., acetyl-Coenzyme A which is formed from acetate and Co A (see p. 86). A source of high energy phosphate, adenosine triphosphate, ATP (see p. 119) is also required. An enzyme (acetyl thiokinase) which catalyzes the formation of acetyl-Co A has been found in pigeon liver. After the active acetate has been formed in the presence of a second enzyme, choline acetylase, **acetylation of choline** by acetyl-Co A occurs.

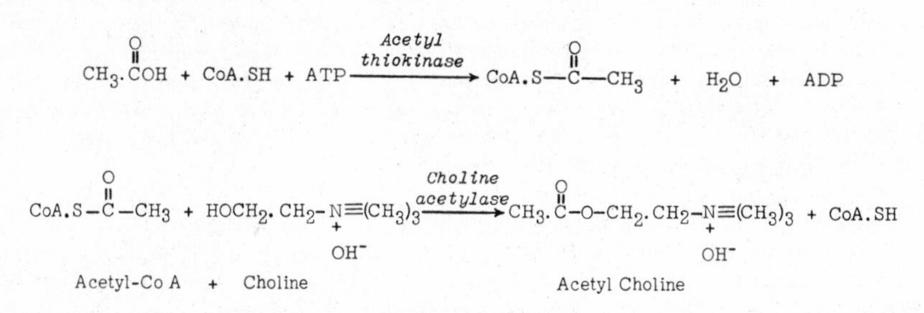

Acetylcholine esterase (ACh-esterase; see p. 109) is an enzyme, present in many tissues, which hydrolyzes acetylcholine to choline and acetic acid. Its importance in nerve activity is discussed on p. 388. It has recently been found that red blood cells can synthesize acetyl choline and that both choline acetylase and acetylcholine esterase are present in red cells. Choline acetylase has also been detected not only in brain and nerve but in skeletal muscle, spleen, and placental tissue as well. The presence of this enzyme in tissues like placenta or erythrocytes which have no nerve supply suggests a more general function for acetylcholine than that in nerve alone. According to Holland and Greig[33], the formation and breakdown of acetylcholine may be related to cell permeability. With respect to red blood cells, they have noted that when the enzyme, choline acetylase, is inactive either because of drug inhibition or because of lack of substrate, the cell loses its selective permeability and undergoes hemolysis.

A chemical method for the determination of choline in the plasma has been developed by Appleton, et al.[36] The free choline level in the plasma of normal male adults averages about 4.4 micrograms per ml. Analyses made over a period of several months indicate that each individual maintains a relatively constant plasma level and that this level is not increased after meals or by the oral administration of large amounts of choline. This suggests that there is a mechanism in the body to maintain the plasma choline at a constant level. Excretion into the urine is a minor factor in this regulatory process, which must therefore be metabolic in origin.

Metabolism of Choline:

A scheme for the biosynthesis of choline has been proposed by Stekol[34], based on experiments with liver slices as well as with the intact animal. According to this scheme, ethanolamine produced by decarboxylation of the amino acid in serine is progressively methylated to choline, but the origin of the first two methyl groups is not the same as that of the third. These first two methyl groups, leading to the production of dimethylamino ethanol, arise by synthesis from one-carbon (hydroxymethyl) moieties through tetrahydro folic acid transport as discussed on p. 93; and only the final methylation - of dimethylaminoethanol to choline - is accomplished by direct transmethylation from methionine. In this latter reaction, folic acid derivatives are not involved. Because of the contributions of the amino acids to synthesis of choline, the quantity of protein in the diet affects the choline requirement; and a choline deficiency is usually coincident with some degree of protein deficiency.

The first reaction in the catabolism of choline is oxidation to betaine aldehyde, which is further oxidized to betaine. This latter compound is an excellent methyl donor and, in fact, choline itself functions as a methyl donor only after oxidation to betaine. After loss of a methyl group by a direct methylation reaction in which a methyl group is transferred to homocysteine to form methionine, or to activated methyl donors, betaine is converted to dimethylglycine. Oxidation of one of the methyl groups on dimethyl glycine produces N-hydroxymethyl sarcosine. The hydroxymethyl group is then lost and sarcosine is formed by transfer of the hydroxymethyl to tetrahydro folic acid. It is by this oxidative reaction and transfer to folic acid derivatives that methyl groups contribute to the one-carbon (formyl) pool (see p. 90). Sarcosine (N-methyl glycine) is now converted to glycine by oxidation and transfer of its methyl group as described above. Glycine is readily converted to serine by addition of a hydroxymethyl group derived from the one-carbon pool of the formylated tetrahydro folic acid derivatives; finally, the decarboxylation of serine to produce ethanolamine starts the cycle of choline synthesis once again. These reactions are diagrammed on p. 100.

It is apparent that dimethyl aminoethanol is the intracellular precursor of choline in man, and there is some evidence that this compound is a more efficient precursor of acetyl choline than choline itself because of limited transport of choline across membranal barriers and because of rapid loss of choline by oxidation to betaine[35]. It follows that dimethyl aminoethanol may also be superior to choline in other reactions requiring the latter compound. In this connection it is of interest that chicks utilize methionine poorly for the synthesis of choline. This is due to limited ability of the chick to synthesize dimethyl aminoethanol. The requirement for choline in the chick is, however, fully satisfied by supplements of dimethyl aminoethanol and methionine.

OTHER VITAMINS

It is very probable that additional vitamins or vitamin-like factors are needed for optimal nutrition of many plant and animal species. Certain microorganisms require growth factors other than those already known, and not infrequently a study of such requirements leads to the discovery of a hitherto unsuspected factor which is needed by higher animals as well. Recent examples are pantothenic acid, lipoic acid, folic and folinic acids, and vitamin B_{12}.

A recently-discovered metabolite essential for microorganisms which may later prove to be of importance in animals is the Lactobacillus bifidus factor described by P. György. Chemically, this substance appears to belong to the nitrogen-containing polysaccharides, such as the hexosamines (see p. 11), which are found in high concentration in the mucoproteins (see pp. 132 and 383). The L. bifidus factor is found in many mucinous secretions, e.g., in hog mucin, which is a good source of the factor. Human milk is also one of the best sources of the L. bifidus factor. The content of protein in human milk (6 to 7% of the total calories) is somewhat lower than the 15 to 20% generally considered as ideal. The high concentration of L. bifidus factor which in rats exerts a growth-promoting effect when the animals are placed on a low-protein diet may provide for such efficient utilization of protein in human milk as to make the lower protein intake adequate.

There is also some evidence that the L. bifidus factor promotes better assimilation of lactose, possibly as a component of a coenzyme involved in galactose utilization.

METABOLISM OF CHOLINE

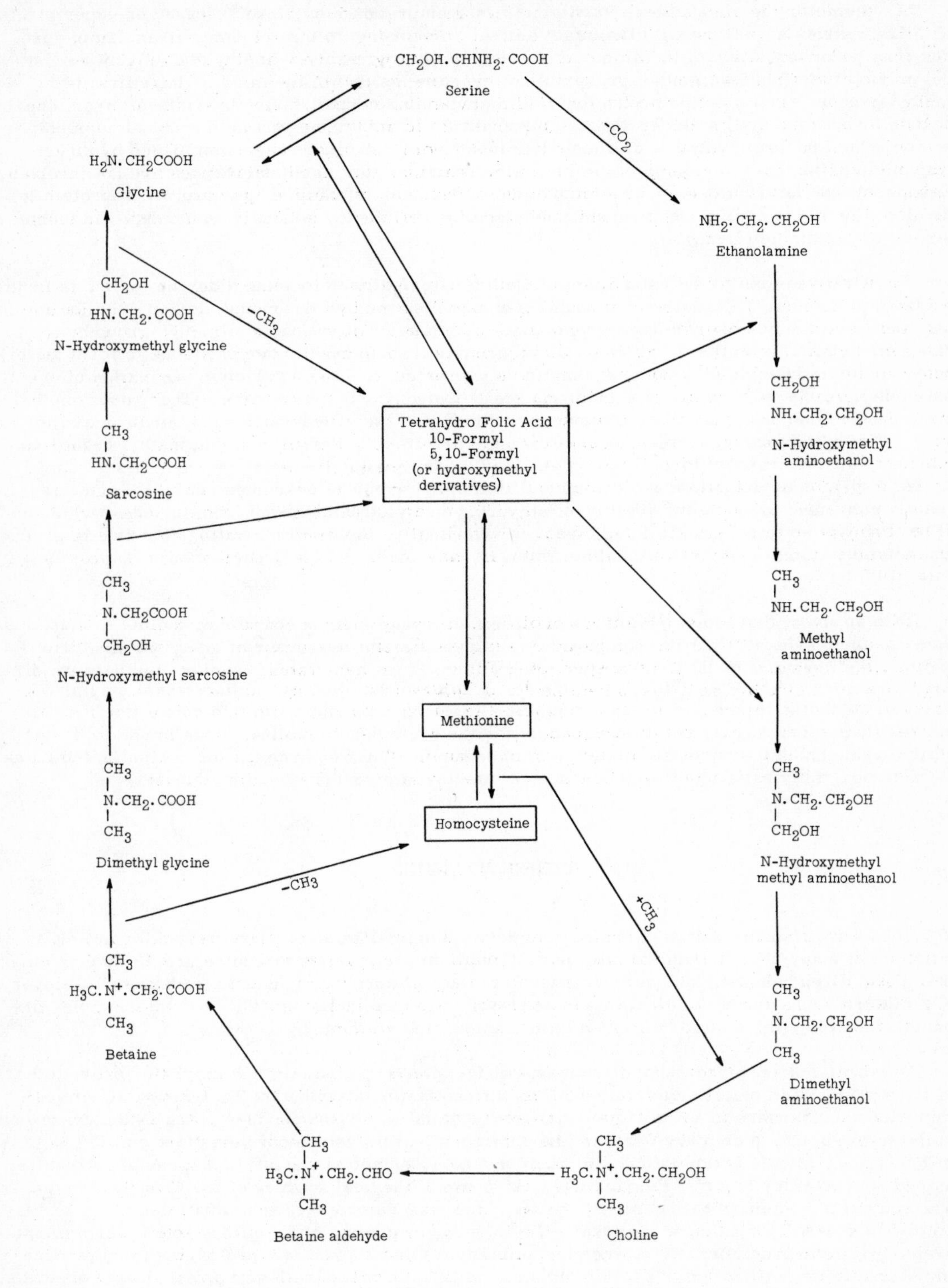

References:

1. a. Sulzberger, M.B., and Lazar, M.P.: J.A.M.A. **146**:788, 1951.
   b. Reinberg, I.E., and Gross, R.J.: Ibid., p. 1222.
   c. Bair, G.: Ibid., p. 1573.
2. Steenbock, H., and Bellin, S.A.: J. Biol. Chem. **205**:985, 1953.
3. Crawford, J.D., Girbetz, D., and Talbot, N.B.: Am. J. Physiol. **180**:156, 1955.
4. Zetterström, R., and Ljunggren, M.: Acta chir. Scandinav. **5**:282, 1951.
5. Nason, A., and Lehman, J.R.: J. Biol. Chem. **222**:511, 1956.
6. Schwartz, K., Ed.: Ann. New York Acad. Sc. **57**:615, 1954.
7. Schwartz, K., Ed.: Ibid., p. 378.
8. Schwartz, K., and Foltz, C.M.: J. Am. Chem. Soc. **79**:3292, 1957.
9. Nair, P., and Magar, N.: J. Biol. Chem. **220**:157, 1956.
10. Jonxis, H.H.P., and Huisman, T.H.J.: Pediatrics **14**:238, 1954.
11. Sealock, R.R., and Goodland, R.L.: Science **114**:645, 1951.
12. Brin, M., Shoshet, S.S., and Davidson, C.S.: J. Biol. Chem. **230**:319, 1958.
13. Teeri, A.E.: J. Biol. Chem. **196**:547, 1952.
14. Suvarnakich, K., Mann, G.V., and Stare, F.J.: J. Nutrition **47**:105, 1952.
15. Singer, T.P., and Kearney, E.B.: Biochim. Biophys. Acta **8**:700, 1952.
16. Fisher, H.F., Conn, E.E., Vennesland, B., and Westheimer, F.H.: J. Biol. Chem. **202**:687, 1953.
17. Pullman, M.E., San Pietro, A., and Colowick, S.P.: J. Biol. Chem. **206**:129, 1954.
18. Reddy, S.K., Reynolds, M.S., and Price, J.M.: J. Biol. Chem. **233**:691, 1958.
19. Christensen, H.N., in "Amino Acid Metabolism," Symposium, McElroy, W.D., and Glass, H.B., Eds. Baltimore, Johns Hopkins Press, 1955.
20. György, P.: J. Clin. Nut. **2**:44, 1954.
21. Wachstein, M., and Lobel, S.: Proc. Soc. Exper. Biol. and Med. **86**:624, 1954.
22. Tower, D.B.: Nutrition Reviews **16**:161, 1958
23. Albers, R.W., and Salvador, R.A.: Science **128**:359, 1958.
24. Nutrition Reviews **16**:10, 1958.
25. Hoagland, M.B., and Novelli, G.D.: J. Biol. Chem. **207**:767, 1954.
26. Moat, A.G., Wilkins, C.N., Jr., and Friedman, H.: J. Biol. Chem. **223**:985, 1956.
27. MacLeod, P.R., Grisolia, S., Cohen, P.P., and Lardy, H.A.: J. Biol. Chem. **180**:1003, 1949.
28. Huennekens, F.M., Osborn, M.J., and Whiteley, H.R.: Science **128**:120, 1958.
29. Eagle, H., Oyama, V.I., Levy, M., and Freeman, A.E.: J. Biol. Chem. **226**:191, 1957.
30. Glass, G.B.J., Boyd, L.J., Gellin, G.S., and Stephanson, L.: Arch. Biochem. and Biophys. **51**:251, 1954.
31. Rosenthal, H.L., and Sarett, H.P.: J. Biol. Chem. **199**:422, 1952.
32. Wagle, S.R., Mehta, R., and Johnson, B.C.: J. Biol. Chem. **233**:619, 1958.
33. Holland, W.C., and Greig, M.E.: Arch. Biochem. and Biophys. **39**:77, 1952.
34. Stekol, J.A.: Am. J. Clinical Nutrition **6**:200, 1958.
35. Pfeiffer, C.C., Jenney, E.N., Gallagher, W., Smith, R.P., Bevan, W., Jr., Killian, K.F., Killian, E.K., and Blackmore, W.: Science **126**:610, 1957.
36. Appleton, B.N. La Du, Jr., Levy, B.B., Steele, J.M., and Brodie, B.B.: J. Biol. Chem. **205**:803, 1953.

Bibliography:

Coward, K.H.: Biological Standardization of the Vitamins. Wood, 1938.
Harris, R.S., Marrian, G.F., and Thimann, K.V., Eds.: Vitamins and Hormones, Vol. 13. Academic, 1955.
Johnson, B.C.: Methods of Vitamin Determination. Burgess, 1948.
Sebrell, W.H., Jr., and Harris, R.S.: The Vitamins. 3 Vols. Academic, 1954.
Annual Reviews of Biochemistry. Annual Reviews, Inc., 1957-58.

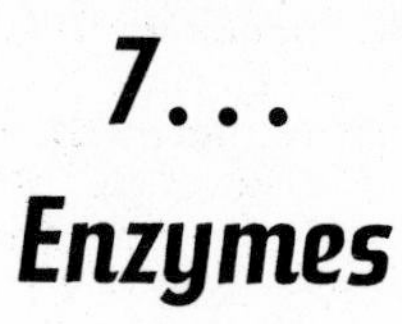

7... Enzymes

An enzyme (ferment, biocatalyst) is an organic chemical substance which is produced by the activity of living organisms - plants, animals, and microorganisms - and which modifies the speed of a reaction without being used up or appearing as one of the reaction products. Most but not all enzymes are thermolabile. Many chemical reactions characteristic of living tissue not only are accelerated by enzymes but probably would not occur at all or at least not to a significant degree in their absence. Although it is true that enzymes are always produced by living cells, they may be isolated from these cells and operate quite independently of them. Every enzyme thus far isolated has been found to be a protein.

The names which were applied to the first enzymes to be described gave no indication of their activity. For example, the designation of enzymes as pepsin, trypsin, or ptyalin suggested nothing of their function. The present convention in enzyme nomenclature is based on a system in which the name of the enzyme is derived from the substrate, i. e., the substance which it attacks, by adding the suffix **-ase**. Thus, we speak of the enzymes that split starch (amylum) as amylases, of those that split fat (lipos) as lipases, of the enzyme that acts on uric acid as uricase, and of the enzyme that acts on urea as urease. Furthermore, groups of enzymes may be designated as carbohydrases, proteases, oxidases, dehydrogenases, glucosidases, etc.

It is estimated that a given cell could contain a thousand different enzymes. There exist enzymes capable of reacting with practically every organic compound which occurs in nature and with many inorganic compounds. These enzymes bring about synthesis, oxidation, hydrolysis, or other chemical changes in the substrate with which they react.

The Genetic Basis for Enzyme Synthesis:

The ability of an organism to manufacture a given enzyme depends on the presence of a particular gene. This important fact has been strikingly demonstrated in the work on the mold, Neurospora crassa. In nature this mold subsists on a very simple medium and is able to synthesize amino acids from an inorganic source of nitrogen such as ammonium salts. Irradiation with x-rays or ultraviolet light, however, brings about genetic changes which result in a loss of the enzymes by which these syntheses are accomplished. Various mutants have thus been prepared which differ from the parent strain in requiring many preformed organic nutrients which could no longer be synthesized in the absence of the genes required for the action or production of the necessary enzymes. For example, seven different mutants requiring the amino acid, arginine, were obtained. This suggests that seven different enzymes are involved in the synthesis of arginine from ammonium salts and carbohydrate.

Factors Which Affect Enzyme Activity:

A. Concentration of the Substrate: If the concentration of the substrate is increased and all other conditions are kept constant, the rate of the enzyme-catalyzed reaction increases to a maximum. Further increase in substrate concentration produces no additional effect, since, presumably, the rate of formation of an enzyme-substrate complex is limited. In this case the enzyme concentration then becomes the limiting factor.

B. Concentration of the Enzyme: Using a highly purified enzyme, the velocity of the reaction is directly proportional to the concentration of the enzyme over a wide range. Eventually the concentration of the substrate limits the rate of the reaction, and accumulation of the products may block the reaction.

C. Effect of pH: Extreme changes in pH cause irreversible destruction of enzymes, probably by denaturation of their protein. When moderate alterations in pH occur, the activity of an enzyme is reversibly altered. There is an optimal pH zone for the activity of each enzyme. When the pH of the medium varies on either side of this zone, the activity gradually decreases until complete inactivation occurs.

D. Temperature: At very low temperatures the effects of an enzyme are greatly decreased. A rise in temperature increases the speed of an enzyme-catalyzed reaction to the point where irreversible destruction of the enzyme occurs. This is probably due to coagulation of the enzyme protein by heat. Destruction of the enzyme by heat serves as simple proof that a reaction is enzymatically catalyzed, since such a reaction will no longer occur (or will occur more slowly) when the mixture has been boiled.

The effect of temperature is well demonstrated by comparing the rate of a reaction at a given temperature with that observed at 10° C. higher or lower. In general, the rate of a reaction is doubled for each 10° C. rise in temperature and reduced by half, for each 10° C. decrease. This is known as the van't Hoff relationship, or the "Q_{10}" temperature coefficient. Many physiologic processes (e.g., the rate of the excised heart) show this relationship to temperature, which suggests that their energy is derived from enzymatically catalyzed reactions.

Enzymes have temperature optima. For enzymes of animal origin this is near body temperature. Some plant enzymes have very high temperature optima; e.g., 65° C. for papain, a proteolytic enzyme derived from the fruit of the papaw.

E. Products of the Reaction: As the products of the reaction become more concentrated, the reaction approaches the point of equilibrium and slows down. In some cases these products actually function as inhibitors of enzyme action.

F. Radiation: In general, enzymes tend to be inactivated by light, although salivary amylase is said to be activated by red and green light. Ultraviolet light is very destructive, perhaps as a result of the denaturing effect of ultraviolet light on proteins. The beta and gamma rays of radium emanations have been shown to be capable of inactivating many enzymes.

G. Oxidation-Reduction Effects: The sulfhydryl (SH) groups of many enzymes are essential to the activity of the enzyme. Oxidation of these groups to an S-S linkage, by removal of the hydrogen, inactivates the enzyme. Reducing agents reactivate the enzyme by restoring SH groups. Glutathione and cysteine are examples of naturally occurring substances which activate these enzymes.

Substrate Induction:

In the intact organism, the injection of a substrate or, in the case of microorganisms, the addition of a substrate to the culture medium causes an increase in the activity of the enzyme which attacks the added substrate. This response to addition of a substrate is called "substrate induction." It is commonly observed in the metabolism of microorganisms, but in animals only a few examples have been demonstrated. These include tryptophan peroxidase-oxidase, ferritin, chick embryo adenosine deaminase, and threonine dehydrase.

Reversibility of Enzymes:

Since enzymes are catalysts, they should be able to influence a reaction in either direction; in other words, enzymatically catalyzed reactions should be reversible. However, the weight of the evidence suggests that in most cases enzyme catalysis occurs in but one direction.

For example, certain esterases (including lipases) may catalyze the synthesis as well as the hydrolysis of an ester. It is more difficult to demonstrate with certainty the reversibility of carbohydrases and proteinases. Since carbohydrate, protein, and fat are broken down by hydrolysis (catalyzed by hydrolases), it is conceivable that their synthesis would be favored by decreasing the amount of water in the presence of the enzymes which would otherwise catalyze their hydrolytic decomposition.

Zymogens:

In some instances an enzyme is manufactured as an inactive precursor called a "zymogen." When this comes in contact with an activator or a kinase, it changes into the active form. For example, trypsinogen of the pancreas is activated in the small intestine by enterokinase, which is elaborated by the duodenal mucosa; pepsin of the stomach, which is produced as pepsinogen, is activated by hydrochloric acid at a pH less than 6.0; bivalent metallic ions, such as cobalt, manganese, nickel, and magnesium, activate some enzymes; and the salivary and pancreatic amylases require chloride.

Coenzymes:

Many enzymes can be shown to consist of two portions. One part, which is heat-labile and nondialyzable, is presumably a protein and is termed the **apoenzyme**. The other portion of the enzyme, relatively heat-stable and dialyzable, is nonprotein in nature. It is attached, as a prosthetic group, to the enzyme protein. This is the **coenzyme**. Unless both components of the enzyme are present, the enzyme remains inactive. The combination of enzyme and coenzyme is called a **holoenzyme**. Many of the B vitamins of the B complex have been shown to function as constituents of coenzymes. Heparin is a coenzyme for lipoprotein lipase ("clearing factor"; see p. 198). Such apoenzyme-coenzyme systems are important in biological oxidation (see p. 115). The hydrolases, such as the enzymes concerned with digestion, have not been shown to be other than simple proteins and do not require coenzymes for their activity.

Antienzymes:

Extracts of the intestinal parasite, Ascaris, contain substances which inhibit pepsin and trypsin. For this reason the parasitic worm escapes digestion in the intestine. Trypsin inhibitors, which appear to be proteins, have been found in pancreas, soybeans, and raw egg white. It is also possible for animals to develop enzyme inhibitors which are antibodies, i.e., substances which are produced in response to the parenteral injection of the enzyme acting as a foreign protein or antigen.

Certain chemical compounds act as specific enzyme inhibitors. An example is the sulfonamide derivative, 2-acetylamino-1, 3, 4-thiadiazole-5-sulfonamide (acetazolamide, **Diamox®**). This compound, although inert as a bacteriostatic agent, is a potent inhibitor of the activity of carbonic anhydrase. Acetazoleamide has found clinical application in the control of water and electrolyte excretion because of the importance of carbonic anhydrase in the activity of the renal tubules (see p. 293).

The Method of Enzyme Action:

Activation of a reactant is a necessary preliminary to a reaction. The energy required for this activation is reduced considerably by an enzyme or other catalyst. In the presence of an enzyme, therefore, a greatly increased rate of reaction per unit of energy in the reactant can take place; this is due to the increase in the number of molecules which can be activated.

It is assumed that a combination of enzyme and substrate is required before the activity of the catalyst can be exerted. The reaction is represented as follows:

$$\text{Enzyme (E)} + \text{Substrate (S)} \underset{(2)}{\overset{(1)}{\rightleftharpoons}} \text{E-S complex} \xrightarrow{(3)} \text{Products of the reaction} + \text{Enzyme}$$

The over-all rate of such a reaction is determined by three individual reactions: (1), (2), and, principally, (3).

Enzyme Specificity:

One of the most noteworthy properties of enzymes is their highly specific action. This may be exhibited in various ways as exemplified below:

A. Absolute Specificity: Pepsin will attack only protein, not fat or carbohydrate; dipeptidase hydrolyzes only dipeptides and not the higher polypeptides.

B. Stereochemical Specificity: Maltase catalyzes the hydrolysis of the α glucosides but not the β glucosides; the β glucosidase of the enzyme, emulsin, will not attack an α glucoside. Both D and L amino acid oxidases are known, each specific for only one or the other isomer.

C. Relative Specificity: Enzymes may have the ability to attack various substrates, but at differing rates.

Theories of Enzyme Activity:

Many enzymes exhibit specificity peculiar to certain linkages in the substrate. Thus, if the amino group in a dipeptide is in the β position, a dipeptidase will not hydrolyze it. From this and other studies of linkage specificity, it is possible to suggest how an enzyme may unite with its substrate. For example, Bergmann[1] has suggested that when a dipeptide is hydrolyzed by a dipeptidase, the enzyme unites with the substrate at two or three points as follows:

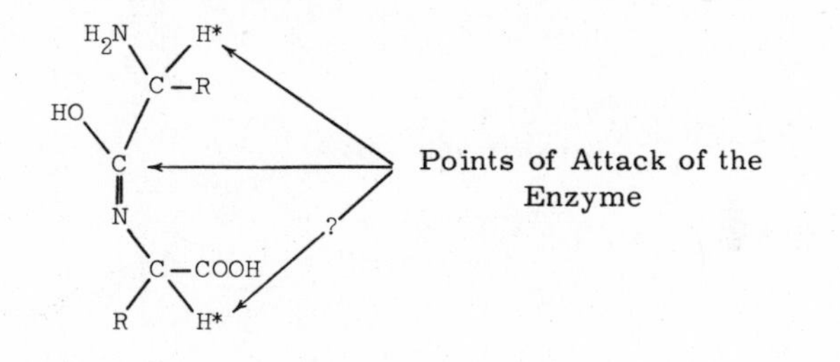

(Enolized form of a polypeptide)

This theory is based on the following experimental evidence:

A. If the dipeptide contains D amino acids, the positions of the H and NH_2 groups on the α carbon atoms would be the reverse of their positions when L amino acids are present in the compound. The dipeptidase will not hydrolyze the dipeptide unless the amino acids are of the L configuration. Therefore, the α H (starred) atoms must confront the enzyme, i.e., act as a point of attachment as shown above.

B. The imino (=N) and the amino groups cannot have any substituent groups in place of the hydrogen if hydrolysis by the dipeptidase is to occur. This suggests that these two positions on the molecule are also points of attachment, since substitution at these positions interferes with enzyme-substrate combination.

C. Esterification of the free carboxyl group does not prevent the action of the dipeptidase, which indicates that this position on the dipeptide molecule may not be a point of attachment for the enzyme.

Two important peptidases are carboxypeptidase in pancreatic juice and aminopeptidase in the intestine. The former attacks a peptide at the point on the molecule where there is a free carboxyl group, e.g.:

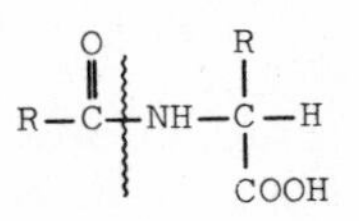

Aminopeptidase removes an amino acid from a peptide at the end of the molecule, where there is a free amino group, e.g.:

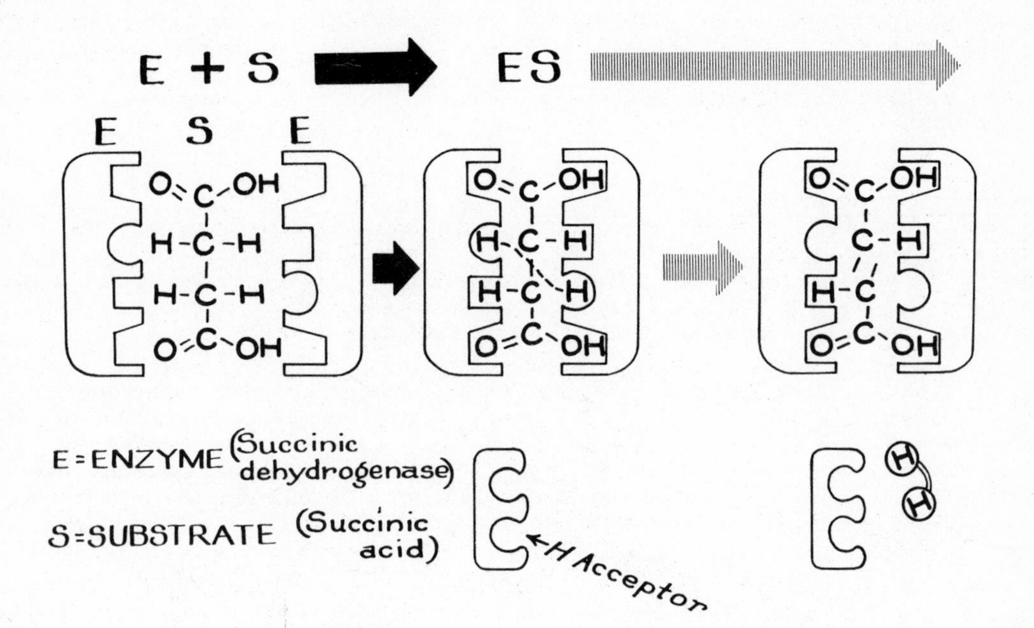

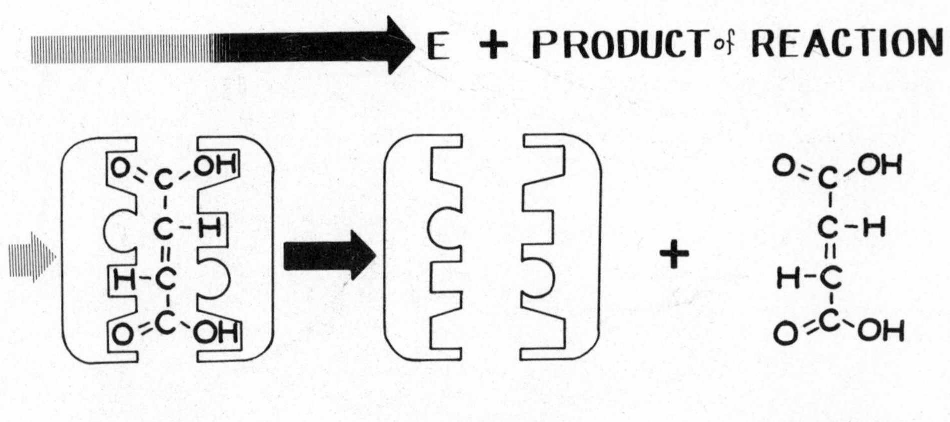

Schematic Representation of Enzyme-Substrate Combination

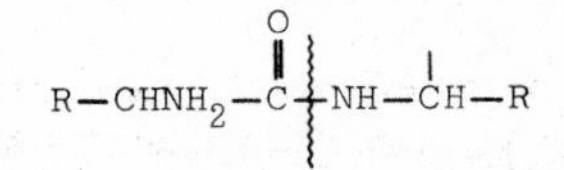

These examples of enzyme specificity support the "lock and key" hypothesis which was advanced long ago to explain the phenomenon. Schematic representations of enzyme-substrate combination and of specificity are illustrated on p. 106.

Competitive and Noncompetitive Inhibition:

It seems likely that compounds other than the natural substrate, particularly those which resemble it closely, might unite with the enzyme in competition with the substrate. However, once combined they would be unable to be activated by the enzyme and the usual reaction catalyzed by the enzyme would not occur. Furthermore, these competitors would prevent the entrance of the natural substrate into the enzyme complex; and if present in sufficient quantity, complete inhibition of activity would result. Such "competitive inhibition" could be reversed by the addition of the natural substrate. In support of this explanation of competitive inhibition is the fact that when a definite ratio between a metabolite (the proper substrate) and its antagonist (the inhibitor) is maintained, a constant degree of inhibition is observed. A further increase in the concentration of one or the other produces an increase or decrease in the enzyme activity, as the case may be. These competitive inhibitors, sometimes designated **antimetabolites** or **metabolic antagonists**, have been shown to inhibit utilization of many B vitamins and some amino acids. An example of this is the action of sulfanilamide on the utilization of *p*-aminobenzoic acid.

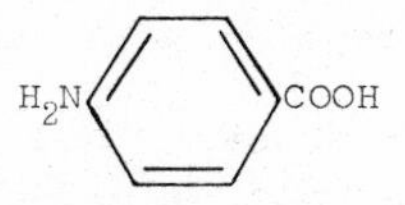

p-Aminobenzoic acid

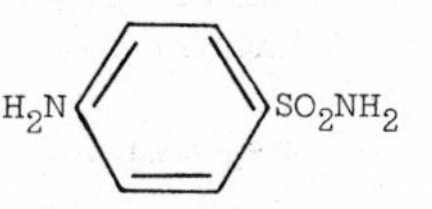

Sulfanilamide

Other antagonists to the B vitamins include pyrithiamin and oxythiamin (antagonists to thiamin), pyridine-3-sulfonic acid (to niacinamide), aminopterin and methopterin (to folic acid), pantoyl taurine and ω-methyl pantothenic acid (to pantothenic acid), deoxypyridoxine (to pyridoxine), and desthiobiotin (to biotin). Bishydroxycoumarin (Dicumarol®; see p. 127) is an antagonist to vitamin K. Purine and pyrimidine antimetabolites have also been prepared and studied as possible chemotherapeutic agents in the treatment of tumors. A recent example is 6-mercaptopurine (Purinethol®, 6-M. P.), which is a hypoxanthine antagonist.

Many other drugs which inhibit enzyme action operate in a similar manner. Eserine, or physostigmine, inhibits the hydrolysis of acetylcholine by cholinesterase by competitive inhibition, probably because it is structurally similar to acetylcholine. D-Histidine inhibits the action of histidase on the natural L-histidine by competitive inhibition.

Noncompetitive inhibition occurs when there is no relationship between the degree of inhibition of an enzyme and the concentration of a substrate, the inhibition depending only on the concentration of the inhibitor. The inhibitory actions of heavy metals, e.g., Ag^{+}, Hg^{++}, and Pb^{++} are examples.

Measurement of Enzyme Activity:

The activity of an enzyme, as determined under specified conditions of substrate concentration, temperature, and pH, is a measure of its concentration. Such activities are expressed in units. For example, the lipase unit is that quantity of lipase which will hydrolyze 24% of 2.5 grams of olive oil in one hour at 30° C. at an initial pH of 8.9 and in the presence of 15 mg. of albumin and 10 mg. of $CaCl_2$. Amylase units are based upon the rate of production of sugar from starch under standardized conditions.

Diagnostic Applications:

The measurement of enzyme activity in body fluids is often of importance in clinical practice. Those enzymes most often measured and their diagnostic significance are listed below.

A. The serum lipase level may be low in liver disease, vitamin A deficiency, malignancy, and in diabetes mellitus. It may be elevated in acute pancreatitis and pancreatic carcinoma[2].

B. The serum amylase level may be low in liver disease, increased in high intestinal obstruction, parotitis, acute pancreatitis, and diabetes[2].

C. Elevations of trypsin in the serum occur during acute disease of the pancreas, with resultant changes in the coagulability of the blood which are reported as antithrombin titers. Direct measurement of the serum trypsin in pancreatic disease has been reported by Nardi[3]. It is stated that elevation in concentration of serum trypsin is a more sensitive and reliable indicator of pancreatic disease than either the serum amylase or lipase.

D. Cholinesterase has been measured in serum in a number of disease states[4]. In general, low levels are found in patients ill with liver disease, malnutrition, chronic debilitating and acute infectious diseases, and anemias. High levels occur in the nephrotic syndrome. A large number of drugs produce a temporary decrease in cholinesterase activity, but the alkyl fluorophosphates (see p. 389) cause irreversible inhibition of the enzyme. Some insecticides in common use depress cholinesterase activity, and tests for the activity of this enzyme in the serum may be useful in detecting over-exposure to these agents.

The content of cholinesterase in young red blood cells is considerably higher than in the adult red blood cells; consequently the cholinesterase titer of erythrocytes in the peripheral blood may be used as an indicator of hematopoietic activity.

E. The blood alkaline phosphatase level may be increased in rickets, hyperparathyroidism, Paget's disease, osteoblastic sarcoma, obstructive jaundice, and metastatic carcinoma.

F. The serum acid phosphatase level may be elevated in prostatic carcinoma with metastases.

G. The transaminase which catalyzes the transfer of an amino group from aspartic acid to ketoglutaric acid to form glutamic and oxaloacetic acids (see p. 226) is designated glutamic-oxaloacetic transaminase (GO-T); that which transfers an amino group from alanine to ketoglutaric acid to form glutamic acid and pyruvic acid (see p. 226) is called glutamic-pyruvic transaminase (GP-T). The levels of these transaminases in normal serum are very low because these enzymes are concentrated within the cells. When there is extensive tissue destruction, the enzymes are liberated into the serum. The myocardium of the heart is rich in transaminase activity; consequently, destruction of heart muscle as occurs in myocardial infarction will be reflected by a rapid and often a striking rise in serum transaminase activity. A test of serum transaminase activity is now used to aid in the diagnosis of heart disease and to evaluate the severity and duration of an attack.

Transaminase activity is also high in liver. The activity of glutamic-pyruvic transaminase in liver is greater than that of glutamic-oxaloacetic transaminase. Both transaminases are elevated in the serum of patients with acute hepatic disease but GP-T is the more specific indicator of liver cell damage. Furthermore, GP-T is not appreciably altered by acute cardiac necrosis[5].

Extensive damage to skeletal muscle, as may occur in severe trauma, will also cause elevated serum transaminase levels.

H. Lactic dehydrogenase (LDH) is an enzyme which can be detected by its ability to catalyze the reduction of pyruvate (see p. 115) in the presence of reduced DPN. In myocardial infarction the concentration of serum LDH rises within 24 hours of the occurrence of the infarct and returns to the normal range within five to six days[6]. High levels of LDH also occur in patients with acute and chronic leukemia in relapse, generalized carcinomatosis, and, occasionally, with acute hepatitis during its clinical peak, but not in patients with jaundice due to other causes. Serum LDH is normal in patients with acute febrile and chronic infectious diseases as well as those with anemia, pulmonary infarction, localized neoplastic disease, and chronic disease processes.

I. Measurement of serum isocitric dehydrogenase, an enzyme of the citric acid cycle, has been found useful, particularly in liver disease[7].

J. Serum copper oxidase activity is associated with the copper-binding serum globulin, ceruloplasmin (see p. 132). A decrease in the activity of this enzyme in the serum is a useful biochemical test to confirm the presence of Wilson's disease (so-called hepatolenticular degeneration).

A number of other enzymes have been studied in connection with diagnosis of disease, particularly as an aid to the diagnosis and measurement of response to therapy in malignant disease. The diagnostic uses of enzymes have been reviewed by Harper[8] and by White[9].

Classification of Enzymes:

A tentative and incomplete classification of enzymes, based on their activity, includes the following major groups:

A. Esterases:
1. Lipase splits fats. The lipases of the intestinal tract are activated by the bile salts.
2. Cholesterol esterase is found in the pancreatic juice and catalyzes the combination of cholesterol with fatty acids to form cholesterol esters. Such esterification of cholesterol is believed to play an important role in its absorption from the intestine.
3. Cholinesterase is an enzyme in serum which hydrolyzes acetylcholine but is not specific for this substrate since it splits other esters as well; it is now termed "pseudo- or nonspecific cholinesterase." Probably most of the so-called lipase activity of normal serum is actually due to this enzyme. The presence of true lipase in the serum can be assumed if the addition of bile salts increases the activity of the serum in the hydrolysis of the substrate. Nonspecific cholinesterase activity is not increased by bile salts.
4. Acetylcholinesterase (ACh-esterase) is specific for the hydrolysis of acetylcholine (see pp. 99 and 388).
5. Liver esterase splits simple esters such as ethyl butyrate (may be the same as serum cholinesterase, above).
6. Phosphatase of bone, kidney, and intestine removes phosphoric acid from monophosphate esters.
7. Pyrophosphatase removes phosphoric acid from di- and triphosphate esters. Examples are adenosine triphosphatase and yeast pyrophosphatase.

B. Carbohydrases:
1. Amylase (also known as diastases) hydrolyzes starch.
2. Cellulase attacks cellulose; cellulases are common in various bacteria and molds.
3. Mucinase (also known as hyaluronidase, "the spreading factor") catalyzes the hydrolysis of hyaluronic acid to acetylated glucosamine and glucuronic acid.
4. Maltase (α-glucosidase) hydrolyzes certain α-glucosidic linkages.
5. β-Glucosidase attacks β-glucosidic linkages. (Emulsin, a crude product from almonds, contains this enzyme.)
6. β-Galactosidase hydrolyzes β-galactosides, such as lactose.
7. β-Glucuronidase is found in many animal tissues. It hydrolyzes glucuronic acid conjugates.

C. Proteolytic Enzymes:
1. Proteinase attacks high molecular weight (native) proteins. Examples include gastric pepsin, pancreatic trypsin, chymotrypsin, and rennin, the milk-coagulating enzyme of the stomach.
2. Peptidase attacks partially digested proteins. Examples include carboxypeptidase, which acts on polypeptides containing a free carboxyl group; aminopolypeptidase, which acts on a polypeptide with a free amino group; and dipeptidase, which splits dipeptides. All these enzymes occur in intestinal juice. The mixture was formerly termed "**erepsin.**"
3. Other proteolytic enzymes -
   a. Papain, occurring in the latex of **Caraca papaya**.
   b. Cathepsin, an enzyme in animal tissues which is concerned with autolysis, occurs in particularly high concentration in liver, kidney, and spleen. Its optimum pH is 4-5, which prevents its action in living tissues.
   c. Thrombin, the factor in blood which converts fibrinogen into fibrin, the substance of the blood clot, occurs in fluid blood as prothrombin.
   d. Fibrinolysins are enzymes which digest fibrin, the substance of the blood clot. **Plasmin,** which occurs as an inactive precursor, **plasminogen,** in Fraction III-3 of blood serum, is a fibrinolysin which causes lysis of the blood clot. Streptokinase (a fibrinolysokinase) is an activator for the conversion of profibrinolysin to fibrinolysin.

D. Phosphorylases: These enzymes have several functions:

1. Catalyze the addition of phosphate to the molecule of another substance, i.e., they catalyze phosphorylysis (analogous to hydrolysis). The general reaction is shown below:

   Example: Conversion of glycogen to glucose monophosphate.

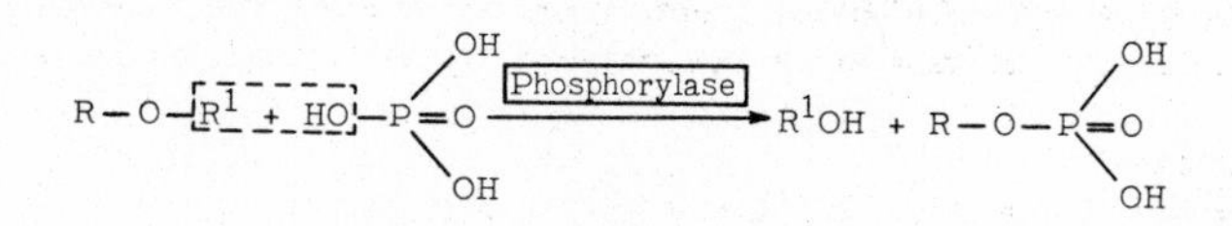

2. Transfer a phosphoric acid group from one compound to another (transphosphorylation). Examples are the enzymes **creatine kinase**, which catalyzes the transfer of phosphate from creatine phosphate to adenosine diphosphate (ADP) to form ATP (see p. 175), and ATP-creatine transphosphorylase, which catalyzes the reverse reaction, i.e., the transfer of phosphate from ATP to creatine, resulting in the formation of creatine phosphate and ADP.
3. Catalyze the transfer of phosphoric acid from one position on a molecule to another point on the same molecule (phosphomutase), or catalyze other alterations in the molecule of a phosphate compound (phosphoisomerases). (Example of the former: glucose-6-phosphate to glucose-1-phosphate. Example of the latter: glucose-6-phosphate to fructose-6-phosphate.)

E. Nucleases: Enzymes which attack nucleic acids or the products derived from them. These include:

1. Polynucleotidases or nucleinases, which convert polynucleotides (nucleic acids) into mononucleotides. **Ribonuclease** attacks ribonucleic acids. **Deoxyribonuclease** attacks deoxyribonucleic acids. These enzymes occur in pancreas, liver, and leukocytes. They are also found in certain microorganisms, e.g., the enzyme **streptodornase,** which contains streptococcal deoxyribonuclease.
2. Mononucleotidases, which convert nucleotides into phosphoric acid and nucleosides. Actually these enzymes are phosphatases. They occur in intestine and liver.
3. Nucleosidases (nucleoside phosphorylases) split nucleosides into their component purine or pyrimidine bases and phosphorylated D-ribose or 2-deoxy-D-ribose. They occur in spleen, lungs, liver, and heart; traces are present in the small intestine.

F. Enzymes Catalyzing Hydrolytic Removal of Ammonia:

1. Deaminases, e.g., enzymes which catalyze the removal of NH_2 from adenine, guanine, or adenylic acid (adenase, guanase, adenylic acid deaminase).
2. Amidases catalyze the removal of NH_2 from amides, e.g., glutaminase or asparaginase, which catalyze the removal of NH_2 from glutamine or asparagine; urease, which converts urea into ammonia and carbon dioxide.
3. Amidinases remove NH_2 groups from a guanidine linkage, e.g., arginase, which converts arginine into ornithine and urea.

G. Carboxylases: Catalyze removal of CO_2 or fixation of CO_2.

1. Carbonic anhydrase - Reaction of CO_2 and H_2O to form H_2CO_3, or the reverse of this reaction, i.e., the removal of CO_2 from H_2CO_3 (see pp. 155 and 293).

$$H_2O + CO_2 \overset{\text{Carbonic Anhydrase}}{\rightleftharpoons} H_2CO_3 \rightleftharpoons H^+ + HCO_3^-$$

2. Pyruvic decarboxylase - Removal of CO_2 from pyruvic acid, converting it into acetaldehyde or acetic acid.
3. Amino acid decarboxylase - Removal of CO_2 from α amino acids to produce primary amines.

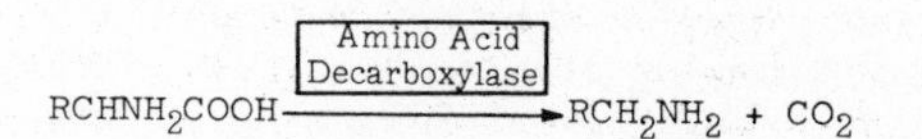

4. Fixation of CO_2, e.g., pyruvic acid to oxaloacetic acid.

$$CH_3.CO.COOH + \overset{*}{C}O_2 \xrightarrow[\text{Fixation of } CO_2]{\boxed{\text{Carboxylase}}} HOO\overset{*}{C}.CH_2.CO.COOH$$

H. Transferring Enzymes:
1. Transaminases transfer NH_2 to α-ketoacids to form α-amino acids.
2. Transamidinases transfer amidine linkages, e.g., from arginine to glycine to form glycocyamine.
3. Transcarbamylases transfer $-\overset{O}{\overset{\|}{C}}.NH_2$ (carbamyl) group from carbamyl phosphate to and from aspartic acid, glutamic acid, and ornithine (see p. 232).
4. Transmethylases transfer CH_3, e.g., from methionine to glycocyamine to form creatine.
5. Transulfurases transfer sulfhydryl groups (SH), e.g., from homocysteine to serine to form cysteine (see p. 248).
6. Transacetylase, e.g., choline acetylase, which transfers acetyl groups to choline from ''active acetate'' (acetyl Coenzyme A). (See p. 98.)
7. Acetylating (activating) enzymes, e.g., acetic thiokinase, which in the presence of ATP, Co A, and acetic acid forms acetyl Co A (active acetate). The enzyme also reacts in a similar manner to activate fatty acids as a necessary preliminary to their oxidation (see p. 209).
8. Thiophorases - Transferring enzymes which act to transfer Co A from acetyl Co A or succinyl Co A to proprionate, acetoacetate, or butyrate to form ''active'' derivatives.

I. Oxidizing and Reducing Enzymes: Include a large group of tissue enzymes which catalyze a variety of energy-yielding reactions. (See Biological Oxidation, p. 112.)
1. Oxidases, e.g., cytochrome oxidase containing iron, ascorbic acid oxidase containing copper.
2. Oxygenases - Enzymes which add both atoms of molecular O_2 to a substrate or add one atom while reducing the other to water.
3. Aerobic dehydrogenases* - Ascorbic acid oxidase containing copper. The D and L amino acid oxidases.
4. Peroxidase, catalase.
5. Anaerobic dehydrogenases, e.g., the succinic and lactic acid dehydrogenases.
6. The flavoprotein enzymes, e.g., the ''yellow enzyme,'' cytochrome reductase, and diaphorase.

J. Isomerases: Catalyze the formation of isomers.
1. Aconitase - Cis-aconitic acid to citric or isocitric acid.
2. Phosphotriose isomerase - Phosphodihydroxyacetone to 3-phosphoglyceraldehyde.
3. Enolase - 2-Phosphoglyceric acid to enolphosphopyruvic acid.

K. Condensing Enzymes: Catalyze the transfer of an acyl group ($CH_3.C=O$) from acetyl Co A (active acetate) to an acceptor such as sulfanilamide (see p. 169), oxaloacetic acid to form citric acid (see p. 181), or choline to form acetylcholine (see p. 98).

• • •

References:

1. Bergmann, M., and Fruton, J.S.: Advances in Enzymology **1**:63, 1941.
2. Dreiling, D.A., and Janowitz, H.D.: Am.J.Med.**21**:98, 1956.
3. a. Nardi, G.L., and Lees, C.W.: New England J.Med.**258**:797, 1958.
   b. Nardi, G.L.: J.Lab. and Clin.Med. **52**:66, 1958.
4. Vorhaus, L.J., and Kark, R.M.: Am.J.Med.**14**:707, 1953.
5. Wroblewski, F., and La Due, B.: Proc.Soc.Exper.Biol. and Med.**91**:569, 1956.
6. Wroblewski, F.: Science **123**:1122, 1956.
7. Sterkel, R.L., Spencer, J.A., Wolfson, S.K., Jr., and Williams-Ashman, H.G.: J.Lab. and Clin.Med.**52**:176, 1958.
8. Harper, H.A.: Ann.Rev.Med.**9**:461, 1958.
9. White, L.P., Ed.: Enzymes in Blood. Ann.N.Y.Acad.Sci.**75**; Art.1, p.1, 1958.

Bibliography:

Baldwin, E.: Dynamic Aspects of Biochemistry. Cambridge, 1952.
Martin, G.J.: Biological Antagonism. Blakiston, 1951.
Sumner, J.B., and Myrbäck, K.: The Enzymes. 2 Vols. Academic, 1950.
McElroy, W.D., and Glass, H.B., Eds.: The Chemical Basis of Heredity. Johns Hopkins Press, 1957.

---

*These dehydrogenases can react directly with molecular oxygen and transfer hydrogen to oxygen without the intervention of an intermediary hydrogen carrier. In this sense they might be included with the oxidases.

8...

Biological Oxidation

All vital processes require energy, which is obtained from chemical reactions carried on in the living cell. The gradual oxidation of various metabolites is the principal mechanism for the liberation of energy. The utilization of oxygen and the production of carbon dioxide by the tissues in the process of cellular respiration is but the final phase of biological oxidation. Many other oxidative changes which involve energy transfer precede this final aspect of cellular respiration. These reactions are discussed in this chapter.

The energy transfers involved in oxidation-reduction systems are measured by differences in the electrical potential of the various systems. The system with the higher potential oxidizes the one with the lower potential, with a consequent liberation of energy for vital processes.

Transfer of electrons is involved in all oxidation and reduction reactions. Oxidation is defined as removal of electrons with an increase in valence; reduction is defined as a gain in electrons with a decrease in valence. Every oxidation must therefore be accompanied by a simultaneous reduction, and the energy required for the removal of electrons in oxidation is supplied by the accompanying reduction.

The oxidation or reduction of iron illustrates these definitions.

$$Fe^{++} \rightleftharpoons Fe^{+++} + 1 \text{ electron}$$

The older concept of oxidation as an addition of oxygen or a removal of hydrogen, while not strictly accurate, is still useful in the case of complex reactions in which electron transfer is difficult to visualize. These oxidations may be exemplified as follows:

Addition of oxygen:

$$CH_3.CH_2OH + O_2 \xrightarrow{\boxed{\text{oxidase}}} CH_3.COOH + H_2O$$

Removal of hydrogen:

$$CH_3.CH_2OH \xrightarrow{\boxed{\text{dehydrogenase}}} CH_3.CHO + 2H$$

<u>Anaerobiosis:</u>

It is characteristic of biological oxidation that it can be accomplished in the absence of oxygen. This anaerobiosis, or life without air, is fundamental to anaerobic organisms. Under certain conditions, aerobic organisms also carry out oxidative changes without oxygen, although oxygen must be present if the physiologic oxidative process in these organisms is to be completed.

<u>The Pasteur Effect:</u>

Anaerobic oxidative reactions are decreased by aeration. This was first observed by Pasteur in studies of the fermentation of glucose by yeast. He also noted that in the presence of oxygen less glucose was broken down by the yeast cells and less alcohol was formed. Under anaerobic conditions, more alcohol was formed and more glucose was fermented. This is now termed the "Pasteur effect." It is explained by the fact that in the presence of oxygen complete breakdown of the glucose occurs and maximum energy is derived from the molecule. In the absence of oxygen, the oxidation of glucose proceeds only to the production of alcohol; and the entire energy content of glucose is not made available. For the same amount of energy, more glucose must therefore be oxidized under anaerobic than under aerobic conditions.

Oxidation by Removal of Hydrogen:

Cellular enzyme systems accomplish oxidation without oxygen by removal of hydrogen from an oxidizable substrate. Such a method of oxidation is extremely common in biological systems.

The concept of removal of hydrogen as a mechanism of biological oxidation was first proposed by Wieland. In the following example, $A.H_2$ is the reduced form of an oxidizable substrate; B is the oxidized form of another metabolite of higher oxidation-reduction potential. According to the Wieland theory, oxidation proceeds as follows:

$$A.H_2 + B \xrightarrow{\text{specific dehydrogenase}} A + B.H_2$$

The reductant, $A.H_2$, is termed a hydrogen donator; and the oxidant, B, is the hydrogen acceptor. In order for this reaction to proceed readily, a catalyst is necessary. This is supplied by the specific enzymatic dehydrogenases of the tissues.

The Dehydrogenases:

The action of the dehydrogenases of the tissues was studied extensively by Thunberg. In the Thunberg technic, methylene blue is used as a hydrogen acceptor; and, since it is reduced to a colorless leuco base when it accepts hydrogen, it acts as a visible indicator of hydrogen transfer. If the dye is added to a suspension of minced tissue such as muscle, kidney, or brain, it is rapidly reduced and loses its color. This reaction indicates that dehydrogenase systems are present in the tissue extracts. The reaction will not occur when the tissue extracts are boiled first. This is evidence that the hydrogen transfer was catalyzed by an enzyme, since it no longer occurs when this enzyme is destroyed by heat. If the tissue is washed, the reduction of the dye will be considerably delayed, because the oxidizable substrates, such as lactate or succinate, which act as hydrogen donators, have been washed out. If the substrate is replaced, rapid reduction will occur again. By such studies, a considerable number of dehydrogenases, specific often to only one substrate, have been identified in the tissues.

Methylene blue is a valuable tool for the study of dehydrogenase systems, but since it is not present in natural materials it can be considered only as an artificial hydrogen acceptor. Other substances in the cells function like hydrogen acceptors in conjunction with the action of dehydrogenases.

The Aerobic Dehydrogenases:

Oxygen is the final acceptor of hydrogen in cellular respiration, and those dehydrogenases which can transfer hydrogen from a metabolite directly to oxygen are known as aerobic dehydrogenases (Example II, p. 116). However, very few of the dehydrogenases react directly with oxygen, and the majority are therefore anaerobic dehydrogenases.

The Oxidases:

The oxidases are often classed with the aerobic dehydrogenases; but the oxidases can use only oxygen as hydrogen acceptor, whereas the aerobic dehydrogenases can utilize methylene blue or other artificial acceptors in addition to oxygen (Examples I and II, p. 116).

Most oxidases are conjugated proteins. Their prosthetic groups contain metals such as copper, iron, or zinc, and also a riboflavin-containing complex.

The transfer of hydrogen to oxygen by an oxidase (or an aerobic dehydrogenase) could form either water or hydrogen peroxide:

$$A.H_2 + 1/2\,O_2 \xrightarrow{\boxed{\text{oxidase}}} A + H_2O$$

$$A.H_2 + O_2 \xrightarrow{\boxed{\text{oxidase}}} A + H_2O_2$$

Hydrogen peroxide is toxic to the tissues, but this toxic effect is prevented by two important enzymes, peroxidase and catalase.

Peroxidase is found only in plant tissues. It combines with hydrogen peroxide, activating it to donate its oxygen to act as a hydrogen acceptor for another substrate molecule:

$$A.H_2 + H_2O_2 \xrightarrow{\boxed{\text{peroxidase}}} A + 2H_2O$$

Catalase occurs both in animals and plants. It is a conjugated protein with an iron-containing prosthetic group identical with that of the blood hemoglobin. The enzyme specifically catalyzes the breakdown of H_2O_2 into water and oxygen, the iron atom undergoing a change in valence in the reaction. The catalytic action of this enzyme is extremely rapid (the most rapid enzyme-catalyzed process yet discovered); 1 mg. produces 2,740 liters of oxygen per hour at 0° C.

The Oxygenases:

Recent studies of oxidative reactions in metabolism, utilizing isotopic oxygen (O^{18}), indicate that molecular oxygen serves not only at the end of the respiratory chain as a means of trapping electrons and of forming water from the transferred hydrogen, but that it may also be transferred directly to various substrates by the catalytic activity of enzymes termed **oxygenases** by Hayaishi, Rothberg, and Mehler[1]. These enzymes should be differentiated from the oxidases, mentioned above, which catalyze transfer of hydrogen directly to oxygen without the need for intervening enzyme systems, as is the case with the dehydrogenases.

Oxygenases may function either by adding both atoms of molecular oxygen to a substrate (S) oe by adding only one atom of oxygen while reducing the other to water. Examples of each of these reactions are as follows:

$$1.\quad S + \underset{(O_2)}{O{=}O} \longrightarrow S\genfrac{}{}{0pt}{}{\leqslant O}{\leqslant O} \qquad \text{(S = Substrate which is oxidized)}$$

$$2.\quad S + O{=}O + DPN.H + H^+ \longrightarrow S{=}O + DPN + H_2O$$

Oxygenase reactions have been found to be particularly important in oxidation of aromatic compounds, such as the hydroxylation of steroids and of phenylalanine to tyrosine. In the metabolism of tryptophan, oxygen is also added by the catalytic activity of oxygenases.

An example of the oxygenase reaction with an aliphatic compound is that of the lactic acid decarboxylase isolated from Mycobacterium phlei:

$$CH_3.CHOH.COOH + O_2^{18} \longrightarrow CH_3COO^{18}H + CO_2 + H_2O^{18}$$

This reaction is of the category shown in Example 2, above, in which one atom of oxygen is accepted by the substrate and the other atom is reduced to water.

The Cytochromes:

It has been noted that most dehydrogenases are anaerobic and will not catalyze the transfer of hydrogen from a substrate directly to oxygen. Consequently, other systems are required which will couple on the one hand with the anaerobic dehydrogenases and, on the other, with molecular oxygen. The iron-containing hemoproteins or **cytochromes**, which are widely distributed in the tissues, perform this function. Various enzyme systems in the cells reduce the cytochromes by transfer of electrons (the cytochromes do not transfer hydrogen). In the presence of oxygen, other cell systems reoxidize the reduced cytochrome, and hydrogen is simultaneously transferred to oxygen (Example III, p. 116). These changes in the cytochromes are easily detected by absorption spectroscopy because in the reduced state the cytochromes have well-defined absorption bands which disappear when the cytochromes are oxidized. Narcotics prevent the reduction of the cytochromes, and such well-known inhibitors of respiration as cyanide, hydrogen sulfide, and carbon monoxide block their reoxidation.

Three different cytochromes which can be differentiated by their absorption bands were described by Keilin in 1925. These were designated cytochromes a, b, and c. Later, Keilin found that heart muscle preparations contain also a cytochrome which resembles cytochrome a with respect to its absorption bands but differs in being sensitive to carbon monoxide. This cytochrome was designated a_3; it is probably identical with cytochrome oxidase in heart muscle, an enzyme necessary for reoxidation of cytochrome c.

Cytochrome c is found in relatively large concentration in heart muscle. It is the only readily soluble cytochrome. It consists of a protein with an iron-containing prosthetic group similar to but not identical with the heme of hemoglobin (see p. 61). In oxidation and reduction, the iron changes in valence. Cytochrome oxidase, as pointed out above, is required for its oxidation, although it is easily reduced by fairly mild reducing agents, including dehydrogenase systems.

Cytochrome a resembles cytochrome a_3 (cytochrome oxidase), and the two appear to be related functionally; but cytochrome a does not react with oxygen or with carbon monoxide, whereas a_3 is auto-oxidizable and forms a definite compound with carbon monoxide, which suggests that a_3 is identical with Warburg's respiratory enzyme or "Atmungsferment."

Cytochrome b, like cytochromes a and a_3, is not readily extractable from tissues. It is slowly auto-oxidizable, but does not react with cyanide or carbon monoxide as does cytochrome a_3. It is believed that cytochrome b acts as an electron carrier in the respiratory chain between the flavoproteins (flavin adenine dinucleotides, FAD; see p. 79) and cytochrome c.

The action of cyanide on the respiration of the cell has been noted above, but cyanide does not completely abolish cellular respiration. It is possible that the residual respiratory activity is made possible by cytochrome b activity, since this substance does not require the cyanide-susceptible cytochrome oxidase. However, the majority of respiration must operate through cytochromes a and c, because a large fraction of cellular respiration is abolished by cyanide inactivation of the oxidase on which these two components depend.

An alternate explanation of the residual respiratory activity of the cyanide-poisoned system is found in the action of the flavin aerobic dehydrogenases, which can transfer hydrogen to molecular oxygen without the cytochrome system (see Example IV, p. 116).

The Pyridine Coenzymes of the Dehydrogenases:

Many dehydrogenases require activators, or coenzymes, which are present in the intact cell. The first such coenzyme to be discovered was found to be required by the dehydrogenase which oxidizes glucose monophosphate to phosphogluconic acid, the hexosemonophosphate dehydrogenase (see p. 187). This coenzyme was called Coenzyme II, or triphosphopyridine nucleotide (TPN), when its structure was revealed. Later, Coenzyme I, diphosphopyridine nucleotide (DPN), was discovered in yeast juice. The structure of this codehydrogenase is given on p. 81; it contains the vitamin, niacinamide. The two coenzymes differ only by one phosphate molecule. As mentioned previously, these coenzymes act as hydrogen and electron transfer agents by reversible oxidation and reduction.

The function of DPN can be exemplified by its role in the oxidation of lactic acid to pyruvic acid. When lactic acid as substrate and the lactic acid dehydrogenase are added to the oxidized form of DPN, pyruvic acid and reduced DPN are formed:

$$\underset{\text{(DPN oxidized)}}{\text{DPN}} + CH_3.CHOH.COOH \xrightleftharpoons{\boxed{\text{lactic acid dehydrogenase}}} CH_3.CO.COOH + \underset{\text{(DPN reduced)}}{\text{DPN.H}} + H^+$$

EXAMPLES OF BIOLOGICAL OXIDATION ENZYME SYSTEMS

Example I

$$M.H_2 + 1/2\ O_2 \xrightarrow{\boxed{\text{oxidase}}} M + H_2O$$

Example II

$$M.H_2 + 1/2\ O_2 \xrightarrow{\boxed{\text{aerobic dehydrogenase}}} M + H_2O$$

(Experimentally, methylene blue is also used as H acceptor)

Example III

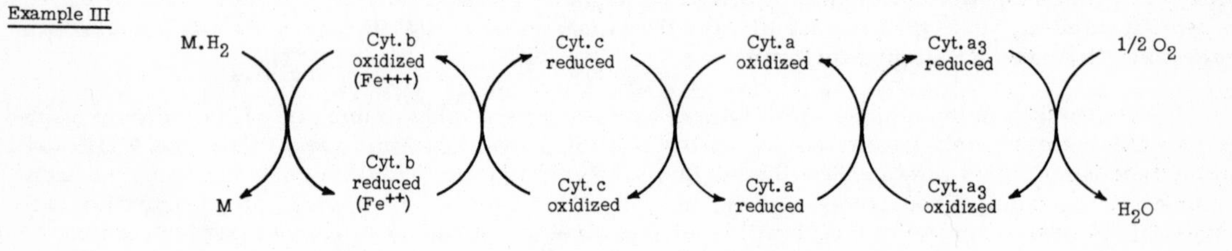

Example IV

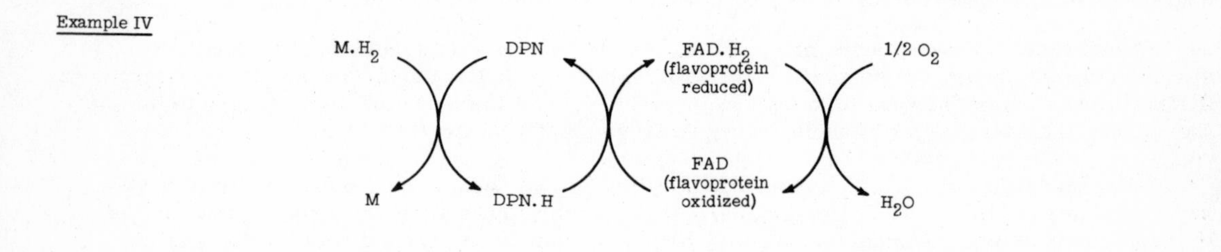

Example V

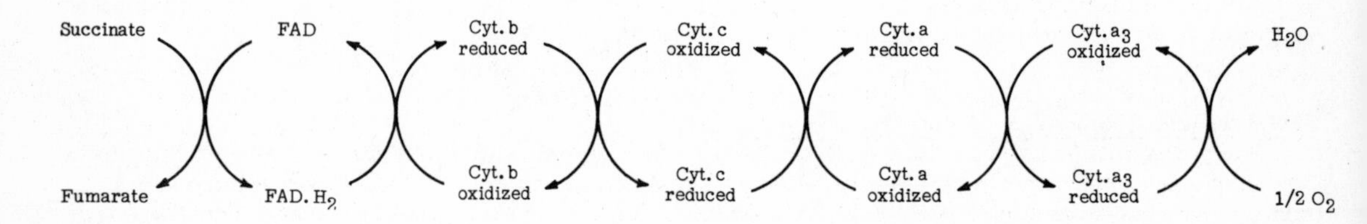

Example VI

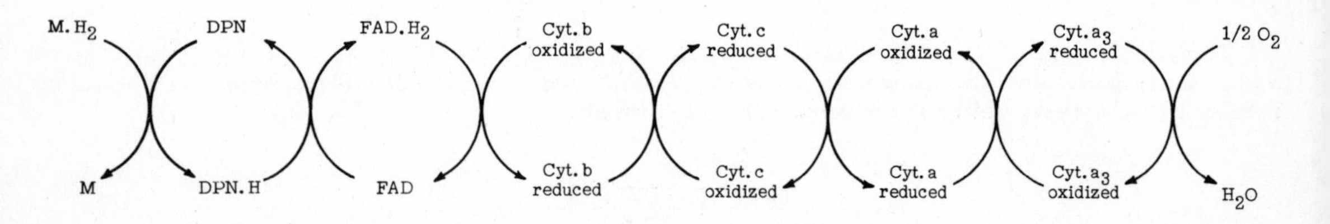

This system is reversible, the direction of the reaction depending on the relative concentrations of the lactic and pyruvic acids.

The Flavoproteins:

Dehydrogenases operating through DPN or TPN require other systems to accomplish the transfer of hydrogen from the reduced coenzyme to the cytochrome system. This is indicated by the observation that purification of natural systems results in failure of the known enzymes to couple with the cytochromes.

The wide distribution of flavoproteins in all tissues, both plant and animal, verifies the suggestion that they act as the link between DPN and TPN and the cytochromes. These flavoproteins act as hydrogen transfer agents in a manner similar to DPN and TPN.

Other enzyme systems which operate between the anaerobic dehydrogenases and the cytochrome system, or molecular oxygen, will probably be discovered in the course of future research. An example is electron-transferring protein (ETF), described below.

In connection with the desaturation of fatty acids (Reaction 2, p. 209), a flavoprotein is associated as a coenzyme (H acceptor) with the specific acyl dehydrogenase which catalyzes the α-β unsaturation of the ''active'' (Co A) fatty acid. The reduced flavoprotein which is formed in the reaction cannot be reoxidized by conventional electron acceptors such as oxygen or cytochrome c. Instead, it requires another flavoprotein which has been designated **electron-transferring flavoprotein (ETF)** by Green (see p. 210). In the reaction, ETF $\rightleftarrows$ ETF. H_2, the original flavoprotein of the acyl dehydrogenase is reoxidized. Reduced ETF (ETF. H_2) is then itself reoxidized by oxygen or with cytochrome c.

The prosthetic group of ETF appears to be flavin adenine dinucleotide (FAD, see p. 79). The riboflavin content of ETF is 0.45%; it has a minimum molecular weight of 83,500.

Structure and Cellular Locale of the Respiratory Chain:

It has long been recognized that the oxidation in tissue preparations of a metabolite by oxygen is dependent upon the structural integrity of an intracellular unit which is actually a packet of many enzymes. For example, the oxidation in rat liver of succinate by oxygen through what is termed the ''succinoxidase system'' is associated with the mitochondria; in heart muscle, the oxidative system is localized in the ''sarcosomes'' which are cytologically analogous to mitochondria in other tissues. The sequence of electron transfer to oxygen in the succinoxidase system within the mitochondria appears to be as follows (see Example V, p. 116):

$$\text{Succinate} \rightarrow \text{Flavoprotein} \rightarrow \text{Cyt. b} \rightarrow \text{Cyt. c} \rightarrow \text{Cyt. a} \rightarrow \text{Cyt. } a_3 \rightarrow O_2$$

Using mitochondrial preparations from liver, reduced DPN can be rapidly oxidized by oxygen. In this case the sequence of electron transfer is as follows (see Example VI, p. 116):

$$\text{Substrate} \rightarrow \text{DPN} \rightarrow \text{Flavoprotein} \rightarrow \text{Cyt. b} \rightarrow \text{Cyt. c} \rightarrow \text{Cyt. a} \rightarrow \text{Cyt. } a_3 \rightarrow O_2$$

Summary:

In summary, biological oxidation involves hydrogen and electron transfer through a series of enzyme systems, much like the transfer of a commodity through various ''middle men'' before the ultimate consumer (oxygen) is reached. The oxidative process is initiated by the action of a dehydrogenase, specific to the metabolite, which catalyzes the removal of hydrogen and thus oxidizes the metabolite. The aerobic dehydrogenases (and the oxidases) transfer hydrogen directly to molecular oxygen (Examples I and II, p. 116). Other dehydrogenases, the anaerobic dehydrogenases, will not react directly with oxygen but require intermediary systems. These intermediary systems include DPN and TPN, the flavoproteins, and the cytochromes. Because of the cyclical character of the reduction and reoxidation of the various components of the oxidation-reduction systems, it is possible for very small quantities of an enzyme, acting as a hydrogen and electron transfer agent, to accomplish the transformation of large amounts of the original substrate.

Types of Compounds of Phosphoric Acid*

<table>
<tr><th>Type</th><th>Structure</th><th>Energy of Hydrolysis</th><th>Examples</th></tr>
<tr><td>Phosphoric ester of simple alcohol</td><td>H R   OH
| | /
—C—C—O—P=O
| | \
    H   OH</td><td>Low</td><td>Glucose-6-phosphate
3-Phosphoglyceraldehyde
2-Phosphoglyceric acid</td></tr>
<tr><td>Phosphoric acid acetal</td><td>   O—   OH
| | /
—C—C—O—P=O
| | \
    H   OH</td><td>Low</td><td>Glucose-1-phosphate</td></tr>
<tr><td>Anhydride of phosphoric acid</td><td>   O   OH
    ‖   /
—O—P—O—P=O
    | \
    OH   OH</td><td>High</td><td>Adenosine triphosphate (ATP)
Adenosine diphosphate (ADP)</td></tr>
<tr><td>Mixed anhydride of phosphoric acid</td><td>   O   OH
| ‖   /
—C—C—O—P=O
| \
              OH</td><td>High</td><td>Acetyl phosphate
1, 3-Diphosphoglyceric acid (the 1-phosphate)</td></tr>
<tr><td>Enol phosphate</td><td>   OH
| /
—C=C—O—P=O
    | \
    C   OH
    |
    R</td><td>High</td><td>Phosphopyruvic acid</td></tr>
<tr><td>N-substituted phosphamic acid</td><td>   OH
| /
—C—N—P=O
| | \
    H   OH</td><td>High</td><td>Phosphocreatine
Phosphoarginine</td></tr>
</table>

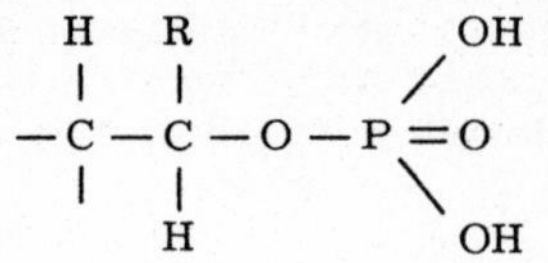

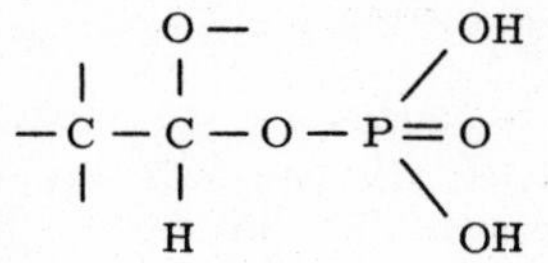

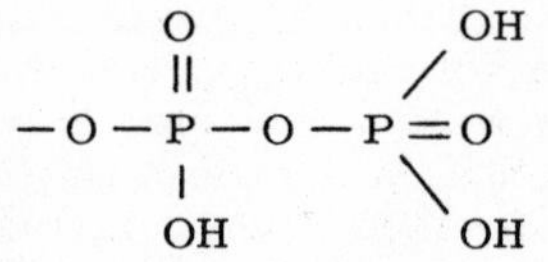

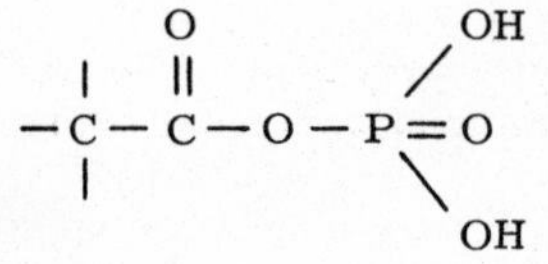

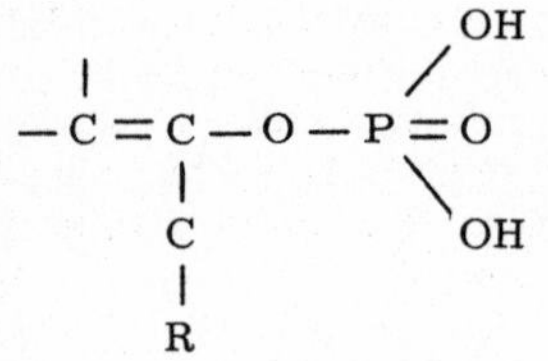

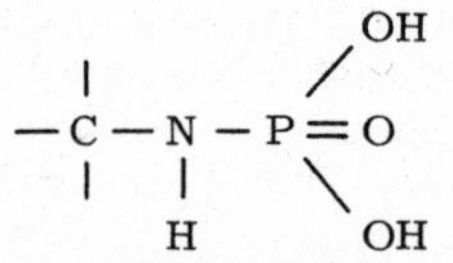

*From Stetten, D.: "Carbohydrate Metabolism," Am. J. Med. 7:571, 1949. Reproduced with permission.

Phosphate Bond Energy:

Combinations of various metabolites with phosphate are common in the tissues. These phosphate compounds are closely connected with the oxidative breakdown of the metabolites and with the exchange of chemical energy developed during the process. Measurement of the energy liberated when these various phosphate esters were hydrolyzed indicated that the amounts of energy were in some cases small and that the reaction was thus easily reversed; in other cases, larger amounts of energy were made available on hydrolysis of the phosphate ester so that it became customary to classify phosphate compounds as high-energy or low-energy in accordance with the energy released on hydrolysis of the phosphate ester. Such a classification is exemplified in the table on p. 118.

In the earlier studies on phosphate bond energy, a value of about 12,000 calories per mol was assigned to the typical "high-energy" bond such as the terminal phosphate bond of adenosine triphosphate (ATP) or of creatine phosphate (CP). More recent work indicates that the value for hydrolysis of ATP is closer to 7600-7800 calories per mol. Because a number of other compounds previously considered "low-energy" may approach this value, a sharp distinction between low-energy and high-energy bonds may not be possible.

It is customary to designate energy-rich phosphate bonds by the symbol ~ and the energy-rich phosphate as Ⓟ. An example of a structure written in this fashion is as follows:

Adenosine triphosphate (ATP): (See also p. 47.)

A—O—P(=O)(OH)—O~Ⓟ(=O)(OH)~O~Ⓟ(=O)(OH)-OH or A—P~Ⓟ~Ⓟ

(A = adenosine: adenine + D-ribose)

Although the concept of the "high-energy bond" is a useful one in understanding the biochemical changes which occur in intermediary metabolism, there are objections to this terminology. It should not be assumed that the energy released on hydrolysis is actually concentrated in one chemical bond, although the use of a special bond symbol as described above might suggest that this is the case. What is really being considered is the change in chemical potential which occurs when certain groups are transferred from one molecule to another, e.g., in the transfer of the terminal phosphate of ATP to water (by hydrolysis) to form ADP and H_3PO_4. The term "**group-transfer potential**" has therefore been suggested as a more accurate term than bond energy.

Exchange of Energy From Phosphate Compounds (The "Biological Storage Batteries"):

The fundamental importance of the high-energy phosphate compounds is their role in energy exchange for cellular activity. Thus the hydrolysis of the terminal phosphate bond of ATP to produce ADP liberates the energy which is apparently universally used by the cells of the body to support their metabolic activities. The resynthesis of the phosphate esters, i.e., the incorporation of inorganic phosphate into a high-energy linkage with an organic compound, requires the simultaneous incorporation of an amount of energy equal to that liberated on hydrolysis of the high-energy bond. This energy is obtained from the oxidative breakdown of various metabolites such as sugars and lipids. In this manner the energy yielded by the breakdown of these metabolites can be stored by the formation of high-energy phosphate bonds. These high-energy bonds, by storing energy and delivering it later by hydrolysis of the bond, may be considered to act as "biological storage batteries" which can be charged and discharged under various conditions within the cell.

Coupling of Oxidation and Phosphorylation:

In the course of the removal and transfer of electrons during the oxidation of a substrate, energy is liberated. This energy may temporarily be stored or "trapped" in a high-energy bond as chemical energy, or it may be liberated as heat. Ultimately, all of the energy produced in the body will be dissipated as heat, so that measurement of heat production can be used to estimate the expenditure of energy by the organism (see p. 370). The storage of energy in the form of a high-energy phosphate bond is limited to one bond for each step in the transfer of electrons. Thus, a maximum of 7600-7800 small calories per mol of substrate oxidized can be stored as a result of a one-step electron transfer. Since the transfer of a pair of electrons from a substrate to oxygen may liberate much more energy than this, it is advantageous to transfer the electrons to oxygen in a series of steps rather than in one step. Each step might then permit the formation of an energy-rich phosphate bond by a process which **couples oxidation with phosphorylation**, so that a large proportion of the liberated energy of the substrate would be stored and little wasted as heat. This process of "oxidative phosphorylation" has been studied in connection with the transfer of electrons over oxidative systems such as are shown in Example VI, p. 116. For each atom of oxygen utilized, three atoms of phosphorus are incorporated into high-energy bonds, giving a P:O ratio for this system of 3:1. This means that as many as 23,400 calories of energy from the original substrate were stored per mol oxidized. In some cases an additional high-energy bond is formed at the first step in oxidation, i.e., at the substrate level, when the electrons are transferred to DPN. Such a stepwise oxidation (e.g., that of α-ketoglutaric acid to succinic acid, reaction 7, p. 181) over the DPN-flavoprotein-cytochrome system would result in the formation of four high-energy bonds per mol of substrate oxidized and a total of 4 × 7800 = 31,200 calories of trapped energy.

The probable sites of phosphorylation coupled to oxidation in the respiratory chain are shown below[2].

$$\text{Substrate} \longrightarrow \text{DPN} \xrightarrow{\text{ATP}} \text{Flavoprotein} \longrightarrow \text{Cyt. b} \xrightarrow{\text{ATP}} \text{Cyt. c} \longrightarrow \text{Cyt. a} \xrightarrow{\text{ATP}} \text{Cyt. } a_3 \longrightarrow O_2$$

According to present knowledge, oxidative phosphorylation is restricted to those oxidations which utilize the complete system from pyridine nucleotide (DPN) or flavoproteins through the cytochromes to oxygen. Oxidative reactions catalyzed by oxidases or by aerobic dehydrogenases (Examples I and II, p. 116) are not phosphate-linked even at the substrate level.

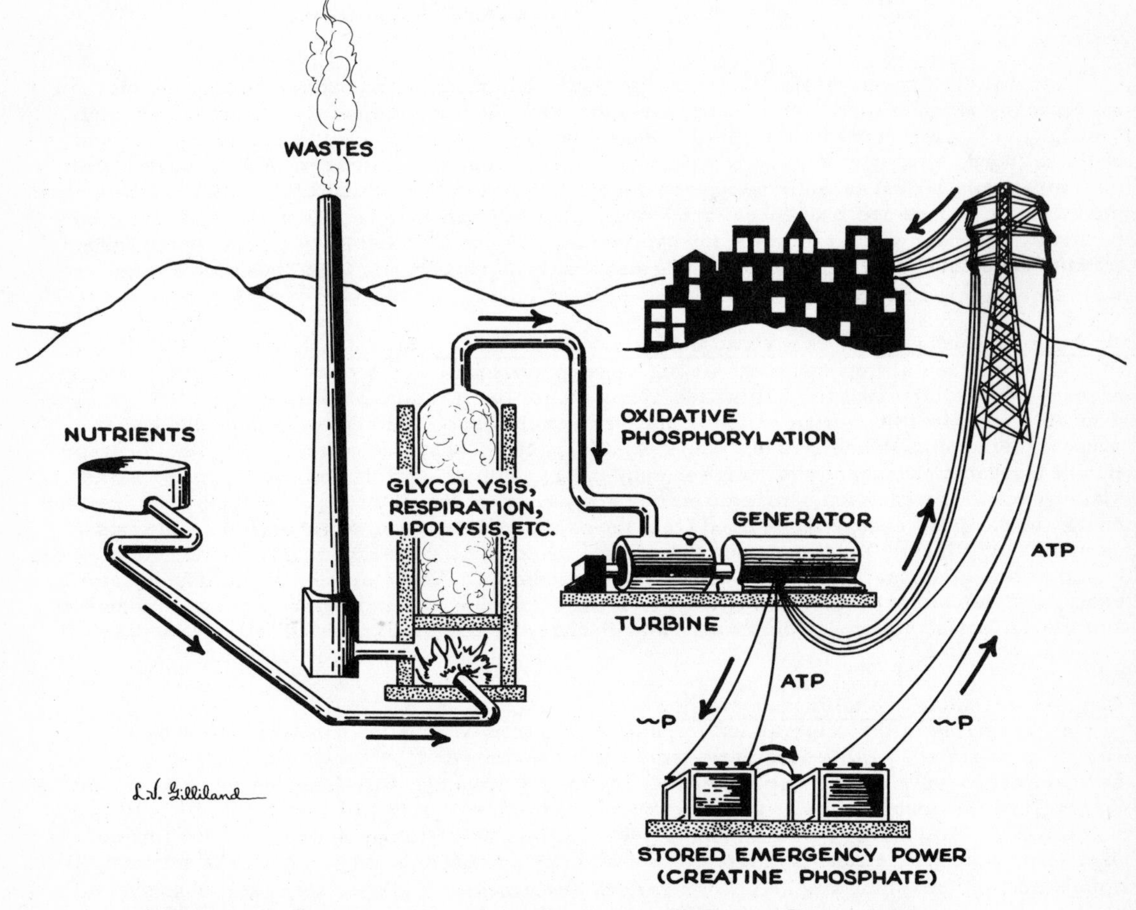

A Metaphoric Representation of the Conversion of Energy Obtained From Nutrients Into "High-energy" Phosphate for Cellular Work

The direct oxidation of TPN.H does not result in the formation of high-energy phosphate, in contrast to the oxidation of DPN.H. However, TPN.H may contribute to the formation of high-energy phosphate indirectly through the mediation of enzymes, **transhydrogenases**, which catalyze the transfer of hydrogen from TPN.H to DPN, thus oxidizing TPN and producing reduced DPN, which then becomes re-oxidized through the respiratory chain. Experiments with soluble enzymes from human placenta have demonstrated that certain steroid hormones, notably estradiol and testosterone, may function as coenzymes in the transfer of hydrogen between the reduced and oxidized forms of di- and triphosphopyridine nucleotides (see p. 359).

The conversion of energy derived from the metabolism of various nutrients into ''high-energy'' phosphate which in turn serves as a direct source of energy for cellular work is metaphorically represented on p. 120. Creatine phosphate is shown as a source of emergency power for use in a tissue such as muscle where a sudden demand for large amounts of energy may exceed the rate at which the tissue can produce it by the usual mechanisms. This reserve source of "power" must of course be recharged during the so-called recovery period of muscular contraction utilizing a transfer of energy from ATP to creatine (see p. 175).

The significance of the stepwise system which is characteristic of biological oxidative processes is clarified by these observations on the coupling of oxidation and phosphorylation. Apparently it functions to permit a gradual rather than an explosive and wasteful evolution of energy from a metabolite. As a result, the energy-yielding process is maintained at a very high level of efficiency wherein as much as 40 to 50% of the output may be recovered as useful work.

Mechanisms of Oxidative Phosphorylation:

The exact mechanisms involved in the coupling of oxidation and phosphorylation are not yet known. It is certain, however, that the process is an important function of the mitochondria. Oxidative phosphorylation is not confined to the phosphorylation of adenine nucleotides (ADP and AMP), although 90% of the activity is in connection with this reaction. Phosphorylation of deoxynucleotides by respiring mitochondria has also been observed, as well as the conversion of UDP (see p. 48) to UTP and of UMP to UDP, probably through ATP.

The amounts of lipid in the mitochondria are relatively large. This lipid is present in lipoproteins which seem to be located between some sections of the respiratory electron transfer chain; in such instances, one component of the chain reacts with the next component only through the lipoprotein segment. Within the lipoprotein there is located a newly-discovered coenzyme which because of its quinone structure has been named Coenzyme Q. The coenzyme which may be derived from vitamin K (see p. 73) is a 2, 3-dimethoxy-5-methylbenzoquinone with a polyisoprenoid side chain at carbon 6.

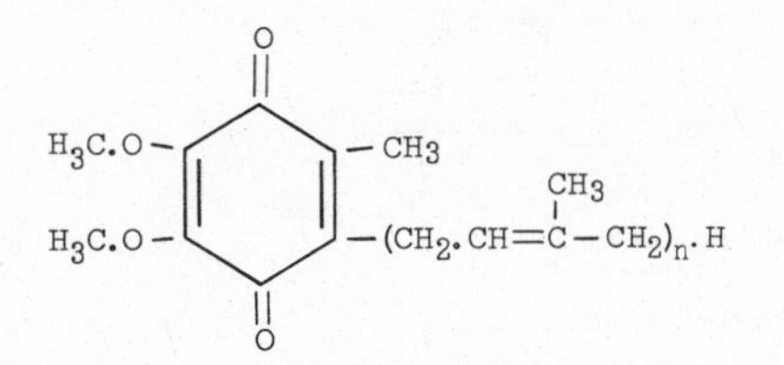

Coenzyme Q

It has been suggested by Green and his co-workers that Coenzyme Q actively participates not only in electron transfer but also in oxidative phosphorylation associated with the transfer of electrons along the respiratory chain. The suggested mechanism is shown on p. 122. Coenzyme Q accepts an electron from a substrate and thus becomes reduced to the semiquinone, which reacts with inorganic phosphate to form the semiquinone phosphate. Release of this phosphate to ADP to form ATP, and oxidation to the quinone, now occur simultaneously. The quinone may then participate once again in the oxidative phosphorylation cycle.

Uncoupling of Phosphorylation from Oxidation:

Under certain conditions, oxidation of a substrate with transfer of electrons to oxygen over a stepwise oxidative system may occur without the formation of high-energy phosphate bonds, in which case much energy is liberated as heat. Thus phosphorylation is not essential to oxidation. This dissociation or ''uncoupling'' of oxidation from phosphorylation may be brought about by 2, 4-dinitrophenol (DNP), azide dyes such as thionine, pentachlorophenol, bilirubin, the antibiotic

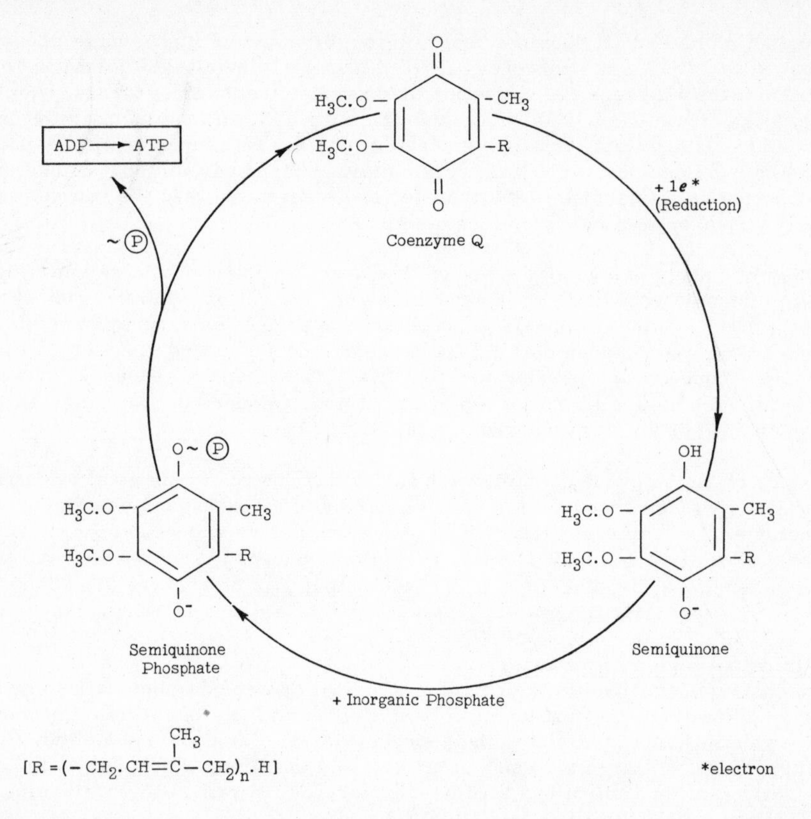

gramicidin, adrenochrome, the hormones thyroxin and triiodothyronine, and the vitamin K antagonist, bishydroxycoumarin, a fact which is of interest in connection with the role of Coenzyme Q (see above). Calcium ion also acts as an uncoupling agent, possibly acting as antagonist to magnesium ion, which is essential in oxidative phosphorylation.

There is evidence that the various uncoupling agents mentioned above do not all exert this action by the same mechanism. In the case of thyroxin and triiodothyronine, the effect may be indirect, the result of changes which these hormones induce in the permeability or structure of the mitochondrial membrane[3]. Using rat liver mitochondria with disrupted membranes, thyroxine or triiodothyronine had no effect on P:O ratios even at concentrations higher than would have been required had intact preparations been used. Dinitrophenol was, however, still effective as an uncoupling agent with the disrupted mitochondrial systems. The uncoupling effect of thyroxine (but not of dinitrophenol) is prevented by glutathione or magnesium or manganous ions. Glutathione overcomes the action of adrenochrome, but magnesium ions or manganous ions are ineffective.

• • •

References:

1. Hayaishi, O., Rothberg, S., and Mehler, A.H.: Abst. Am. Chem. Soc., 130th Meeting, p. 53C, Sept., 1956.
2. Lehninger, A.L., Wadkins, C.L., Cooper, C., Devlin, T.M., and Gamble, J.M., Jr.: Science 128:450, 1958.
3. Tapley, D.F., and Cooper, C.: J. Biol. Chem. 222:341, 1956.

Bibliography:

Baldwin, E.: Dynamic Aspects of Biochemistry. Cambridge, 1952.
Green, D.E.: Mechanisms of Biological Oxidation. Cambridge, 1940.
Klotz, I.M.: Energetics in Biochemical Reactions. Academic, 1957.
Neilands, J.B.: Biological Oxidations. Ann. Rev. Biochem. 27:455, 1958.
Lardy, H.A.: Respiratory Enzymes. Burgess, Rev. Ed., 1950.
McElroy, W.D., and Glass, H.B., Eds.: Symposium, Phosphorus Metabolism, 2 vols. Johns Hopkins Press, 1951-52.

9...

The Blood, Lymph and Cerebrospinal Fluid

BLOOD

Blood is a tissue which circulates in what is virtually a closed system of blood vessels. It consists of solid elements - the red and white blood cells and the platelets - suspended in a liquid medium, the plasma.

The Functions of the Blood:

1. Respiration - Transport of oxygen from the lungs to the tissues and of carbon dioxide from the tissues to the lungs.
2. Nutrition - Transport of absorbed food materials.
3. Excretion - Transport of metabolic wastes to the kidneys, lungs, skin, and intestines for removal.
4. Maintenance of normal acid-base balance in the body.
5. Regulation of water balance through the effects of blood on the exchange of water between the circulating fluid and the tissue fluid.
6. Regulation of body temperature by the distribution of body heat.
7. Defense against infection in the white cells and the circulating antibodies.
8. Transport of hormones; regulation of metabolism.

Packed Cell Volume:

When blood which has been prevented from clotting by the use of a suitable anticoagulant is centrifuged, the cells will settle to the bottom of the tube while the plasma, a straw-colored liquid, will rise to the top. Normally the cells comprise about 45% of the total volume. This reading (45%) is a normal hematocrit, or packed cell volume, for males; for females, the normal packed cell volume is about 41%.

The specific gravity of whole blood varies between 1.054 and 1.060; the specific gravity of plasma is about 1.024 to 1.028.

The viscosity of blood is about 4.5 times that of water. Viscosity of blood varies in accordance with the number of cells present and with the temperature and degree of hydration of the body. Because these three factors are relatively constant under normal conditions, the viscosity of the blood does not ordinarily influence the physiology of the circulation.

Blood Volume:

The volume of the blood can be measured by several procedures, although the results obtained will vary somewhat with the method used. The method of Gregersen, which is employed frequently in clinical practice, utilizes T-1824, a blue dye. The dye is injected intravenously, and its concentration in the plasma is determined after sufficient time has elapsed to allow for adequate mixing (usually ten minutes). The plasma volume thus obtained is then converted to total blood volume by the following formula:

$$\text{Total blood volume (ml.)} = \frac{\text{Plasma volume (ml.)} \times 100}{100 - \text{hematocrit}}$$

Presumably the dye is bound to the plasma protein; for this reason it remains in the circulation for some time. Actually some of the dye does escape into the extracellular fluid; therefore the results obtained by this method are somewhat high. Normal volumes in males as found by the T-1824 method are given as follows: plasma volume, 45 ml./Kg. body weight; blood volume, 85 ml./Kg. The corresponding values in the female are somewhat lower[1].

I^{131}-labelled human serum albumin is also used to determine plasma volume. A carefully measured dose is injected, and the dilution of the label is then obtained.

There are objections to measurements of plasma volume as a means of determining whole blood volume. The blood volume is calculated from the plasma volume by the use of the hematocrit, which, since peripheral blood is used, may not represent the ratio of cells to plasma throughout the circulation. Consequently, a direct measurement of the whole blood volume by the use of labelled red cells is preferred when greater accuracy is required, particularly in states where there is an impairment of circulatory efficiency (e.g., shock or cardiac failure). P^{32} and radio-iron have both been used to label the red cells for the purpose of measuring whole blood volume.

Blood Osmotic Pressure:

The osmotic pressure of the blood is kept relatively constant mainly by the kidney. Osmotic pressure can be determined by measurement of the freezing point depression. −0.537°C. has been established as the average delta value for whole blood. This corresponds to an osmotic pressure of seven to eight atmospheres at body temperature. A solution of sodium chloride containing 0.9 Gm. per 100 ml. has an osmotic pressure equal to that of whole blood. Such a saline solution is termed "isotonic" or "physiologic" saline. Actually these sodium chloride solutions are not physiologic, since additional ions are lacking which are necessary for the function of the tissues. Other solutions, which are not only isotonic but which contain these ions in proper proportions, are more appropriate from a physiologic standpoint. Examples of these balanced ionic solutions are Ringer's, Ringer-Locke's, and Tyrode's solutions. The formulas are given below:

Composition of Some Saline Solutions Isotonic With Blood

	Saline (%)	Mammalian Ringer (%)	Ringer-Locke (%)	Tyrode (%)
NaCl	0.9	0.9	0.9	0.9
$CaCl_2$	-	0.026	0.048	0.02
KCl	-	0.03	0.042	0.02
$NaHCO_3$	-	-	0.01-0.03	0.1
Glucose	-	-	0.10-0.2	0.1
$MgCl_2$	-	-	-	0.01
NaH_2PO_4	-	-	-	0.005

THE CLOTTING OF BLOOD

When blood is drawn and allowed to clot, a clear liquid (serum) exudes from the clotted blood. Plasma, on the other hand, separates from the cells only when blood is prevented from clotting.

The blood clot is formed by a protein **(fibrinogen)** which is present in soluble form in the plasma and which is transformed to an insoluble network of fibrous material (**fibrin**, the substance of the blood clot) by the clotting mechanism.

According to the original **Howell theory** of blood coagulation, the change of fibrinogen into fibrin is caused by **thrombin**, which in fluid blood exists as **prothrombin.** The conversion of prothrombin to thrombin depends on the action of **thromboplastin** and calcium. These steps in the clotting process may be diagrammed as follows:

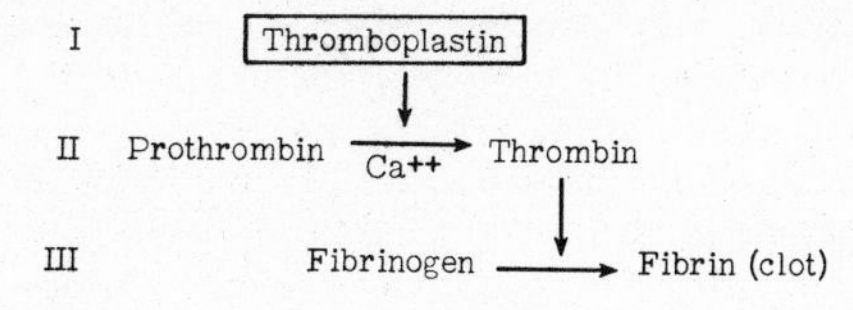

In recent years it has become evident that a number of factors are concerned with each of the three essential steps in the simplified scheme given above. These will be discussed separately.

I. Origin of Thromboplastin:

Substances with thromboplastic activity are contributed by the plasma, the platelets, and the tissues.

A. From the Plasma:
1. Plasma thromboplastic factor A (PTF-A) - Synonyms: anti-hemophilic factor (AHF), anti-hemophilic globulin, thromboplastic plasma component, thrombokinase. This factor is relatively heat-stable, but labile on storage. A deficiency of this factor is the cause of the classical type of hemophilia. It occurs in Cohn Fraction I (see p. 131).
2. Plasma thromboplastic factor B (PTF-B)[2] - Synonyms: plasma thromboplastin component (PTC), Christmas factor. This factor is found in both serum and plasma; it is absolutely necessary for the production of thromboplastin. The factor occurs in the β-2 globulin fraction in a concentration of less than 1 mg./100 ml. plasma. Activity of this factor is reduced in vitamin K deficiency[3].
3. Plasma thromboplastic factor C (PTF-C) - Synonyms: plasma thromboplastin antecedent (PTA). A factor described by Rosenthal, et al.[4, 5]
4. Plasma thromboplastic factor D (PTF-D) - A fourth plasma factor necessary for thromboplastic activity described by Spaet, Aggeler, and Kinsell[6].

Deficiencies of each of the four thromboplastic factors have been described. The clinical picture in each case is similar, i.e., a hemophilioid disease, but the defect is a distinct one in each instance. However, the occurrence of hemophilioid states traceable to deficiencies of clotting factors other than PTF-A, the anti-hemophilic globulin, are rare. In classical hemophilia (PTF-A deficiency) and in PTF-B (PTC) deficiency, the extent of the deficiency varies. In the most severe cases the factor is completely absent, whereas moderate or mild cases may have from 3 to 16% of the normal content of the factor. As much as 33% of normal may be found in the subhemophilic cases. These latter individuals may not show symptoms of hemophilia except under conditions of stress such as excessive hemorrhage.

B. From the Platelets: The platelet thromboplastic factor (thromboplastinogenase) is obtained upon disintegration of the platelets, as by contact with a rough surface. Thrombin catalyzes the formation of the platelet factor.

C. From the Tissues: Thromboplastin precursors are also supplied by the tissues. These may be very important in initiating coagulation reactions because they are supplied from outside the circulation itself.

II. Conversion of Prothrombin to Thrombin:

The prothrombin of the blood exists in an inactive as well as an active form.

A. The conversion of the inactive form of prothrombin to the active form is catalyzed by:
1. Prothrombin conversion accelerator I*, which occurs in both plasma and serum but in a more active form in the serum. Synonyms: labile component, labile factor, proaccelerin, accelerin, factor V, factor VI, plasma or serum accelerator globulin; and by
2. Thrombin, which also accelerates the change of inactive to active prothrombin.

---

*According to Lewis and Ware[7] there may actually be only **one** accelerator system in human plasma. The evidence strongly suggests that plasma Ac-globulin (II, A, 1 above) is converted by thrombin to the much more active serum Ac-globulin, most of which is then rapidly inactivated by an inhibitor. Not all of the plasma Ac-globulin is converted during the process of spontaneous coagulation.

B. Active prothrombin is converted to thrombin under the influence of thromboplastin. This is accelerated by:
1. Calcium ions.
2. An accelerator from the platelets (platelet accelerator factor I).
3. Prothrombin conversion accelerator I* (same as A, 1 above).
4. Prothrombin conversion accelerator II*, which occurs in both plasma and serum but in a more active form in serum. Activity of this factor is reduced in vitamin K deficiency. Synonyms: serum prothrombin conversion accelerator, stable component, stable factor, convertin, co-thromboplastin, factor VII.

III. Formation of Fibrin From Fibrinogen:

The protein, fibrinogen, loses one or more peptides under the influence of thrombin, which is itself a proteolytic enzyme. A factor from the platelets (platelet accelerator factor II) also catalyzes the conversion of fibrinogen to fibrin. The result of these reactions is the formation, first, of activated fibrinogen (F′), which then undergoes a spontaneous but reversible polymerization resulting in the production of fibrin, a protein of much larger molecular weight than the original fibrinogen.

Other Aspects of the Clotting Mechanism:

A. Autocatalysis: In the first few seconds following an injury, little or nothing observable happens with respect to the clotting of shed blood. Clotting then begins suddenly, and the reactions become accelerated with the passage of time. This acceleration phenomenon is caused by **autocatalysis**, whereby certain products formed in the coagulation process actually catalyze the reactions by which they themselves were formed. The principal autocatalyst is thrombin. It was noted above that thrombin catalyzes the formation of the platelet thromboplastic factor (I, B) and the conversion of inactive to active prothrombin (II, A, 2).

B. Vasoconstrictor Action: In addition to the coagulation reactions at the site of the injury other factors may aid in securing hemostasis. These include:
1. A prompt reflex vasoconstriction in the region of the injury.
2. Compression of the vessels in the area of injury by the mass of clotted blood in the tissues. The blood vessels become so compressed that their endothelial linings may actually adhere to one another.
3. Liberation of a vasoconstrictor principle upon lysis of the platelets. This may be **serotonin** (hydroxytryptamine; see p. 257), which is known to be adsorbed and concentrated in the platelets although it is not manufactured there.

C. Inhibitors of Prothrombin Activation and Conversion: The clotting of blood may be prevented by the action of substances which interfere with the conversion of prothrombin to thrombin. The best-known inhibitor of this reaction is **heparin**, a water-soluble, thermostable compound which is extremely potent. As little as 1 mg. can prevent the clotting of 100 ml. or more of blood. Heparin occurs in liver and certain other tissues such as lung.

Another inhibitor is referred to as "anti-thromboplastin," although it has not been proved that this substance acts specifically against thromboplastin. All that is known is that it inhibits the first phase of the process by which prothrombin is activated.

Plasma contains **antithrombic activity**, since it will cause the destruction of large quantities of thrombin by irreversible conversion to **metathrombin.** The antithrombin activity of the plasma is not influenced by heparin.

Calcium is also necessary for the conversion of prothrombin to thrombin. Citrates and oxalates which are commonly used as anticoagulants are effective because they remove calcium from the blood by the formation of insoluble citrate or oxalate salts of calcium. If calcium is added in excess, the clotting power of the blood is restored.

Blood may also be prevented from clotting by defibrination, i.e., removal of fibrin by allowing it to form around a glass rod with which the blood has been stirred or on glass beads shaken in the flask with the blood. This technic is often used in preparing blood for use in blood media in bacteriology.

*See Note (*) on p. 125.

The Fibrinolytic System:

In addition to the mechanisms for the formation of a clot in the blood, there is also a mechanism which is concerned with the lysis of the clot. In plasma and serum there is a substance called **profibrinolysin (plasminogen)** which becomes activated to **fibrinolysin (plasmin)**, the enzyme which lyses the clot. As was the case with prothrombin, profibrinolysin must be activated before it is converted to fibrinolysin. An activator has been found in many tissues; it is called **fibrinolysokinase (fibrinokinase)**. Attempts to prepare potent concentrates of the activator have not yet been successful, so artificial substances have been used instead. For example, shaking with chloroform will activate profibrinolysin, and enzymes found in certain bacteria are also effective. Such enzymes are **staphylokinase** and **streptokinase**. The ability of some bacteria to lyse fibrin clots is undoubtedly due to the presence of such activating enzymes.

The essentials of the fibrinolytic system are diagrammed below. Note the analogy to the clotting system as it pertains to prothrombin activation and conversion, and to the action of thrombin and antithrombin.

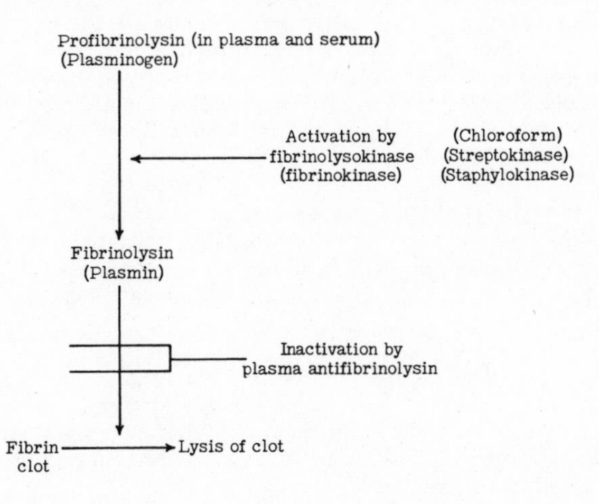

Prothrombin Production:

Prothrombin is manufactured exclusively by the liver, and vitamin K (see p. 72) is necessary for prothrombin production by this organ. A deficiency of prothrombin results when liver damage is so extensive as to interfere with prothrombin synthesis or when the absorption of vitamin K from the intestine is impaired - particularly when the flow of bile to the intestine is prevented, as in obstructive jaundice.

Bishydroxycoumarin (Dicumarol®), acting as an antagonist to vitamin K, produces hypoprothrombinemia. This was the first drug to be used clinically to reduce the clotting time of blood, but a number of other antagonists (anticoagulant drugs) are also available to produce hypoprothrombinemia.

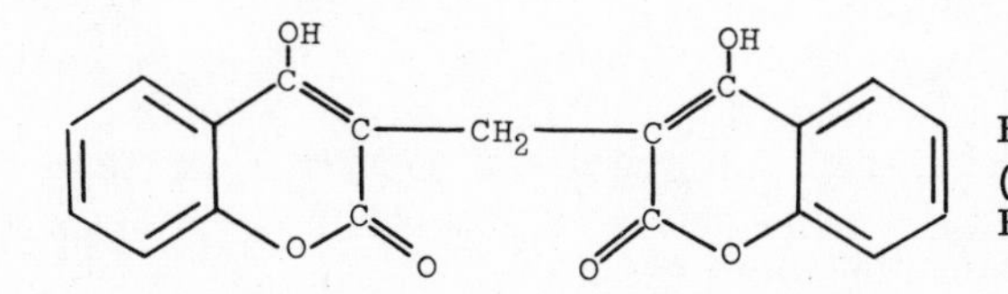

Bishydroxycoumarin (Dicumarol®), 3, 3-Methylene Bishydroxycoumarin

Dicumarol® is used in the treatment of thrombosis, or intravascular clotting. To control the dosage, frequent determinations of the **prothrombin level** of the blood are necessary. One technic for making these measurements is the method of Quick, in which an excess of thromboplastic substance (obtained from rabbit brain) and calcium are added to diluted plasma and the clotting time noted. A comparison of the clotting time of the patient with that of a normal control is always required for a valid test.

Various preparations to aid blood coagulation are also used. These include preparations of thrombin for topical use and "fibrin foam," which is applied as a hemostatic agent directly to a bleeding surface. These materials are of particular value in specialized surgical procedures.

THE PLASMA PROTEINS

The total protein of the plasma is about 7 to 7.5 Gm. per 100 ml. Thus, the plasma proteins comprise the major part of the solids of the plasma. The proteins of the plasma are actually a very complex mixture which includes not only simple proteins but also mixed or conjugated proteins such as glycoproteins and various types of lipoproteins.

The separation of individual proteins from a complex mixture is frequently accomplished by the use of various solvents and/or electrolytes to remove different protein fractions in accordance with their solubility characteristics. This is the basis of the so-called "salting-out" methods commonly utilized in the determination of protein fractions in the clinical laboratory. Thus it is customary to separate the proteins of the plasma into three major groups (fibrinogen, albumin, and globulin), by the use of varying concentrations of sodium or ammonium sulfate. Since it is likely that the subsequent analysis of the protein fractions will require a nitrogen analysis, sodium sulfate is preferred to ammonium sulfate.

Fibrinogen is the precursor of fibrin, the substance of the blood clot. It resembles the globulins in being precipitated by half-saturation with ammonium sulfate; it differs from them in being precipitated in a 0.75 molar solution of Na_2SO_4 or by half saturation with NaCl. In a quantitative determination of fibrinogen, these reactions are used to separate this protein from other closely related globulins.

Fibrinogen is a large, asymmetric molecule (see diagrams on next page) which is highly elongated, having an axial ratio of about 20:1. The molecular weight is between 350,000 and 450,000. It normally constitutes 4 to 6% of the total proteins of the plasma. This protein is manufactured in the liver; in any situation where excessive destruction of liver tissue has occurred, a sharp fall in blood fibrinogen results.

The serum proteins include mainly the albumin and globulin fractions of the plasma since most of the fibrinogen is removed in the clotting process, which is incident to the preparation of the serum. These two fractions may be separated by the use of a 27% solution of sodium sulfate, which precipitates the globulins and leaves the albumins in solution. By analysis of the nitrogen* in the filtrate following such a separation, a measure of the serum albumin concentration is obtained.

A similar analysis of the total protein of the serum, when corrected for the nonprotein nitrogen, may be used to estimate the total of the albumin and globulin. The globulin concentration is obtained by subtracting the albumin concentration (determined by direct analysis) from the total. The concentration of these two major protein fractions is often expressed as the ratio of albumin to globulin (A/G ratio). If proper separation of albumin and globulin fractions is accomplished†, the normal value for this ratio is about 1.2:1. In many clinical situations, this ratio is reversed or "inverted."

---

*It is customary to convert the nitrogen into protein by the use of the factor 6.25 ($N \times 6.25$ = protein). However, the average nitrogen factor found by analyses of dried proteins from pooled human plasma was reported as 6.73 by Armstrong, et al.[8]

---

†A clinical method for separating the albumins from the globulins of the serum was introduced by Howe in 1921. In this procedure, 1 ml. of serum was treated with 30 ml. of 22.2% Na_2SO_4, giving a final salt concentration of 21.5%. It has subsequently been found that this does not remove all of the globulins, so that the albumin values obtained by this method are too high and consequently the globulins are too low, i.e., the A/G ratio is too high. This method has not yet been modified in many clinical laboratories; it is therefore necessary to consider the method of separation used in serum protein analyses when the results are considered for diagnosis and treatment of disease.

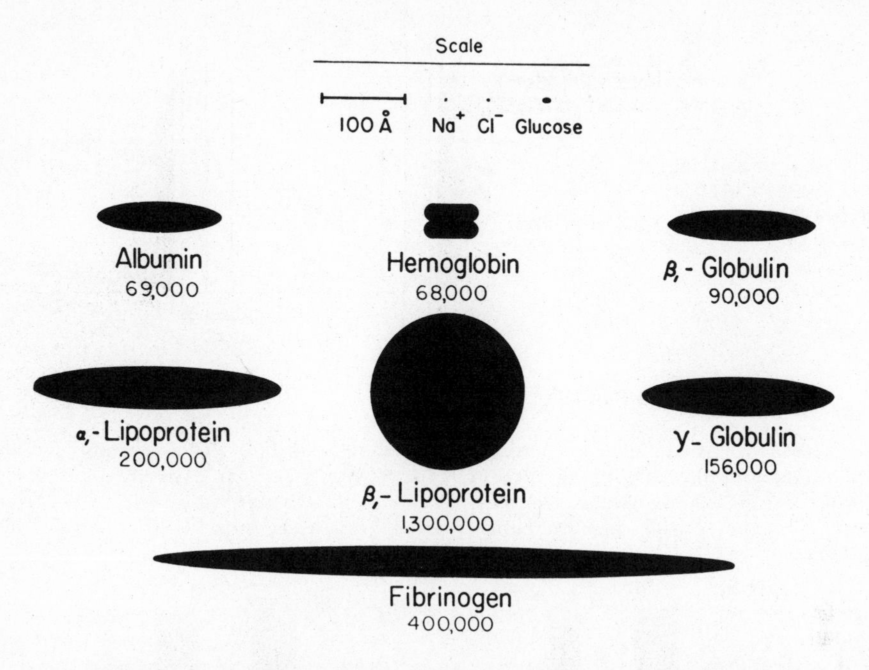

Relative Dimensions of Protein Molecules in the Blood (Oncley)

Electrophoretic Determination of Serum Proteins:

Electrophoresis is the migration of charged particles in an electrolytic solution which occurs when an electric current is passed through the solution. Various protein components of a mixture, such as plasma, at pH values above and below their isoelectric points will migrate at varying rates in such a solution because they possess different surface charges. The proteins will thus tend to separate into distinct layers. Tiselius has applied this principle to the analysis of plasma proteins. The sample for analysis is dissolved in a suitable buffer (for plasma, usually 0.1 N sodium diethylbarbiturate at pH 8.6). This mixture is then placed in the U-shaped glass cell of the Tiselius electrophoresis apparatus, and positive and negative electrodes are connected to each limb of the cell. When the current is applied, migration of the protein components begins. The albumin molecules, which are smaller and more highly charged, exhibit the most rapid rate of migration, followed by various globulins. After a time, boundaries between the separate fractions can be detected because of differences in the index of refraction due to variations in concentrations of protein. A photographic record of these variations constitutes what is termed an electrophoretic pattern. The diagram on p. 130 illustrates typical patterns as seen in each limb of the cell; these are called descending or ascending patterns in accordance with the direction of protein migration.

In normal human plasma, six distinct moving boundaries have been identified. These are designated in order of decreasing mobility as albumin, $alpha_1$ and $alpha_2$ globulins, beta globulin, fibrinogen, and gamma globulin. The distribution of electrophoretic components of normal human plasma is reported by Armstrong, et al., as follows[8]:

Albumin	55.2% (of total plasma protein)
Globulin	44.9
α_1 globulin	5.3
α_2 globulin	8.7
β globulin	13.4
γ globulin	11.0
Fibrinogen	6.5
A/G ratio	1.2

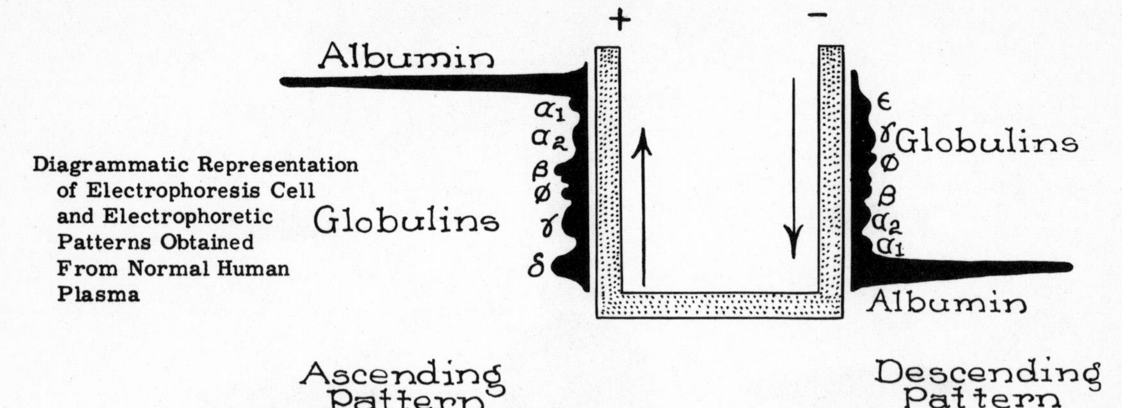

Diagrammatic Representation of Electrophoresis Cell and Electrophoretic Patterns Obtained From Normal Human Plasma

The electrophoresis of proteins on filter paper is also possible. This method, which requires relatively simplified and inexpensive equipment, is now widely used in clinical medicine. A detailed review of the clinical significance of analyses of serum proteins by paper electrophoresis has been prepared by Jencks, Smith, and Durrum[9]. Owen[10] has also reviewed this subject in a more extensive fashion.

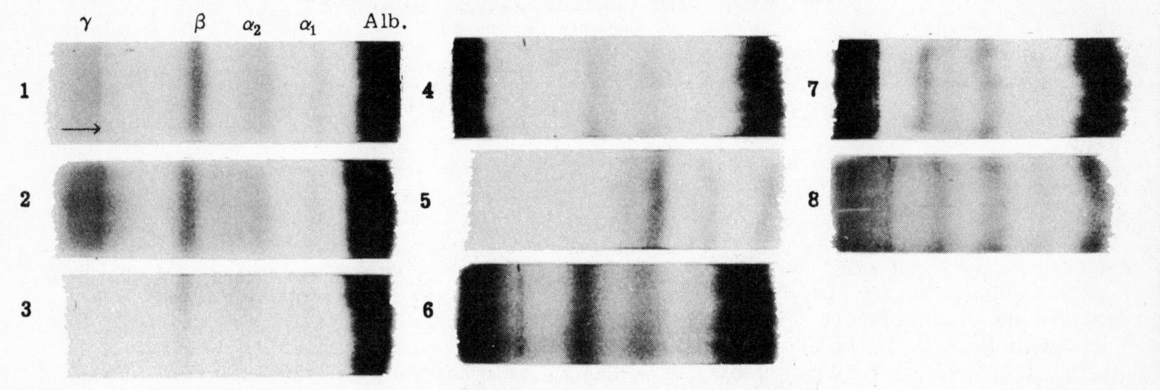

Examples of Abnormalities in Serum Protein Distribution Which Are Evident on Inspection of Electrophoretic Patterns Obtained by Electrophoresis on Filter Paper. The direction of migration is indicated by the arrow. (1) Normal. (2) Infectious mononucleosis. (3) Hypogammaglobulinemia. (4) Leukemia (type undetermined). (5) Nephrotic syndrome. (6) Infectious hepatitis. (7) Multiple myeloma. (8) Sarcoidosis. (Reproduced, with permission, from Jencks et al.: Am. J. Med. **21**:387, 1956.)

Other Methods of Separation of Proteins in Plasma:

The separation of proteins by electrophoretic analysis depends on a single property, the mobility of the proteins in an electrical field. It is known that some plasma proteins which differ in size, shape, composition, and physiologic functions may nonetheless have identical or nearly identical mobilities under the usual conditions of electrophoretic analysis. Thus the conventional electrophoretic fractions are by no means single protein components. Other methods of analysis, such as ultracentrifugation, alcohol precipitation, or immunologic analysis, reveal a considerable number of individual entities within each electrophoretic component.

E. J. Cohn and his collaborators have developed methods for the fractionation of plasma proteins which are particularly useful for the isolation in quantity of individual components. Their method is carried out at low temperatures and with low salt concentrations. Differential precipitations of the proteins is accomplished by variation of the pH of the solution and the use of different concentrations of ethyl alcohol.

The results of fractionation of pooled normal human plasma by the method of Cohn is shown on p. 131. Five major fractions are obtained. These account for the vast majority of the total proteins of the plasma. The supernatant, after removal of Fraction V (Fraction VI), contains less than 2% of the total protein. Fractions II and V are relatively homogeneous; the other fractions

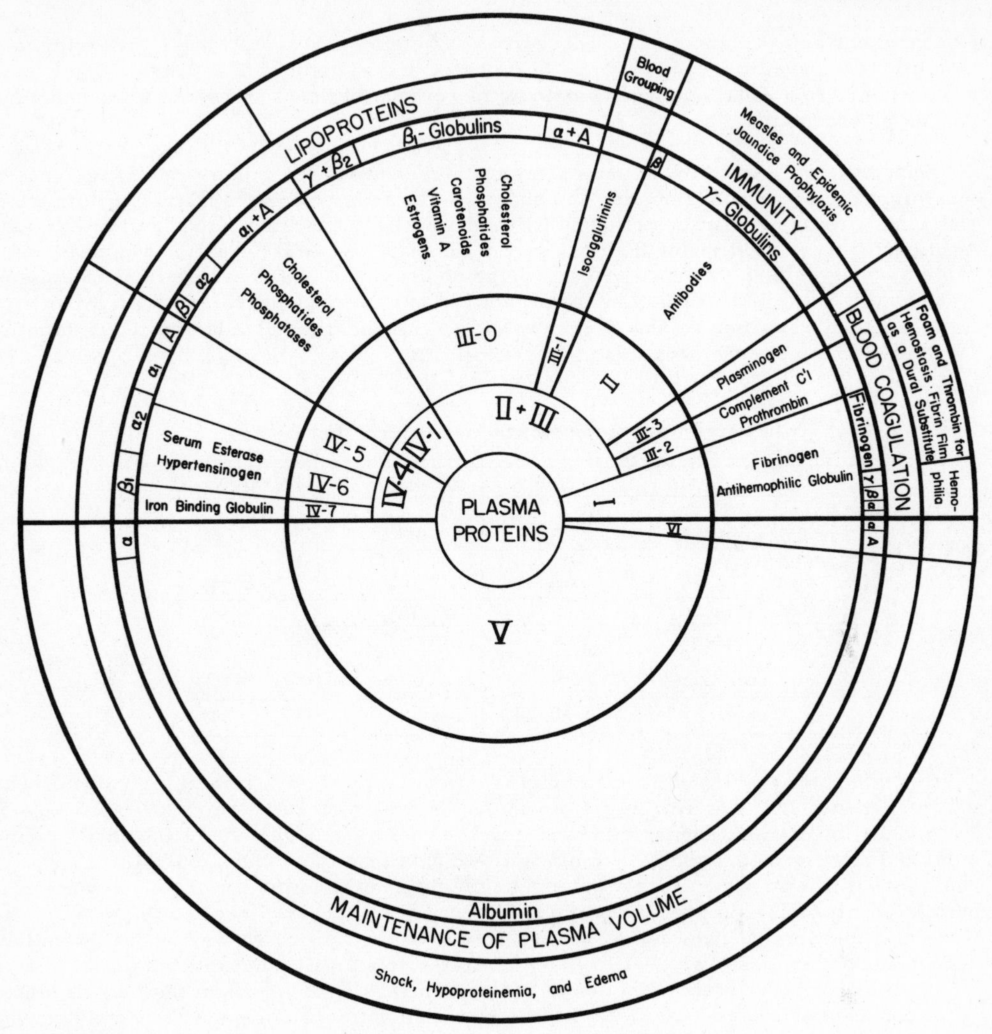

Plasma Proteins - Their Natural Functions and Clinical Uses and Separation into Fractions. (Revised by L. E. Strong from Figure 1, in Cohn, E. J., "Blood Proteins and Their Therapeutic Value," Science **101**:54, 1945.)

are very complex mixtures; subfractionation has revealed more than 30 components. From Fraction V by reprecipitation, albumin which is electrophoretically 97 to 99% homogeneous can be prepared. This is the salt-poor human serum albumin which is used clinically. Fraction II is almost pure gamma globulin. It is rich in antibodies and has thus found application in prophylaxis and modification of measles and infectious (viral) hepatitis (epidemic jaundice).

Albumin. Although this fraction of the serum proteins is not absolutely homogeneous, one component accounts for about two-thirds of the total. This fraction, **mercaptalbumin,** contains one free SH group per mol; it can be separated from the total serum albumins by crystallization as the mercuric salt. The remaining serum albumins apparently have no free SH groups.

Glycoproteins have also been found in the albumin fraction.

Globulins. The globulin fraction of the serum proteins is a very complex mixture. Certain components of particular interest will be described.

A. Mucoproteins and Glycoproteins: These are combinations of carbohydrate (hexosamine) moieties with globulin, found principally in the α_1 and α_2 globulin fractions. Meyer defines **mucoproteins** (mucoids) as those containing more than 4% hexosamine and those containing less as **glycoproteins**.

B. The Lipoproteins: About 3% of the plasma protein consists of combinations of lipid and protein migrating with the α globulins and about 5% of similar mixtures migrating with the β globulins. Using the ultracentrifuge, Hillyard, et al.[11] separated human serum into various fractions so as to account for the total serum lipoproteins. Each fraction was analyzed for protein, phospholipid, free and esterified cholesterol, and triglycerides. The results of these analyses are shown below. Fraction A contains the β lipoproteins, with densities less than 1.063; fraction B, the α_2 lipoproteins, with densities of 1.063 to 1.107; fraction C, the α_1 lipoproteins, with densities of 1.107 to 1.220.

Percentage Composition of Lipoproteins in Man

	Fraction A	Fraction B	Fraction C
Lipids			
Phospholipid	21%	29%	20%
Cholesterol			
Free	8%	7%	2%
Esterified	29%	23%	13%
Triglyceride	25%	8%	6%
Total Lipid	83%	67%	41%
Protein	17%	33%	59%

It will be noted that the β lipoproteins (Fraction A, above) are rich in fat and, consequently, low in protein. They are very large molecules, having molecular weights in the range of 1,300,000. In contrast, the lipoproteins migrating with the α globulins have smaller amounts of fat and more protein, and their molecular weights tend to be much lower (in the range of 200,000). It is also apparent that the higher the fat and the lower the protein content of a lipoprotein, the lower is its specific gravity. As the fat content declines and the protein rises, there is a concomitant rise in the specific gravity of the lipoprotein so that the so-called high-density lipoproteins (sp. gr. > 1.220) contain relatively small amounts of lipid.

The lipoproteins probably function as major carriers of the lipids of the plasma since most of the plasma fat is associated with them. Such combinations of lipid with protein provide a vehicle for the transport of fat in a predominantly aqueous medium such as plasma.

C. Metal Binding Proteins: Globulins which combine stoichiometrically with iron and copper comprise about 3% of the plasma protein. **Siderophilin (trans-ferrin)**, found in Cohn Fraction IV-7, is an example of a protein in the plasma which binds iron. The main function of this protein is to transport iron in the plasma. In states of iron deficiency or in pregnancy, there is a significant increase in the concentration of this metal-binding protein in the plasma. In disease, such as in pernicious anemia, chronic infections, or liver disease, there is a reduction in the amounts of this protein. A blue-green copper binding protein has been isolated from normal plasma. This protein, ceruloplasmin (molecular weight, 150,000), contains about 0.34% copper.

D. Gamma Globulins: This fraction of the serum proteins is the principal site of the circulating antibodies. Gamma globulin preparations which are electrophoretically better than 98% homogeneous may be found to contain as many as 20 distinct antibodies, when investigated by immunochemical technics.

Origin of the Plasma Proteins:

The liver is the sole source of fibrinogen, prothrombin, and albumin. Most of the α and β globulins are also of hepatic origin, but the γ globulins originate from other tissues. In plasmapheretic studies*, Whipple and co-workers found that dogs on whom an Eck fistula had been

*Plasmapheresis is a technic for depleting the plasma proteins by withdrawal of blood and reinjection of washed cells suspended in Ringer's solution.

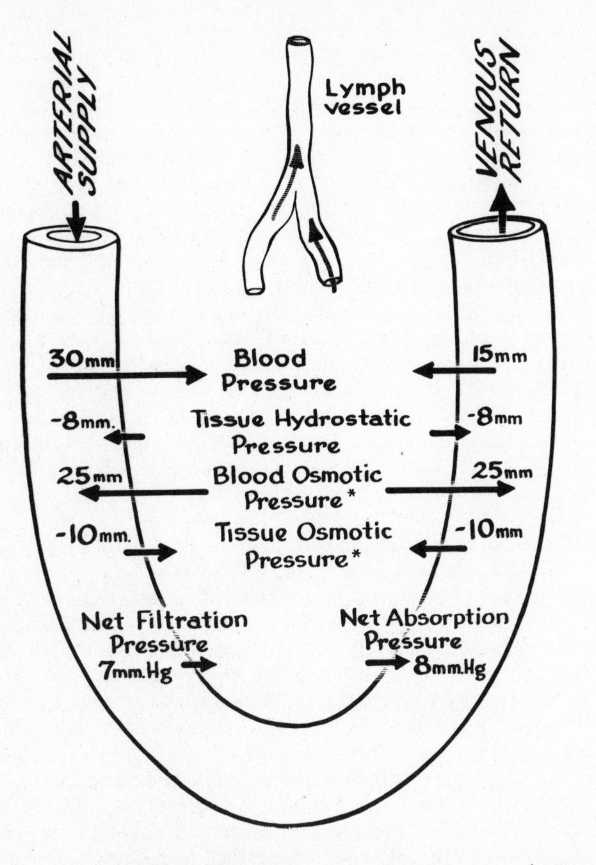

*These osmotic pressures are actually due only to the protein content of the respective fluids. They do not represent the total osmotic pressure (see Water Exchange, below).

Capillary Filtration and Reabsorption (''Starling Hypothesis'')

performed (portal blood diverted to the vena cava) were able to regenerate plasma proteins at a rate only 10% of that of control dogs. The Eck operation results in progressive impairment of liver function. A similar decline in liver function occurs in chronic liver disease (cirrhosis), and here also low plasma protein levels (particularly albumin levels) are very characteristic.

There is good evidence that the reticuloendothelial system participates in the formation of antibodies. This relates this system to the production of gamma globulin.

Dietary protein serves as a precursor of plasma protein. Many experiments have demonstrated a direct relationship between the quantity and quality of ingested protein and the formation of plasma protein, including also antibody formation. All dietary proteins are not equally effective in supplying materials for the regeneration of plasma protein. In studies on the dog after plasmapheresis, fresh and dried beef serum and lactalbumin, a protein of milk, were most effective. Egg white, beef muscle, liver, casein, and gelatin follow in that order.

<u>Functions of the Serum Proteins:</u>

A. Water Exchange: An important function of the serum proteins is the maintenance of osmotic relations between the circulating blood and the tissue spaces. The concentration of the electrolytes and of the organic solutes in plasma and tissue fluids is substantially the same; therefore the osmotic pressures due to these substances are practically identical. However, the total osmotic pressure of the plasma, which exceeds 6.5 atmospheres (4940 mm. Hg), is due not only to inorganic electrolytes and organic solutes but also to the plasma protein. These blood proteins are responsible for about 25 mm. of the total osmotic pressure of the plasma. Because there is also a small amount of protein in the tissue fluids which exerts an osmotic pressure of about 10 mm. Hg, the **effective** osmotic pressure of the blood over that of the tissue fluid is 15 (25–10) mm. Hg. This has the effect of attracting fluid and dissolved substances into the circulation from the tissue spaces. Opposing this force is the hydrostatic pressure of the blood, which tends to force fluids out of the circulation and into the tissue spaces. On the arterial side of the capillary loop, the hydrostatic pressure may be considered to be about 30 mm. Hg, and as the blood flows farther from the heart this pressure gradually decreases until it has fallen as low as 15 mm. Hg in the venous capillaries and even lower in the lymphatics. The hydrostatic pressures of the capillary are opposed by approximately 8 mm. Hg hydrostatic pressure in the tissue spaces. The effective hydrostatic pressure in the arterial capillary is therefore 22 mm. Hg (30-8); in the venous capillary, 7 mm. Hg (15-8). On the arterial side, the net result of these opposing pressures is a 7 mm. Hg (22-15) excess of hydrostatic pressure over osmotic pressure; this favors filtration of materials outward from the capillary to the tissue spaces. On the venous side, the 8 mm. Hg difference (15-7) is in favor of reabsorption of materials, because the intravascular osmotic pressure now predominates. Reabsorption is also aided by the lymphatics. This explanation of the mechanism of exchange of fluids and dissolved materials between the blood and tissue spaces is called the "Starling hypothesis." It is diagrammed on the preceding page.

The accumulation of excess fluid in the tissue spaces is termed "edema." Any alteration of the balance described above may result in edema. Decreases in serum protein concentration or increases in venous pressure, as in heart disease, are examples of pathologic processes which by altering the balance between osmotic and hydrostatic pressure would foster edema.

Each gram percent of serum albumin exerts an osmotic pressure of 5.54 mm. Hg, whereas the same quantity of serum globulin exerts a pressure of only 1.43 mm. Hg. This is due to the fact that the albumin fractions consist of proteins of considerably lower molecular weight than those of the globulin fractions. Albumin is of major importance in maintaining serum osmotic pressure. One gram of albumin will hold 18 ml. of fluid in the blood stream. Concentrated albumin infusions (25 Gm. in 100 ml. of diluent) are equivalent in osmotic effect to 500 ml. of citrated plasma. This effect may be beneficial in treatment of shock or in any situation where it is desired to remove fluid from the tissues or to increase the blood volume.

B. Blood Buffers: The serum proteins, like other proteins, are amphoteric and can combine with acids or bases. At the normal pH of the blood, the proteins act as an acid and combine with alkali cations (mainly sodium). The buffer pair which is formed constitutes a relatively small fraction of the total blood buffers, since only about 16 mEq. of sodium per liter are combined with protein anions.

C. A Reserve of Body Protein: Serum albumin, when administered parenterally, is effective as a source of protein in hypoproteinemic patients. It is well assimilated and is not excreted unless a proteinuria already exists. Plasma is also effective as a source of nutrient protein.

The circulating plasma protein is not static; it constantly interchanges with a labile tissue reserve equal in quantity to the circulating protein. The term "dynamic equilibrium" has been applied to this interchange. In protein starvation, the body draws upon this tissue reserve as well as upon plasma protein for its metabolic needs.

D. Other Functions of Serum Proteins: These include the transport of lipids, including fat-soluble vitamins and steroid hormones, antibodies, and possibly certain carbohydrates as well.

Plasma Protein Changes in Disease:

The albumin component of the serum is either unchanged or, more usually, lowered in pathologic states. Albumin levels do not rise above normal except in the presence of hemoconcentration or dehydration. A decline in serum albumin levels (hypoalbuminemia) follows prolonged malnutrition or the loss of protein either in the urine as in the nephrotic syndrome, or by extravasation as in burns. Inability to synthesize albumin is a prominent feature of chronic liver disease (cirrhosis); thus hypoalbuminemia is a characteristic finding in this type of liver disease.

Changes in the globulin fractions in disease usually involve increases in many fractions. The gamma globulins will be expected to rise as a result of antigenic stimulation; thus infectious processes usually produce a rise in this component.

The alterations which occur in the α and β components are also of considerable interest in disease. An increase in α globulins, particularly in the glycoproteins and mucoproteins, is a noteworthy feature of acute febrile disease. This seems to be related to inflammation or tissue destruction. It is also noted in moderate to advanced tuberculosis and in advanced carcinoma where tissue wasting is also occurring. In many diseases there is a constant association between decreased albumin and increased α globulin; nephrosis and cirrhosis are examples; acute infections such as pneumonia, acute rheumatic fever, and typhus are other examples.

As noted above, the level of mucoproteins and glycoproteins associated with the α-globulins of the serum is increased above normal in many diseases. However, it has been reported that the mucoprotein levels fall below normal in most patients with infectious or homologous serum hepatitis or with portal cirrhosis[12]. In normal men the mucoprotein (M) levels ranged from 48 to 75 mg./100 ml. serum; in women, from 40 to 70 mg. In a high percentage of cases of hepatitis or of portal cirrhosis, subnormal M levels were found. In contrast, in only three of 125 patients with obstructive biliary tract disease were the M levels low. Even in advanced biliary cirrhosis, M levels did not fall. These authors suggest this test as an additional aid in the differential diagnosis of jaundice.

An abnormal constituent in the α-globulin fraction of human serum is the so-called **C-reactive protein** (CRP)[13]. This protein is formed by the body in response to an inflammatory reaction. It is called C-reactive protein because it forms a precipitate with the somatic C-polysaccharide of the pneumococcus. Small amounts of this protein may be detected in human serum by a precipitin test using a specific antiserum from rabbits hyperimmunized with purified C-reactive protein.

Increases in β globulins are often associated with accumulations of lipids. The relation of the lipoproteins to the genesis of atherosclerosis is under study at the present time. Changes in the lipoproteins of the β globulin component of the serum are responsible for positive thymol turbidity tests (see p. 281). This test is positive not only in hepatitis and cirrhosis but also in many other diseases in which increased serum lipids are a feature.

Multiple myeloma is characterized by hyperproteinemia, which is caused by extremely high globulin levels. Smith et al.[14] state that individual serum globulins from different patients with myeloma may differ markedly from one another. The electrophoretic pattern of the serum is usually normal except for a superimposed protein boundary which generally possesses the mobility of a γ or a β globulin, although there have been a few reports of excessive α globulins. These authors have concluded that the serum globulins in myeloma are not abnormal globulins but probably individual normal γ globulins which arise in abnormal amounts because of excessive proliferation of plasma cells. A similar hypothesis was advanced by Putnam and Udin[15], who believe that a disturbance in protein synthesis occurs in myeloma which results in the massive production of one globulin randomly selected from the family of normal globulins.

The proteins described above as occurring in large amounts in the serum of the myeloma patient have molecular weights in excess of 160,000. However, in about 30% of cases of this disease an additional group of peculiar proteins of lower molecular weights (24,000 to 90,000) may be detected in the blood and, because of their low molecular weights, may be excreted into the urine. These are the so-called Bence Jones proteins (see p. 304). Bence Jones proteins occur only in multiple myeloma and in a few cases of myeloid leukemia or other diseases which extensively involve the bone marrow. It has now been established that they are not degradation products of serum or tissue proteins but are synthesized from free amino acids. The Bence Jones proteins seem to be components of larger normal protein molecules which are produced as a result of disorders of protein synthesis. According to Deutsch[16], it is possible that in the course of accelerated synthesis of certain globulins a step in the normal synthetic mechanism is missing. Such an impairment in the synthetic pathway may result in the production of only a fragment of the large protein molecule which would normally result. Analyses of individual Bence Jones proteins have often indicated that the amino acid, methionine, is present in low amounts or not at all. It is possible that the synthetic defect may be caused by failure to incorporate methionine adequately under conditions of accelerated globulin synthesis.

The Bence Jones proteins formed in any given interval are almost entirely excreted within 12 hours. As a result, a patient exhibiting Bence Jones proteinuria may excrete as much as one-half his daily nitrogen intake as Bence Jones protein.

Congenital Deficiencies of Plasma Protein Fractions[17]:

A. Thromboplastic Factors: The best example of a disease entity which is caused by a congenital lack of an essential factor in the clotting process is classical hemophilia. This hemorrhagic disease is due to a deficiency in the plasma content of antihemophilic globulin, a component of Cohn Fraction I (see p. 131). The defect is believed to be inherited as a sex-linked recessive trait, transmitted exclusively through the females of an affected family although the females do not exhibit the bleeding tendency. The males, on the other hand, exhibit the disease but do not transmit the defect. The abnormality is characterized by a marked prolongation of the coagulation time of the blood with no abnormality in the prothrombin time. When antihemophilic globulin is added to hemophilic blood, clotting of such blood becomes normal. In the presence of hemorrhage, the hemophilic patient may be treated by injections of Cohn Fraction I or by transfusions of fresh blood or plasma from normal donors.

Other hemophilioid states have also been discovered (see p. 125), although these are less common than classical hemophilia. Each is due to a specific deficiency of a plasma thromboplastic factor.

Congenital as well as acquired deficiencies of the accelerator factors involved in conversion of prothrombin to thrombin (see p. 125) have also been described.

B. Afibrinogenemia and Fibrinogenopenia: Afibrinogenemia is another congenital hemorrhagic disease which in its clinical manifestations superficially resembles hemophilia. It is a hereditary disease which is characterized by the absence or near-absence of fibrinogen. It is transmitted as a non-sex-linked recessive trait, although it occurs slightly more frequently in males. In a typical case, the clotting time and the prothrombin time are prolonged indefinitely. All of the clotting factors of the blood other than fibrinogen are present in normal amounts. In case of injury, death will occur from uncontrollable hemorrhage unless fibrinogen is supplied. Administered fibrinogen is lost from the body by normal decay in 12 to 21 days; therefore, the protein must be replaced every ten to 14 days in quantities sufficient to maintain the fibrinogen level above 50 mg. per 100 ml. of plasma.

Fibrinogenopenia and afibrinogenemia may also be acquired, most commonly as a complication of pregnancy where a long-standing intrauterine death of the fetus has occurred or where there has been premature separation of the placenta or the occurrence of amniotic fluid embolism following administration of pitocin. It is thought that acute depletion of fibrinogen is brought about by release of thromboplastin-like substances from placenta and amniotic fluid, which results in extensive intravascular clotting and defibrination of the blood.

C. Agammaglobulinemia and Hypogammaglobulinemia: Another apparently sex-linked recessive factor is involved in the congenital transmission of a defect in plasma protein production which is characterized by the complete or near-complete absence of γ globulin from the serum. The cases reported so far have all been among males. Acquired forms of this disease have also been found; these occur in both sexes.

Patients afflicted with this disorder of protein formation exhibit a greatly increased susceptibility to bacterial infection and an absence of γ globulin from the serum and of circulating antibodies in the blood and tissues. Furthermore, there is a complete failure to produce antibodies in response to antigenic stimulation. Either the zinc turbidity test or a direct turbidimetric measurement of γ globulin (see pp. 281 and 282) may be used to detect these cases. Injections of γ globulin may be used to aid in controlling bacterial infections in these patients. Peculiarly, their resistance to viral diseases seems normal.

In a study of several patients with agammaglobulinemia, some acquired and others congenital in origin, no abnormalities in the plasma clotting factors were found[18]. This indicates that none of the factors involved in clotting mechanisms are gamma globulins. It is probable that congenital agammaglobulinemia is the result of an isolated deficiency of protein synthesis resulting from a lack of a single enzyme system. Furthermore, the deficiency of gamma globulin synthesis in these patients does not involve the liver but depends rather upon an anomaly of protein metabolism existing elsewhere in the reticuloendothelial system.

It is reported that in hypogammaglobulinemia there is a disturbance in the architecture of the lymphoid follicles, a lack of plasma cells, and a failure to form plasma cells after antigenic stimulation. There may also be a deficiency of at least two β globulins which are immunochemically unrelated to γ globulin.

HEMOGLOBIN

The red coloring matter of the blood is a conjugated protein, hemoglobin. The normal concentration of hemoglobin is 14 to 16 Gm. per 100 ml. of blood, all confined to the red cell. It is estimated that there are about 750 Gm. of hemoglobin in the total circulating blood of a 70 Kg. man and that about 6.25 Gm. (90 mg./Kg.) are produced and destroyed each day.

Dilute acid will readily hydrolyze hemoglobin into the protein, globin, which belongs to the class of proteins known as histones (see p. 26), and its prosthetic group, **heme** (hematin). Crystals of the hydrochloride of heme, **hemin**, can be easily prepared and its chemical structure then determined. Heme is an iron porphyrin. The formation of the porphyrins and of heme is described in Chapter 5 on p. 57.

The average life of a red blood cell in the body is about 120 days. When the red cells are destroyed the porphyrin moiety of hemoglobin is broken down to form the bile pigments, biliverdin and bilirubin, which are carried to the liver for excretion into the intestine by way of the bile. The details of this process are described in Chapter 5 on p. 64.

The most characteristic property of hemoglobin is its ability to combine with oxygen to form **oxyhemoglobin**. This combination is reversed merely by exposing the oxyhemoglobin to a low oxygen tension. The absorption spectra which are obtained when white light is passed through hemoglobin solutions or closely related derivatives are useful in distinguishing these compounds from one another. Oxyhemoglobin or diluted arterial blood shows three absorption bands: a narrow band of light absorption at a wavelength of $\lambda = 578\ m\mu$, a wider band at $\lambda = 542\ m\mu$, and a third with its center at $\lambda = 415\ m\mu$ in the extreme violet end of the spectrum. Reduced hemoglobin, on the other hand, shows only one broad band with its center at $\lambda = 559\ m\mu$.

When blood is treated with either ozone, potassium permanganate, potassium ferricyanide, chlorates, nitrites, nitrobenzene, pyrogallic acid, acetanilid, or certain other substances, methemoglobin is formed. In this compound the iron, which is in the ferrous (Fe^{++}) state in hemoglobin, is oxidized to the ferric (Fe^{+++}) state. In acid solution, methemoglobin has one absorption band with its center at $\lambda = 634\ m\mu$.

Carbon monoxide combines with hemoglobin even more readily than does oxygen. The carbon monoxide hemoglobin shows two absorption bands, the middle of the first at $\lambda = 570\ m\mu$ and the second at $\lambda = 542\ m\mu$. Combinations of hemoglobin with hydrogen sulfide or hydrocyanic acid also give characteristic absorption spectra. This is a valuable means of detecting these compounds in the blood of individuals suspected of having been poisoned with H_2S or HCN.

Abnormal Hemoglobins:

The red blood cell of an individual may contain two or three different molecular species of hemoglobin. The fetus of a given animal species produces a hemoglobin which is different from that of the adult of the same species; in certain anemic states it has been observed that the anemic patient may still be producing this fetal (hemoglobin F) type of hemoglobin at an age when it has disappeared from normal adult individuals.

The mechanism for the synthesis of hemoglobin is inherited from each parent in a manner similar to that of other genetically controlled characteristics. Most individuals have inherited a normal mechanism from each parent, and their red blood cells contain only normal adult (A) hemoglobin, except for the first few months of postnatal life, when some fetal hemoglobin is still present.

The identification of some of the abnormal hemoglobins in the blood may be accomplished by the use of paper electrophoresis since there is a variation in the mobilities of the various hemoglobins on the paper strip. The electrophoretic identification of a number of hemoglobins is illustrated on p. 138. The presence of abnormal hemoglobins in the blood is often associated with abnormalities in red cell morphology and definite clinical manifestations. Each abnormality appears to be transmitted as a Mendelian-recessive characteristic; if heterozygous, i.e., inherited from only one parent, and associated with normal hemoglobin inherited from the other parent, the patient has only a so-called "trait" (e.g., sickle-cell trait) and may be free of clinical findings, but the presence of the abnormality can be detected electrophoretically.

At the present time, in addition to hemoglobin F, several other abnormal hemoglobins have been described, usually in association with various atypical anemias. These are designated hemoglobins S (sickle cell), C, D, E, G, H, I, J, and K[19,20]. Hemoglobins H, I, and J, when analyzed electrophoretically at pH 8.6, exhibit a migration faster than that of hemoglobin A. This is in contrast to the other abnormal hemoglobins, which have slower mobilities as noted on p. 138.

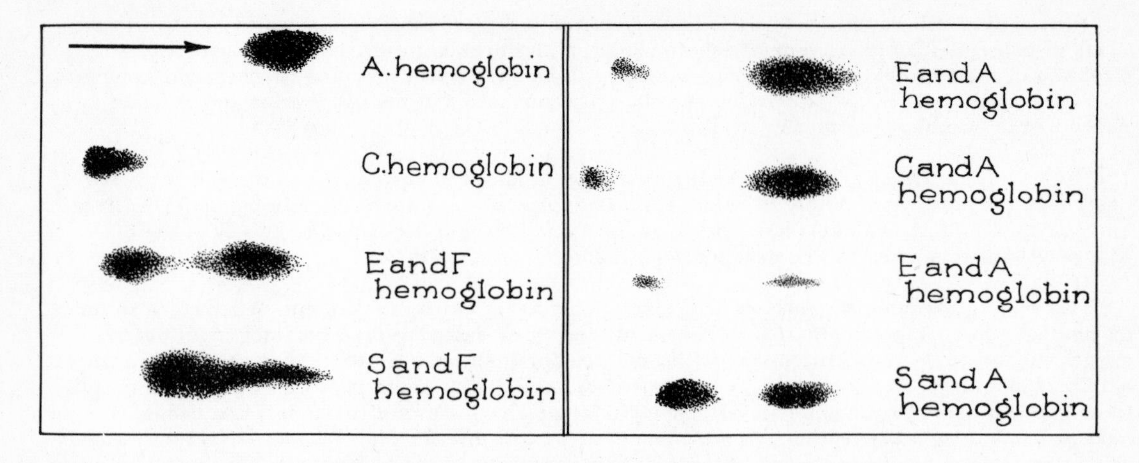

Paper Electrophoresis of a Number of Hemoglobin Specimens Containing Various Types of Hemoglobins. (Redrawn from Chernoff and Minnich, "Hemoglobin E, a Hereditary Abnormality of Human Hemoglobin," Science **120**:605, 1954.)

The appearance of sickle cell instead of normal hemoglobin results from a single mutation. Ingram[21] has shown that hemoglobin from patients with homozygous (SS) sickle cell disease differs from normal hemoglobin simply by replacement of one glutamic acid residue by one valine residue in the polypeptide chain. This is the only difference among 300 amino acid molecules in the polypeptide chain.

Discussions of the abnormal hemoglobins may be found in reviews by Itano[22] and by Chernoff[23].

THE ANEMIAS

The concentration of hemoglobin in the blood may be measured by a number of methods, including iron analysis, oxygen combining power, and, most commonly, by the color of the acid heme formed when a measured quantity of blood is treated with acid or alkali. **Anemia** exists when the hemoglobin content of the blood falls below normal.

Anemia may result from a decreased rate of production or from an increased loss or destruction of red blood cells. This may occur in acute or chronic hemorrhages, or may be produced by toxic factors (poisons or infections) which cause hemolysis and increased erythrocyte destruction. Decreased production of blood may be due to destruction or loss of function of the blood-forming tissue, as in the leukemias, Hodgkin's disease, multiple myeloma, and aplastic anemia. Certain drugs (benzene, gold salts, arsphenamine), chronic infections, and radiations may also lead to severe anemias because of their destructive or suppressive effect on erythrogenic tissue.

Failure of erythrocyte production may also be caused by a lack of iron and protein in the diet. These nutritional or hypochromic anemias are common in infancy and childhood as well as in pregnancy and in chronic blood loss where the iron and/or protein intake is inadequate.

Pernicious anemia is due to a failure in red cell production occasioned by a lack of a factor (or factors) necessary to erythrocyte maturation. In uncomplicated pernicious anemia this deficiency is completely corrected by vitamin B_{12} (see p. 95). According to Castle, two factors are necessary for the formation of red blood cells: (1) the **extrinsic factor** found in meat, yeast, liver, rice-polishings, eggs, and milk; and (2) the **intrinsic factor**, produced by the gastric mucosa and possibly also by the duodenal mucosa. Vitamin B_{12} is the extrinsic factor. It is now apparent that the function of the intrinsic factor is to assure absorption of vitamin B_{12} from the intestine. It has no direct effect on erythrocyte production, since parenterally administered vitamin B_{12} is itself sufficient to correct pernicious anemia (see p. 96).

Anemias are sometimes classified according to the predominating size of the erythrocytes, e.g., macrocytic (large cell), microcytic (small cell), or normocytic (no significant alteration).

There are three red cell indices which are useful in the differential diagnosis of the anemias:

1. Mean corpuscular volume (MCV) - The range of normal MCV is 80-94 cu.μ; average, 87 cu.μ. MCV is calculated from the packed cell volume (PCV) and red blood count. Example: PCV = 45%. RBC = 5,340,000/cu.mm.

$$\frac{\text{Vol. r.b.c. in ml./liter blood}}{\text{RBC} \times 10^6\text{/cu.mm. blood}} = \frac{450}{5.34} = 84.2 \text{ cu.}\mu$$

2. Mean corpuscular hemoglobin (MCH) - This expresses the amount of hemoglobin per red blood cell. It is reported in micro-micrograms ($\mu\mu$g.). The normal range is 27-32 $\mu\mu$g.; average, 29.5 $\mu\mu$g. For children, the range is 20-27 $\mu\mu$g. MCH is calculated from the red blood count and the hemoglobin concentration. Example: Hemoglobin = 15.6 Gm./100 ml. RBC = 5,340,000.

$$\frac{\text{Hemoglobin in Gm./liter blood}}{\text{RBC} \times 10^6\text{/cu.mm. blood}} = \frac{156}{5.34} = 29.2\ \mu\mu\text{g.}$$

3. Mean corpuscular hemoglobin concentration (MCHC) - This is the amount of hemoglobin expressed as percent of the volume of a red blood cell. The normal range is 33-38%; average, 35%. It is calculated from the hemoglobin concentration and the packed cell volume. Example: Hemoglobin = 15.6 Gm./100 ml. blood. PCV = 45%.

$$\frac{\text{Hemoglobin in Gm./liter blood}}{\text{Packed Cell Volume}} \times 100 = \frac{15.6}{45} \times 100 = 34.7\%$$

The characteristic values for the above indices in three types of anemia are shown below:

Characteristic Values in Three Types of Anemia*

Type of Anemia	M.C.V. (cu. μ)	M.C.H. ($\mu\mu$g.)	M.C.H.C. (Concen. %)
Normal Blood†	84-95	28-32	33-38
Macrocytic	95-160	30-52	31-38
Microcytic	72-79	22-26	31-38
Hypochromic	50-71	14-21	21-29

* Modified and reproduced, with permission, from Krupp et al., Physician's Handbook, 10th Ed., 1958. Lange Medical Publications, Los Altos, California.

†Or normocytic anemia.

BLOOD CHEMISTRY

Determination of the content of various compounds in the blood is of increasing importance in the diagnosis and treatment of disease. The blood not only reflects the overall metabolism of the tissues but affords the most accessible method for the sampling of body fluids.

Many of the methods of blood chemistry require the preparation of a protein-free filtrate which is then analyzed for those constituents which remain. The most common technic for the preparation of a protein-free filtrate is the method of Folin and Wu, which utilizes sodium tungstate and sulfuric acid to make tungstic acid for the precipitation of the plasma proteins. Other protein precipitants include trichloroacetic acid or picric acid and a mixture of sodium hydroxide and zinc sulfate (the Somogyi precipitants). Urea, nonprotein nitrogen, uric acid, sugar, and

BLOOD, PLASMA, OR SERUM VALUES*

Determination	Material Analyzed	Amt. Required† (Fasting State)	Normal Values (Values vary with procedure used.)
Acetone bodies	Plasma	2 ml.	0.8-5.5 mg./100 ml.
Amino acid nitrogen	Plasma	2 ml. fasting	3.0-5.5 mg./100 ml.
Ammonia	Blood	2 ml.	40-70 mcg./100 ml.
Amylase	Serum	2 ml.	80-180 Somogyi units/100 ml.
Ascorbic acid	Plasma White cells (of whole blood)	1 ml. fasting 10 ml. fasting	0.4-1.5 mg./100 ml. (fasting) 25-40 mg./100 ml.
Base, total serum	Serum	2 ml.	145-160 mEq./L.
Bilirubin	Serum	2 ml.	Direct, 0.1-0.4 mg./100 ml. Indirect, 0.2-0.7 mg./100 ml.
Calcium	Serum	2 ml. fasting	9.0-11.0 mg./100 ml.; 4.5-5.5 mEq./L. (varies with protein concentration)
Carbon dioxide Content Combining power	Serum or plasma	1 ml.	24-28 mEq./L. 55-65 Vol. % 55-75 Vol. %
Carotenoids	Serum	2 ml. fasting	100-300 I. U./100 ml.
Vitamin A	Serum	2 ml. fasting	40-100 I. U./100 ml.; 40-60 mcg./100 ml.
Chloride	Serum	1 ml.	100-106 mEq./L.; 350-370 mg./100 ml. (as chloride)
Cholesterol	Serum	1 ml.	150-280 mg./100 ml.
Cholesterol esters	Serum	1 ml.	65% of total cholesterol
Copper	Serum	5 ml.	130-230 mcg./100 ml.
Creatinine	Blood or serum	1 ml.	1-2 mg./100 ml.
Glucose (Folin) Glucose (true) Glucose tolerance	Blood Blood Blood	0.1-1.0 ml. fasting 0.1-1.0 ml. fasting (See p. 173)	80-120 mg./100 ml. (fasting) 60-100 mg./100 ml.
Hemoglobin	Blood	0.05 ml.	12-16 Gm./100 ml.
Iodine, protein-bound (thyroid hormone)	Serum	5 ml.	4-8 mcg./100 ml.
Iron	Serum	5 ml.	50-200 mcg./100 ml.
Lipase	Serum	2 ml.	0.2-1.5 units (ml. of 0.1 N NaOH)
Lipids, total	Serum	5 ml.	500-600 mg./100 ml.
Magnesium	Serum	2 ml.	1-2 mEq./L.; 1-3 mg./100 ml.
Nonprotein nitrogen‡	Serum or whole blood	1 ml.	15-35 mg./100 ml.
Oxygen Capacity	Blood	5 ml.	19-22 Vol. % (varies with Hgb. concentration)
Arterial content	Blood	5 ml.	18-21 Vol. % (varies with Hgb. content)
Arterial % saturation	-		94-96% of capacity
Venous content	Blood	5 ml.	10-16 Vol. %
Venous % saturation	-		60-85% of capacity
pH (reaction)	Arterial plasma	1 ml.	7.35-7.45
Phosphatase, acid	Serum	2 ml.	1-5 units (King-Armstrong); 0.5-2.0 units (Bodansky); 0.5-2.0 units (Gutman); 0.0-1.1 units (Shinowara)
Phosphatase, alkaline	Serum	2 ml.	5-13 units (King-Armstrong); 2.0-4.5 units (Bodansky); 3.0-10.0 units (Gutman); 2.2-8.6 units (Shinowara)
Phosphorus, inorganic	Serum	1 ml. fasting	3.0-4.5 mg./100 ml. (children, 4-7 mg.); 0.9-1.5 mm./L.
Phospholipid	Serum	2 ml.	145-200 mg./100 ml.
Potassium	Serum	1 ml.	3.5-5.0 mEq./L.; 14-20 mg./100 ml.
Protein Total	Serum	1 ml.	6.0-8.0 Gm./100 ml.
Albumin§	Serum	1 ml.	4.5-5.5 Gm./100 ml.
Globulin§	Serum		1.5-3.0 Gm./100 ml.
Fibrinogen	Plasma	1 ml.	0.2-0.6 Gm./100 ml.
Prothrombin clotting time	Plasma	2 ml.	By control.
Pyruvic acid	Blood	2 ml.	0.7-1.2 mg./100 ml.
Sodium	Serum	1 ml.	136-145 mEq./L.; 310-340 mg./100 ml.
Sulfate	Plasma or serum	2 ml.	0.3-1.0 mEq./L.
Urea nitrogen‡	Serum or whole blood	1 ml.	10-20 mg./100 ml.
Uric acid	Serum	1 ml.	3-6 mg./100 ml.
Specific gravity	Blood	0.1 ml.	1.056 (varies with Hgb. and protein concentration)
	Serum	0.1 ml.	1.0254-1.0288 (varies with protein concentration)

Blood volume (Evans blue dye method): Adults, 2990-6980 ml. (Women, 46.3-85.5 ml./Kg.; Men, 66.2-99.7 ml./Kg.)

*Reproduced, with permission, from Krupp, et al., Physician's Handbook, 10th Ed., 1958. Lange Medical Publications, Los Altos, California.

†Minimum amount required for any procedure.

‡Do not use anticoagulant containing ammonium oxalate.

§Albumin and globulin values obtained by use of 22% sodium sulfate; not in agreement with electrophoretic data. See p. 128.

creatinine are examples of blood constituents commonly determined in protein-free filtrates. Many other substances are determined directly, using oxalated whole blood, blood serum, or blood plasma without prior removal of protein.

The normal ranges in concentration of many of the important constituents of the blood are listed in the table on p. 140. (From Krupp, Sweet, Jawetz, and Armstrong, Physician's Handbook, 10th Ed., Lange Medical Publications, Los Altos, Californa, 1958.)

LYMPH

The lymphatic fluid is a transudate formed from the plasma by filtration through the wall of the capillary. It resembles plasma in its content of substances which can permeate the capillary wall, although there are some differences in the electrolyte concentrations. The distribution of the nonelectrolytes such as glucose and urea is about equal in plasma and lymph, but the protein concentration of the lymph is definitely lower than that of plasma.

In its broadest aspects, the term "lymph" includes not only the fluid in the lymph vessels but also the fluid which bathes the cells, the "tissue," or "interstitial fluid." The chemical composition of lymph would therefore be expected to vary with the source of the sample investigated. Thus, the fluid from the leg contains 2 to 3% protein, whereas that from the intestines contains 4 to 6% and that from the liver 6 to 8%.

The lymphatic vessels of the abdominal viscera, the "lacteals," absorb the majority of fat from the intestine. After a meal, this chylous fluid, milky-white in appearance because of its high content of neutral fat, can be readily demonstrated. Except for this high fat content, the chyle is similar in chemical composition to the lymph in other parts of the body.

CEREBROSPINAL FLUID

The cerebrospinal fluid is formed as an ultrafiltrate of the plasma by the choroid plexuses of the brain. The process is not one of simple filtration, since active secretory processes are required. The normal fluid is water-clear, with a specific gravity of 1.006 to 1.008. Normally, the protein content is low, about 25 mg. per 100 ml., with an albumin-globulin ratio of 3:1, but in disease an increase in protein, particularly in globulin, is characteristic. In inflammatory meningitis, for example, the protein may rise as high as 125 to over 1 Gm. per 100 ml.

Various diseases of the brain (neurosyphilis, encephalitis, abscess, tumor) show protein elevations of 20 to 300 mg. per 100 ml. The Pandy globulin test and the Lange colloidal gold test are diagnostic tests based on changes in the spinal fluid proteins.

The sugar in spinal fluid is somewhat less than in blood, 40 to 70 mg. per 100 ml. in the fasting adult. It is raised in encephalitis, central nervous system syphilis, abscesses, and tumors. It is decreased in purulent meningitis.

The chloride concentration is normally 720 to 750 mg. per 100 ml. (expressed as NaCl) or 123 to 128 mEq. per liter. It is generally decreased in meningitis and unchanged in syphilis, encephalitis, poliomyelitis, and other diseases of the central nervous system. The chloride is especially low in tuberculous meningitis.

The concentration of calcium in normal human cerebrospinal fluid is 2.43 mEq. per liter ± 0.05, that of magnesium, 2.40 mEq. per liter ± 0.14. Thus the calcium content of the cerebrospinal fluid is considerably less than that of the serum, whereas that of magnesium is slightly higher; the ratio of calcium to magnesium was found to be 1.01 ± 0.06.

References:

1. New and Nonofficial Remedies. J. Am. Med. Assoc. **150**:1486, 1952.
2. White, S.G., Aggeler, P.M., and Glendening, M.B.: Blood **8**:101, 1953.
3. Naeye, R.L.: Proc. Soc. Exper. Biol. and Med. **91**:101, 1956.
4. Rosenthal, R.L., Dreskin, O.H., and Rosenthal, N.: Proc. Soc. Exper. Biol. and Med. **82**:171, 1953.
5. Ramot, B., Angelopoulos, B., and Singer, K.: Arch. Int. Med. **95**:705, 1955.
6. Spaet, T.H., Aggeler, P.M., and Kinsell, B.G.: J. Clin. Invest. **33**:1095, 1954.
7. Lewis, M.L., and Ware, A.G.: Blood **9**:520, 1954.
8. Armstrong, S.H., Jr., Budka, M.J.E., and Morrison, K.C.: J. Am. Chem. Soc. **69**: 416, 1947.
9. Jencks, W.P., Smith, E.R.B., and Durrum, E.L.: Am. J. Med. **21**:387, 1956.
10. Owen, J.A.: "Paper Electrophoresis of Proteins and Protein-Bound Substances in Clinical Investigations," Advances in Clinical Chemistry, I., 238, Academic, 1958.
11. Hillyard, L.A., Entenman, C., Feinberg, H., and Chaikoff, I.L.: J. Biol. Chem. **214**: 79, 1955.
12. Greenspan, E.M., and Dreiling, D.A.: Arch. Int. Med. **91**:474, 1953.
13. Abernethy, T.J., and Avery, O.T.: J. Exper. Med. **73**:173, 1941.
14. Smith, E.L., Brown, D.M., McFadden, M.L., Buettner-Janusch, V., and Jager, B.V.: J. Biol. Chem. **216**:601, 1955.
15. Putnam, F.W., and Udin, B.: J. Biol. Chem. **202**:727, 1953.
16. Deutsch, H.F.: J. Biol. Chem. **216**:97, 1955.
17. Gitlin, D., and Janeway, C.A.: Pediatrics, 21:1034, 1958.
18. Frick, P.G., and Good, R.A.: Proc. Soc. Exper. Biol. and Med. **91**:169, 1956.
19. Thorup, O.A., Itano, H., Wheby, M., and Leavell, B.S.: Science **123**:889, 1956.
20. Battle, J.D., Jr., and Lewis, L.: J. Lab. and Clin. Med. **44**:765, 1954.
21. Ingram, V.M.: Nature **180**:326. 1957.
22. Itano, H.: Arch. Int. Med. **96**:287, 1955.
23. Chernoff, A.I.: New Eng. J. Med. **253**:322, 365, 416, 1955.

Bibliography:

Albritton, E.C., Ed.: Standard Values in Blood. Saunders, 1952.

Best, C.H., and Taylor, N.B.: Physiological Basis of Medical Practice, 6th Ed. Williams and Wilkins, 1955.

Hepler, O.E.: Manual of Clinical Laboratory Methods, 4th Ed. Thomas, 1951.

Peters, J.P., and Van Slyke, D.D.: Quantitative Clinical Chemistry, Vol. II, Methods. Williams and Wilkins, 1932. Reprinted 1956.

Roughton, F.J.W., and Kendrew, J.C., Eds.: Haemoglobin. Interscience, 1949.

10 . . .

The Chemistry of Respiration

PHYSICAL EXCHANGE OF GASES

The term ''respiration'' is applied to the interchange of the two gases, oxygen and carbon dioxide, between the body and its environment.

Composition of Atmospheric Air:

The atmospheric air which we inhale has the following composition: oxygen, 20.96%; carbon dioxide, 0.04%; and nitrogen, 79%. Other gases are present in trace amounts but are not of physiologic importance.

Composition of Expired Air:

The expired air contains the same amount of nitrogen as the inspired air, but the oxygen has been reduced to about 15% and the carbon dioxide increased to about 5%. About one-fourth of the oxygen of the inspired air has passed into the blood and has been replaced in the expired air by an equal amount of carbon dioxide which has left the blood.

Partial Pressure of Gases:

In the mixture of gases in air, each gas exerts its own partial pressure. For example, the partial pressure of oxygen at sea level would be 20% of the total pressure of 760 mm. Hg; i.e., the partial pressure of oxygen (pO_2) = 760 × 0.20 = 152 mm. Hg. In the alveoli of the lung the oxygen content is 15%. The total pressure after correction for the vapor pressure of water in the alveolar air (47 mm. Hg at 37° C.) is 760 − 47 = 713 mm. Hg. The partial pressure of oxygen in the lung is therefore about 107 mm. Hg (713 × 0.15 = 107); that of carbon dioxide, 36 mm. Hg (713 × 0.05 = 36).

Diffusion of Gases in the Lungs:

When the gases of the inspired air come in contact with the alveolar membrane of the lung, it is assumed that the exchange of gases takes place in accordance with the usual physical laws of diffusion. Thus, the gas passes through the membrane and into the blood, or in the reverse direction, in accordance with the difference in the pressure of that particular gas on either side of the membrane. The gas pressures in the blood are usually expressed as gas ''tensions''; for example, the CO_2 ''tension'' (pCO_2) is the pressure of the dry gas (mm. Hg) with which the dissolved carbonic acid in the blood is in equilibrium; similarly pO_2 (oxygen tension) is the pressure of the dry gas with which the dissolved oxygen in the blood is in equilibrium.

The exchange of gases between the alveoli and the blood is illustrated by the following:

Oxygen tension in alveolar air - 107 mm. Hg
Oxygen tension in venous blood - 40 mm. Hg

A pressure difference of 67 mm. Hg serves to drive oxygen from the alveoli of the lung into the blood.

Carbon dioxide tension in alveolar air - 36 mm. Hg
Carbon dioxide tension in venous blood - 46 mm. Hg

A relatively small difference of 10 mm. Hg is sufficient to drive carbon dioxide from the blood into the lung. This small difference in pressure is adequate because of the rapidity of the diffusion of carbon dioxide through the alveolar membrane. In the resting state, a difference of as little as 0.12 mm. Hg in carbon dioxide tension will still provide for the elimination of this gas.

The tension of nitrogen is essentially the same in both venous blood and lung alveoli (570 mm. Hg). This gas is therefore physiologically inert.

After this exchange of gases has occurred, the blood becomes arterial (in a chemical sense). Arterial blood has an oxygen tension of about 100 mm. Hg and a carbon dioxide tension of 40 mm. Hg. The nitrogen tension is, of course, unchanged (570 mm. Hg). These gases are dissolved in the blood in simple physical solution, and the quantity of each gas which might be carried in the blood in this manner can be calculated according to Henry's law from their absorption coefficients.

It is of interest to compare the quantities of each of these gases which could be dissolved (under physiologic conditions of temperature and pressure) with the actual quantities found in the blood.

Comparison of the Calculated Content with the Actual Content of Oxygen, Carbon Dioxide, and Nitrogen in the Blood

	O_2 ml./100 ml.	CO_2 ml./100 ml.	N_2 ml./100 ml.
Calculated content in blood	0.393	2.96	1.04
Actually present - arterial blood	20.0	50.0	1.70
Actually present - venous blood	14.0	56.0	1.70

It is apparent that considerable quantities of oxygen and carbon dioxide are carried in the blood in other than simple solution. The mechanisms by which these increased amounts of oxygen and carbon dioxide are transported will now be investigated.

THE TRANSPORT OF OXYGEN BY THE BLOOD

Function of Hemoglobin:

The transportation of oxygen by the blood from the lungs to the tissues is due mainly to the ability of hemoglobin to combine reversibly with oxygen. This may be represented by the equation:

$$Hb + O_2 \rightleftharpoons HbO_2$$

(Hb = reduced hemoglobin; HbO_2 = oxyhemoglobin)

The combination of hemoglobin and oxygen is best thought of as a loose affinity rather than as a chemical combination such as an oxide. The degree of combination or of its reversal, i.e., dissociation of oxyhemoglobin to release oxygen, is determined by the tension of the oxygen in the medium surrounding the hemoglobin. At a tension of 100 mm. Hg or more, hemoglobin is completely saturated. Under these conditions, approximately 1.34 ml. of oxygen are combined with each gram of hemoglobin. Assuming a hemoglobin concentration of 14.5 Gm. per 100 ml. of blood, the total oxygen which would be carried as oxyhemoglobin would be 14.5×1.34, or 19.43 ml. per 100 ml. (19.43 Vol. %). To this may be added the 0.393 ml. physically dissolved; the total, approximately 20 Vol. %, is the oxygen capacity of blood which contains 14.5 Gm. of hemoglobin per 100 ml. It is apparent that the oxygen carrying power of the blood (the oxygen content) is largely a function of the hemoglobin (red cell) concentration.

Dissociation of Oxyhemoglobin:

The important relationship between the saturation of hemoglobin and the oxygen tension may be perceived by an examination of the dissociation curve of oxyhemoglobin, in which the percent saturation is plotted against the oxygen tension (see below). The shape of the dissociation curve varies with the tension of CO_2. The curve drawn with CO_2 at a tension of 40 mm. Hg is to be considered as representative of the normal physiologic condition. It will be noted that at the oxygen tension which exists in arterial blood (100 mm. Hg), the hemoglobin is 95 to 98% saturated; that is, almost complete formation of oxyhemoglobin has occurred. A further increase in oxygen tension has only a slight effect on the saturation of hemoglobin.

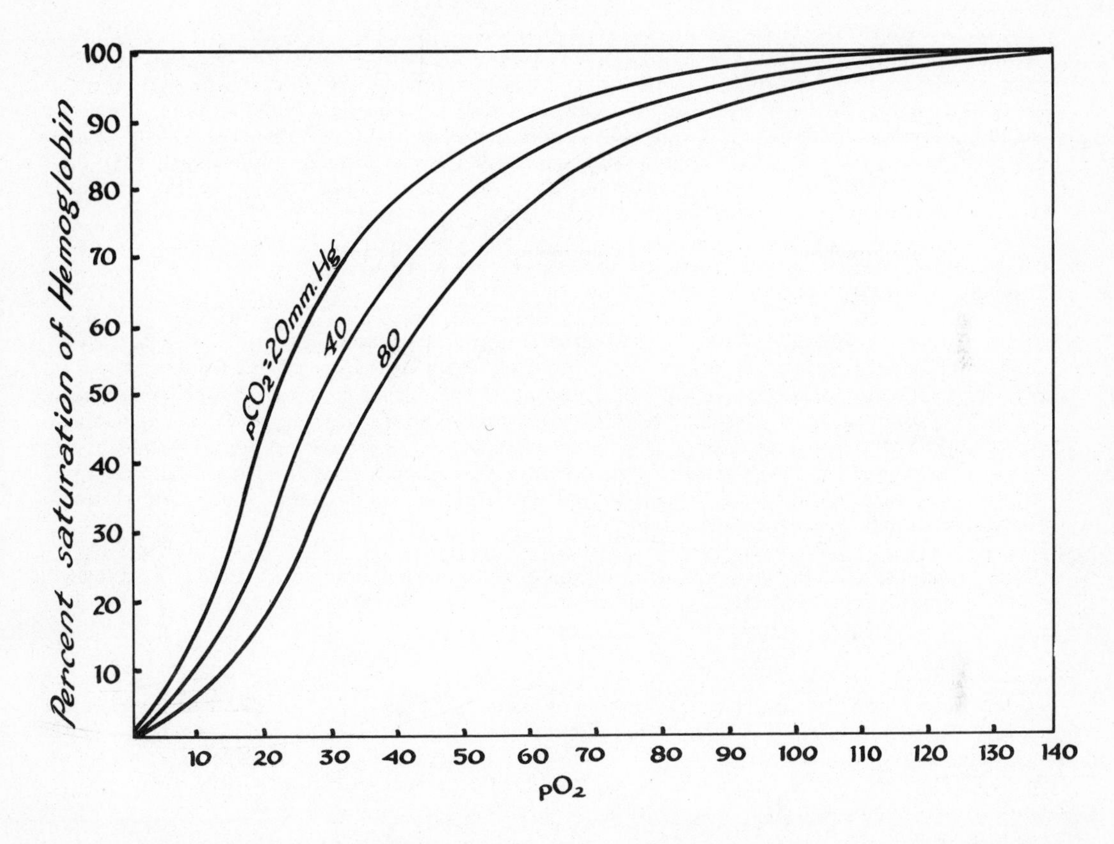

The Dissociation Curves of Hemoglobin at 38° C. and at Partial Pressures of Carbon Dioxide Equal to 20, 40 and 80 mm. Hg. (Redrawn from Davenport, the ABC of Acid-Base Chemistry, The University of Chicago Press.)

As the oxygen tension falls, the saturation of hemoglobin declines slowly until the oxygen tension drops to about 50 mm. Hg, at which point a rapid evolution of oxygen occurs. This is the "unloading tension" of hemoglobin. This initial lag in dissociation of oxyhemoglobin provides a fairly wide margin of safety which permits the oxygen tension in the lung to fall as low as 80 mm. Hg before any decrease in the oxygenation of hemoglobin occurs.

In the tissues, where the oxygen tension is about 40 mm. Hg (approximately the unloading tension of hemoglobin), oxyhemoglobin dissociates and oxygen is readily made available to the cells. In the course of a single passage of the blood through the tissues, the oxygen content of the blood falls only from 20 to about 15 Vol. %. This provides a considerable reserve of oxygenated blood in the event of inadequate oxygenation at the lung.

Clinical Signs of Variation in Hemoglobin Saturation:

The red color of reduced hemoglobin is darker than the bright red of oxyhemoglobin. For this reason arterial blood is always brighter than venous blood. A decrease in normal oxygenation of the blood, with a consequent increase in reduced hemoglobin, gives a characteristic bluish appearance to the skin. This is spoken of as cyanosis. It is characteristic of cyanide poisoning, where respiration is also impaired. A cyanotic appearance is dependent on the presence of at least 5 Gm. of **reduced hemoglobin** per 100 ml. of capillary blood (Lundsgaard). In severe anemia, the concentration of hemoglobin may be so low as to make cyanosis impossible even though the oxygen content of the blood is reduced. In carbon monoxide poisoning, the formation of the cherry-red carbon monoxide–hemoglobin produces a very characteristic ruddy appearance, particularly noticeable in the lips.

Several Factors Affect the Dissociation of Oxyhemoglobin:

A. Temperature: A rise in temperature decreases hemoglobin saturation. For example, at an oxygen tension of 100 mm. Hg, hemoglobin is 93% saturated at 38° C. but 98% saturated at 25° C. If the saturation of hemoglobin is measured at 10 mm. Hg oxygen tension, these differences are even greater; at 25° C., hemoglobin is still 88% saturated, whereas at 37° C. it is only 56% saturated. This last observation is of physiologic interest since it indicates that in the warm-blooded animal hemoglobin gives up oxygen more readily when passing from high to low oxygen tensions (as from lungs to tissues) than it does in cold-blooded animals.

B. Electrolytes: At low oxygen tensions, oxyhemoglobin gives up oxygen more readily in the presence of electrolytes than it does in pure solution.

C. Effect of Carbon Dioxide: This effect is illustrated in the graph on p. 145, in which the curves for percentage saturation of hemoglobin at various tensions of oxygen are shown to vary with different tensions of CO_2. It is probable that the influence of CO_2 on the shape of the dissociation curve is actually the effect of carbonic acid formation, with consequent lowering of the pH of the environment. The increase in acidity, by altering the pH of the medium to the acid side of the iso-electric point of hemoglobin, apparently facilitates the dissociation of oxyhemoglobin. The ability of CO_2 to shift the slope of the oxyhemoglobin dissociation curve to the right is known as the **Bohr effect.**

Under physiologic circumstances, this action of the electrolytes and of carbon dioxide on the liberation of oxygen from oxyhemoglobin is important in the delivery of oxygen to the tissues.

Carboxyhemoglobin:

As has been noted, hemoglobin combines with carbon monoxide even more readily than with oxygen (210 times as fast) to form cherry-red carboxyhemoglobin. This reduces the amount of hemoglobin available to carry oxygen. When the carbon monoxide in the inspired air is as low as 0.02%, headache and nausea occur. If the carbon monoxide concentration is only 1/210 that of oxygen in the air (approximately 0.1% carbon monoxide), unconsciousness will occur in one hour and death in four hours.

THE TRANSPORT OF CARBON DIOXIDE IN THE BLOOD

Carbon dioxide is carried by the blood both in the cells and in the plasma. The carbon dioxide content of the arterial blood is from 50-53 Vol. %, and in venous blood it is from 54-60 Vol. % (ml. carbon dioxide per 100 ml. = volumes percent). In accordance with the data on the solubility of carbon dioxide, 100 ml. of blood at 37° C., exposed to a carbon dioxide tension of 40 mm. Hg - as it is in alveolar air or arterial blood, for example - would dissolve only about 2.9 ml. It is obvious that, as was shown for oxygen, the large majority of the blood carbon dioxide is not physically dissolved in the plasma but must exist in other forms. These comprise three main fractions: (1) a small amount of carbonic acid; (2) the "carbamino-bound" carbon dioxide, which is transported in combination with proteins (mainly hemoglobin); and (3) that carried as bicarbonate in combination with the cations, sodium or potassium.

The carbamino-bound carbon dioxide, although it constitutes only about 20% of the total blood CO_2, is important in the exchange of this gas because of the relatively high rate of the reaction:

$$Hb.NH_2 + CO_2 \rightleftharpoons HbNHCOOH$$

where $Hb.NH_2$ represents a free amino group of hemoglobin (or other blood protein) which is capable of combination with carbon dioxide to form the carbamino compound.

The amount of carbon dioxide physically dissolved in the blood is not large, but it is important because any change in its concentration will cause the following equilibrium to shift:

$$\uparrow CO_2 + H_2O \rightleftharpoons H_2CO_3 \rightleftharpoons H^+ + HCO_3^-$$

Carbonic Anhydrase:

The rate at which equilibrium in the above reaction is attained is almost 100 times too slow to account for the amount of carbon dioxide which is eliminated from the blood in the one second allowed for passage through the pulmonary capillaries. Nevertheless, about 70% of the carbon dioxide is derived from that fraction which is carried as bicarbonates in the blood. This apparent inconsistency is explained by the action of an enzyme, **carbonic anhydrase,** which is associated with the hemoglobin in the red cells (never in the plasma). The enzyme has been isolated in highly purified form and shown to be a zinc protein complex. It specifically catalyzes the removal of carbon dioxide from H_2CO_3. The reaction is, however, reversible. At the tissues, the formation of H_2CO_3 from CO_2 and H_2O is also accelerated by carbonic anhydrase. Small amounts of carbonic anhydrase are also found in muscle tissue, in the pancreas, and in spermatozoa. Much larger quantities occur in the parietal cells of the stomach, where, it is believed, the enzyme is involved in the secretion of hydrochloric acid. Carbonic anhydrase also occurs in the tubules of the kidney; here its function is also involved in hydrogen ion secretion (see p. 293).

Effect of Carbon Dioxide on Blood pH:

Although it is true that the CO_2 evolved from the tissues will form carbonic acid, as shown in the above reaction, very little CO_2 can actually be carried in this form because of the effect of carbonic acid on the pH of the blood. It is estimated that in 24 hours the lungs remove the equivalent of 20 to 40 liters of 1-normal acid as carbonic acid. This large acid load is successfully transported by the blood with hardly any variation in the blood pH, since most of the carbonic acid formed is promptly neutralized by the alkali cations of the blood (represented by B^+ in the equation):

$$H_2CO_3 \rightleftharpoons H^+ + HCO_3^- + B^+ \rightleftharpoons BHCO_3 + H^+ \text{ (See text below)}$$

At the pH of blood (7.40), a ratio of 20 to 1 must exist between the bicarbonate and carbonic acid fractions. This ratio is calculated from the Henderson-Hasselbalch equation as follows:

$$7.4 = \text{pH of blood}$$
$$6.1 = \text{pKa, } H_2CO_3$$

$$\text{pH} = \text{pKa} + \log \frac{\text{salt}}{\text{acid}}$$

$$7.40 = 6.10 + \log \frac{S}{A}\left(\frac{BHCO_3}{H_2CO_3}\right)$$

$$1.30 = \log \frac{BHCO_3}{H_2CO_3}$$

$$\text{Antilog } 1.3 = 20$$

$$\text{Therefore, } \frac{20}{1} = \frac{BHCO_3}{H_2CO_3}$$

Any increase or decrease in H^+ ion activity will be met by an adjustment in the above reaction. **As long as this ratio is maintained, the pH of the blood will be normal.** Any alteration in the ratio will disturb the acid-base balance of the blood in the direction of acidemia or alkalemia.

THE BUFFER SYSTEMS OF THE BLOOD

Since most of the CO_2 is carried in the blood as bicarbonate, it is apparent that the transport of this gas demands a considerable source of alkali. Furthermore, it is important that additional supplies of alkali be made available to the venous blood, which is carrying more carbon dioxide than the arterial blood. This extra alkali must be suitably neutralized when the extra carbon dioxide is eliminated in the lung. These requirements are so well met that the pH of venous blood differs from that of the arterial blood by only 0.01 to 0.03 units, i.e., pH 7.40 to pH 7.43.

The maintenance of the blood pH depends ultimately on the activity of the blood buffers, which include the plasma proteins, hemoglobin, and oxyhemoglobin as well as bicarbonate and inorganic phosphate. The small change in pH which is allowed to occur when carbon dioxide enters the venous blood at the tissues will cause a shift in the ratio of acid to salt in all of these buffer pairs, thus liberating alkali for the neutralization of the incoming CO_2. In this respect the plasma phosphates and bicarbonates play a minor role. The buffer effect of the plasma proteins is of greater importance since they release sufficient alkali to carry about 10% of the total CO_2. The red cell phosphates are responsible for about 25% of the total CO_2 carried. But most important of all is the buffering capacity of hemoglobin, which accounts for 60% of the CO_2 capacity of whole blood. This role of hemoglobin is, of course, in addition to the part it plays in the carriage of the carbamino-bound CO_2.

The Hemoglobin Buffers:

The remarkable buffering capacity of hemoglobin is due to the fact that oxyhemoglobin is a stronger acid than reduced hemoglobin, as is shown by their respective dissociation constants:

$$K, \text{ oxyhemoglobin} = 2.4 \times 10^{-7}$$

$$K, \text{ reduced hemoglobin} = 6.6 \times 10^{-9}$$

At a given pH, more alkali must therefore be combined with oxyhemoglobin than with reduced hemoglobin; and when some of the oxyhemoglobin changes to reduced hemoglobin (the isohydric change) at the tissues, there is a corresponding release of alkali which is then utilized to neutralize the carbonic acid resulting from the evolution of CO_2 by the tissues. This function of hemoglobin directly or indirectly supplies sufficient alkali for the carriage of over half of the total CO_2 of the blood.

When the blood returns to the lungs, exactly the opposite situation prevails. As CO_2 leaves the blood, some alkali will become excess. However, oxyhemoglobin is being formed simultaneously, and this alkali is promptly utilized to neutralize it. These changes are illustrated in the following diagram:

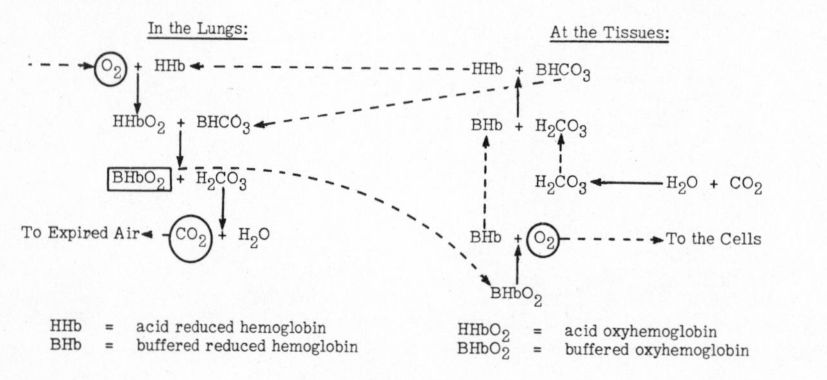

The Buffering Action of Hemoglobin

It must be recalled that there is never a **complete** change of **all** of the oxyhemoglobin to reduced hemoglobin, so that these reactions actually represent a variation in the concentrations of the hemoglobin buffer pairs -

$$\frac{BHbO_2}{HHbO_2} \text{ and } \frac{BHb}{HHb}$$

However, alkali is made available for the neutralization of incoming CO_2, to the extent that more of the reduced hemoglobin system is produced by dissociation of oxyhemoglobin.

The Chloride Shift:

It has been shown above that hemoglobin is responsible for about 60% of the buffering capacity of the blood. The cell phosphates contribute another 25%. Thus some 85% of the CO_2 carrying power of the blood resides within the cell. It is for this reason that the buffering power of whole blood greatly exceeds that of plasma or serum. However, it is also true that most of the buffered CO_2 is carried as bicarbonate in the plasma. These observations pose the question of how it is possible for the majority of the buffer capacity of the blood to reside in the cells but to be exerted in the plasma.

CO_2 reacts with water to form carbonic acid, mainly inside the red cell since the catalyzing enzyme, carbonic anhydrase, is found only within the erythrocyte. The carbonic acid is then buffered by the intracellular buffers, phosphate and hemoglobin, combining in this case with potassium. Bicarbonate ion also returns to the plasma and exchanges with chloride, which shifts into the cell when the tension of CO_2 increases in the blood. The process is reversible, so that chloride leaves the cells and enters the plasma when the CO_2 tension is reduced. This fact is confirmed by the finding of a higher chloride content in arterial plasma than in venous plasma.

It is considered that under normal conditions the red cell is virtually impermeable to sodium or potassium. But since it is permeable to hydrogen, bicarbonate, and chloride ions, intracellular sources of alkali cation (potassium) are indirectly made available to the plasma by chloride (anion) exchange. This permits the carriage of additional CO_2 (as sodium bicarbonate) by plasma.

The reactions of the chloride shift as they occur at the tissues between the plasma and red cells are summarized in the following diagram:

The Chloride Shift

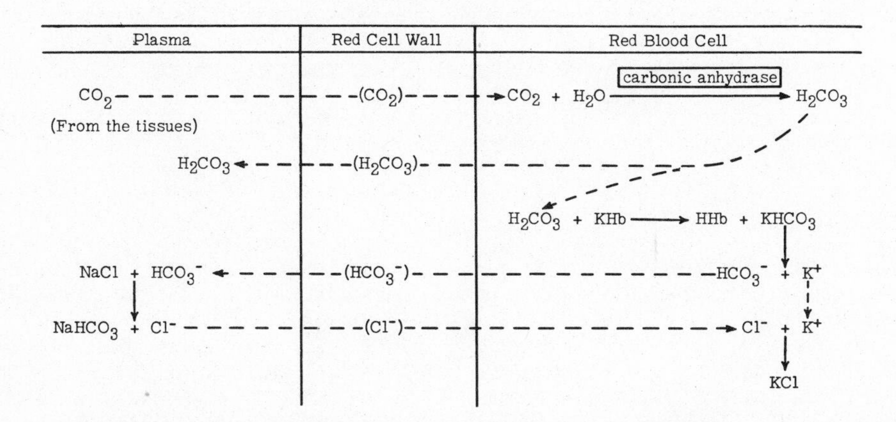

The CO_2 entering the blood from the tissues passes into the red cells, where it forms carbonic acid, a reaction catalyzed by carbonic anhydrase. Some of the carbonic acid returns to the plasma. The remainder reacts with the hemoglobin buffers to form bicarbonate, which then returns to the plasma where it exchanges with chloride. Sodium bicarbonate is formed in the plasma; and the chloride which the bicarbonate has replaced enters the red cells, where it is neutralized by potassium.

All of these reactions are reversible. At the lung, when the blood becomes arterial, chloride shifts back into the plasma, thus liberating intracellular potassium to buffer the newly-formed oxyhemoglobin and, in the plasma, neutralizing the sodium liberated by the removal of CO_2 during respiration.

ACID-BASE BALANCE

It has been noted above that as long as the ratio of carbonic acid to bicarbonate in the blood is 1:20, the pH of the blood remains normal; and that any alteration in this ratio, which was calculated from the Henderson-Hasselbalch equation at the normal pH of the blood (see p. 147), will disturb the acid-base balance of the blood and tissues in the direction of acidosis or alkalosis.

The content of H_2CO_3 in the blood is under the control of the respiratory system because of the dependence of carbonic acid on the pCO_2, which in turn is controlled by the organs of respiration. In consequence, disturbances in acid-base balance which are due to alterations in content of H_2CO_3 of the blood are said to be respiratory in origin. Thus **respiratory acidosis** will occur when circumstances are such as to cause an accumulation of H_2CO_3 in the blood; and **respiratory alkalosis** will occur when the rate of elimination of CO_2 is excessive, so that a reduction of H_2CO_3 occurs in the blood. In either instance, the normal 1:20 ratio of H_2CO_3 to bicarbonate is disturbed, and the pH of the blood will fall or rise in accordance with the retention or the excessive elimination of CO_2. If, however, the bicarbonate content of the blood can be adjusted to restore the 1:20 ratio between carbonic acid and bicarbonate, the pH will once more return to normal. Such an adjustment can be accomplished by the kidneys (see p. 293) - in respiratory acidosis by reabsorption of more bicarbonate in the renal tubules, and in respiratory alkalosis by permitting more bicarbonate to escape reabsorption and thus to be excreted into the urine. The respiratory acidosis or alkalosis is then said to be **compensated,** which means that even though the amounts of H_2CO_3 and of bicarbonate in the blood are abnormal, the pH is normal because the ratio of the two has been restored to the normal 1:20. It follows from the above discussion that the carbon dioxide content (see p. 152) of the plasma, which is a measure of both carbonic acid and bicarbonate, will be higher than normal in compensated respiratory acidosis and lower than normal in compensated respiratory alkalosis.

Disturbances in acid-base balance which are due to alterations in the content of bicarbonate in the blood are said to be metabolic in origin. A deficit of bicarbonate without any change in H_2CO_3 will produce a **metabolic acidosis**; an excess of bicarbonate, a **metabolic alkalosis**. Compensation will occur by adjustments of the carbonic acid concentrations, in the first instance by elimination of more CO_2 (hyperventilation) and in the latter instance by retention of CO_2 (depressed respirations). The CO_2 content of the plasma will obviously be lower than normal in metabolic acidosis and higher than normal in metabolic alkalosis.

The biochemical changes which occur in the various types of acidosis and alkalosis, both uncompensated and compensated, are shown on p. 151.

Causes of Disturbances in Acid-Base Balance:

A. Acidosis:

1. Metabolic acidosis is caused by a decrease in the bicarbonate fraction, with either no change or a relatively smaller change in the carbonic acid fraction. This is the most common, classical type of acidosis. It occurs in uncontrolled diabetes with ketosis, in some cases of vomiting when the fluids lost are not acid, in renal disease, poisoning by an acid salt, excessive loss of intestinal fluids (particularly from the lower small intestine and colon, as in diarrheas or colitis), and whenever excessive losses of electrolyte have occurred. Increased respirations (hyperpnea) may be an important sign of an uncompensated acidosis.
2. Respiratory acidosis is caused by an increase in carbonic acid relative to bicarbonate. This may occur in any disease which impairs respiration, such as pneumonia, emphysema, congestive failure, asthma, or in depression of the respiratory center (as by morphine poisoning). A poorly functioning respirator may also contribute to respiratory acidosis.

Biochemical Changes in Acidosis and Alkalosis

Vol. % / mEq./L. scale: $H.HCO_3$ 3 — 1.35; $B.HCO_3$ 60 — 26; 120 — 52

	Normal	Acidosis: Metabolic U*	Acidosis: Metabolic C*	Acidosis: Respiratory U*	Acidosis: Respiratory C*	Alkalosis: Metabolic U*	Alkalosis: Metabolic C*	Alkalosis: Respiratory U*	Alkalosis: Respiratory C*
Serum CO_2 content (Vol. %)	63	33	31.5	66	126	93	94.5	61.5	31.5
Serum pCO_2	→	→	↓	↑	↑	→	↑	↓	↓
pH	→	↓	→	↓	→	↑	→	↑	→
Ratio of H_2CO_3 to $B.HCO_3$	1:20	<1:20	1:20	<1.20	1.20	>1:20	1:20	>1:20	1:20

*U = Uncompensated. C = Compensated.

B. Alkalosis:

1. Metabolic alkalosis occurs when there is an increase in the bicarbonate fraction, with either no change or a relatively smaller change in the carbonic acid fraction. A simple alkali excess leading to alkalosis is produced by the ingestion of large quantities of alkali, such as might occur in patients under treatment for peptic ulcer. But this type of alkalosis occurs much more commonly as a consequence of high intestinal obstruction (as in pyloric stenosis), after prolonged vomiting, or after the excessive removal of gastric secretions containing hydrochloric acid (as in gastric suction). The elevated blood pH of an uncompensated alkalosis often leads to tetany, possibly by inducing a decrease in ionized serum calcium. This is sometimes referred to as gastric tetany, although its relation to the stomach is, of course, incidental. The common denominator in this form of alkalosis is a chloride deficit caused by the removal of gastric secretions which are low in sodium but high in chloride (i.e., as hydrochloric acid). The chloride ions which are lost are then replaced by bicarbonate. This type of metabolic alkalosis is aptly termed ''hypochloremic'' alkalosis. The frequent association of potassium deficiency with hypochloremic alkalosis is discussed in Chapter 19, p. 320. Hypochloremic alkalosis also occurs in Cushing's disease and during corticotropin or cortisone administration.

   In all types of uncompensated alkalosis, the respirations are slow and shallow; the urine may at first be alkaline in reaction, but, usually because of a concomitant deficit of sodium and potassium, the urine will give an acid reaction even though the blood bicarbonate is elevated. This paradox is attributable in part to the fact that the excretion of the excess bicarbonate by the kidney will require an accompanying loss of sodium which under the conditions described (low sodium) cannot be spared. Thus the kidney defers to the necessity for maintaining sodium concentrations in the extracellular fluid at the expense of acid-base balance. However, an equal if not, in the usual situations, a more important cause of the excretion of an acid urine in the presence of an elevated plasma bicarbonate is the effect of a potassium deficit on the excretion of hydrogen ions by the kidney as described on p. 296. Alkalosis as usually encountered clinically is almost always associated with a concomitant deficiency of potassium.

2. Respiratory alkalosis occurs when there is a decrease in the carbonic acid fraction with no corresponding change in bicarbonate. This is brought about by hyperventilation, either voluntary or forced. Examples are hysterical hyperventilation, central nervous system disease affecting the respiratory system, the early stages of salicylate poisoning (see below), the hyperpnea observed at high altitude, or injudicious use of respirators. Respiratory alkalosis may also occur in patients in hepatic coma.

Measurement of Acid-Base Balance:

The existence of uncompensated acidosis or alkalosis is most accurately determined by measurement of the pH of the blood. In respiratory acidosis or alkalosis, blood pH determination is essential to a satisfactory biochemical diagnosis. However, determination of the pH of the blood is often not feasible clinically. Furthermore, it is necessary to know to what extent the electrolyte pattern of the blood is disturbed in order to prescribe the proper corrective therapy. For these reasons, a determination of the CO_2 derived from a sample of blood plasma after treatment with acid (CO_2 capacity or CO_2 combining power) is used instead. This measures essentially the total quantity of H_2CO_3 and of bicarbonate in the plasma but gives no information as to the ratio of distribution of the two components of the bicarbonate buffer system (and hence of the blood pH). It should also be noted that such a single determination also fails to take into account the concentration of other buffer systems such as hemoglobin (both oxygenated and reduced), serum protein, and phosphates. In disease, these may be notably altered and thus exert important effects on acid-base balance. However, the total blood CO_2 determination is reasonably satisfactory when taken in association with clinical observations and the history of the case. In addition, as noted above, it yields information on the degree of depletion or excess of bicarbonate so that the proper correction may be instituted.

Measurement of the CO_2, carbonic acid, and bicarbonate of plasma derived from blood collected under oil to prevent loss of gases to the air gives what is designated as **CO_2 content.** It is reported as volumes of CO_2 per 100 ml., at standard conditions of temperature and pressure, since all of the bicarbonate and carbonic acid are converted to CO_2 by acidification and by the imposition of a vacuum in the apparatus used to measure the gas. The CO_2 content of venous blood is naturally higher than that of arterial blood.

If the plasma is first equilibrated with normal alveolar air (CO_2 tension, 40 mm. Hg) before it is measured, the **CO_2 capacity** (or CO_2 combining power) is obtained. Ordinarily, the CO_2 content and the CO_2 combining power are practically identical; but if the CO_2 tension in the alveolar air of the patient is less than 40 mm., CO_2 capacity will be greater than the CO_2 content.

In a clinical appraisal of the severity of a metabolic acidosis or alkalosis, the bicarbonate fraction of the blood is of primary interest. The plasma bicarbonate is sometimes designated the **alkali reserve** because it is this fraction of the plasma electrolyte which is used to neutralize all acidic compounds entering the blood and tissues. In this capacity, the plasma bicarbonate constitutes a sort of first line of defense. As a result, any threat to the acid-base equilibrium of the body will be reflected in a change in this component of the electrolyte structure. The concentration of the plasma bicarbonate, which is used to measure the alkali reserve, can be obtained from the CO_2 combining power. For this purpose, it is assumed that a 20:1 ratio exists between bicarbonate and carbonic acid; by dividing the CO_2 combining power (expressed in volumes percent) by 2.24, plasma bicarbonate concentration in milliequivalents per liter is derived. A reduction in the plasma bicarbonate is usually sufficient to make a diagnosis of acidosis, although this may be erroneous since the ratio of carbonic acid to bicarbonate, which determines the blood pH, is not known.

An important illustration of the limitations of CO_2 content determinations is found in the chemical imbalance which prevails in salicylate poisoning. In the initial stages, respiratory alkalosis occurs because of hyperventilation induced by the toxic effect of the drug on the respiratory center. Compensation produces a decrease in carbon dioxide content which by itself suggests the presence of metabolic acidosis; however, the blood pH will be found to be elevated above normal, which confirms the presence of an uncompensated respiratory alkalosis. Later, because of renal failure and other metabolic disturbances, a metabolic acidosis supervenes. This is of course aggravated by the lowered alkali reserve brought about by attempts at compensating for the pre-existing respiratory alkalosis. The time at which the metabolic acidosis occurs can be detected only by measurement of the blood pH, an observation which is of great importance in respect to electrolyte therapy. The sequence of events described above as characteristic of salicylate poisoning also occurs consequent to respiratory alkalosis from any other cause.

Fixed Base:

The term "fixed base" refers to the nonvolatile alkali of the blood and tissues. It is equal to the sum of the cations. In the extracellular fluid the fixed base is equal to 150 to 155 mEq. per liter. Most of this is sodium; the remainder consists of smaller amounts of potassium, calcium, and magnesium. The fixed base can be measured by the use of an electrical conductivity apparatus as well as by analysis for the individual cations.

The Role of the Kidney in Acid-Base Balance:

In addition to carbonic acid, which is eliminated by the respiratory organs as carbon dioxide, other acids, which are not volatile, are produced by metabolic processes. These include lactic and pyruvic acids and the more important inorganic acids, hydrochloric, phosphoric, and sulfuric. About 50 to 150 ml. of a normal solution of these inorganic acids are eliminated by the kidneys in a 24-hour period. It is of course necessary that these acids be partially neutralized by alkali; but in the distal tubules of the kidney some of this fixed base is reabsorbed (actually exchanged for hydrogen ion), and the pH of the urine is allowed to fall. This acidification of the urine in the distal tubule is a valuable function of the kidney in conserving the reserves of fixed base in the body.

Another device used by the kidney to buffer acids and thus to conserve fixed base (cation) is the production of ammonia from amino acids. The ammonia is substituted for alkali cations, and the amounts of ammonia mobilized for this purpose may be markedly increased when the production of acid within the body is excessive (e.g., as in metabolic acidosis such as occurs as a result of the ketosis of uncontrolled diabetes).

When alkali is in excess, the kidney excretes an alkaline urine to correct this imbalance. The details of the renal regulation of acid-base equilibrium will be discussed in Chapter 18, p. 293.

In kidney disease, glomerular and tubular damage results in considerable impairment of these important renal mechanisms for the regulation of acid-base balance. Tubular reabsorption of sodium in exchange for hydrogen ion is poor, and excessive retention of acid catabolites, such as phosphates and sulfates occurs because of decreased glomerular filtration. In addition, the mechanism for ammonia production by the tubules is inoperative. As a result, acidosis is a common finding in nephritis.

• • •

Bibliography:

Best, C.H., and Taylor, N.B.: Physiological Basis of Medical Practice, 6th Ed. Williams and Wilkins, 1955.

Davenport, H.W.: The ABC of Acid-Base Chemistry, 4th Ed. University of Chicago Press, 1958.

Weisberg, H.F.: Water, Electrolyte and Acid-Base Balance. Williams and Wilkins, 1953.

11...

Digestion and Absorption from the Gastrointestinal Tract

Most foodstuffs are ingested in forms which are unavailable to the organism, since they cannot be absorbed from the digestive tract until they have been reduced to smaller molecules. This breakdown of the naturally-occurring foodstuffs into assimilable forms is the work of digestion.

The chemical changes incident to digestion are accomplished with the aid of the enzymes of the digestive tract. These enzymes catalyze the hydrolysis of native proteins to amino acids, of starches to monosaccharides, and of fats to glycerol and fatty acids. It is probable that in the course of these digestive reactions, the minerals and vitamins of the foodstuffs are also made more assimilable. This is certainly true of the fat-soluble vitamins, which are not absorbed unless fat digestion is proceeding normally.

DIGESTION IN THE MOUTH

Constituents of the Saliva:

The oral cavity contains saliva secreted by three pairs of salivary glands: the parotid, submaxillary, and sublingual. The saliva consists of about 99.5% water, although the content varies with the nature of the factors exciting its secretion. The saliva acts as a lubricant for the oral cavity, and by moistening the food as it is chewed it reduces the dry food to a semisolid mass which is easily swallowed. The saliva is also a vehicle for the excretion of certain drugs (e.g., alcohol and morphine) and of certain inorganic ions such as K^+, Ca^{++}, HCO_3^-, iodine, and thiocyanate (SCN).

The pH of the saliva is usually slightly on the acid side, about 6.80, although in various individuals it varies on either side of neutrality.

Salivary Digestion:

The saliva is relatively unimportant in digestion. It contains a starch-splitting enzyme, a salivary amylase known as **ptyalin**; but the enzyme is readily inactivated at pH 4 or less, so that digestive action on food in the mouth will soon cease in the acid environment of the stomach. Although saliva is capable of bringing about the hydrolysis of starch to maltose in the test tube, this actually is of little significance in the body because of the short time it can act on the food. Furthermore, other amylases of the intestine are capable of accomplishing complete starch digestion. In many animals, a salivary amylase is entirely absent.

DIGESTION IN THE STOMACH

Stimulation of Gastric Secretion:

Gastric secretion is initiated by nervous or reflex mechanisms. The effective stimuli for these reflexes are similar to those which operate in salivary secretion. The continued secretion of gastric juice is, however, due to a hormonal stimulus, **gastrin** (gastric secretin). This chemical stimulant is produced by the gastric glands and absorbed into the blood, which carries it back to the stomach where it excites gastric secretion. Histamine, produced by decarboxylation of the amino acid, histidine, also acts as a potent gastric secretagogue.

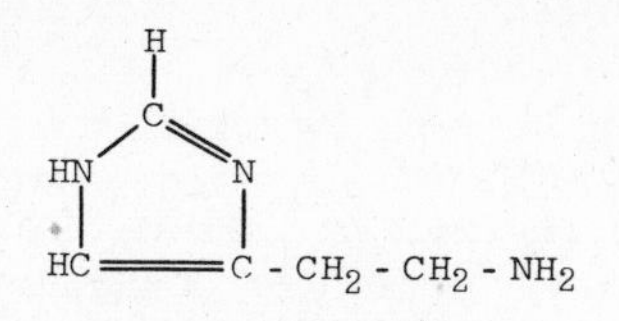

Histamine

Gastric Constituents and Gastric Digestion:

In the mucosa of the stomach wall, two types of secretory glands are found: those exhibiting a single layer of secreting cells (the chief cells), and those with cells arranged in layers (the parietal cells), which secrete directly into the gastric glands. The mixed secretion is known as gastric juice. It is normally a clear, pale yellow fluid of high acidity, 0.2 to 0.5% HCl, with a pH of about 1.0.

A. Hydrochloric Acid: The parietal cells are the sole source of gastric hydrochloric acid. HCl is said to originate from NaCl and carbonic acid according to the following reaction:

$$1.\quad H_2O + CO_2 \longrightarrow H_2CO_3 \rightleftharpoons H^+ + HCO_3^-$$

$$2.\quad H^+ + HCO_3^- + Na^+ + Cl^- \longrightarrow HCl + NaHCO_3$$

HCl ↓ (secreted into gastric juices); $NaHCO_3$ ↓ (reabsorbed into the blood)

Reaction 1 is catalyzed by carbonic anhydrase.

An alkaline urine often follows the ingestion of a meal ("alkaline tide"), presumably as a result of the formation of extra sodium bicarbonate in the process of hydrochloric acid secretion by the stomach in accordance with the above reaction.

The gastric juice is 97 to 99% water. The remainder consists of mucin and inorganic salts, the digestive enzymes (pepsin and rennin), and a lipase.

B. Pepsin: The chief digestive function of the stomach is the partial digestion of protein. Gastric pepsin is produced in the chief cells as the inactive zymogen, **pepsinogen**, which is activated to pepsin by the action of HCl and, autocatalytically, by itself, i.e., a small amount of pepsin can cause the activation of the remaining pepsinogen. The enzyme transforms native protein into proteoses and peptones which are still reasonably large protein derivatives.

C. Rennin (Chymosin) (Rennet): This enzyme causes the coagulation of milk. It is important in the digestive processes of infants because it prevents the rapid passage of milk from the stomach. In the presence of calcium, rennin changes irreversibly the casein of milk to a paracasein which is then acted on by pepsin. This enzyme is said to be absent from the stomachs of adults.

D. Lipase: The lipolytic action of gastric juice is not important, although a gastric lipase capable of mild fat-splitting action is found in gastric juice.

PANCREATIC AND INTESTINAL DIGESTION

The stomach contents, or **chyme,** which are of a thick creamy consistency, are intermittently introduced during digestion into the duodenum through the pyloric valve. The pancreatic and bile ducts open into the duodenum at a point very close to the pylorus. The high alkaline content of pancreatic and biliary secretions neutralizes the acid of the chyme and changes the pH of this material to the alkaline side; this shift of pH is necessary for the activity of the enzymes contained in pancreatic and intestinal juice.

Stimulation of Pancreatic Secretion[1]:

Like the stomach, the pancreas secretes its digestive juice almost entirely by means of hormonal stimulation. The hormones are secreted in the duodenum and upper jejunum as a result of stimulation of hydrochloric acid, fats, proteins, carbohydrates, and partially digested foodstuffs, and are carried by the blood to the pancreas, liver, and gallbladder after absorption from the small intestine through the portal blood. The active hormonal components of the duodenum (originally termed ''secretin'' by Bayliss and Starling) have now been separated into five separate factors: (1) secretin (high in bicarbonate but low in enzyme content), which stimulates the production of a thin, watery fluid by the pancreas; (2) pancreozymin (low in bicarbonate but high in enzyme content), which stimulates excretion of a viscous pancreatic juice; (3) hepatocrinin, which causes the liver to secrete a thin, salt-poor type of bile; (4) cholecystokinin, which induces contraction and emptying of the gallbladder; and (5) enterocrinin, which induces the flow of succus entericus (intestinal juice).

Constituents of Pancreatic Secretion:

Pancreatic juice is a nonviscid watery fluid which is similar to saliva in its content of water and contains some protein and other organic and inorganic compounds, mainly Na^+, K^+, HCO_3^-, and Cl^-. Ca^{++}, Zn^{++}, $HPO_4^=$, and SO_4 are present in small amounts. The pH of pancreatic juice is distinctly alkaline, 7.5 to 8.0 or higher.

The enzymes contained in pancreatic juice include trypsin, chymotrypsin, and peptidases, α-amylase, lipase, cholesterol esterase, ribonuclease, deoxyribonuclease, and collagenase. Some of these enzymes are secreted as inactive precursors (zymogens) such as trypsinogen or chymotrypsinogen, but are activated on contact with the intestinal mucosa. The activation of trypsinogen is attributed to **enterokinase,** which is produced by the intestinal glands. A small amount of active trypsin then autocatalytically activates additional trypsinogen and chymotrypsinogen.

A. Trypsin and Chymotrypsin: The protein-splitting action of pancreatic juice (proteolytic action) is due to trypsin and chymotrypsin, which attack native protein, proteoses, and peptones from the stomach to produce polypeptides. Chymotrypsin has more coagulative power for milk than trypsin, and, as previously noted, it is activated not by enterokinase but by active trypsin.

B. The ''Peptidases'': The further attack on protein breakdown products, i.e., on the polypeptides, is accomplished by a mixture of **carboxypeptidase**, a zinc-containing enzyme of the pancreatic juice, and **aminopeptidase** and **dipeptidase** of the intestinal juices. Such a mixture was formerly termed ''erepsin.'' The carboxy- and aminopeptidases, as their names imply, attack polypeptides either at the end of the amino chain, where a free carboxyl group exists, or at the amino linkages of the peptide chain. This system of intestinal proteoses converts food proteins into their constituent amino acids for absorption by the intestinal mucosa and transfer to the circulation.

C. Amylase: The starch-splitting action of pancreatic juice is due to a pancreatic α-amylase (amylopsin). It is similar in action to salivary ptyalin, hydrolyzing starch to maltose at an optimum pH of 7.1.

D. Lipase: Fats are hydrolyzed by a pancreatic lipase (steapsin) to fatty acids, glycerol, and monoglycerides. This is an important enzyme in digestion. It is possibly activated by bile salts.

E. Cholesterol Esterase: Esterification of cholesterol with fatty acids is believed to play an important part in the absorption of cholesterol from the intestine. Cholesterol esterase, an enzyme found in pancreatic juice, together with bile salts, catalyzes the esterification of cholesterol and therefore its absorption from the lumen of the intestine for transfer to the lymphatics.

F. Ribonuclease (RNase) and deoxyribonuclease (DNase) have been prepared from pancreatic tissue (see p. 43).

G. Collagenase[2] is a recently described pancreatic enzyme which acts on collagen.

Constituents of Intestinal Secretions:

The intestinal juice secreted, under the influence of enterocrinin (see p. 156), by the glands of Brunner and of Lieberkühn also contains digestive enzymes. In addition to the proteolytic enzymes already mentioned, these include the following:

A. The specific disaccharidases, i.e., sucrase, maltase, and lactase, which convert sucrose, maltose, or lactose into their constituent monosaccharides for absorption.

B. A phosphatase, which removes phosphate from certain organic phosphates such as hexosephosphates, glycerophosphate, and the nucleotides derived from the diet.

C. Polynucleotidases (nucleinases) (phosphodiesterases), which split nucleic acids into nucleotides.

D. Nucleosidases (nucleoside phosphorylases), one of which attacks only purine-containing nucleosides, liberating adenine or guanine and the pentose sugar which is simultaneously phosphorylated. The pyrimidine nucleosides (uridine, cytidine, and thymidine) are broken down by another enzyme which differs from the purine nucleosidase.

E. Lecithinase: The intestinal juice is also said to contain a lecithinase, which attacks lecithins to produce glycerol, fatty acids, phosphoric acid, and choline.

The Results of Digestion:

The final result of the action of the digestive enzymes already described is to reduce the foodstuffs of the diet to forms which can be absorbed and assimilated. These end-products of digestion are, for carbohydrates, the monosaccharides (principally glucose); for proteins, the amino acids; and for fats, the fatty acids, glycerol, and the monoglycerides, although some unhydrolyzed fat is probably also absorbed.

THE BILE

In addition to many functions in intermediary metabolism, the liver, by producing bile, plays an important role in digestion. The **gallbladder**, a saccular organ attached to the hepatic duct, stores a certain amount of the bile produced by the liver between meals. During digestion, the gallbladder contracts and supplies bile rapidly to the small intestine by way of the common bile duct. The pancreatic secretions mix with the bile, since they empty into the common duct shortly before its entry into the duodenum.

Composition of Bile:

The composition of hepatic bile differs from that of gallbladder bile, as shown in the following table:

The Composition of Hepatic and of Gallbladder Bile

	Hepatic Bile (as secreted)		Bladder Bile
	% of total bile	% of total solids	% of total bile
Water	97.00	-	85.92
Solids	2.52	-	14.08
Bile acids	1.93	36.9	9.14
Mucin and pigments	0.53	21.3	2.98
Cholesterol	0.06	2.4	0.26
Fatty acids and fat	0.14	5.6	0.32
Inorganic salts	0.84	33.3	0.65
Specific gravity	1.01	-	1.04
pH	7.1-7.3	-	6.9-7.7

SUMMARY OF DIGESTIVE PROCESSES

Source of Enzyme and Stimulus for Secretion	Enzyme	Method of Activation and Optimum Conditions for Activity	Substrate	End Products or Action
Salivary Glands of the Mouth. Secrete saliva in reflex response to presence of food in the mouth.	Ptyalin	Chloride ion necessary. pH - 6.6-6.8.	Starches	Maltose
Stomach Glands. Chief cells and parietal cells secrete gastric juice in response to reflex stimulation and the chemical action of gastrin.	Pepsin	Pepsinogen converted to active pepsin by HCl and by active pepsin. pH - 1.0-2.0.	Protein	Proteoses Peptones
	Rennin	Calcium necessary for activity - pH - 4.0.	Casein of milk.	Coagulates milk.
Pancreas. Presence of acid chyme from the stomach activates duodenum to produce (a) secretin which hormonally stimulates **flow** of pancreatic juice. (b) pancreozymin which stimulates the **production** of enzymes.	Trypsin	Trypsinogen converted to active trypsin by enterokinase of the intestine at pH 5.2-6.0. Active trypsin converts trypsinogen at pH 7.9.	Protein Proteoses Peptones	Polypeptides Dipeptides
	Chymotrypsin	Secreted as chymotrypsinogen and converted to active form by trypsin. pH - 8.0.	Same as trypsin.	Same as trypsin. More coagulating power for milk.
	Carboxypeptidase	Same as trypsin?	Peptides with free carboxyl groups.	Lower peptides. Free amino acids.
	Amylase (Amylopsin)	pH - 7.1.	Starch	Maltose
	Lipase (Steapsin)	Activated by bile salts? pH - 8.0.	Primary ester linkages of fats.	Fatty acids Monoglycerides Glycerol
	Ribonuclease		Ribonucleic ac.	Nucleotides
	Deoxyribonuclease		Deoxyribonucleic acids.	Nucleotides
	Cholesterol esterase	Activated by bile salts.	Free cholesterol.	Esters of cholesterol with fatty acids.
Liver and Gallbladder.	Bile salts and alkali.	Cholecystokinin and hepatocrinin - hormones from the intestine; stimulate gallbladder and liver to secrete bile.	Fats - also neutralize acid chyme.	Fatty acid-bile salt conjugates and finely emulsified neutral fat.
Small Intestine. Secretions of Brunner's glands of the duodenum and the glands of Leiberkühn. Enterocrinin induces flow of succus entericus.	Aminopeptidase		Amino groups in peptide linkages.	Lower peptides. Free amino acids.
	Dipeptidase		Dipeptides	Amino acids.
	Sucrase	pH - 5.0-7.0.	Sucrose	Fructose, glucose.
	Maltase	pH - 5.8-6.2.	Maltose	Glucose
	Lactase	pH - 5.4-6.0.	Lactose	Glucose, galactose.
	Phosphatase	pH - 8.6.	Organic phosphates.	Free phosphate.
	Polynucleotidase		Nucleic acid.	Nucleotides
	Nucleosidases		Purine or Pyrimidine nucleotides.	Purine or Pyrimidine bases Pentose-phosphate
	Lecithinase		Lecithin	Glycerol Fatty acids Phosphoric acid Choline

Stimulation of Gallbladder and Bile Formation:

Contraction of the gallbladder and relaxation of its sphincter are initiated by a hormonal mechanism. This hormone, **cholecystokinin** (see p. 156), is secreted by the intestine in response to the presence of foods, mainly meats and fats. Bile salts and various other chemical substances, such as calomel, act as stimulants to bile flow. These biliary stimulants are known as **cholagogues**. A hormone of duodenal mucosal origin (hepatocrinin; see p. 156), is also active in stimulation of hepatic secretion of bile.

Bile Acids:

Because of their detergent and emulsifying effects on fats, the conjugated bile acids may be important in digestion. Four bile acids have been isolated from human bile. Cholic acid, whose structure is shown below, is the acid which is found in the largest amounts in the bile itself. Its structure is that of a completely saturated sterol, having three OH groups on the nucleus (at positions 3, 7, and 12). The other bile acids are deoxycholic acid, which lacks the OH group at position 7; chenodeoxycholic acid, lacking an OH at position 12; and lithocholic acid, which has only one OH group (at position 3). Deoxycholic acid is the main bile acid found in the feces of normal adult human beings[3].

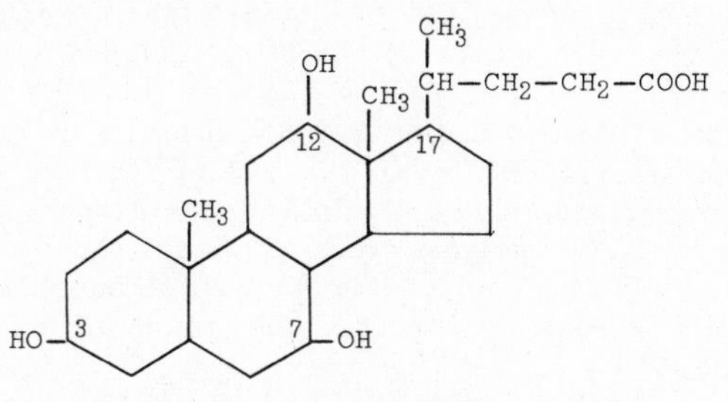

Cholic Acid

The bile acids are the major end products of the metabolism of cholesterol since the cholesterol which the liver removes from the blood is excreted into the bile not as cholesterol but as cholic acid and the other bile acids. Measurement of the output of bile acids is thus the most accurate means of estimating cholesterol excretion.

The bile acids are not excreted in the bile in the free state but are conjugated by the liver with glycine or taurine (a cystine derivative; see p. 250). In this conjugated form the bile acids are water-soluble. The conjugation with glycine or taurine occurs through the carboxyl group on the side chain. Because of the alkalinity of the bile, the conjugated bile acids may also be largely neutralized with sodium or potassium to form the glycocholates or taurocholates. These are the so-called bile salts; they exert a powerful detergent effect within the intestine and thus aid in the digestion of fats.

The conjugation of cholic acid has been found to require preliminary activation with coenzyme A (see p. 85). The activating enzyme[6] which occurs only in the microsomes of the liver catalyzes the following reaction:

$$\text{Cholic acid} + \text{Co A} + \text{ATP} \xrightarrow{Mg^{++}} \text{Cholyl Co A}$$

A second enzyme in the liver catalyzes conjugation of cholyl Co A with taurine.

Functions of the Bile System:

A. Emulsification: The bile salts have considerable ability to lower the surface tension of water. This enables them to emulsify fats in the intestine and to dissolve fatty acids and water-insoluble soaps. The presence of bile in the intestine is necessary to accomplish the digestion and absorption of fats as well as the absorption of the fat-soluble vitamins A, D, E, and K. When fat digestion is impaired, other foodstuffs are also poorly digested, since the fat covers the food particles and prevents enzymes from attacking them. Under these conditions, the activity of the intestinal bacteria causes considerable putrefaction and production of gas.

B. Neutralization of Acid: In addition to its functions in digestion, the bile is a reservoir of alkali, which helps to neutralize the acid chyme from the stomach.

C. Excretion: Bile is also an important vehicle of excretion. It removes many drugs, toxins, bile pigments, and various inorganic substances such as copper, zinc, and mercury. Cholesterol, either derived from the diet or synthesized by the body, is eliminated almost entirely in the bile. Free cholesterol is not soluble in water but is emulsified by the bile salts. Very often it precipitates from the bile and forms stones in the gallbladder or ducts. These stones, or calculi, may also contain a mixture of cholesterol and calcium, although pure cholesterol stones are the most common, probably because of the inability of the gallbladder to cope with an excess of cholesterol.

D. Bile Pigment Metabolism: The origin of the bile pigments from hemoglobin is discussed in Chapter 5 (see p. 64). Further consideration of bile pigment metabolism will be given in the discussion of liver function (see p. 273).

INTESTINAL PUTREFACTION AND FERMENTATION

Most ingested food is absorbed from the small intestine. The residue passes into the large intestine. Here considerable absorption of water takes place, and the semi-liquid intestinal contents gradually become more solid. During this period, considerable bacterial activity occurs. By fermentation and putrefaction, the bacteria produce various gases, such as carbon dioxide, methane, hydrogen, nitrogen, and hydrogen sulfide, as well as acetic, lactic, and butyric acids. The bacterial decomposition of lecithin may produce choline and related toxic amines such as neurine and muscarine.

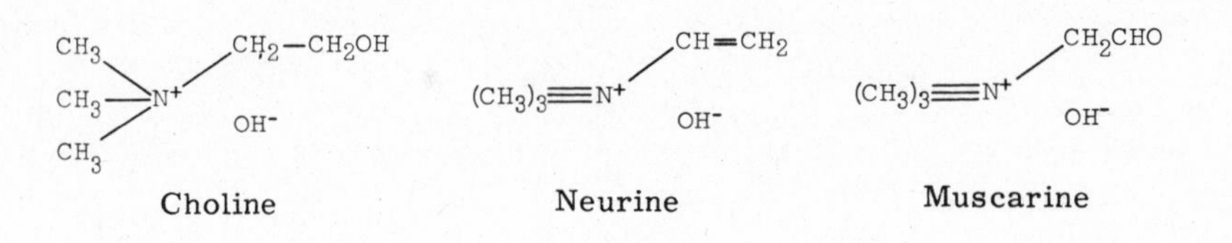

Choline Neurine Muscarine

Fate of Amino Acids:

Many amino acids undergo decarboxylation as a result of the action of intestinal bacteria to produce toxic amines, or ptomaines.

$$RCHNH_2\boxed{COO}H \xrightarrow{\boxed{\text{Bacterial Decarboxylase}}} \underset{\text{(ptomaine)}}{RCH_2NH_2} + CO_2$$

Such decarboxylation reactions produce cadaverine from lysine; agmatine from arginine; tyramine from tyrosine; putrescine from ornithine; and histamine from histidine. Many of these amines are powerful vasopressor substances.

The amino acid, tryptophan, undergoes a series of reactions to form indole and methylindole (skatole), the substances particularly responsible for the odor of the feces.

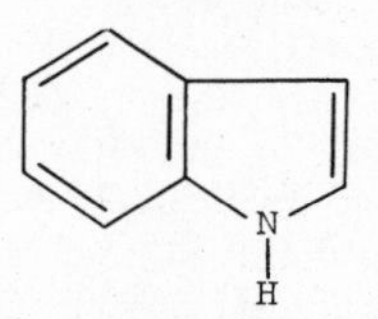

Indole

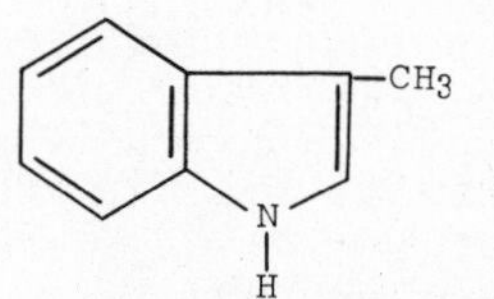

Skatole

The sulfur-containing amino acid, cystine, undergoes a series of transformations to form mercaptans such as ethyl and methyl mercaptan, as well as H_2S.

$$\begin{array}{c} CH_3 \\ | \\ CH_2SH \end{array} \qquad\qquad CH_3SH \xrightarrow{+2H} CH_4 + H_2S$$

Ethyl Mercaptan   Methyl Mercaptan

The large intestine is a source of considerable quantities of ammonia, presumably as a product of the putrefactive activity on nitrogenous substrates by the intestinal bacteria. This ammonia is absorbed into the portal circulation, but under normal conditions it is rapidly removed from the blood by the liver. In liver disease, this function of the liver may be impaired, in which case the concentration of ammonia in the peripheral blood will rise to toxic levels. It is believed that ammonia intoxication may play a role in the genesis of hepatic coma in some patients. In dogs on whom an Eck fistula has been performed (complete diversion of the portal blood to the vena cava), the feeding of large quantities of raw meat will induce symptoms of ammonia intoxication (meat intoxication) accompanied by elevated levels of ammonia in the blood. The oral administration of neomycin has been shown to reduce the quantity of ammonia delivered from the intestine to the blood[5]. It is possible that the feeding of high-protein diets to patients suffering from advanced liver disease may also contribute to the development of ammonia intoxication.

Intestinal Bacteria:
The intestinal flora may comprise as much as 25% of the dry weight of the feces. In herbivora, whose diet consists largely of cellulose, the intestinal bacteria are essential to digestion, since they decompose this polysaccharide and make it available for absorption. In addition, the intestinal bacteria may accomplish the synthesis of protein precursors for these animals. In man, although the intestinal flora is not as important as in the herbivora, nevertheless some nutritional benefit is derived from bacterial activity in the synthesis of certain vitamins, particularly vitamin K, and possibly certain members of the B complex, which are made available to the body. Information gained from current experiments with animals raised under strictly aseptic conditions should help to define further the precise role of the intestinal bacteria.

ABSORPTION FROM THE GASTROINTESTINAL TRACT

There is little absorption from the stomach, even of foodstuffs like glucose which can be absorbed directly from the intestine. Although water is not absorbed to any extent from the stomach, considerable gastric absorption of alcohol is possible.

The small intestine is the main digestive and absorptive organ. About 90% of the ingested foodstuffs is absorbed in the course of passage through the approximately 25 feet of small intestine. Water is absorbed from the small intestine at the same time. Considerably more water is absorbed after the foodstuffs pass into the large intestine, so that the contents, which were fluid in the small intestine, gradually become more solid in the colon.

There are two general pathways for the transport of materials absorbed by the intestine: the veins of the portal system, which lead directly to the liver; and the lymphatic vessels of the intestinal area (the lacteals), which eventually lead to the blood by way of the lymphatic system and the thoracic duct.

Absorption of Carbohydrates:
Carbohydrates are absorbed from the intestine into the blood of the portal venous system in the form of monosaccharides, chiefly the hexoses, glucose, fructose, mannose, and galactose, although the pentose sugars, if present in the food ingested, will also be absorbed. The passage of the hexose sugars across the intestinal barriers occurs at a fixed rate even against an osmotic gradient and independent of their concentration in the intestinal lumen. Furthermore, they are absorbed much faster than the pentose sugars. The constant and rapid absorption of the hexoses

indicates that a mechanism other than simple diffusion is operating in the intestinal absorption of these sugars. This mechanism is an enzymatic phosphorylation process which catalyzes the formation of a hexose phosphate as a preliminary to absorption. It requires an enzyme, which is termed a **hexokinase**, and also a source of high energy phosphate which is presumably adenosine triphosphate (ATP). The intestinal hexokinase differs from those described on page 176 which catalyze the formation of hexose phosphates as a preliminary to glycogenesis or to glycolysis. However, a similar if not identical hexokinase is found in the renal tubules where it catalyzes the phosphorylation of glucose incident to its reabsorption from the tubular filtrate (see p. 288). The pentoses, on the other hand, are apparently not phosphorylated. If phosphorylation is prevented by poisoning with monoiodoacetic acid or phlorhizin, the hexose sugars can now be absorbed only by diffusion; and the rate of absorption of these sugars decreases to that of the pentose sugars.

The rate of absorption of different hexoses is not the same. If the rate of absorption of glucose is taken as 100, the absorption rates of certain other sugars are as follows: (Data from Cori, using the rat as experimental animal.)

Galactose	-	110	Fructose	-	43	Xylose	-	15
Glucose	-	100	Mannose	-	19	Arabinose	-	9

Absorption of Fats:

The complete hydrolysis of fats (triglycerides) produces glycerol and fatty acids. However, the first, second, and third fatty acids are hydrolyzed from the triglycerides with increasing difficulty, the removal of the last fatty acid requiring special conditions. Mattson and Beck[6] have presented evidence that pancreatic lipase is specific for the hydrolysis of primary ester linkages; for this reason the digestion of a triglyceride by lipase would proceed first, by removal of a terminal fatty acid to produce a 1, 2-diglyceride, and then the other terminal fatty acid would be removed to produce a 2-monoglyceride. Since this last fatty acid is linked by a secondary ester group, its removal would require isomerization to a primary ester linkage. This is a relatively slow process; as a result, monoglycerides are the major end products of fat digestion and less than one-third of the ingested fat is completely broken down to glycerol and fatty acids. The digestion of a triglyceride as described above may be illustrated as follows:

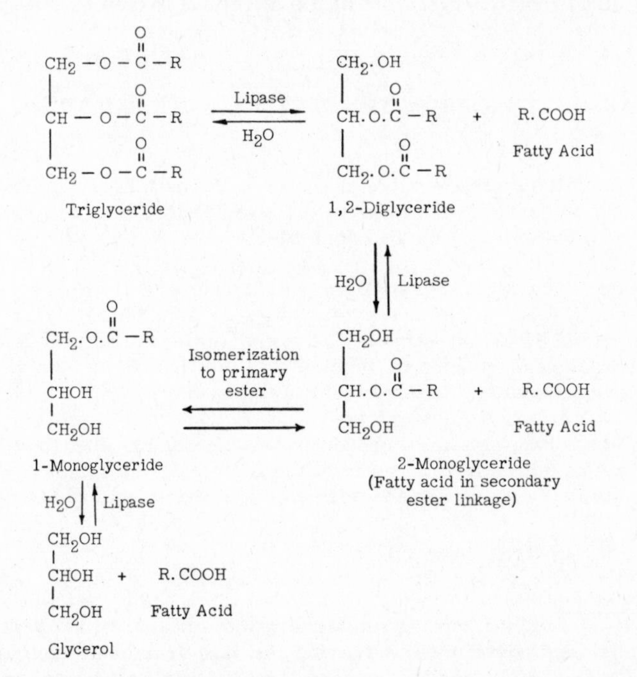

Glycerol, because it is water-soluble, is easily absorbed from the intestine; but the fatty acids and monoglycerides, since they are only sparingly soluble in an aqueous medium, require some auxiliary mechanism for their absorption. It is believed that a combination of the bile salts with the fatty acids produces a soluble product (the hydrotropic complexes) which can be absorbed from the intestinal lumen into the mucosal cells which line the intestinal cavity. After these fatty acid–bile salt complexes and glycerol have passed into the mucosal cells, the fatty acid is released from the bile salt and recombined with a glycerol moiety to resynthesize a neutral fat. The bile salts are then absorbed into the portal circulation and pass to the liver, where they are re-excreted in the bile. This is the so-called enterohepatic circulation of the bile salts.

There is some evidence to indicate that the glycerol originally obtained from the digestion of the fat may not be used when the fat is re-formed within the intestinal cells. This was suggested by the experiments of Reiser[7], in which C^{14}-labeled glycerol obtained from the hydrolysis of a fat in the intestine did not appear in the triglycerides of the lymph. Instead, a three-carbon compound derived from glycolysis, dihydroxyacetone, was found to function as a source of glycerol for the esterification of fatty acids and the formation of neutral fat.

Neutral fat, having been resynthesized in the intestinal mucosa after absorption as described above, is not transported to any extent in the portal venous blood. Instead the great majority of absorbed fat appears as particles of lipid associated with only small quantities of protein, the **chylomicrons**, in the lymphatic vessels of the abdominal region (the lacteals). The majority of absorbed fatty acids of more than ten carbon atoms in length are resynthesized with glycerol in the intestinal mucosa into new triglycerides which are then transported via the lymph. Fatty acids with carbon chains shorter than ten to 12 carbons are transported in the portal venous blood as unesterified (free) fatty acids. This was demonstrated in experiments in which labeled fatty acids were fed either as free acids or in neutral fat. About 90% of labeled palmitic (C_{16}) acid was found in the lymph after it was fed to experimental animals. Stearic (C_{18}) acid was poorly absorbed. Myristic (C_{14}) acid was well absorbed mostly into the lymph; however, while almost all of the lauric (C_{12}) acid or decanoic (C_{10}) acid was absorbed, these were not found to any extent in the lymph. Presumably they passed into the portal blood. These two fatty acids are, however, not important constituents of fats ordinarily taken in the diet except in the fats of milk.

Phosphate has been shown to accelerate fat absorption, and under certain conditions inhibition of phosphorylation by the use of phlorhizin or monoiodoacetate hinders the absorption of fat. These facts suggest that phosphorylation may be involved in fat absorption. Such phosphorylation may require the formation of a phospholipid (a lecithin), which increases the solubility of the fat and thus promotes its absorption. The process may occur in the intestinal mucosa after absorption of glycerol and fatty acids, as described above. However, the increase in the phospholipid content of the blood and lymph during fat absorption is very slight.

There is evidence that unhydrolyzed fat can be absorbed if it is dispersed in very fine particles (not over 0.5 micron in diameter). A combination of bile salts, fatty acids, and a monoglyceride will bring about this fine degree of dispersion of neutral fats. Certain synthetic "wetting agents" such as Tween® 80 (sorbitan monooleate) have a similar effect and are used therapeutically to promote fat absorption. Since bile salts play such an important role in the absorption of fats, it is obvious that fat absorption is seriously hampered by a lack of bile in the intestine such as results when the bile duct is completely obstructed.

The absorption of cholesterol is facilitated by esterification with fatty acids, a process catalyzed by **cholesterol esterase** of pancreatic and intestinal secretions. The plant sterols (sitosterols) are esterified by the same enzyme, but these sterols are not absorbed from the intestine. It has been reported that the feeding of soybean sterols to rats resulted in a lowering of cholesterol in the blood and lymph; the decrease was mainly in the ester fraction. It is suggested that this effect of the soybean sterols may be explained as an interference by the plant sterol with the action of cholesterol esterase, possibly by competing with cholesterol for the enzyme as well as for bile salts and fatty acids.

Summary of Fat Absorption:

The dietary fat is digested, by the action of the lipase present in the intestine, to glycerol, fatty acids, and partially split products such as monoglycerides and diglycerides. With the aid of the bile salts, these products of fat digestion enter the mucosal cells of the small intestine where a resynthesis of fat occurs, possibly utilizing three-carbon compounds derived from glycolysis as a source of the glycerol moiety. The resynthesized fat then passes into the lymphatics (the lacteals) of the abdominal cavity and thence by way of the thoracic duct to the blood, where it may be detected as lipid-protein particles about one micron in diameter, the so-called **chylomicrons.** The bile salts are carried by the portal blood to the liver and excreted in the bile back to the intestine; this is the so-called enterohepatic circulation of the bile salts.

Some phospholipid may be formed in the course of the absorption of fat, although there is no evidence that the amount or turnover rate of phospholipid in the intestinal wall is increased during fat absorption. Phospholipid obtained from the diet may be absorbed as such because of its hydrophilic nature; it may then travel by way of the portal blood directly to the liver.

Cholesterol is absorbed into the lymphatics. Its absorption is facilitated by esterification with fatty acids; both free cholesterol and cholesterol esters are found in the blood. Of the plant sterols (phytosterols), none is absorbed from the intestine except ergosterol, which is absorbed after it has been converted by irradiation to vitamin D.

Absorption of Protein:

It is probable that under normal circumstances the dietary proteins are completely digested to their constituent amino acids and that these end products of protein digestion are then rapidly absorbed from the intestine into the portal blood. The amino acid content of the portal blood rises during the absorption of a protein meal, and the rise of individual amino acids in the blood a short time after their oral administration can be readily detected. Animals may be successfully maintained in a satisfactory nutritional state with respect to protein when a complete amino acid mixture is fed to them. This indicates that intact protein is not necessary.

A difference in the rate of absorption from the intestine of the two isomers of an amino acid has been reported. The natural (L) isomer is apparently absorbed more rapidly than the D isomer. This suggests that amino acids may not be absorbed by a simple diffusion process but that some specific coupling mechanism which permits stereochemical differentiation is involved. When groups of amino acids are fed, there is some evidence of competitive effects whereby one amino acid fed in excess can retard the absorption of another. These observations are similar to those made with respect to reabsorption of amino acids by the renal tubules.

The absorption of small peptide fragments from the intestine is undoubtedly possible, and it is very likely that this normally occurs. During the digestion and absorption of protein, an increase in the peptide nitrogen of the portal blood has in fact been found.

A puzzling feature of protein absorption is that in some individuals sensitivity to protein (in the immunologic sense) results when they eat certain proteins. It is known that a protein is antigenic, i.e., able to stimulate an immunologic response, only if it is in the form of a relatively large molecule; the digestion of a protein even to the polypeptide stage destroys its antigenicity. Those individuals in which an immunologic response to ingested protein occurs must therefore be able to absorb some unhydrolyzed protein. This is not entirely undocumented, since the antibodies of the colostrum are known to be available to the infant.

• • •

References:

1. Dreiling, D.A., and Janowitz, H.D.: Am.J.Med. **21**:98, 1956.
2. Ziffrin, S.E., and Hosie, R.T.: Proc.Central Soc.Clin.Research **28**:86, 1955.
3. Carey, J.B., and Watson, C.J.: J.Biol.Chem. **216**:847, 1955.
4. Siperstein, M.D., and Murray, A.W.: Science **123**:377, 1956.
5. Silen, W., Harper, H.A., Mawdsley, D.L., and Weirich, W.L.: Proc.Soc.Exper.Biol. and Med. **88**:138, 1955.
6. Mattson, F.H., and Beck, L.W.: J.Biol.Chem. **214**:115, 1955; ibid., **219**:735, 1956.
7. Reiser, R., and Williams, M.C.: J.Biol.Chem. **202**:815, 1953.

12...

Detoxication

The term ''detoxication'' has been applied to chemical reactions in the body which serve to convert a toxic substance into a nontoxic form for removal by some excretory route. The detoxication of indole is an example. This substance is produced in the intestine from tryptophan by bacterial action (see p. 160). Most of the indole is eliminated in the feces; but that which may be absorbed from the intestine is detoxified by oxidation to **indoxyl**, which is then combined with sulfate to form **indican** for excretion into the urine. The quantity of indican in the urine may be a measure of the extent of intestinal putrefaction. The formation of indican from indole is illustrated below:

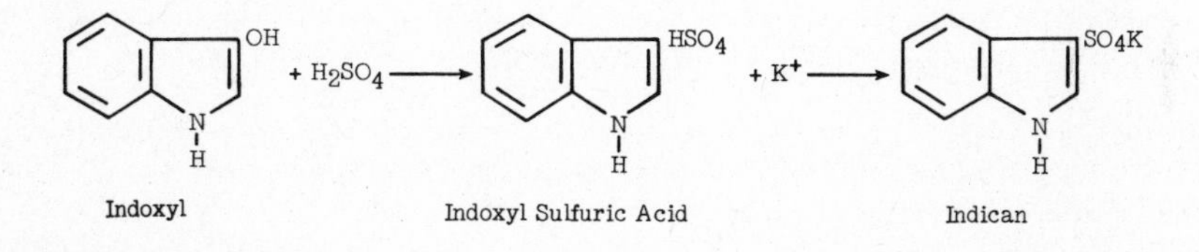

Indoxyl   Indoxyl Sulfuric Acid   Indican

Later studies have shown that these reactions are rather common in the body and do not necessarily involve a toxic substance. The term ''protective synthesis'' was suggested as more descriptive of the process. Most of the reactions occur in the liver, but the kidney and other tissues may also participate.

The most important of these protective reactions involve (1) oxidation, (2) reduction, or (3) conjugation. Each of these will be illustrated in the following sections.

OXIDATION

Aliphatic compounds and aliphatic side chains of aromatic derivatives are more easily oxidized than the aromatic derivatives themselves, although there are instances of the oxidation of aromatic rings. The metabolic degradation of the aromatic amino acids, tyrosine and phenylalanine, for example, involves opening of the aromatic nucleus (see p. 254).

Some of the benzene is also converted to phenol and further metabolized. Benzene derivatives like toluene, ethylbenzene, benzaldehyde, and benzyl alcohol may be oxidized to benzoic acid. Phenylethyl alcohol gives phenylacetic acid. *m*-Xylene (a dimethyl benzene) or trimethyl benzene may have only one methyl group oxidized to benzoic acid.

The aromatic amines are usually only partially oxidized, whereas the aliphatic amines are completely oxidized and destroyed in the body. Thus benzylamine ($C_6H_5.CH_2.NH_2$) yields benzoic acid. Aniline ($C_6H_5.NH_2$) is oxidized to *p*-aminophenol, and the amino group remains unchanged. Acetanilid, a common ingredient of sedative drugs, is similarly oxidized to form *p*-acetyl aminophenol.

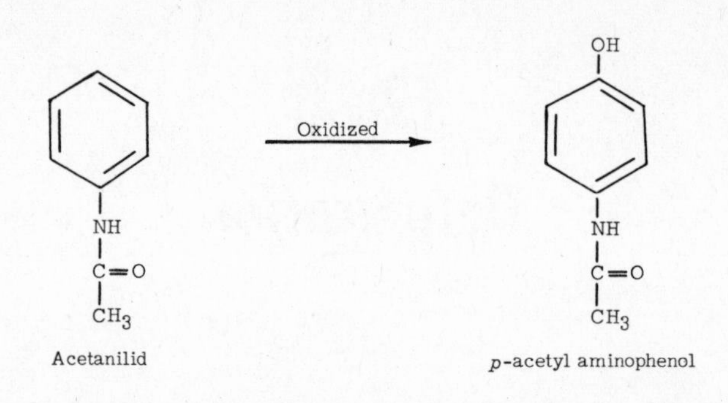

Acetanilid *p*-acetyl aminophenol

Ring structures more complex than benzene are oxidized by the addition of hydroxy groups at various positions on the ring. Such ring structures are exemplified by anthracene or its isomer, phenanthrene, which is the parent structure of the sterid nucleus.

Most of the compounds formed by oxidative changes undergo still further alteration, as will be described in the following sections.

REDUCTION

Protective synthesis (detoxication) by reduction is less common than detoxication by oxidation. Examples are (1) the conversion of picric acid to picramic acid:

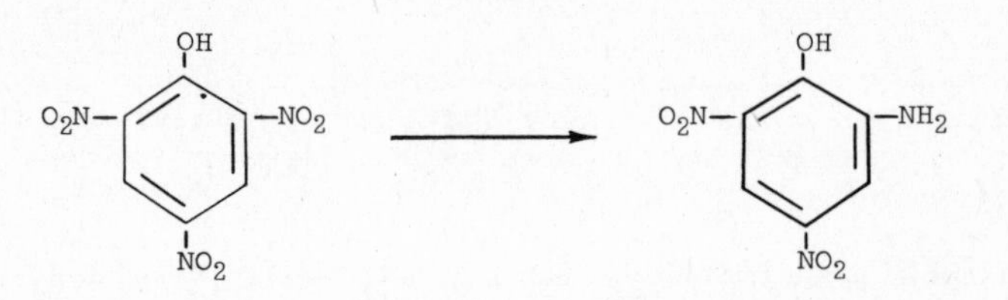

and (2) the simultaneous reduction and oxidation of the same compound, as in the conversion of *p*-nitrobenzaldehyde to *p*-aminobenzoic acid, or of nitrobenzene to *p*-aminophenol:

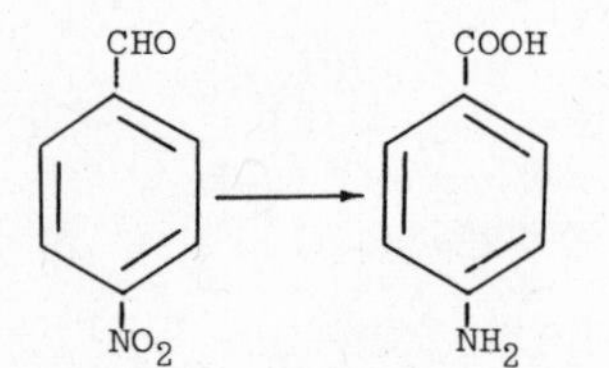

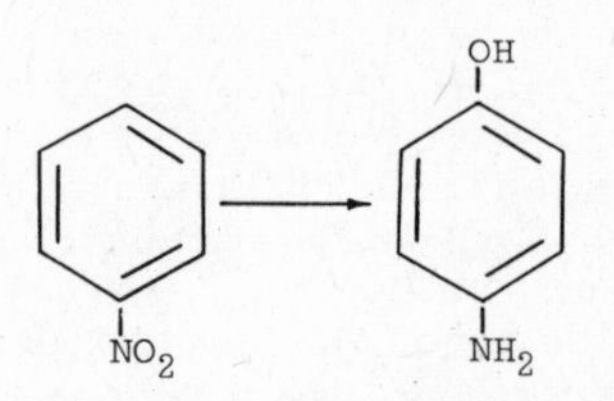

p-nitrobenzaldehyde **p-aminobenzoic acid** nitrobenzene *p*-aminophenol

CONJUGATION

Conjugation reactions involve the combination of a metabolite with some other substance followed by elimination of the resulting conjugate. Conjugation often occurs in association with oxidation or reduction. Substances used for conjugation with the metabolite to be eliminated include the amino acids, glycine, glutamine, ornithine, and cysteine; and sulfuric, glucuronic, and acetic acids as well as the methyl group (methylation).

Examples of Amino Acid Conjugation:

A. Glycine:

1. Combines with benzoic acid to produce benzoyl glycine (hippuric acid) (see also p. 235).

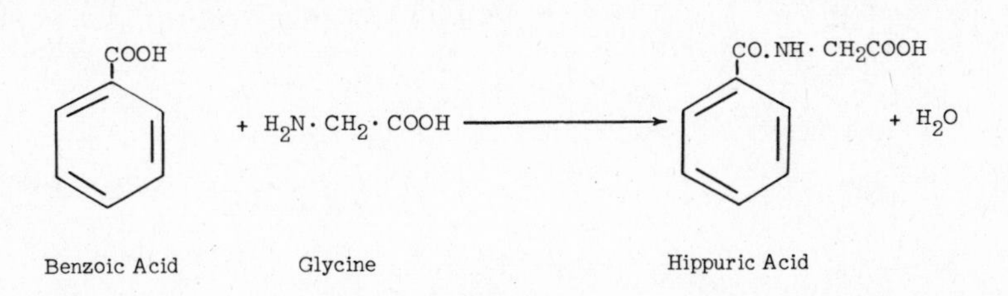

Benzoic Acid　　Glycine　　Hippuric Acid

A similar reaction occurs with *p*-aminobenzoic acid.

The formation of hippuric acid after the administration of benzoate is used as a test of liver function (see p. 279). Schachter and Taggart[1] have presented evidence that activation of benzoic acid by formation of the coenzyme A derivative (with Co A and ATP) is a necessary preliminary to conjugation with glycine. Their studies were carried out with a soluble pig kidney enzyme preparation, and the essential factors in the condensation reaction were found to be limited to benzoyl-coenzyme A, glycine, and the kidney enzyme.

2. Combines with salicylic acid (*o*-hydroxybenzoic acid) to form salicyluric acid.
3. Combines with nicotinic acid to form nicotinuric acid.

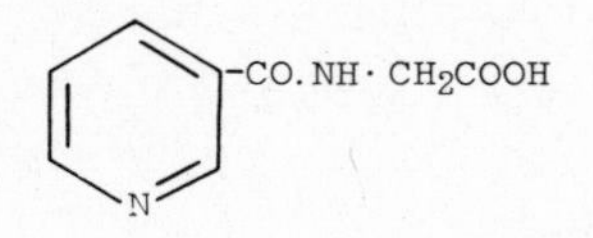

B. Cysteine: This conjugation is demonstrated by the feeding of aromatic halogen derivatives such as brombenzene. Acetylation of the amino group of cysteine also occurs in this reaction, and mercapturic acids are formed.

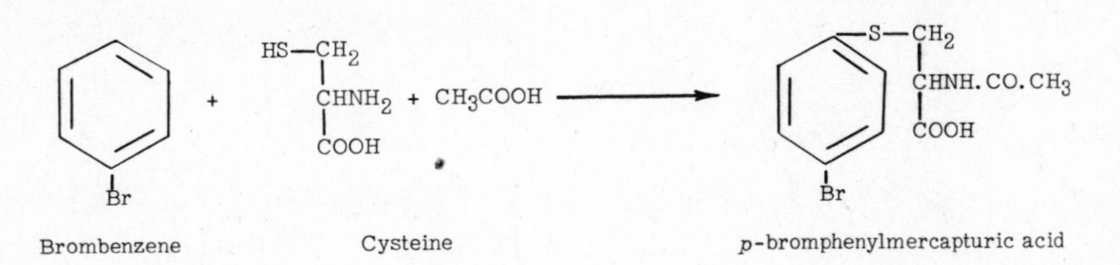

Brombenzene　　Cysteine　　*p*-bromphenylmercapturic acid

Brombenzene may also be oxidized to bromphenol and excreted in combination with sulfuric or glucuronic acid.

Naphthalene, anthracene, and possibly phenanthrene also may yield mercapturic acids.

Conjugations With Sulfuric Acid:

A number of compounds may be conjugated with sulfate. These include indole, phenols, the amino acid tyrosine being an example, and the steroid hormones such as the 17-ketosteroids.

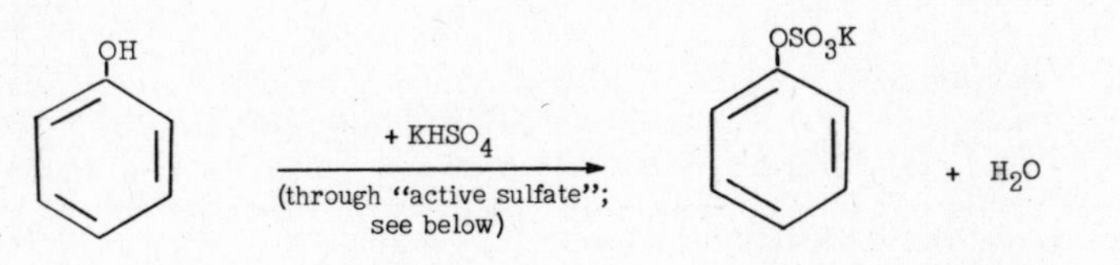

The amino acids methionine and cysteine (see p. 249) serve as the sources of sulfate used in these conjugations. The sulfate conjugates excreted in the urine constitute the so-called "ethereal fraction" of the total urinary sulfur.

Conjugation of phenol as well as steroids with sulfate has been shown to require an "active sulfate" intermediate formed with the aid of ATP[2]. The reactions are as follows:

1. ATP + $SO_4^{=}$ $\xrightleftharpoons{Mg^{++}}$ Adenosine phosphosulfate (APS) + 2Ⓟ

2. APS + ATP ⟶ Adenosine-3'-phospho-5'-phosphosulfate (PAPS) + ADP ("active sulfate")

Adenine — ribose — Ⓟ — O — S(=O)(=O) — O⁻ (with P on ribose)

Structure of Phospho-adenine Phosphosulfate (PAPS) ("Active Sulfate")

The actual transfer of sulfate from "active sulfate" to the substrate (e.g., phenol) is catalyzed by enzymes designated **sulfokinases.**

Sulfate is bound mostly in ester linkages in many compounds of biological importance, e.g., sulfated mucopolysaccharides like chondroitin sulfuric acid. In the formation of these compounds, the active sulfate system is also essential.

Biological sulfate activation and transfer has recently been reviewed by Lipmann[3].

Conjugation With Glucuronic Acid:

This glucose derivative is a common conjugate, particularly with compounds containing carboxyl or hydroxy groups.

Benzoic acid combines with glucuronic acid to form 1-benzoylglucuronic acid. Phenol forms phenolglucuronic acid.

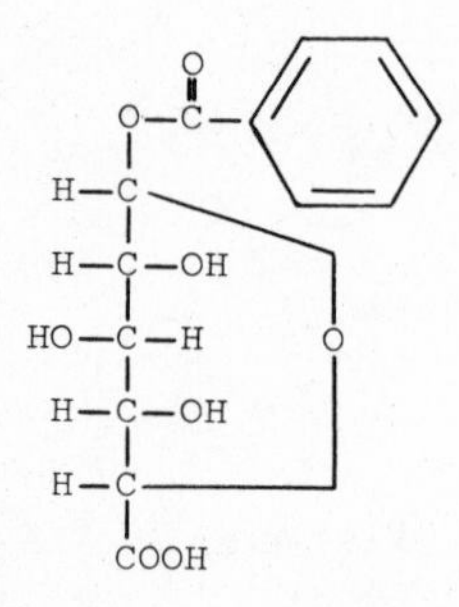

Benzoylglucuronic Acid

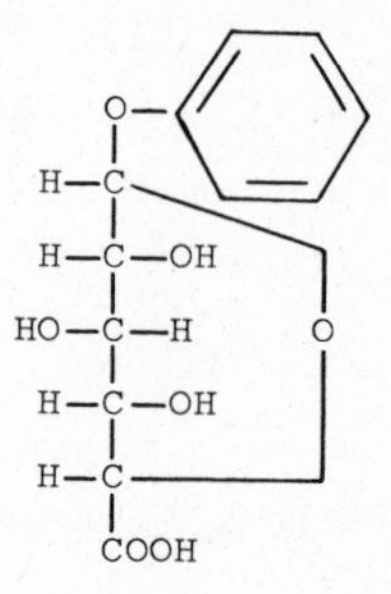

Phenolglucuronic Acid

Certain hormone derivatives (see p. 353) are excreted partially as glucuronides. Phenolphthalein and anthracene may also form glucuronides.

When camphor is reduced, a secondary alcohol, borneol, is produced. This is exclusively eliminated in combination with glucuronic acid.

Bilirubin is conjugated in the liver with glucuronic acid and thus converted into a water-soluble form for excretion in the bile. The conjugated bilirubin, which is identical with "direct-reacting" bilirubin (see p. 275), is a diglucuronide in which the two mols of glucuronic acid are conjugated to the carboxyl groups of the propionic acid residues on Rings III and IV of bilirubin.

In conjugation reactions with glucuronic acid, the glucuronyl moiety must be in the form of the nucleotide, uridine diphosphate glucuronide[4]. This compound, which may be thought of as "active" glucuronide, is formed from glucose in the uronic acid pathway as shown on p. 192. A specific transferring enzyme in liver, glucuronyl transferase, is also required to complete the reaction whereby conjugations of glucuronic acid with various substrates may be accomplished.

Beta-glucuronidase is an enzyme found in many animal tissues which hydrolzes glucuronic acid conjugates. A preparation of this enzyme obtained from the spleen of the calf is used to hydrolyze urinary steroid glucuronide conjugates as a preliminary to the measurement of these hormones in the urine.

Hydrolysis of the glucuronide linkage produces free glucuronic acid, which is a reducing substance. If this occurs in a urine where there happens to be a considerable quantity of these conjugates, false-positive glucose tests may be obtained.

Conjugation With Acetic Acid:

Acetylation of amino groups is a frequent "detoxication" mechanism. The role of acetylation in the formation of mercapturic acids has already been mentioned. *p*-Aminobenzoic acid and sulfonamide derivatives are largely acetylated before excretion, although phenolic compounds are probably also formed and detoxified with sulfate.

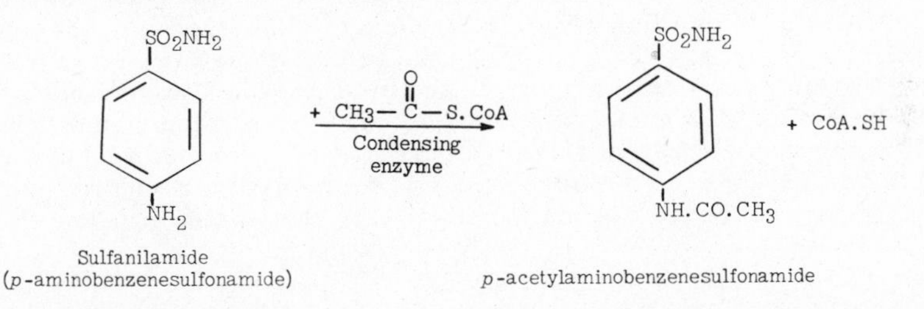

The acetylation of sulfanilamide is often used as a classical illustration of acetylation reactions in general and, in fact, as a means of testing for the occurrence of acetylation in experimental systems. The enzyme system catalyzing this and other acetylations requires coenzyme A and ATP to form acetyl-Co A ("active acetate"), which serves as the acetate donor. Condensing enzyme (see p. 111) then catalyzes the transfer of the acetyl group to the acceptor, in this case sulfanilamide.

Conjugation by Methylation:

Although reactions involving methylation of natural metabolites occur frequently in the body, the methylation of a foreign substance, pyridine, was the first to be detected. The methylated hydroxypyridine, after formation probably by the liver, is then excreted in the urine. The various derivatives of nicotinic acid, which is also a pyridine, are also largely excreted as methylated compounds (see p. 81).

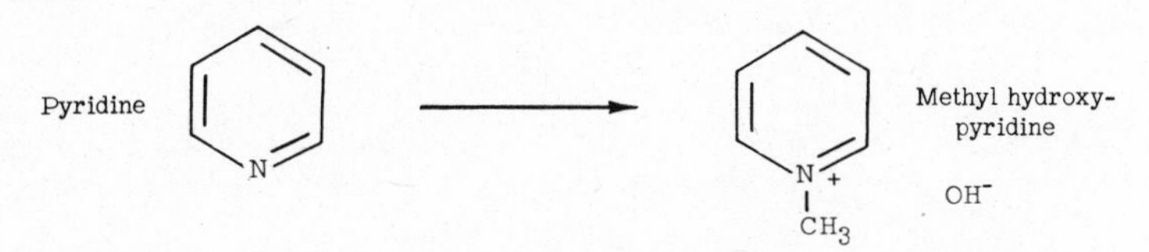

Sources of the labile methyl groups required for methylation include methionine, choline, and betaine. The body can also synthesize methyl groups. The important physiologic role of methyl groups is further discussed in connection with transmethylation on p. 206.

• • •

References:

1. Schachter, D., and Taggart, J. V.: J. Biol. Chem. **203**:925, 1953.
2. Brunngraber, E.: J. Biol. Chem. **233**:472, 1958.
3. Lipmann, F.: Science **128**:575, 1958.
4. Dutton, G. J., and Storey, I. D. E.: Biochem. J. **53**:xxxvii, 1953.

Bibliography:

Williams, R. T.: Detoxication Mechanisms. Wiley, 1948.

13...
Metabolism of Carbohydrate

In the average diet carbohydrate comprises more than half of the total caloric intake. However, only a limited amount of this dietary carbohydrate can be stored as such (see table below). It is now known that the unstored portion of the ingested carbohydrate is converted to fat by the metabolic processes of lipogenesis. In the form of fat, the extra carbohydrate is stored and utilized. It should be emphasized, however, that this conversion of carbohydrate and its utilization as fat requires the simultaneous metabolism of a minimum of carbohydrate (see p. 211).

Lipogenesis (the conversion of glucose to fat) takes place both in the liver and in the peripheral tissues, including adipose tissue[1]. In fact, the ability of adipose tissue to convert glucose to fatty acid actually exceeds that of the liver[2]. Thus, although the liver plays a major role in lipogenesis from glucose, the action of adipose tissue in this respect may be more important. In addition to its other effects on carbohydrate metabolism, as will later be described, insulin is essential to lipogenesis both in liver and in adipose tissue. Diet also exerts an influence on lipogenesis[3]. Both lipogenesis and glucose oxidation are diminished in adipose tissue by fasting or by a high-fat diet; dietary carbohydrate, on the other hand, stimulates lipogenesis and glucose utilization.

Storage of Carbohydrate in Normal Adult Man (70 Kg.)

Liver glycogen	6.00% = 108 Gm. (Liver weight, 1800 Gm.)
Muscle glycogen	0.70% = 245 Gm. (Muscle mass, 35 Kg.)
Extracellular sugar	0.10% = 10 Gm. (Total Volume, 10 liters)
TOTAL	363 Gm. × 4 = 1452 Calories

Not all carbohydrates can be utilized for the nutrition of the cell; the three principal monosaccharides utilized are glucose, fructose, and galactose; of these, only glucose is of major importance. The disaccharides, maltose, lactose, and sucrose, are converted to these three monosaccharides in the course of digestion. Galactose is important as a constituent of lactose, the sugar of milk. It is also found in the cerebrosides, the carbohydrate-containing fats of nervous tissue. Both galactose and fructose are convertible to glucose in the body.

Pentose sugars, such as xylose and arabinose, may be obtained from the diet; but all of the details of their metabolism are not yet known. The pentoses, D-ribose and 2-deoxy-D-ribose, are synthesized in the body for incorporation into nucleoprotein and other important metabolites.

THE BLOOD SUGAR

Sources of Blood Sugar:

The sugar of the blood is glucose. It is obtained from three sources.

A. Carbohydrates by Digestion: The main sources of glucose are the carbohydrates of the diet. The digestion of starches and sugars produces glucose, which is absorbed from the intestine into the blood.

B. Various Glycogenic Compounds by **Gluconeogenesis**: A second source results from gluconeogenesis, the conversion of noncarbohydrate precursors into glucose. These precursors include:
1. The glycogenic amino acids (see p. 230).
2. A number of compounds which are formed in the metabolic breakdown of glucose. These intermediates may produce glucose because many of the reactions by which they are formed are reversible. Examples of such glycogenic metabolites are the dicarboxylic acids - succinic, fumaric, etc. - of the citric acid cycle (see p. 181), and lactic and pyruvic acids.
3. Glycerol obtained by hydrolysis of neutral fat. Odd-chain fatty acids such as propionic acid are also glycogenic. The ability of long-chain, even-numbered fatty acids to be converted to glucose is disputed.

C. Liver Glycogen: A third source of blood glucose is liver glycogen; glucose is formed by **glycogenolysis**, the hydrolysis of glycogen stored in the liver.

In experiments with hepatectomized dogs it has been shown that the kidneys may contribute sugar to the blood. The extent of the renal contribution to the blood sugar in the intact animal is not known.

Concentration of Sugar in the Blood:

When the quantity of glucose in the blood is measured after a period of fasting, it is found to be between 70 and 100 mg. per 100 ml. of blood. Shortly after the ingestion of carbohydrate, the blood sugar rises to 120 to 130 mg. per 100 ml. or higher. Under normal conditions, the glucose level soon declines from these peak values so that after one and one-half to two hours the fasting levels are once again approached. The blood sugar level remains constant between meals despite the fact that glucose is constantly used by the tissues of the body. This regulation of the blood sugar depends upon the ability of the liver to remove glucose from the blood, when the normal level is exceeded, by the various processes mentioned above. A portion of the blood glucose is converted by the liver to glycogen, sometimes termed animal starch, by the process of hepatic glycogenesis, and some is directly broken down and utilized for energy. However, the majority of the glucose derived from the digestion of dietary carbohydrate is converted to fat by the liver and the peripheral tissues, including adipose tissue, and stored in the fat depots throughout the body for later use. Between meals, the glycogen stores of the liver maintain the blood sugar by re-forming glucose as required. This process, **glycogenolysis**, is augmented when necessary by **gluconeogenesis.**

The liver (and, to a slight extent, the kidney) is the only source of sugar for the blood during periods of fasting. Muscle, although it stores glycogen, does not contribute glucose to the blood. The unique position of the liver and the kidney as sources of glucose derived from the breakdown of glycogen is due to the presence in these organs of a specific enzyme, a **phosphatase** (glucose-6-phosphatase), which can split glucose-6-phosphate to free glucose.

Regulation of the Blood Sugar; Effects of Hormones:

The maintenance of normal levels of sugar in the blood is one of the most finely regulated of all the homeostatic mechanisms and one in which the liver plays an essential role. The liver functions both in the removal of sugar from the blood and in the addition of sugar to the blood. During periods of fasting, the liver contributes sugar by glycogenolysis and gluconeogenesis. When blood sugar levels are raised, there is an immediate inhibition of the hepatic production of sugar, and sugar is removed from the blood and stored as liver glycogen. If the increase in the blood sugar level is too great to be controlled by hepatic factors alone, storage in other tissues occurs (e.g., muscle). There is also an increase in the rate of utilization of sugar by the peripheral tissues. In the experiments of Soskin and Levine the utilization of dextrose in totally abdominally eviscerated dogs at normal blood sugar levels (80 to 100 mg. per 100 ml.) was about 224 mg. per Kg. per hour, whereas an elevation of the blood sugar to 500 mg. per 100 ml. was accompanied by gradual increases in peripheral utilization of glucose to a maximum of about 550 mg. per Kg. per hour.

The activity of the liver in maintaining normal levels of glucose in the blood is influenced by various hormones. These include the following:

1. **Insulin**, produced by the beta cells of the **islets of Langerhans** in the pancreas, fosters glycogenesis as well as lipogenesis from carbohydrate. It may also inhibit hepatic production of glucose. The effect of this hormone is, therefore, to promote a **reduction** in the blood sugar by its influence on those metabolic processes which remove sugar from the blood.
2. The **anterior pituitary** gland secretes hormones which tend to **elevate** the blood sugar by acting as antagonists to the action of insulin. These are pituitary growth hormone, ACTH (corticotropin), and a third "diabetogenic" principle which is believed to be secreted by the pituitary.
3. The **adrenal cortex** secretes a number of steroid hormones which have profound effects on metabolism. Those which have an oxygen on position 11 of the steroid nucleus (the 11-oxysteroids; S hormones of Albright) stimulate gluconeogenesis, the formation of glucose from the amino acids of protein precursors. These hormones are also insulin antagonists. The diabetogenic action of ACTH is probably through its action on the production of the 11-oxysteroids by the adrenal cortex.
4. **Epinephrine**, secreted by the adrenal medulla, stimulates hepatic glycogenolysis, the formation of glucose from the glycogen of the liver. This hormone causes an elevation in the blood sugar if the liver is filled with glycogen. In experiments with rat liver slices epinephrine partially inhibited the incorporation of acetate into fatty acids. If it exerts a similar action in the intact animal, this would inhibit the disposal of carbohydrate by interference with lipogenesis, thus producing a rise in blood sugar.
5. **Glucagon**, the hormone of the alpha cells of the pancreas, also raises the blood sugar by increasing glycogenolysis. It is therefore often referred to as the "hyperglycemic-glycogenolytic factor," **HGF**. Like epinephrine, the action of glucagon on the blood sugar depends upon the presence of glycogen in the liver.
6. **Thyroid** hormone should also be considered as affecting the blood sugar. In severe thyrotoxic states, symptoms of mild diabetes may be noted. These disappear when the disease has been successfully treated. There is also a complete absence of glycogen from the livers of thyrotoxic animals. It is likely that the effects of thyroid hormone on carbohydrate metabolism are related to its general accelerating effects on metabolism; a specific connection with the diabetic state is not yet apparent.

Because of its immediate effect on the blood sugar, insulin is of specific value in the treatment of diabetes. The 11-oxysteroids of the adrenal cortex, Compounds A, B, E (cortisone), and F (hydrocortisone) and various hormones of the anterior pituitary are diabetogenic. Their effect is opposite to that of insulin. Thus, whereas insulin reduces the blood sugar and induces **glycogenesis** and **lipogenesis** (formation of fat from carbohydrate), these diabetogenic hormones induce hyperglycemia and possibly glycosuria; they also exert an inhibitory action on lipogenesis. The metabolic effects on the utilization of carbohydrate of these diabetogenic hormones are therefore opposed to the effects of insulin. Under normal conditions the utilization of carbohydrate is regulated by a balance between the two groups of hormones. Diabetes could therefore be caused either by a lack of insulin in the presence of normal quantities of pituitary and adrenocortical hormones or by an excess of these latter hormones in the presence of a normal insulin output. It is for this reason that one may refer to pancreatic, pituitary, or steroid diabetes as separate entities. In pituitary insufficiency or in adrenal insufficiency (such as in Addison's disease), patients exhibit abnormal sensitivity to insulin as evidenced by a tendency to decreased levels of sugar in the blood, i.e., hypoglycemia. This is obviously the direct opposite of the diabetic state.

The Renal Threshold for Glucose:

When the blood sugar rises to relatively high levels, the kidney also exerts a regulatory effect. Glucose is continually filtered by the glomeruli but is ordinarily returned completely to the blood by the reabsorptive system of the renal tubules. The reabsorption of glucose is effected by phosphorylation in the tubular cells, a process which is similar to that utilized in the absorption of this sugar from the intestine. The phosphorylation reaction is enzymatically catalyzed, and the capacity of the tubular system to reabsorb glucose is limited by the concentration of the enzymatic components of the tubule cell to a rate of about 350 mg. per minute. When the blood levels of glucose are elevated, the glomerular filtrate may contain more glucose than can be reabsorbed; the excess passes into the urine to produce a **glycosuria**.

In normal individuals, glycosuria occurs when the venous blood sugar exceeds 170 to 180 mg. per 100 ml. This level of the venous blood sugar is termed the **renal threshold** for glucose. Since the maximal rate of reabsorption of glucose by the tubule (**Tm_G** - the tubule maximum for glucose) is a constant, it is a more accurate measurement than the renal threshold, which varies with changes in the glomerular filtration rate.

Glycosuria may be produced in experimental animals with phlorhizin, which inhibits the glucose reabsorptive system in the tubule. This is known as **renal glycosuria** since it is caused by a defect in the renal tubule and may occur even when blood glucose levels are normal. Glycosuria of renal origin is also found in human subjects. It may result from congenital defects in the kidney (see p. 289) or as a result of disease processes.

Carbohydrate Tolerance:

The ability of the body to utilize carbohydrates is termed **carbohydrate tolerance.** It is determined by the nature of the blood sugar curve following the administration of carbohydrate. **Diabetes mellitus** ("sugar" diabetes) is characterized by decreased tolerance to carbohydrate. This is manifested by elevated blood sugar levels (hyperglycemia) and glycosuria and may be accompanied by defects in fat metabolism. Tolerance to carbohydrate is decreased not only in diabetes but also in conditions where the liver is damaged and in some infections as well. It would also be expected to occur in the presence of hyperactivity of the pituitary or adrenal cortex because of the antagonism of the hormones of these endocrine glands to the action of insulin.

Insulin, the hormone of the islets of Langerhans of the pancreas, increases tolerance to carbohydrate. Injection of insulin lowers the content of the sugar in the blood and increases the utilization of glycogen and its storage in the liver and muscle. An excess of insulin may lower the blood sugar level to such an extent that severe hypoglycemia occurs which results in convulsions and even in death unless glucose is administered promptly. In man, hypoglycemic convulsions may occur when the blood sugar is lowered to about 20 mg. per 100 ml. or less. Increased tolerance to carbohydrate is also observed in pituitary or adrenocortical insufficiency; presumably this is attributable to a decrease in the normal antagonism to insulin which results in a relative excess of that hormone.

Measurement of Glucose Tolerance:

The glucose tolerance test is a valuable diagnostic aid. Glucose tolerance (ability to utilize carbohydrate) is decreased in diabetes, increased in hypopituitarism, hyperinsulinism, and adrenocortical hypofunction (such as in Addison's disease).

A. Standard Test: After a 12-hour fast, the patient is given 0.75 to 1.5 Gm. of glucose per Kg. (or a standard dose of between 50 and 100 Gm. of glucose may be used). Specimens of blood and urine are taken before the administration of the sugar and at intervals of one-half or one hour thereafter for three to four hours. The concentration of sugar in the blood is measured and plotted against time. Qualitative tests for sugar in the urine are also made for detection of glycosuria of renal origin. As a result of the administration of glucose as described, the blood sugar in normal individuals increases in one hour from about 100 mg. per 100 ml. to about 150 mg. per 100 ml.; at the end of two to two and one-half hours, a return to normal levels occurs. In a diabetic patient, the increase in the blood sugar level is greater than in normal subjects and a much slower return to the pre-test level is observed; i.e., the glucose tolerance curve is typically higher and more prolonged than normal.

B. Exton-Rose Test: 100 Gm. of glucose are dissolved in 650 ml. of water and the solution divided into two equal parts. After a 12-hour fast, samples of blood and urine are collected; the first dose of the glucose solution (50 Gm.) is administered to the patient by mouth. Thirty minutes later, a second blood sample is collected and the remainder of the glucose solution is given. At the end of an hour, blood and urine samples are again collected. In the normal person, the blood sugar rises to a maximum of 75 mg. per 100 ml. above the fasting level in the first half hour; in the blood sample taken one hour after the administration of the first dose (30 minutes after the second dose) of glucose, the sugar is either lower than in the 30-minute blood sample or not more than 5 to 10 mg. per 100 ml. above it. In the diabetic both the first and second blood specimens show an elevation of the blood sugar above the values obtained in normal subjects at the same periods. All urine specimens in the normal subject are negative when tested for glucose.

C. An **intravenous** test is preferred if abnormalities in absorption of glucose from the intestine, as might occur in hypothyroidism or in sprue, are suspected. A 20% solution of glucose (0.5 Gm. per Kg.) is given intravenously at a uniform rate over a period of one-half hour. A control (fasting) blood specimen is taken, and additional blood samples are obtained one-half, one, two, three, and four hours after the glucose injection. In normal individuals, the control specimen of blood contains a normal amount of glucose; the concentration does not exceed 250 mg. per 100 ml. after the infusion has been completed; by two hours the concentration of sugar in the blood has fallen below the control level, and between the third and fourth hours it has returned to the normal fasting level.

GLYCOLYSIS

The breakdown of carbohydrate to pyruvic and lactic acids is termed **glycolysis.** This series of reactions is an important mechanism for the production of energy in the body: The process was originally studied in muscle, although it occurs in many other tissues. It is likely that glycolysis takes place by almost identical mechanisms in all tissues.

When a muscle contracts in a medium from which oxygen is excluded, anaerobic glycolysis takes place. Glycogen disappears, and pyruvic and lactic acids appear as the principal end products. When oxygen is admitted, aerobic recovery takes place; glycogen reappears, while pyruvic and lactic acids disappear. It is possible, then, to separate muscle activity into **anaerobic contraction** and **aerobic recovery** phases. Studies of oxygen consumption during the aerobic phase have shown that about one-fifth of the lactic acid produced is oxidized further, presumably to carbon dioxide and water. The remaining four-fifths is then reconverted to glycogen. These events are represented as in the diagram below.

The Lactic Acid Cycle

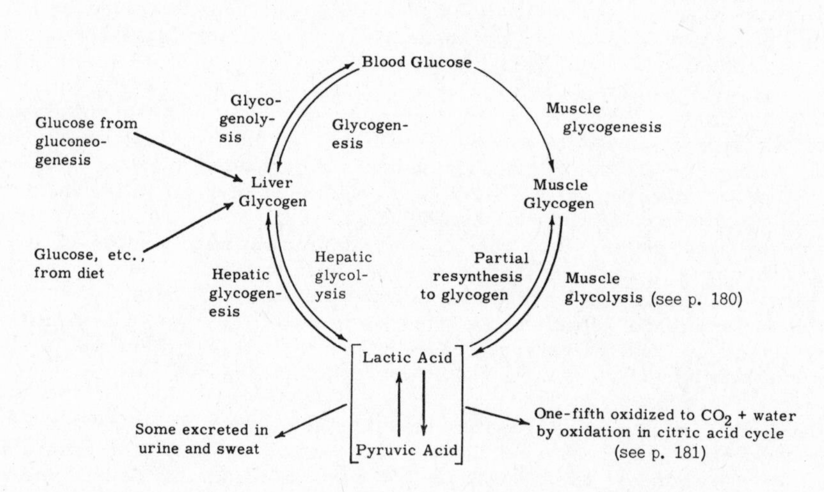

Energy for Muscular Contraction:

The energy obtained on hydrolysis of the terminal phosphate bond of ATP (see p. 119) to produce ADP is believed to be the immediate source of the energy of muscular contraction. The contractile system in the individual muscle fibril, the myofibril, is essentially composed of the major muscle protein, myosin. When the muscle is in an extended state, it is high in potential energy. Contraction of the muscle is accompanied by the liberation of kinetic energy, some of which is used for work. The remainder is liberated as heat. Relaxation and restoration of the fibril to the high-energy state requires the re-formation of the high-energy phosphate bond, which on hydrolysis originally supplied the energy of contraction. A supply of energy for this purpose is available as long as the concentration of ATP within the muscle is adequate. Creatine phosphate (or phosphocreatine, PC) acts as a source of high-energy phosphate to accomplish the prompt resynthesis of ATP. In the resting state, mammalian muscle contains four to six times as much PC as ATP. Thus muscle creatine phosphate is a reservoir of high-energy phosphate to assure the maintenance of a maximum supply of ATP. The transfer of high-energy phosphate from creatine phosphate to ADP, resulting in the formation of ATP, is known as the Lohmann reaction; it is catalyzed by an enzyme in muscle, **creatine kinase.** The reaction is reversible so that phosphate may be transferred from ATP to creatine to form creatine phosphate. This is catalyzed by an enzyme designated ATP-creatine transphosphorylase by Kirby et al.[4], who isolated the crystalline enzyme from rabbit muscle.

$$C \sim \textcircled{P} + A - \textcircled{P} \sim \textcircled{P} \underset{\text{ATP-creatine transphosphorylase}}{\overset{\text{(Lohmann reaction)} \atop \text{creatine kinase}}{\rightleftharpoons}} A - \textcircled{P} \sim \textcircled{P} \sim \textcircled{P} + C$$

The synthesis of creatine phosphate is made possible by the transfer of high-energy phosphate from ATP, which is formed during oxidative phosphorylation incident to glycolysis and the aerobic oxidation of lactic and pyruvic acids. (See below and pp. 119 and 120.)

Rephosphorylation of ATP and PC can be blocked by poisoning the tissue with iodoacetic acid. When so poisoned, a muscle may contract only about 100 times and then cease because of the depletion of these energy reservoirs. From this experiment and others, in which the breakdown of muscle glycogen has been prevented, it appears that the energy required for the regeneration of high-energy phosphate bonds is supplied by the glycolysis of glycogen to lactic and pyruvic acids and the further oxidation of pyruvic acid in the citric acid cycle. The sequence of events by which glycogen is degraded to supply this energy is described in the following sections.

In summary, the nerve impulse to the muscle initiates muscle contraction; the breakdown of ATP to ADP supplies the energy of this contraction. The resynthesis of ATP from ADP and phosphate (as well as that of PC) is accomplished with the aid of energy obtained from glycolytic mechanisms, which will be described later; the hydrolysis of phosphocreatine is also available to assure the prompt resynthesis of ATP. Obviously, the maintenance of the activity of the muscle is dependent upon a steady supply of high-energy phosphate in the form of ATP and PC.

Phosphorylation of Sugars:

The simple sugars, glucose, fructose, and galactose, enter into the glycolytic pathways only after each has been phosphorylated. For each sugar a specific enzyme is required to catalyze the phosphorylation, and ATP acts as phosphate donor. Thus **galactokinase** catalyzes phosphorylation of galactose to galactose-1-phosphate; **fructokinase,** the phosphorylation of fructose to fructose-1-phosphate; and **glucokinase,** the phosphorylation of glucose to glucose-6-phosphate. These reactions in liver tissue are diagrammed on p. 176.

The glucokinase - and perhaps also the galactokinase - reactions are affected by insulin and by the hormones which are antagonistic to the action of insulin. According to Price, Cori, and Colowick, glucokinase activity is inhibited by an extract of the anterior pituitary gland and this inhibition is reversed by insulin. It has also been shown that adrenocortical extracts augment the inhibition of glucokinase by anterior pituitary extracts and that insulin reverses the combined inhibition. These observations on hormonal effects at the very first step in the utilization of glucose are of considerable significance as a means of explaining one aspect of the role of these hormones in carbohydrate metabolism. According to this hypothesis, at least one function of insulin is to provide for the phosphorylation of glucose to glucose-6-phosphate, a reaction which is essential to the further metabolism of this sugar. The hormones which act antagonistically to insulin would then produce their "diabetogenic" effect by blocking the phosphorylation step. Fructokinase is not affected by insulin or the antagonistic hormones. If given intravenously to diabetics, it will be found to disappear normally from the blood without the need for insulin. The details of fructose metabolism are considered on p. 194.

Pathways in the Liver of the Metabolism of Glucose and Related Hexose Sugars

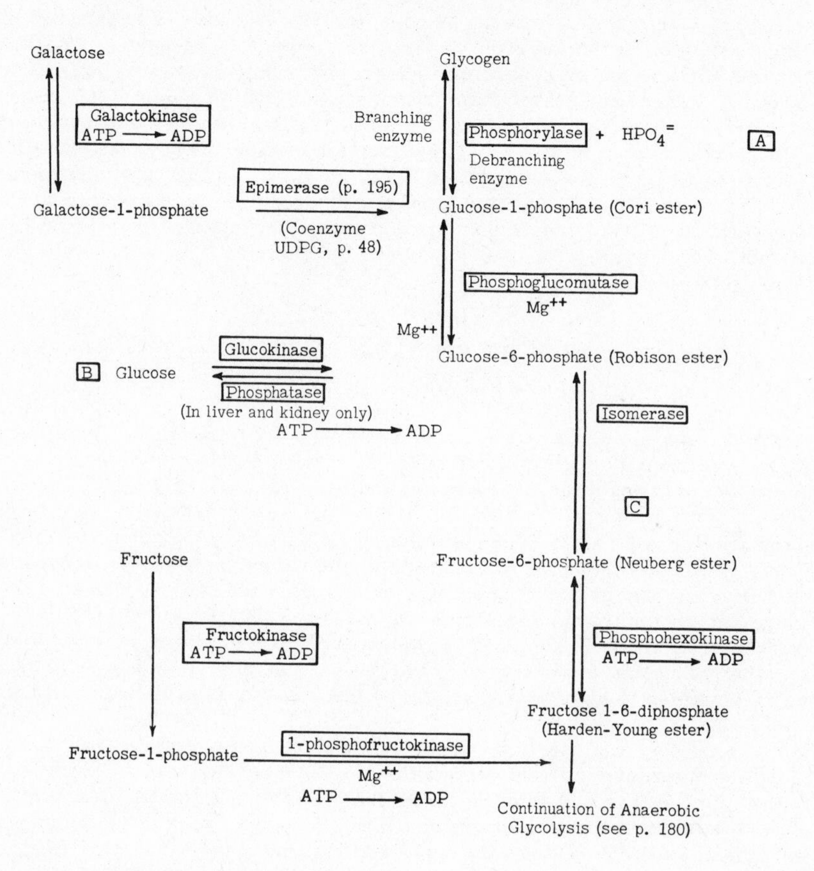

Formation and Degradation of Glycogen:

The formation of glycogen occurs in practically every tissue of the body but chiefly in liver and muscle. When glucose is converted to glycogen, glucose-6-phosphate is first transformed to glucose-1-phosphate in a reaction catalyzed by the enzyme, **phosphoglucomutase.** Glucose-1-phosphate is then polymerized to glycogen as described below.

Glycogen or animal starch is a branched polysaccharide composed entirely of α-D-glucose units[5]. These glucose units are connected to one another by glucosidic linkages between the first and the fourth carbon atoms of adjacent glucosyl moieties, except at branch points where the linkages are between carbons 1 and 6. The molecular weight of glycogen may vary from one to four million or more. (See p. 177.)

STRUCTURE OF GLYCOGEN

The first step in glycogen synthesis is removal of the phosphate group from glucose-1-phosphate under the catalytic action of **phosphorylase,** and the dephosphorylated glucose is then attached by its now free first carbon atom to the fourth carbon of a glucose residue on some preexisting glycogen. This action of phosphorylase has been studied in vitro using a purified crystalline phosphorylase obtained from rabbit muscle. The reaction can be followed by measurement of the amounts of inorganic phosphate liberated. When the enzyme and glucose-1-phosphate are

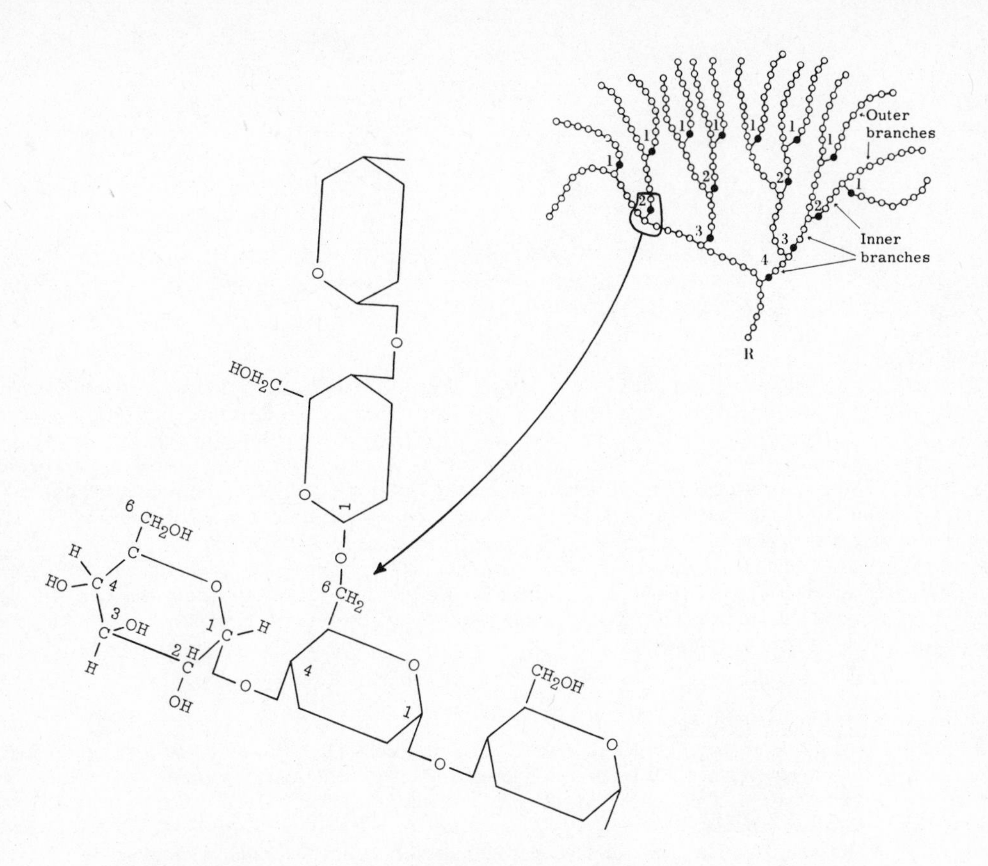

The Glycogen Molecule as Depicted by G. T. Cori. Open circles are alpha 1-4 glucosidic links and solid circles are 1-6 links. Segment at the branch point is enlarged to show structure.

incubated together, no reaction takes place until a little glycogen is added to the system, at which point synthesis begins at once and proceeds with increasing rapidity up to a certain limit as more glycogen is added. It is concluded from these observations that phosphorylase cannot begin the formation of polysaccharide chains simply by condensing two or more molecules of glucose-1-phosphate. Apparently pre-formed long chains are required as "primers" to get the reaction started.

The addition of a glucose residue to a glycogen chain occurs at the non-reducing outer end of the molecule so that the "branches" of the glycogen "tree" thus become elongated as successive 1-4 linkages occur. When the chain has been lengthened to eight glucose residues, a second enzyme, the "**branching enzyme,**" acts on the glycogen. This enzyme transfers a part of the 1-4 chain to a neighboring chain but by a 1-6 linkage, thus establishing a branch point in the molecule. The branching enzyme is thus acting as a **transglucosidase.**

The action of this transglucosidase has been studied in the living animal by feeding C^{14}-labeled glucose and examining the liver glycogen at intervals thereafter. At first only the outer branches of the chain are labeled, indicating that the new glucose residues are added at this point. Later, some of these outside chains are transferred to the inner portion of the molecule, appearing as labeled 1-6 linked branches. Thus, under the combined action of phosphorylase and transglucosidase (branching enzyme), the glycogen molecule grows like a tree.

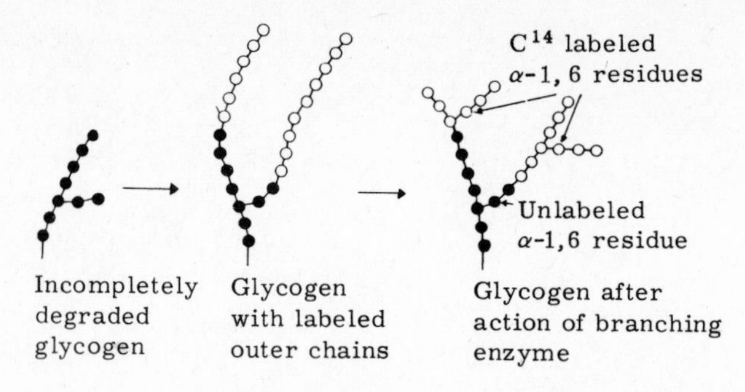

The activity of branching enzyme is not reversible. Consequently, another enzyme is required for degradation of the glycogen (glycogenolysis). This is the **debranching enzyme** which has been isolated from muscle and liver. It causes hydrolytic breakdown at the 1-6 linkage, but only after phosphorylase, in the presence of inorganic phosphate (which favors glycogen breakdown), has removed the 1-4 glucose residues from the outermost chains, converted them to glucose-1-phosphate, and thus exposed the 1-6 branch points to the action of debrancher. The glucose-1-phosphate residues are themselves converted to glucose-6-phosphate because mutase action is reversible. In liver and kidney (but not in muscle), there is a specific enzyme, **glucose-6-phosphatase**, which removes phosphate and frees the glucose to diffuse from the cell into the extracellular spaces, including the blood. This is the final step in glycogenolysis, which is reflected by a rise in the blood sugar.

The Diseases of Glycogen Storage:

Abnormal metabolism of glycogen is a feature of the so-called diseases of glycogen storage. The subject has been reviewed by Recant[6]. Glycogen storage disease is a generic term for a group of hereditary disorders characterized by the deposition of abnormally large amounts of glycogen in the tissues. In hepato-renal glycogenosis (Von Gierke's disease), both the liver cells and the cells of the renal convoluted tubules are characteristically loaded with glycogen. However, these glycogen stores seem to be metabolically unavailable, as evidenced by the occurrence of hypoglycemia and a lack of glycogenolysis under stimulus by epinephrine. Ketosis and hyperlipemia are also present in these patients, as would be characteristic of an organism deprived of carbohydrate. It has recently been found[7] that the activity of glucose-6-phosphatase is either extremely low or that no phosphatase activity at all can be detected in the liver tissue obtained from these patients, so that this enzymatic defect may be one important etiologic factor in the disease.

In other types of glycogen storage disease which are less frequently encountered than the hepato-renal type, either decreased brancher or decreased debrancher activity has been detected in the liver. On the other hand, glycogen storage disease of the heart has not so far been associated with any detectable abnormality of carbohydrate metabolism.

Pathways for Metabolism of Glucose-6-phosphate:

Glucose-6-phosphate is a key compound in the metabolism of the hexose sugars. Its formation from glucose and the related sugars of the diet is illustrated on p. 176. Once formed glucose-6-phosphate may be stored by conversion to glycogen or reconversion to free glucose, or may be further metabolized by several pathways. The most important of these from the standpoint of energy metabolism is the glycolytic pathway to pyruvic and lactic acids in a series of reactions now referred to as the Embden-Meyerhof (E-M) pathway. The end-product of the metabolism of glucose via the E-M pathway is pyruvic acid, which is oxidatively decarboxylated to acetic acid. This latter two-carbon compound (as ''active acetate''), derived not only from the metabolism of sugars but from fatty acid metabolism and some amino acids as well, enters the aerobic, citric acid cycle of Krebs, where it is completely degraded to carbon dioxide and water. Thus, by the combined action of the Embden-Meyerhof-Krebs pathway, hexose sugars are completely metabolized for the principal purpose of producing energy, as high-energy phosphate, for cellular metabolism.

The biochemical separation of the reactions for the metabolism of glucose into anaerobic and aerobic phases has its parallel in the cytologic distribution of the enzymes which comprise each phase. In rat liver, for example, the glycolytic enzymes are located free in the soluble non-particulate portion of the cytoplasm, whereas the enzymes of the citric acid cycle are contained within the organized mitochondria, in close association with the respiratory chain (see p. 117).

A second pathway for the metabolism of glucose-6-phosphate is the Direct Oxidative pathway (also termed the phosphogluconate oxidative pathway) or the Hexose Monophosphate shunt (HMP shunt) pathway. The first reaction of the shunt pathway is the oxidation of glucose-6-phosphate to 6-phosphogluconic acid; hence the term "direct oxidative" in describing this pathway. The details and significance of the shunt pathway will be discussed on p. 186.

Glucose-6-phosphate may also be converted to glucuronic acid by way of glucose-1-phosphate. This so-called "uronic acid" pathway may be considered a third route for the metabolism of glucose. The reactions of the uronic acid pathway are described on p. 191.

ANAEROBIC METABOLISM OF CARBOHYDRATE
(Embden-Meyerhof Pathway of Glycolysis)

The first reaction in the metabolism of glucose-6-phosphate through the Embden-Meyerhof pathway of glycolysis is its conversion to fructose-6-phosphate in a reaction catalyzed by **isomerase** (phosphohexoseisomerase).

The Fate of Fructose-6-phosphate:

The formation of fructose-6-phosphate from glucose-6-phosphate is followed by the addition of another phosphate linkage on position 1, for which ATP also acts as phosphate donor. The resultant compound is fructose-1, 6-diphosphate, the Harden-Young ester. Phosphohexokinase is the catalyzing enzyme.

The Fate of Fructose-1, 6-diphosphate: (See p. 180.)

The fructose diphosphate splits at the midpoint to yield two triose phosphates, glyceraldehyde-3-phosphate and dihydroxyacetone phosphate. This reaction is catalyzed by zymohexase (aldolase).

Under the action of a phosphotriose isomerase, the dihydroxyacetone phosphate is changed to another molecule of glyceraldehyde-3-phosphate.

Oxidation of Glyceraldehyde-3-phosphate to 1, 3-Diphosphoglyceric Acid:

It has been shown by Krimsky and Racker[8] that glutathione (see p. 235) is a prosthetic group for glyceraldehyde-3-phosphate dehydrogenase, and Racker and Krimsky[9] suggest that an SH group on the dehydrogenase actually participates in the reaction by which the glyceraldehyde is oxidized. According to this scheme, the aldehyde group combines with the SH group on the enzyme. Oxidation then occurs, in the typical anaerobic manner, by removal of hydrogen which is transferred to DPN. Finally, by phosphorylysis (addition of phosphoric acid), 1, 3-diphosphoglyceric acid is formed together with liberation of the free enzyme.* These reactions may be represented as shown on p. 182.

The glyceraldehyde dehydrogenase is readily inhibited by iodoacetate.

It is known that the phosphate on position 1 of 1, 3-diphosphoglyceric acid is of the high-energy variety and that the oxidation of the aldehyde to the acid is an exothermic reaction. Thus in the above scheme it is likely that a high-energy sulfur bond is created and that this energy is incorpo-

*The method for oxidation of glyceraldehyde described above may represent a general scheme for the oxidation of other aldehydes. Thus acetaldehyde (derived from oxidation of ethyl alcohol through its specific dehydrogenase) will also act as a substrate for glyceraldehyde-3-phosphate dehydrogenase and the final product of the oxidation is acetyl phosphate.

A SUMMARY OF ANAEROBIC GLYCOLYSIS [P = -P(OH)(=O)(OH)]

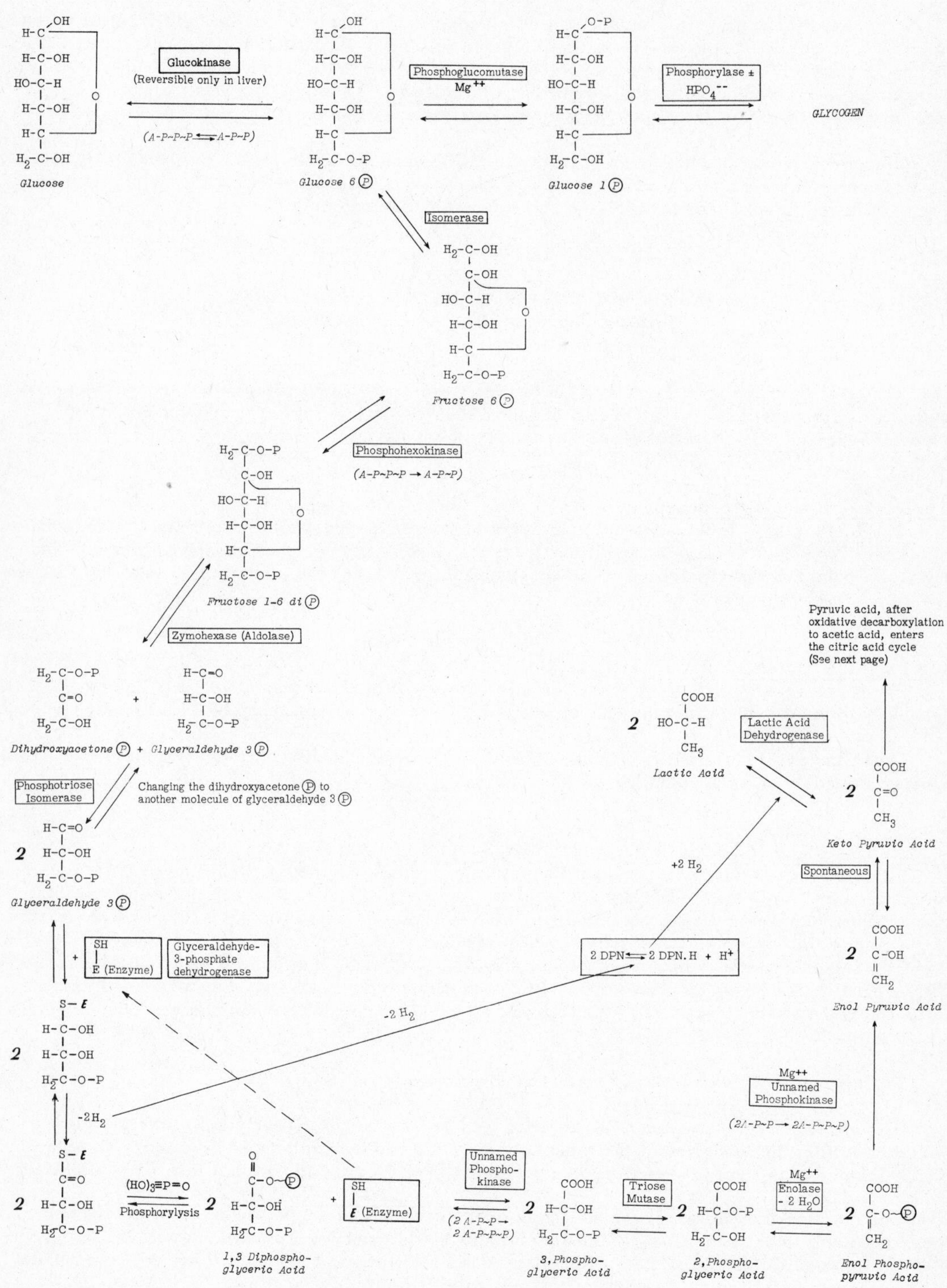

THE CITRIC ACID CYCLE
(Aerobic Metabolism of Carbohydrate)

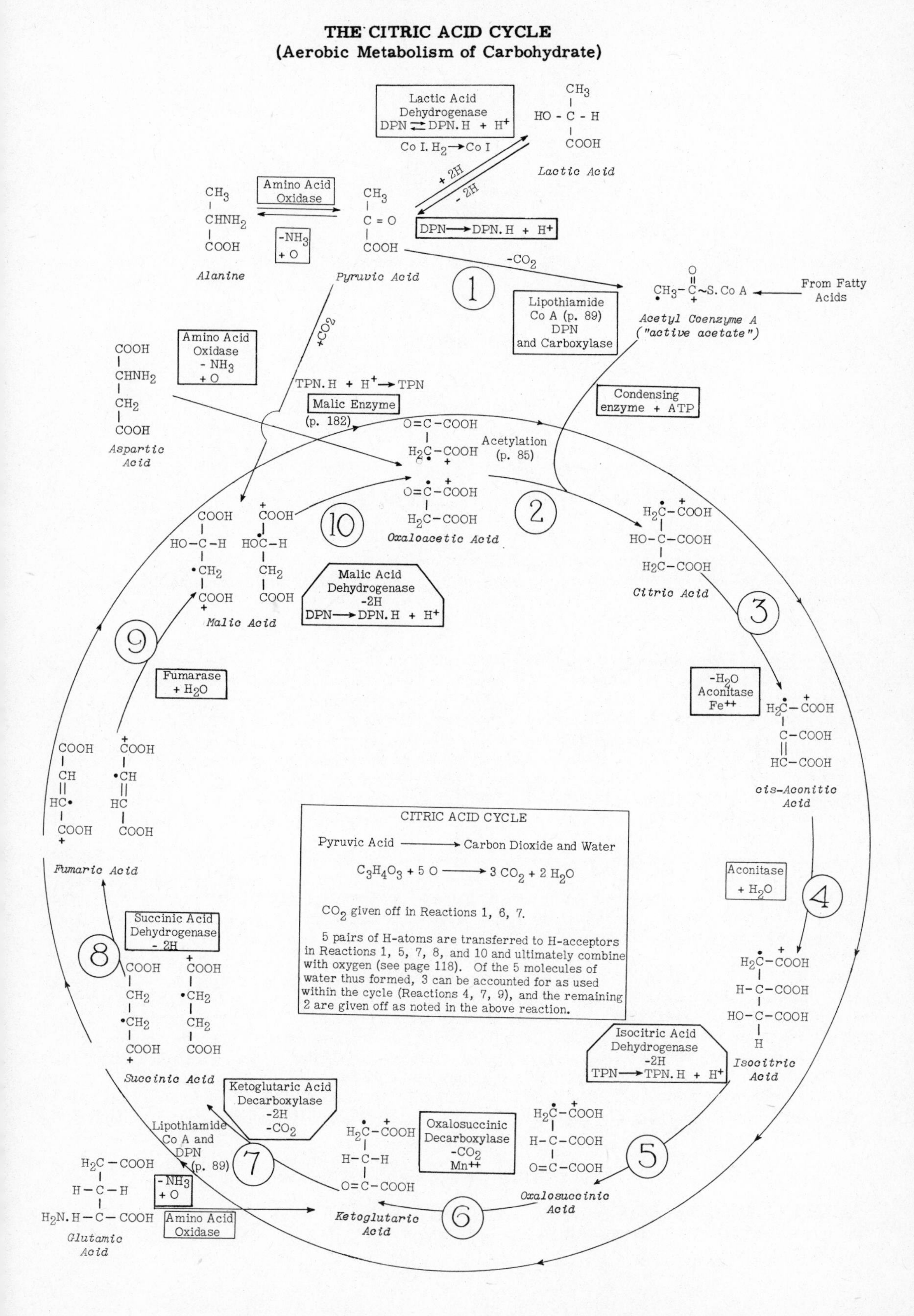

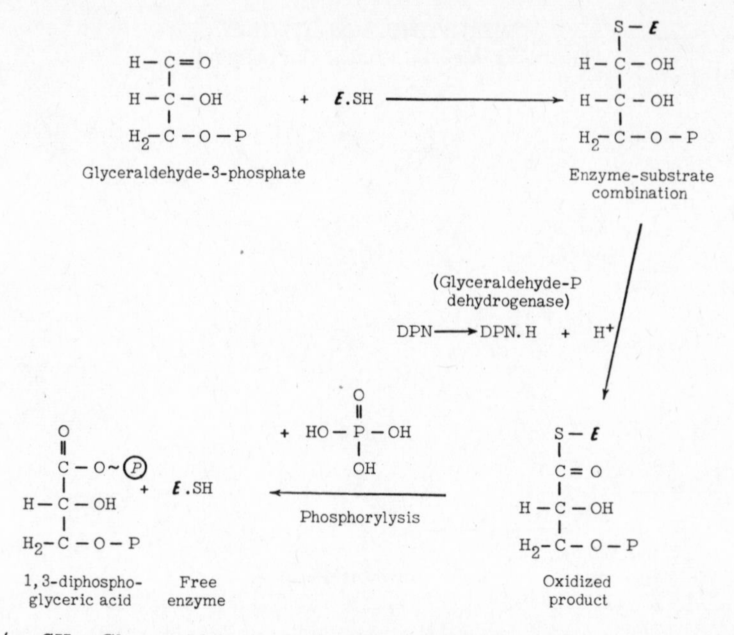

(E .SH = Glyceraldehyde-3-phosphate dehydrogenase-glutathione)

rated after phosphorylysis into the phosphate on position 1. This high-energy phosphate is now transferred to ADP to produce two molecules of high-energy ATP, and 3-phosphoglyceric acid remains.

Metabolism of Phosphoglyceric Acid; Formation of Pyruvic Acid:

The 3-phosphoglyceric acid is next converted to a 2-phosphoglyceric acid by the action of a triose mutase. The enzyme enolase* catalyzes the dehydration of 2-phosphoglyceric acid to a phosphoenol-pyruvic acid. In the dehydration step there is a redistribution of energy within the molecule to raise the phosphate on position 2 to the high-energy state. The subsequent dephosphorylation to enol pyruvic acid permits the transfer of the phosphate removed to ADP, forming more high-energy ATP. Thus in the course of anaerobic glycolysis four high-energy bonds are created per mol of glucose broken down. Two are required to "pay back" those required for the formation of glucose-6-phosphate and fructose-1, 6-diphosphate; the other two represent the net gain of this phase of glycolysis.

Enol pyruvic acid readily changes to the keto form. If anaerobic conditions prevail, keto pyruvic acid is reduced to lactic acid, utilizing the hydrogen of reduced DPN which was produced above in the oxidation of glyceraldehyde to glyceric acid. The reduction of pyruvic acid to lactic acid serves to regenerate oxidized DPN, which then participates again as a hydrogen acceptor in the oxidation of glyceraldehyde to glyceric acid as described above. Under aerobic conditions, reduced DPN can transfer its hydrogen through a riboflavin-cytochrome system to oxygen (see p. 116). The formation of lactic acid is then not required, and in this case pyruvic acid becomes the final product of glycolysis. The reoxidation of DPN by the riboflavin-cytochrome hydrogen transfer system may be adequate when the muscle is not working extensively. When muscular effort is increased, reduced DPN is produced at a more rapid rate; an additional mechanism for its reoxidation becomes necessary. This mechanism is provided by the reduction of pyruvic acid to lactic acid. The additional quantities of lactic acid thus produced may be detected in the tissues and in the blood and urine. The formation of lactic acid is therefore essentially a mechanism for the reoxidation of DPN.

A summary of the reactions in anaerobic muscle glycolysis is shown on p. 180.

*Enolase is inhibited by sodium fluoride. The inhibition of glycolysis by fluoride is the reason for the addition of this ion to anticoagulant mixtures which are to be used when blood is collected for sugar determinations.

THE AEROBIC METABOLISM OF CARBOHYDRATE
(Citric Acid or Krebs Cycle)

The complete oxidative breakdown of glucose to carbon dioxide and water takes place in anaerobic and aerobic phases. The anaerobic phase of glycolysis has just been discussed. The aerobic phase is concerned with the conversion of lactic and pyruvic acids, the end products of anaerobic glycolysis, to CO_2 and water.

The various reactions involved in the conversion of pyruvic acid to CO_2 and water comprise the citric acid cycle (Krebs cycle, tricarboxylic acid cycle). As will become evident in further studies of intermediary metabolism, this integrated series of reactions is actually a common channel not only for the oxidation of the products of glycolysis but also for the oxidation of fatty acids and of many amino acids as well. In other words, the citric acid cycle constitutes a final common pathway in metabolism.

Pyruvic acid is first converted to acetyl coenzyme A (''active acetate'') by an oxidative decarboxylation which is catalyzed by a decarboxylase requiring lipothiamide pyrophosphate (LTPP) as coenzyme as well as coenzyme A. The details of this important reaction are shown on p. 88.

After the formation of citrate by the acetylation of oxaloacetate, a series of reactions follows in the course of which one mol of pyruvic acid disappears and oxaloacetic acid is regenerated to start a new cycle. The details of the reactions of the citric acid cycle are shown on p. 181. In order to follow the passage of acetate through the cycle, the two carbon atoms of acetate are shown labeled with isotopic carbon on the carboxyl carbon [using the designation (+)] and on the methyl carbon [using the designation (•)]. Note that although two carbon atoms are lost (reactions 6 and 7) in one passage of acetate through the cycle, the acetate carbons remain, and labeled CO_2 would not be expected to appear until the labeled oxaloacetate, which results from the introduction of labeled acetate into the cycle, has re-entered the cycle at reaction 2.

In reaction 7 ''randomization'' occurs, resulting in the formation of succinic acids labeled as shown. The further progress of these succinic acids through the cycle ultimately produces oxaloacetic acids labeled either on the carboxyl and keto carbons or on the opposite carboxyl carbon and its adjacent carbon atom. The conversion of these labeled oxaloacetate molecules to pyruvic acid by loss of CO_2 (see p. 187) would then produce two types of labeled pyruvic acid: one labeled on the methyl carbon only and the other doubly labeled - on the carboxyl and α-keto carbons.

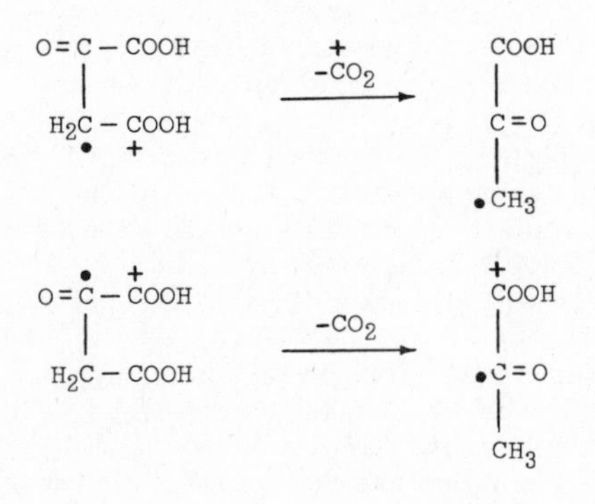

When these labeled pyruvates form glucose, the distribution of the isotopic carbons in the glucose molecules can be expected to be as follows:

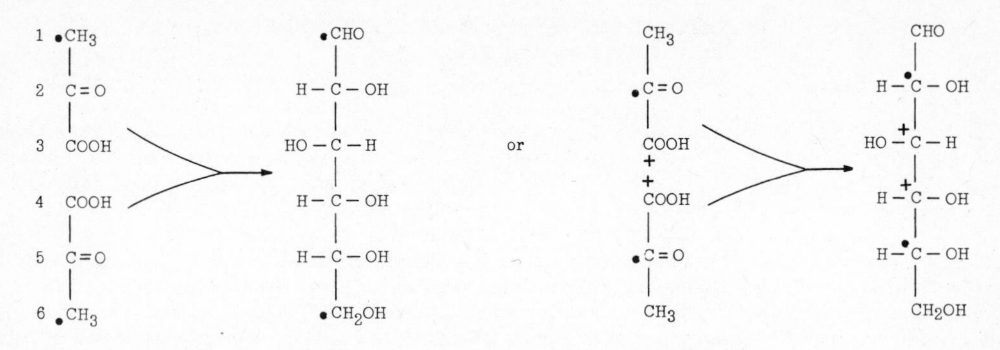

From these data the contribution of each of the two carbon atoms of acetate to the carbon skeleton of glucose can be deduced. It is apparent that the carboxyl carbon of acetate [designated (+)] contributes to carbon atoms 3 and 4 of glucose, whereas the methyl carbon [designated (•)] contributes to carbon atoms 1, 2, 5, and 6.

Oxaloacetic acid, the compound essential to the commencement of the citric acid cycle, can also be formed by the addition of CO_2 to pyruvic acid (carboxylation). This reaction was first studied in microorganisms, where it is referred to as the **Wood-Werkman reaction** or **fixation of CO_2**. In animal tissue (liver) the formation of oxaloacetic acid by CO_2 fixation has been found to utilize not pyruvic acid but **enol-phosphopyruvic acid** as the CO_2 acceptor. The details of this very important reaction involving the liver enzyme, oxaloacetic carboxylase, are shown on p. 187.

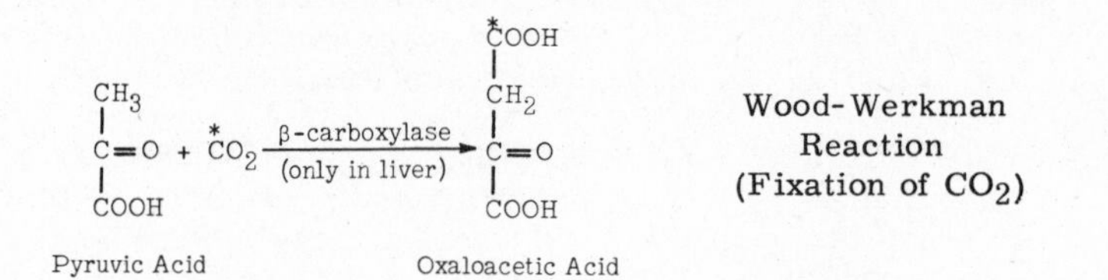

This reaction is undoubtedly of major importance in the maintenance of the citric acid cycle; for, although oxaloacetic acid is theoretically regenerated completely, it must be assumed that this process is not entirely efficient and that some losses of this compound do occur. It therefore becomes necessary to replenish and maintain the supplies of oxaloacetic acid if the acetate derived from pyruvate (and from fat) is to enter the citric acid cycle and complete its metabolism to CO_2 and water. Carbohydrate (i. e., pyruvate) may not only form acetate for utilization within the citric acid cycle; by CO_2 fixation (carboxylation) of pyruvate, it also assures the production of the compound (oxaloacetate) with which the acetate must be combined. On the other hand, other metabolites, notably fat, present the cycle with acetate but cannot provide oxaloacetate. It is for these reasons that at least a minimum amount of carbohydrate must be metabolized to maintain the citric acid cycle and thus provide for the complete utilization of other metabolites. This will be discussed further in connection with the ketolytic action of carbohydrate (see p. 212).

When pyruvate enters the citric acid cycle through carboxylation to oxaloacetic acid, thus forming a dicarboxylic acid, it may be considered to be increasing the amounts of the citric cycle intermediates. Under these circumstances, some of the cycle intermediates can be drained off and used in synthetic pathways. Examples are the conversion of oxaloacetic acid or of ketoglutaric acid to aspartic or glutamic acid, respectively, by amination of these keto acid citric cycle intermediates. On the other hand, when pyruvate, after decarboxylation, enters the cycle as acetyl-Coenzyme A, there is no net increase in the intermediates of the cycle, so that the reaction can be used for energy purposes only. These considerations indicate that alterations in the metabolic pathway for utilization of pyruvate by the citric acid cycle may be a means to influence the proportions of carbohydrate which are catabolized for energy or are diverted to synthetic pathways. Freedman and Graff[10] have found that in the fasted rat, carbohydrate is used by the citric acid cycle in the liver primarily as a source of dicarboxylic acids. Under these conditions, the principal source of energy for the liver is probably fat. When the rat is given glucose, a significant amount of pyruvate entering the citric acid cycle does so by decarboxylation to a two-carbon fragment (acetate).

Influence of Toxic Agents on the Citric Acid Cycle:

By the use of certain poisons, the reactions of the citric acid cycle can be blocked at various points. This blockage causes the accumulation of the products immediately preceding the blocked reaction and permits isolation of the intermediates for identification. Examples of such poisons are fluoroacetic acid, which blocks the utilization of citric acid so that this compound accumulates; arsenite, which blocks reaction No. 7, in which case ketoglutaric acid accumulates; and malonate, which blocks reaction No. 8 by inhibiting succinic acid dehydrogenase, in which case succinic acid accumulates.

Vitamins in Utilization of Carbohydrate:

Many of the B vitamins have a specific function in the aerobic and anaerobic breakdown of carbohydrate. Niacin (in DPN and TPN) and riboflavin play essential roles in oxidation by acting as hydrogen transfer agents; thiamine, lipoic acid, and pantothenic acid are involved in the conversion of pyruvic acid to acetic acid and of ketoglutaric acid to succinic acid. These three vitamins are therefore essential at the gateway of the citric acid cycle. The role of biotin in CO_2 fixation to convert pyruvic acid to oxaloacetic acid has also been mentioned.

The Energy Aspect of Carbohydrate Oxidation:

In the course of the complete oxidative breakdown of glucose to CO_2 and water, a net gain of at least 38 high-energy phosphate bonds may occur. Assuming 7600 small calories per bond (see p. 119), the 38 new bonds represent 288,800 calories. Since one molecule of glucose is known to contain about 686,000 calories, the "capturing" of 288,800 calories represents an efficiency of 288,800/686,000 or 42%. The formation of these high-energy bonds occurs through the process of phosphorylation coupled to oxidation (see p. 119) as well as by direct transfer of high-energy phosphate from a substrate to ADP. Such "substrate phosphorylation" occurs in glycolysis to pyruvate in the conversion of 1, 3-diphosphoglyceric acid to 3-phosphoglyceric acid and in the dephosphorylation of enol-phosphopyruvic acid to enol-pyruvic acid. In addition, in the oxidation of glyceraldehyde, reduced DPN is formed; its reoxidation through the flavoprotein-cytochrome system is associated with the formation of three high-energy bonds (P:O ratio = 3:1; see p. 120). Thus, when the complete oxidation of one mol of glucose to pyruvate occurs ten high-energy bonds are created, since glucose breaks down into two triose molecules and each of the above oxidative and substrate phosphorylations is associated with one triose molecule. The net gain, however, is only eight bonds, since two were used at the start of glycolysis (in the formation of glucose-6-phosphate and of fructose-1, 6-diphosphate).

In the course of the complete oxidation of pyruvate to CO_2 and water through the reactions of the citric acid cycle, 15 high-energy bonds may be created for each mol of pyruvate oxidized. Considering that one mol of glucose gives rise to two mols of pyruvate, a total of 30 high-energy phosphate bonds may be considered to result from this phase of the oxidation of glucose. The sum of eight from glycolysis to pyruvate and 30 from complete oxidation of two mols of pyruvate to CO_2 and water through the citric acid cycle equals the 38 high-energy bonds which may be derived from the complete oxidation of one mol of glucose, as mentioned above.

The points of formation of the 15 high-energy bonds within the citric acid cycle are as follows (numbers refer to the reactions shown on p. 181):

Reaction No. 1 = 3 (oxidative through DPN-flavoprotein-cytochrome)

5 = 3 (oxidative through TPN-flavoprotein-cytochrome)

7 = 4 (1 at the substrate level; 3 oxidative through DPN-flavoprotein-cytochrome)

8 = 2 (oxidative through flavoprotein)

10 = 3 (oxidative through DPN-flavoprotein-cytochrome)

In reaction 5 (p. 181), TPN participates in oxidative phosphorylation through a transhydrogenase reaction (see p. 121) with DPN, which is actually the pyridine component of the respiratory chain directly involved in the process of oxidative phosphorylation. Reaction 8 is listed as forming only two high-energy bonds since the succinoxidase system is but a two-step phosphorylation process not involving DPN (see p. 117).

It is obvious that the aerobic phase of carbohydrate oxidation (citric acid cycle) contributes much more energy than does anaerobic glycolysis to pyruvate and lactate. This emphasizes the importance of adequate supplies of oxygen to the cells to permit the most efficient acquisition of energy for the performance of cellular work. (See also discussion of the Pasteur effect, p. 112.)

Integration of Metabolic Processes:

It is now apparent that each of the major classes of nutrients - carbohydrate, fat, and protein - contributes to the citric acid cycle. It is therefore no longer accurate to differentiate rigidly between the metabolism of these substances. The conversion of carbohydrate to fat (or vice versa) is now interpreted as due to a reversal of these integrated actions.

Reversibility of Glycolysis; Action of Oxaloacetic Carboxylase:

The constituents of the glycolytic systems can all be shown to form glycogen. However, Utter and Kurahashi[11] found that the conversion of enol-phosphopyruvic acid to enol-pyruvic acid is not reversible. This indicates that pyruvic and lactic acid as well as the constituents of the citric acid cycle must form glycogen by a method other than a simple reversal of the reactions by which they were formed.

In their studies of the action of oxaloacetic carboxylase of liver, Utter and Kurahashi[12] have indicated how enol-phosphopyruvic acid can be formed from lactic or pyruvic acids.

Lactic acid is converted to pyruvic acid by the well-known reaction involving lactic acid dehydrogenase; then, through the action of the "malic" enzyme (Ochoa) together with reduced TPN carboxylation and reduction of pyruvic acid to malic acid occur. Oxidation of malic acid to oxaloacetic acid follows, and, through the action of the oxaloacetic carboxylase and a source of high-energy phosphate such as ATP, decarboxylation to enol-phosphopyruvic acid occurs. It is of interest that inosine triphosphate (ITP), which has a structure similar to ATP except that hypoxanthine rather than adenine is the purine component, or guanosine triphosphate (GTP), is an even more efficient source of high-energy phosphate for this reaction. The action of oxaloacetic carboxylase depends also on the presence of SH groups such as glutathione or cysteine. The reactions just discussed are shown on the next page.

Reaction (§) (see p. 187) is readily reversible if a phosphate **acceptor** is present. ADP or, preferably, IDP or GDP is suitable. It is by this very important reaction, shown below, that the citric acid cycle can be maintained by the products of glycolysis, since it provides a method for the synthesis of oxaloacetic acid (see p. 184).

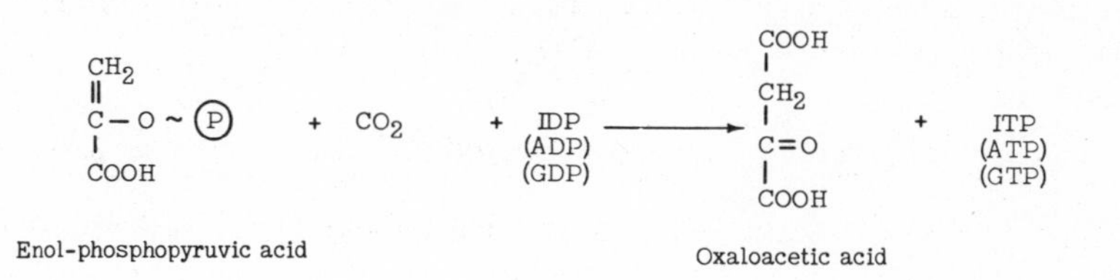

ALTERNATE PATHWAYS OF CARBOHYDRATE OXIDATION; THE DIRECT OXIDATIVE PATHWAY

The first step in the direct oxidative pathway is the oxidation of glucose-6-phosphate to 6-phosphogluconic acid in a TPN-dependent reaction catalyzed by **glucose-6-phosphate dehydrogenase** (the zwischenferment of Warburg). The product of the oxidation, 6-phosphogluconic acid

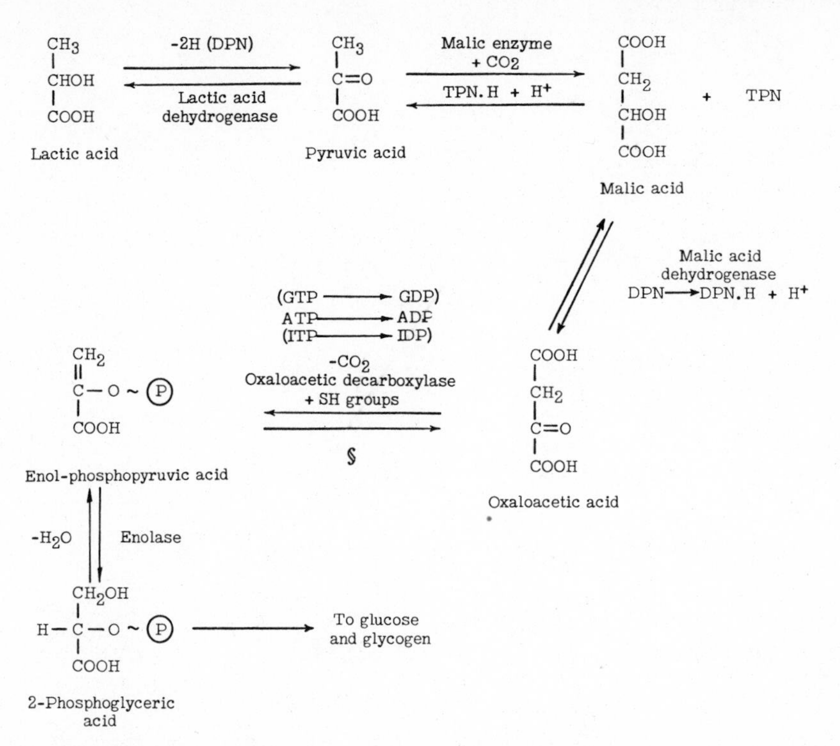

(lactone), is then oxidized at carbon 3 to 3-ketogluconic acid–6-phosphate, the reaction being catalyzed by another TPN-dependent enzyme, **6-phosphogluconic acid dehydrogenase.** Decarboxylation of 3-ketogluconic acid–6-phosphate produces the ketopentose, ribulose-5-phosphate, which through formation of an enediol intermediate, results in a balanced mixture of ribulose and ribose; the latter is the sugar of ribonucleic acids. Through the action of a **phosphoribomutase,** ribose-1-phosphate may be formed from the 5-phosphate. These reactions, which are shown on p. 188, lead to the production of pentoses important in nucleic acid metabolism and also to the production of reduced TPN (TPN.H). The formation of these two compounds is apparently a very important function of the shunt pathway. (See p. 191.)

The further stages of the direct oxidative pathway involve the metabolism of the pentoses, ribulose and ribose-5-phosphate. Through the catalytic action of **phosphopentose epimerase,** some of the ribulose-5-phosphate is epimerized to D-xylulose-5-phosphate: This compound then serves as a source of a two-carbon moiety (sometimes referred to as "active glycolaldehyde") which can be transferred to ribose-5-phosphate, forming a seven-carbon sugar, sedoheptulose-7-phosphate. The transfer of the two-carbon moiety from xylulose is called transketolation, and the enzyme which catalyzes such a transfer is termed **transketolase.** This latter enzyme requires thiamine diphosphate as a cofactor. Glyceraldehyde phosphate remains as the product after removal of the two-carbon moiety from xylulose. (See p. 188.)

Sedoheptulose-7-phosphate then reacts with the glyceraldehyde phosphate and, in a reaction of transaldolation, a three-carbon moiety is transferred from sedoheptulose to glyceraldehyde, forming fructose-6-phosphate and a tetrose, D-erythrose-4-phosphate. The transaldolase enzyme is not known to require a cofactor.

Finally, another reaction of transketolation takes place between xylulose as donor of the two-carbon fragment and erythrose as receptor. The products of this reaction are fructose and glyceraldehyde. (See p. 189.)

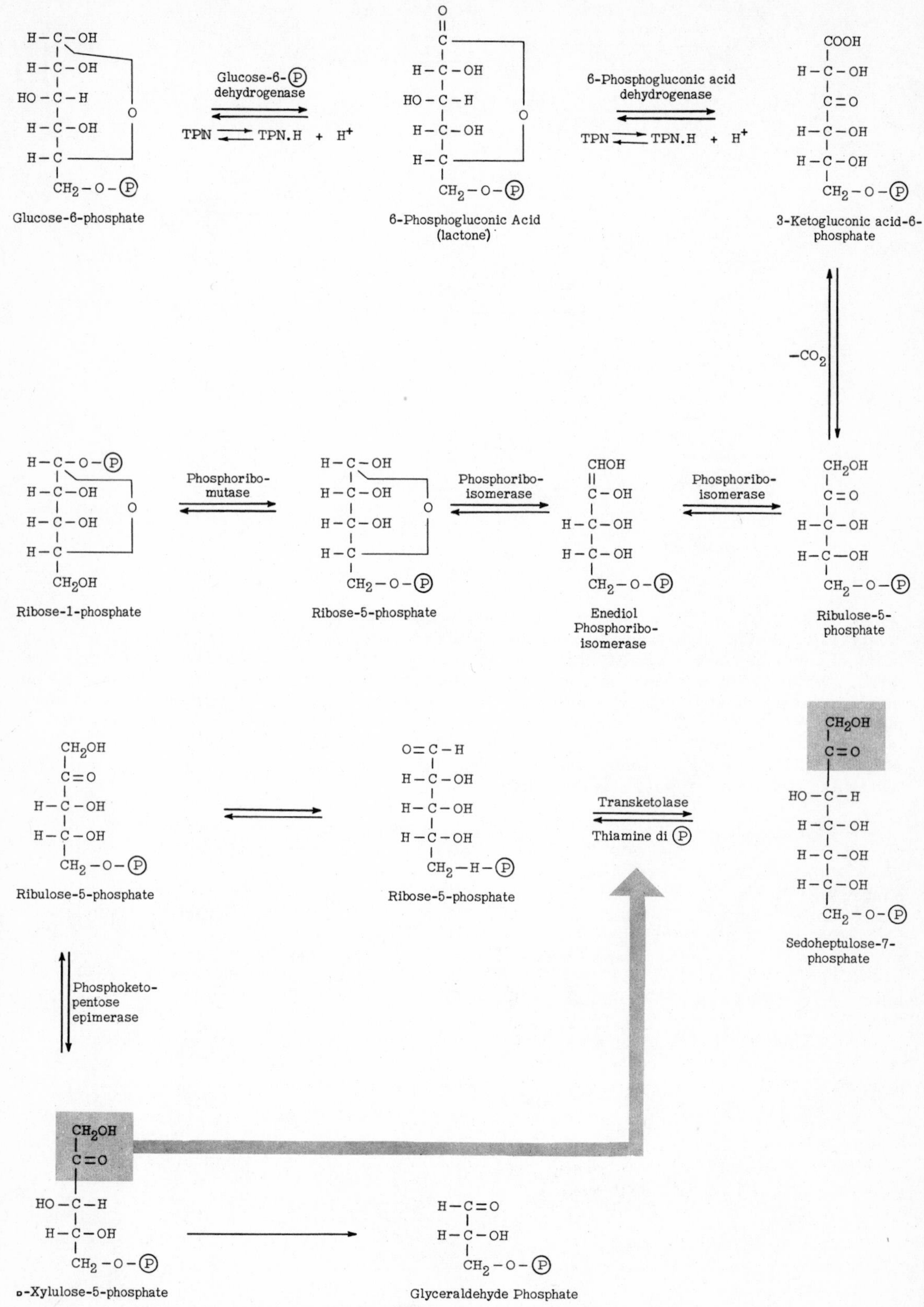

Glucose-6-phosphate
Glucose-6-(P) dehydrogenase
TPN ⇌ TPN.H + H+
6-Phosphogluconic Acid (lactone)
6-Phosphogluconic acid dehydrogenase
TPN ⇌ TPN.H + H+
3-Ketogluconic acid-6-phosphate
$-CO_2$
Ribose-1-phosphate
Phosphoribo-mutase
Ribose-5-phosphate
Phosphoribo-isomerase
Enediol Phosphoribo-isomerase
Phosphoribo-isomerase
Ribulose-5-phosphate
Ribulose-5-phosphate
Ribose-5-phosphate
Transketolase
Thiamine di (P)
Sedoheptulose-7-phosphate
Phosphoketo-pentose epimerase
D-Xylulose-5-phosphate
Glyceraldehyde Phosphate

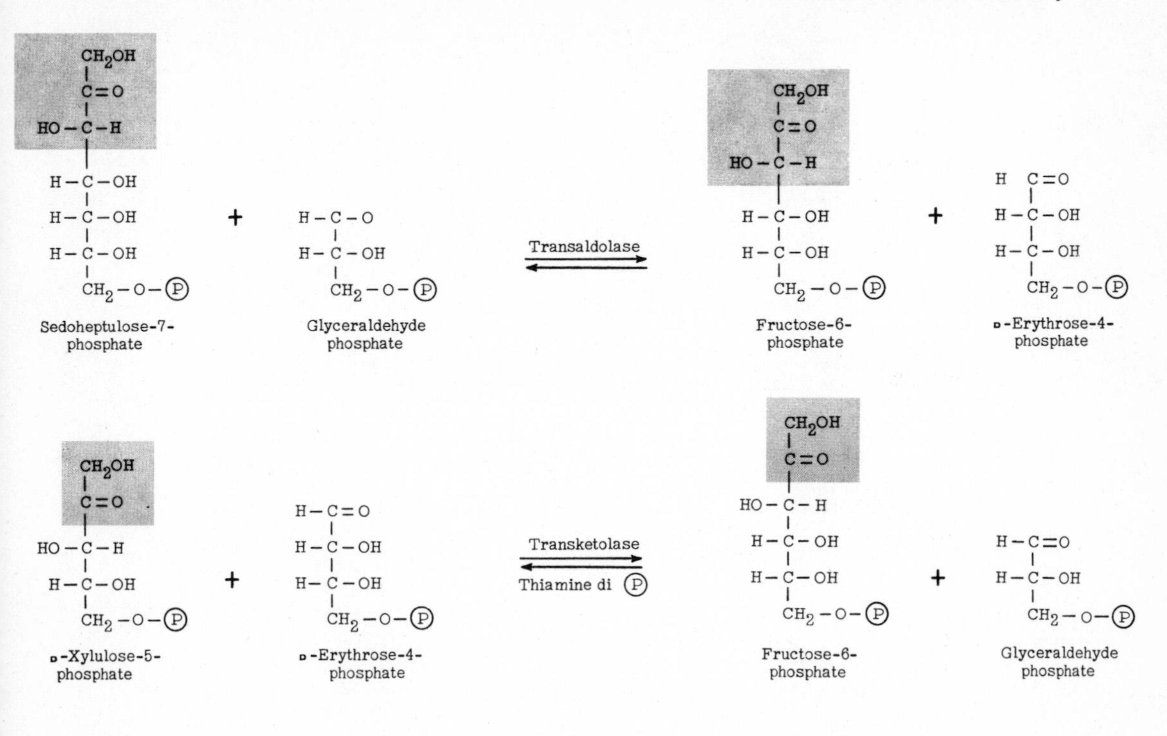

A summary of the reactions of the direct oxidative pathway is given below.

1. 2-Glucose-6-(P) + O_2 ⟶ Ribose phosphate + Xylulose phosphate + CO_2
2. Ribose phosphate + Xylulose phosphate ⟶ Sedoheptulose phosphate + Glyceraldehyde phosphate
3. Sedoheptulose phosphate + Glyceraldehyde phosphate ⟶ Fructose phosphate + Erythrose phosphate
4. Erythrose phosphate + Xylulose phosphate ⟶ Fructose phosphate + Glyceraldehyde phosphate

It is obvious that the direct oxidative pathway is markedly different from the Embden-Meyerhof pathway of glycolysis. Oxidation occurs very early, and carbon dioxide, which is not produced at all in the E-M pathway, is a characteristic product of the shunt mechanism obtained by decarboxylation of phosphogluconic acid to the pentoses. In this reaction the CO_2 is derived from the original carbon-1 of glucose-6-phosphate. This fact is used in experiments designed to measure the relative proportions of carbohydrate metabolized in a given tissue by the classical (E-M) pathway versus the direct oxidative pathway. Most of such studies have been based on measurement of differences in the rate of liberation of labeled CO_2 from glucose-1-C^{14} and from glucose-6-C^{14}. In the glycolytic pathway, carbons 1 and 6 of glucose are both converted to the methyl carbon of pyruvic acid and therefore metabolized in the same manner. In the direct oxidative pathway, carbons 1 and 6 of glucose are treated differently, carbon 1 being promptly eliminated as labeled CO_2 when the C-1 labeled glucose is metabolized via the shunt pathway.

Except for the two oxidative steps which occur in the conversion of glucose-6-phosphate to 6-phosphogluconic acid and thence to 3-ketogluconic acid phosphate, the remaining reactions of the direct oxidative pathway proceed anaerobically by way of group transfer reactions catalyzed by transketolase and transaldolase. The end products of the shunt pathway are fructose and glyceraldehyde. In the sense that some of the glyceraldehyde can be readily converted to dihydroxy acetone and condensation between this compound and the remaining glyceraldehyde will also produce fructose, it can be said that fructose is the actual sole end product of the shunt pathway. Fructose can be converted to glucose, which may then reenter the shunt pathway; in this manner, glucose could be completely oxidized independently of the E-M pathway.

A comparison of the E-M pathway and the shunt pathway is shown on p. 190.

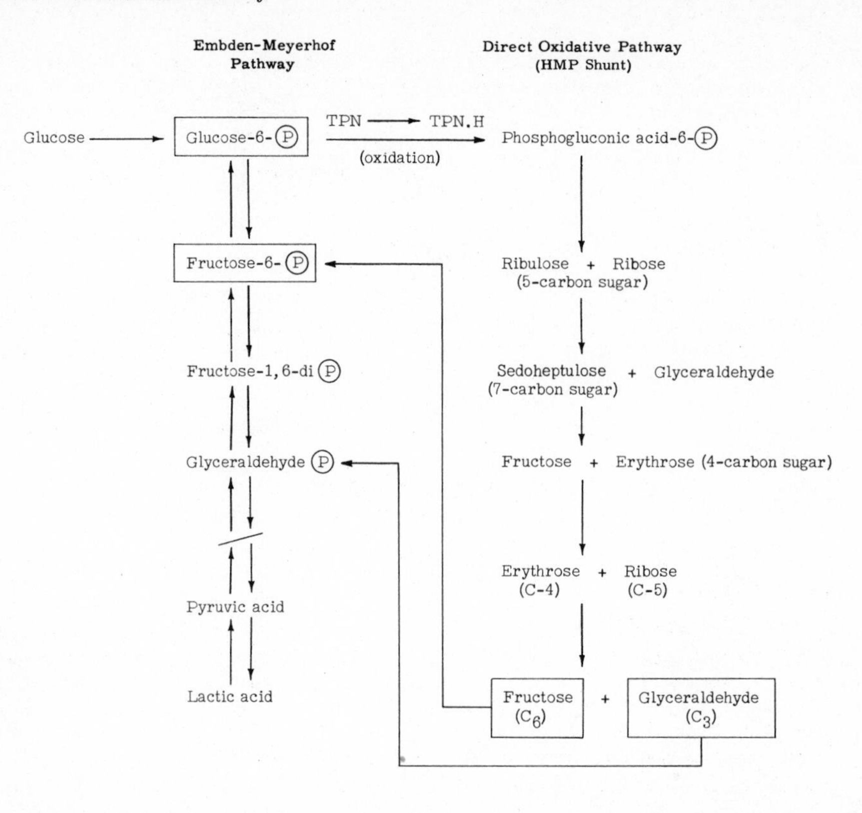

The extent to which the classical glycolytic pathway or the shunt pathway is utilized in various tissues is now under study. In muscle only the glycolytic pathway is utilized, not merely because this tissue is chiefly concerned with the breakdown of carbohydrate for production of energy but mainly because the enzymes necessary for the first two - oxidative-steps of the shunt pathway (glucose-6-phosphate dehydrogenase and 6-phosphogluconic acid dehydrogenase) are not present in muscle. However, the enzymes necessary for the isomerization of the pentose phosphates as well as transketolase and transaldolase are present in muscle, and in this tissue ribose phosphate is rapidly converted to hexose monophosphate.

In liver about 90% of glucose is metabolized by reactions of the glycolytic (E-M) pathway. In the metabolism of rabbit lens tissue, Kinoshita and Wachtl[13] found that most of the glucose utilized was converted to lactic acid; but at least 10% was metabolized via the shunt pathway, which appeared to contribute significantly to the CO_2 produced from the small amount of glucose oxidized in the lens. There is, however, evidence for the fact that the direct oxidative pathway may be of greater significance than the glycolytic pathway in some mammalian cells and tissues, as for example lactating mammary gland, leukocytes, and adrenal cortex. The shunt pathway is also present in red blood cells.

Although under ordinary circumstances liver is not believed to utilize the shunt pathway to an important extent, there is experimental evidence for the idea that this may be altered by certain metabolic factors, specifically the amount of available oxidized pyridine nucleotide (TPN). Brin and Yonemoto have made the interesting observation that methylene blue added to human erythrocytes stimulates the metabolism of glucose by the direct oxidative pathway[14]. It is believed to do so by serving as electron acceptor for reduced TPN (TPN.H), which thus brings about reoxidation of TPN and facilitates glucose oxidation through the shunt pathway. In similar experiments with liver slices from normal and diabetic rats, Cahill et al.[15] found that oxidation-reduction mediators preferentially increased oxidation of carbon 1 of glucose. These authors concluded from their data that the rate of oxidation of a reduced coenzyme may act as a factor controlling a metabolic pathway, and in liver the extent to which the direct oxidative pathway is favored may be determined by metabolic factors tending to produce oxidized TPN.

In tissue homogenates oxidation of glucose via the E-M glycolytic pathway is enhanced by the addition of DPN, whereas oxidation by the shunt pathway is stimulated by TPN, which is additional evidence for the idea that the availability of pyridine nucleotides is a factor influencing the pathway of glucose catabolism.

Metabolic Significance of the Direct Oxidative Pathway:

The E-M glycolytic pathway and the citric acid cycle of Krebs function to provide the major sources of energy (ATP), provide intermediates for lipogenesis (acetyl-Coenzyme A and glycerophosphate) and for the synthesis of some amino acids (pyruvate, oxaloacetate, and ketoglutarate to form, after amination, alanine, aspartic acid, and glutamic acid, respectively).

The shunt pathway provides pentoses for nucleic acid and nucleotides; most importantly, however, it provides reduced TPN, which is essential in several reductive biosynthetic processes such as the reduction of crotonyl Co A to butyryl Co A[16] in synthesis of fatty acids.

$$\underset{\text{Crotonyl Co A}}{CH_3.CH{=}CH{-}\overset{O}{\overset{\|}{C}}{-}S.Co\ A} \xrightarrow{TPN.H \rightarrow TPN} \underset{\text{Butyryl Co A}}{CH_3.CH_2.CH_2{-}\overset{O}{\overset{\|}{C}}{-}S.Co\ A}$$

TPN.H is specifically required in reductive carboxylation of pyruvate to malate (see p. 187), the important reaction by which regeneration of 4-carbon dicarboxylic acids of the citric acid cycle may be carried out. Reduced TPN is also an essential co-factor in several steps in steroid synthesis as well as in hydroxylation of steroids.

Formation of TPN.H seems to be an important function of the operation of the shunt pathway in red blood cells, and a direct correlation has been found between the activity in the erythrocytes of the direct oxidative enzymes, particularly glucose-6-phosphate dehydrogenase, and the fragility of red cells (susceptibility to hemolysis) especially when the cells are subjected to the toxic effects of certain drugs (primaquine[17], acetylphenylhydrazine) or the susceptible individual has ingested fava beans, Vicia fava (favism). The majority of patients whose red cells are readily hemolyzed by the toxic agents mentioned have been found to possess a hereditary deficiency in the oxidative enzymes of the shunt pathway which are normally found in the red blood cell.

It has been mentioned above that the direct oxidative pathway is an important route for metabolism of glucose in lactating mammary gland. This may well be related to the high rate of synthesis of fatty acids in this organ wherein there would be a correspondingly high requirement for reduced TPN produced in the shunt pathway.

Insulin is known to be essential in promotion of lipogenesis, i.e., synthesis, of fatty acids from carbohydrate. Winograd and Renold[18] have found in in vitro experiments that insulin added to adipose tissue (epididymal fat pad of rat) strikingly increases oxidation of C-1 of glucose and a corresponding striking increase in lipogenesis. These findings support the idea that insulin stimulates operation of the shunt pathway, thus increasing production of reduced TPN, which acts as the stimulus to lipogenesis. The action of insulin in promoting operation of the shunt pathway may be related to its influence in production of glucose-6-phosphate.

THE URONIC ACID PATHWAY

Glucuronic acid is formed from glucose by the reactions shown on p. 192. Glucose-6-phosphate is converted to glucose-1-phosphate, which then reacts with uridine triphosphate (UTP) to form the active nucleotide, uridine diphosphate glucose, UDPG (see p. 48). This latter reaction is catalyzed by the enzyme, **UDPG pyrophosphorylase.** After the glucose moiety has been incorporated into the nucleotide structure, it is oxidized at carbon 6 by a two-step process to glucuronic acid. The product of the oxidation which is catalyzed by a DPN-dependent **UDPG dehydrogenase** is, therefore, UDP-glucuronic acid.

Galacturonic acid is an important constituent of many natural products such as the pectins. It may be formed from UDP-glucuronic acid by inversion around carbon 4, as occurs when UDP-glucose is converted to UDP-galactose (see p. 195).

UDP-glucuronic acid is the "active" form of glucuronic acid for reactions involving incorporation of glucuronic acid into chondroitin sulfate or for reactions in which glucuronic acid is conjugated to such substrates as steroid hormones, certain drugs (see p. 168), or bilirubin (formation of "direct" bilirubin; see p. 273).

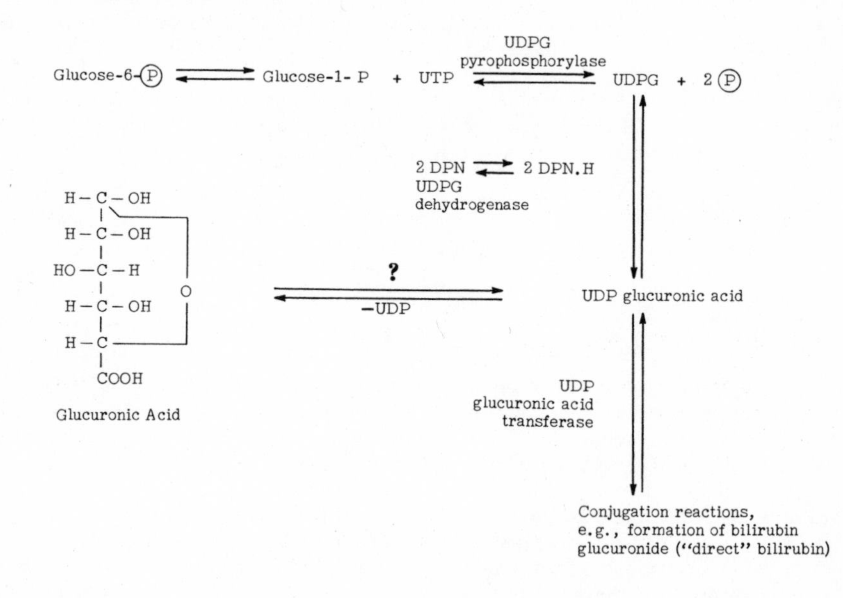

The further metabolism of glucuronic acid is shown on p. 193.[19] In a TPN.H-dependent reaction, glucuronic acid is reduced to L-gulonic acid, which is oxidized in a DPN-dependent reaction to 3-keto-L-gulonic acid. This latter compound is the direct precursor of ascorbic acid in those animals which are capable of synthesizing this vitamin. In man and guinea pigs ascorbic acid cannot be synthesized, and keto gulonic acid is decarboxylated to the pentose, L-xylulose.

Xylulose is a constituent of the direct oxidative pathway; but in the reactions shown on p. 193, the unnatural, L-isomer of xylulose is formed from keto gulonic acid. It is therefore necessary to convert L-xylulose to the natural, D-isomer. This is accomplished by a TPN.H-dependent reduction to xylitol, which is itself then oxidized in a DPN-dependent reaction to D-xylulose; this latter compound, after conversion to D-xylulose-5-phosphate with ATP as phosphate donor, is further metabolized in the direct oxidative pathway.

In the hereditary disease termed "essential pentosuria," considerable quantities of L-xylulose appear in the urine. It is now believed that this may be explained by the absence in such pentosuric patients of the enzyme necessary to accomplish reduction of L-xylulose to xylitol, and hence inability to convert the unnatural L form of the pentose to the natural L form.

The metabolism of glucuronic acid through the pentoses is shown on p. 193.

METABOLISM OF FRUCTOSE

The metabolism of the keto-hexose sugar, fructose, differs in certain respects from that of glucose. Its entrance into the glycolytic cycle and its conversion to glucose is diagrammed on p. 194.

URONIC ACID PATHWAY

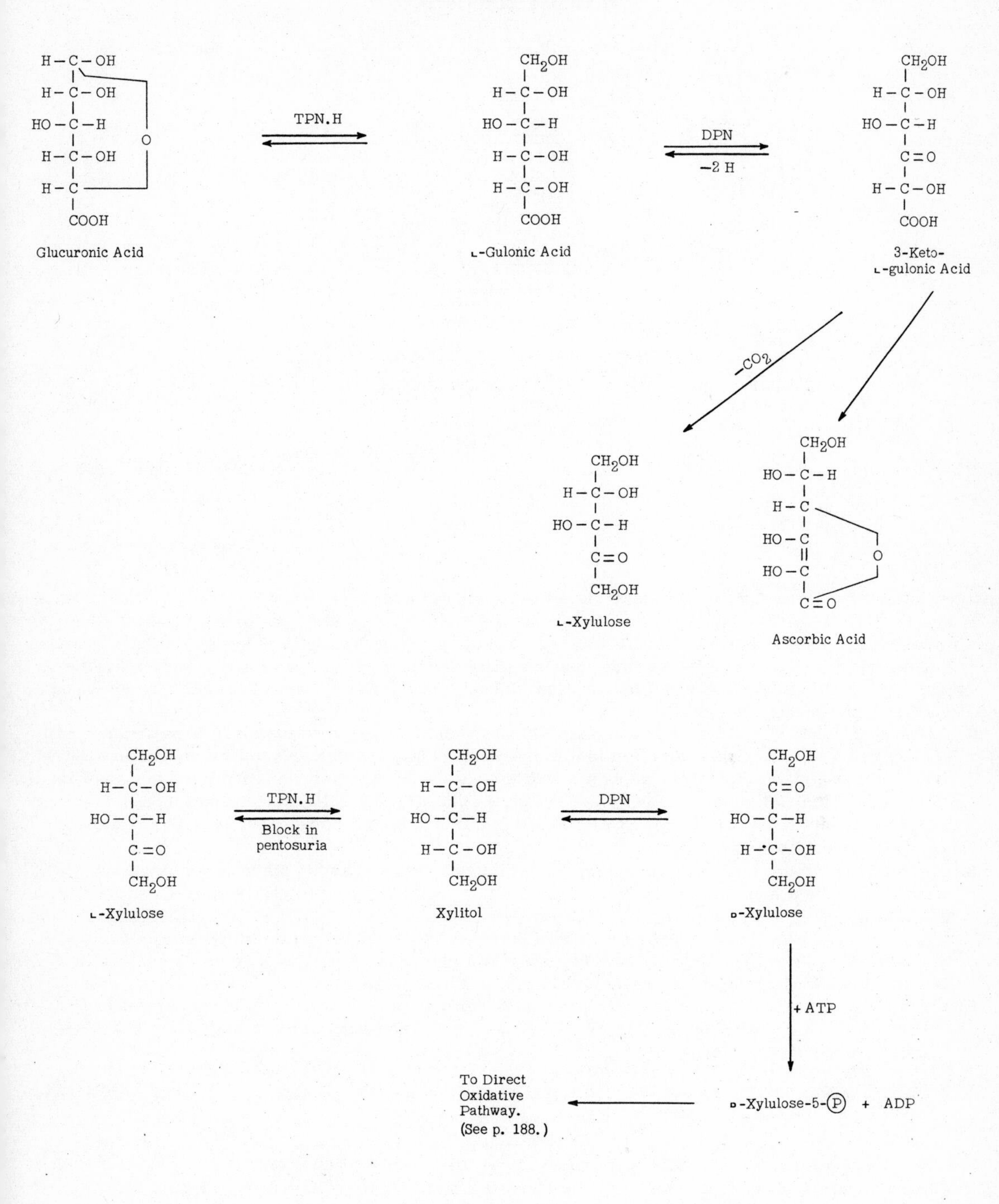

To Direct Oxidative Pathway. (See p. 188.) ← D-Xylulose-5-(P) + ADP

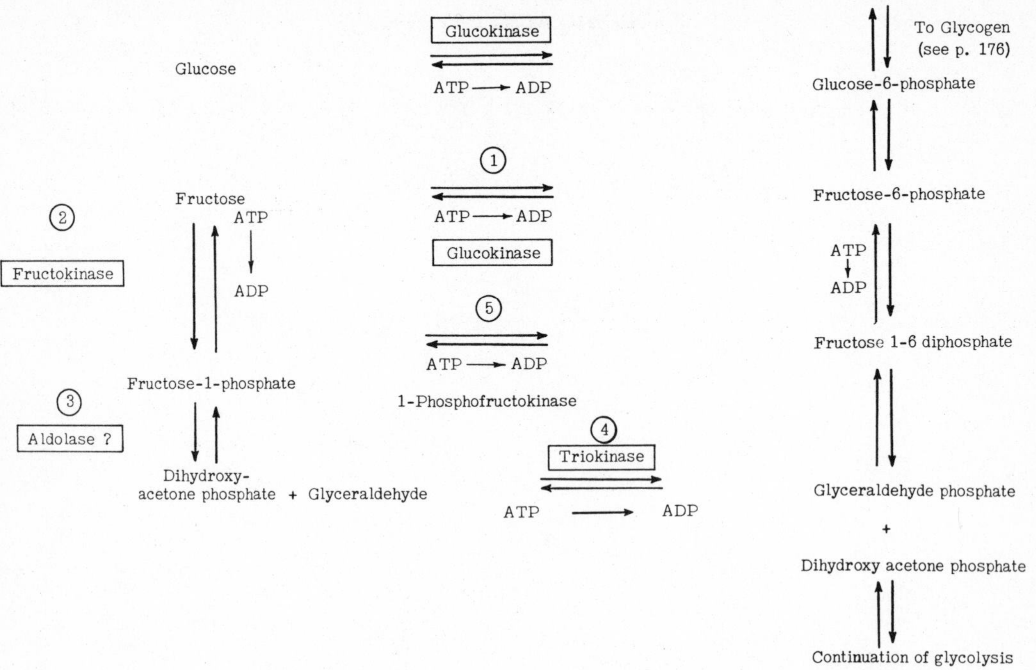

Reaction 1: Fructose may be phosphorylated to form fructose-6-phosphate, with ATP as phosphate donor. However, the reaction is catalyzed by the same hexokinase (glucokinase) which accomplishes the phosphorylation of glucose or of mannose. As a result there is competition between glucose and fructose for this enzyme, and the enzyme has a much greater affinity for glucose than for fructose. This pathway for fructose utilization is very likely therefore a minor one in the intact organism.

Reaction 2: Fructose may be phosphorylated in liver and in muscle to form fructose-1-phosphate. The enzyme catalyzing this reaction is a specific fructokinase which is not competitively inhibited by glucose. It is therefore probable that this is the major route of initial phosphorylation of fructose. Fructokinase, unlike glucokinase, is not affected by insulin (see p. 338), which may explain why fructose disappears from the blood of diabetic patients or of alloxan-diabetic rats at a normal rate.

Reaction 3: Fructose-1-phosphate is then split into two three-carbon compounds, dihydroxyacetone phosphate and glyceraldehyde. The enzyme catalyzing this split is closely related to aldolase.

Reaction 4: Glyceraldehyde is phosphorylated with the aid of the enzyme, triokinase. The two triose phosphates, dihydroxyacetone phosphate and glyceraldehyde phosphate, then enter the Embden-Meyerhof scheme and are metabolized as shown on p. 180.

Reaction 5: Fructose-1-phosphate may be phosphorylated on position 6 to form fructose-1,6-diphosphate, which is in the direct route of glycolysis. The enzyme catalyzing this reaction is called 1-phosphofructokinase.

The utilization of fructose in the phosphogluconate oxidation pathway (see p. 186) should also be recalled.

The metabolism of fructose as described above takes place almost entirely in the liver and, to a lesser extent, in the intestines. In the liver the metabolism of fructose occurs almost entirely by means of reactions 2, 3, and 4. If the liver and intestines of an experimental animal are removed, the conversion of injected fructose to glucose does not take place and the animal succumbs to hypoglycemia unless glucose is administered. It appears that brain and muscle can utilize significant quantities of fructose only after its conversion to glucose. Consequently the adequacy of fructose as a source of energy depends upon competent hepatic function.

METABOLISM OF GALACTOSE

Galactose is derived from the hydrolysis in the intestine of the disaccharide, lactose, the sugar of milk. It is readily convertible in the liver to glucose and further metabolized in this way. The ability of the liver to accomplish this conversion is measured as a test of hepatic function in the galactose tolerance test (see p. 279). The disappearance of galactose from the blood is dependent upon insulin, as is true also of glucose (see p. 338). The pathway by which galactose is converted to glucose is diagrammed below[20].

The Pathway For Conversion of Galactose to Glucose

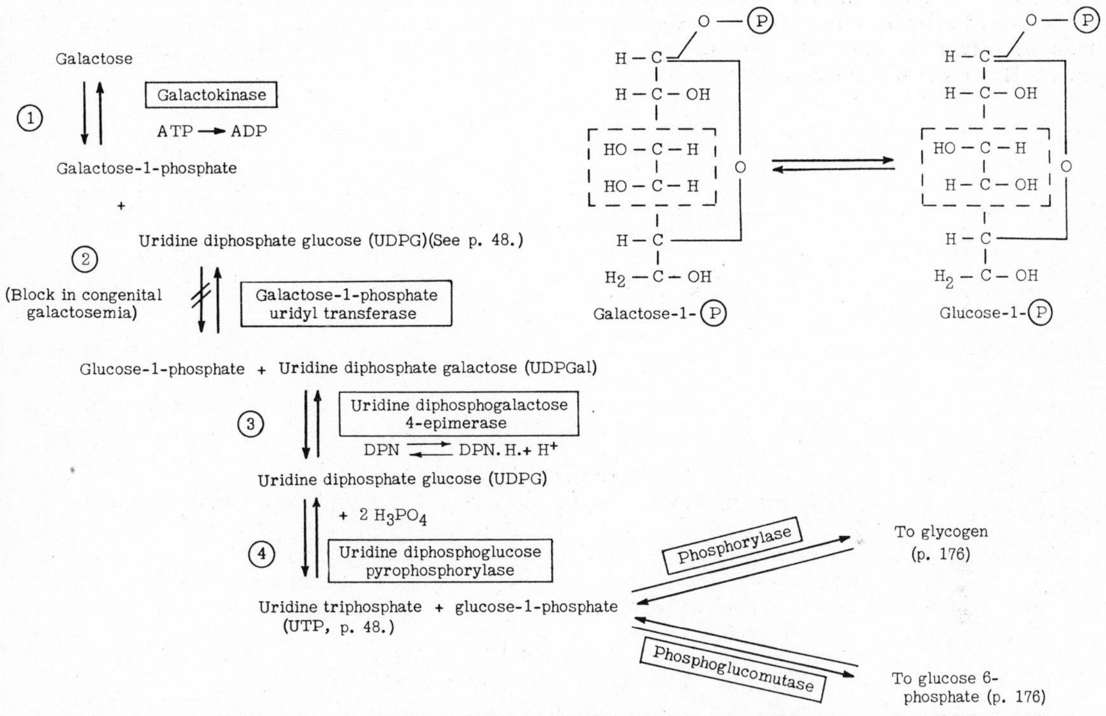

In reaction 1, galactose is phosphorylated with the aid of galactokinase, using ATP as phosphate donor. The product, galactose-1-phosphate, then reacts with the important coenzyme, uridine diphosphate glucose (see p. 48), to form uridine diphosphate galactose and glucose-1-phosphate. In this step (reaction 2), which is catalyzed by an enzyme called phosphogalactose uridyl transferase, galactose is simply transferred to a position on the coenzyme, replacing glucose. The conversion of galactose to glucose now takes place (reaction 3) in a reaction of the galactose-containing coenzyme which is catalyzed by an epimerase. The product is uridine diphosphate glucose. Finally (reaction 4), glucose is liberated from the coenzyme as glucose-1-phosphate with concomitant formation of uridine triphosphate (UTP), a high-energy phosphate analogous to ATP.

Reaction 3 is freely reversible. In this manner glucose can be converted to galactose, so that preformed galactose is not essential in the diet. It will be recalled that galactose is required in the body not only in the formation of milk but also as a constituent of galactolipins (cerebrosides; see p. 24), chondromucoids (see p. 382), and mucoproteins.

Inability to metabolize dietary galactose occurs in galactosemia, a congenital syndrome in which galactose accumulates in the blood and spills over into the urine when this sugar or lactose is ingested. However, there is also marked accumulation of galactose-1-phosphate in the red blood cells of the galactosemic individual, which indicates that there is no deficit of galactokinase (reaction 1).

Recent studies suggest that a congenital lack of phosphogalactose uridyl transferase is responsible for galactosemia. As a result reaction 2 (see p. 195) is blocked. The epimerase (reaction 3) is, however, present in adequate amounts, so that the galactosemic individual can still form galactose from glucose. This explains how it is possible for normal growth and development of these children to occur on the galactose-free diets which are used to control the symptoms of the disease.

Metabolism of Propionate:

Propionic acid, which is formed in metabolism, is known to be glycogenic, but the route by which it is converted to carbohydrate has been in doubt. Direct conversion of propionate to succinate by CO_2 fixation with the aid of soluble enzyme preparations from mammalian liver has now been demonstrated. The reaction involves formation of propionyl Co A and of succinyl Co A, requiring therefore Co A and ATP and possibly cocarboxylase as well as magnesium and manganese ions. As noted on p. 91, biotin is believed to function as a cofactor in CO_2 fixation. It is therefore of interest that enzyme preparations from biotin-deficient rats were much less effective than those from normal rats in catalyzing the carboxylation (CO_2 fixation) of propionate. This metabolic defect was repaired by the injection of biotin.

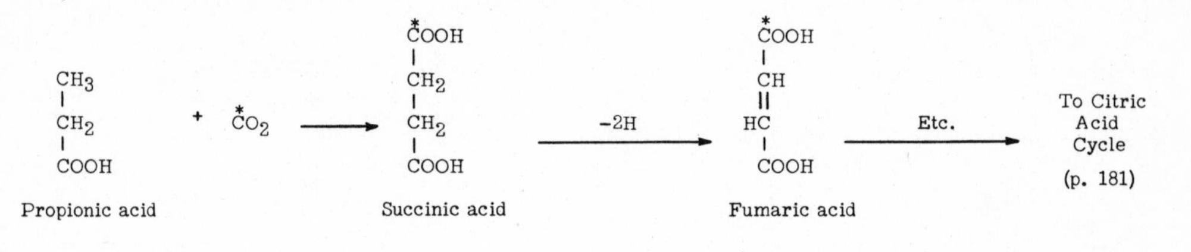

• • •

References:

1. Shapiro, B., and Wertheimer, E.: J. Biol. Chem. **173**:725, 1948.
2. Hausberger, F.X., Milstein, S.W., and Rutman, R.J.: J. Biol. Chem. **208**:431, 1954.
3. Nutrition Reviews, **14**:81, 1956.
4. Kirby, S.A., Noda, L., and Lardy, H.A.: J. Biol. Chem. **209**:191, 1954.
5. Cori, C.F.: Symposium on Clinical and Biochemical Aspects of Carbohydrate Utilization in Health and Disease, p. 3; Najjar, V.A., Ed. Johns Hopkins Press, 1952.
6. Recant, L.: Am. J. Med. **19**:610, 1955.
7. Cori, G.T., and Cori, C.F.: J. Biol. Chem. **199**:661, 1952.
8. Krimsky, I., and Racker, E.: J. Biol. Chem. **198**:721, 1952.
9. Racker, E., and Krimsky, I.: J. Biol. Chem. **198**:731, 1952.
10. Freedman, A.D., and Graff, S.: J. Biol. Chem. **233**:292, 1958.
11. Utter, K., and Kurahashi, M.F.: J. Am. Chem. Soc. **75**:758, 1953.
12. Utter, K., and Kurahashi, M.F.: J. Biol. Chem. **207**:821, 1954.
13. Kinoshita, J.H., and Wachtl, C.: J. Biol. Chem. **233**:5, 1958.
14. Brin, M., and Yonemoto, R.H.: J. Biol. Chem. **230**:307, 1958.
15. Cahill, G.F., Jr., Hastings, A.B., Ashmore, J., and Zottu, S.: J. Biol. Chem. **230**:125, 1958.
16. Brady, R.O., Mamoon, A.M., and Stadtman, E.R.: J. Biol. Chem. **222**:795, 1956.
17. Carson, P.E., Flanagan, C.L., Ickes, C.E., and Alving, A.S.: Science **124**:484, 1956.
18. Winograd, A.I., and Renold, A.E.: J. Biol. Chem. **233**:267, 1958.
19. Horecker, B.L., and Hiatt, H.H.: New Eng. J. Med. **258**:225, 1958.
20. Kalckar, H.M.: Science **125**:105, 1957.

Bibliography:

Annual Review of Biochemistry. Annual Reviews, Inc.
Soskin, S., and Levine, R.: Carbohydrate Metabolism. U. of Chicago Press, 2nd Ed., 1952.

14...
Metabolism of Fat

The fats (lipids) of the body include the neutral fats (triglycerides), the phospholipids, and the sterols. Fat is the principal and most efficient source of stored energy because of its high fuel value (9.3 Calories per gram). These reserves of fat are not stored passively but are constantly being broken down and resynthesized.

Certain unsaturated fatty acids which are essential constituents of the diet are described on p. 203. There is evidence that for optimal nutrition a certain amount of fat should be included in all diets in addition to these essential fatty acids. Furthermore, there is increasing evidence that fat is a major source of energy for many tissues and perhaps the major source of energy for the body as a whole, a role which formerly was attributed to the carbohydrates. It must be pointed out, however, that at least minimal quantities of carbohydrate must be available for the proper utilization of fat as an energy source (see p. 211). This amount of carbohydrate has been estimated to be about 5 Gm. per 100 Calories, i.e., at least 20% of the total calories of the diet should be derived directly from carbohydrates.

TRANSPORT OF FAT FROM THE INTESTINE

After absorption from the intestine (see p. 162), fat is transported in several forms and by various routes. The lymphatic vessels of the abdominal area (the lacteals) transport most of the absorbed fat in the form of fat droplets about one micron in diameter (chylomicrons). The chylomicrons are probably formed in the intestinal mucosa. The lymphatic system transfers the chylomicrons to the blood when the thoracic lymph duct empties into the left subclavian vein. After a fatty meal, the fat content of the blood is substantially increased; this postabsorptive lipemia begins one to three hours after the meal, persists for six to seven hours, and then decreases to the fasting level. In contrast to neutral fat, which utilizes the lymphatic system, the phospholipids may also enter the blood vessels of the portal system and pass directly into the liver.

The Partition of Lipids in the Blood Plasma

Lipid Fraction	Mean (mg./100 ml.)	Range (mg./100 ml.)
Total lipid	570	360-820
Triglyceride	142	80-180
Fatty acids (as stearic)*	340	200-800
Total phospholipid†	215	123-290
Lecithins		50-200
Cephalins		50-130
Sphingomyelins		15-35
Total cholesterol	200	107-320
Free (nonesterified) cholesterol	55	26-106

*Total in all fats; 45% esterified in triglycerides, 35% esterified in phospholipids, 15% esterified with chol esterol, 5% nonesterified and carried with serum albumin.
†Analyzed as lipid phosphorus; mean lipid P = 9.2 mg./100 ml. (range, 6.1 to 14.5). Lipid P × 25 = phospholipid as lecithin (4% P). Serum also contains amall amounts of phosphatidyl serine.

In addition to triglyceride and phospholipids, the blood also contains cholesterol, both free and esterified with fatty acids; other sterids; and free (nonesterified) fatty acids, all protein-bound as lipoproteins. The triglycerides, cholesterol esters, and phospholipids account for 80% of the total blood fat. The quantities of the various lipids in the blood plasma are given in the table on p. 197.

The Composition of Lipids in the Plasma:

As noted above, the various lipid constituents of the plasma are all protein-bound. About 3% of the total plasma protein consists of lipoproteins in the α globulin fraction and about 5% of lipoproteins in the β globulin fraction. The α lipoproteins contain relatively smaller amounts of fat and hence more protein than do the β lipoproteins. The higher the fat and the lower the protein content of a lipoprotein, the lower is its specific gravity (density); thus the terms "low density" or "very low density" may be used to refer to various β lipoproteins (densities less than 1.063), whereas the α lipoproteins, with higher protein and lower fat contents, are called "high density" lipoproteins (densities greater than 1.063) (see also p. 132).

Using ultracentrifugal technics, Gofman[1] and his associates have further characterized the β lipoproteins. Because of their high lipid content, these proteins have a lower density than ordinary proteins. When centrifuged, they can therefore be made to rise to the surface rather than fall out. The measurements of flotation are made in a solution of sodium chloride at a known temperature, and the rate of upward migration of the fat molecule is expressed in Svedberg units of flotation (S_f units). One S_f unit is equal to 10^{-13} cm. per second per dyne per gram at 26° C. in a NaCl solution having a specific gravity of 1.063. A lipoprotein with a flotation rate of 12 S_f units is therefore spoken of as an "S_f 12 lipoprotein."

The approximate percentage composition of several lipoprotein fractions is given below.*

Fraction	Protein	Triglyceride	Cholesterol	Phospholipid
Chylomicrons	2	81	9	7
Very low density lipoproteins ($S_f > 10$)	7	52	22	18
Low density lipoproteins (S_f 0-10)	21	9	47	23
High density lipoproteins	46	8	19	26

*Bragdon et al.: J. Lab. & Clin. Med. **48**:36, 1956.

These data indicate that the triglycerides, which are derived mainly from alimentary fat, are carried principally in the chylomicrons and the very low density lipoproteins. Cholesterol migrates mainly with the very low density and the low density lipoproteins (S_f 0-10 and 10-400) and, to a lesser extent, with the high density α lipoproteins. Consequently, when the serum cholesterol is elevated, the increase will be reflected by a parallel increase in the low or very low density lipoproteins. Phospholipid is concentrated chiefly in the high density fraction and, to a lesser extent, in the low and very low density fractions. These relationships are illustrated on p. 199.

Action of Clearing Factor (Lipoprotein Lipase):

The turbid appearance (lactescence) of the serum during the period immediately following a meal is due to an increased concentration of triglyceride which occurs in the chylomicrons and in the very low density lipoproteins of the higher S_f range. Whenever the triglyceride content of the blood exceeds 20 mEq./liter (590 mg./100 ml.), lactescence will be noted. The chylomicrons leave the blood a relatively short time after they have entered from the lymph, and this is accompanied by gradual clearing of the serum. There is evidence that the chylomicrons disappear into various organs, where hydrolysis of triglyceride then occurs, forming glycerol and free (nonesterified) fatty acids. The nonesterified fatty acids may then return to the plasma to be transported with the albumin fraction of the serum proteins to various tissues for oxidation or, after reconversion to lipid, for storage.

Distribution of Lipids in Serum

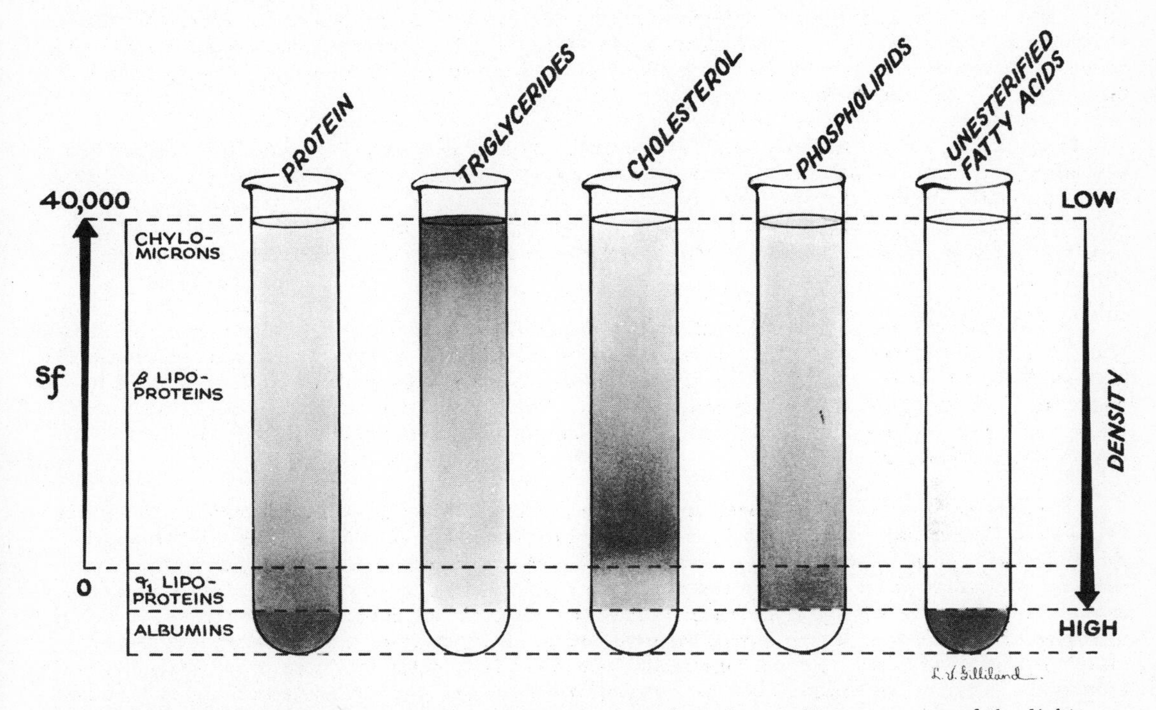

Clearing of post-alimentary lipemic plasma is therefore due to disappearance of the light-scattering chylomicrons and very low density lipoproteins from the plasma. According to Korn[2], the clearing factor is an enzyme, **lipoprotein lipase**, whose substrates are normally the triglycerides of the chylomicrons and of the very low density lipoproteins, although lipoprotein lipase catalyzes the hydrolysis of triglycerides in any lipoprotein. The reaction proceeds rapidly if there is adequate receptor (albumin) to bind the nonesterified fatty acids which are produced by hydrolysis of the triglycerides.

Heparin, or a similar compound, is an intimate part of the lipoprotein lipase molecule, seemingly acting as an essential co-factor for the action of the enzyme.

Although some clearing takes place in the plasma itself, it is believed that lipoprotein lipase activity within the tissues is of far greater importance. Lipoprotein lipase is present in large quantities in adipose tissue and in moderate amounts in heart muscle. It occurs in small amounts in lungs and in skeletal muscle. None has been found in liver or in several other tissues examined. The action of the enzyme in the tissues is a mechanism for release, after hydrolysis, of nonesterified fatty acids stored in the triglycerides of the depot fats.

The clearing action of lipoprotein lipase in vitro may be inhibited by the heparin antagonist, protamine, as well as by diisopropylfluorophosphate (DFP, see p. 389) and by sodium cholate.

Metabolism of the Nonesterified (Free) Fatty Acids of the Plasma:

The nonesterified fatty acid fraction (NEFA) of the plasma lipids is composed chiefly of palmitic, stearic, and oleic acids. The concentration of NEFA in the plasma after an overnight fast is about 300-600 μEq./liter (0.3-0.6 mEq.), but any metabolic situation which increases the demand for fat, such as fasting, leads to an increase in nonesterified fatty acids in the plasma; on the other hand, the ingestion of fat or of any other source of calories produces a rapid decrease in plasma NEFA. It now seems clear that the nonesterified fatty acids of the plasma represent the form in which lipid is transported from the depots to the tissues for oxidation to supply energy for tissue activities.

The "turnover" rate of the NEFA in the plasma is very high, as evidenced by a half-time of only a few minutes. The NEFA are evidently the metabolically active forms of lipid. Apparently all tissues of the mammalian organism so far examined (liver, kidney, heart, brain, pancreas, muscle, testis, and lung) can oxidize fatty acids completely to carbon dioxide and water. This direct and complete oxidation of long-chain fatty acids can be a major source of energy for tissues surviving in vitro. In some tissues, fatty acids seem to be the preferred fuel, e.g., in the myocardium of the heart, where in the fasting subject as much as 70% of the calories being used by this organ may be derived from nonesterified fatty acids.

In the fasting state, the fatty acids are taken up from the plasma by a variety of tissues and also slowly added to the plasma. Adipose tissue itself, perhaps with the aid of lipoprotein lipase, is exceptionally active in furnishing fatty acids to the plasma. In fact, the evidence is gradually accumulating to support the concept that adipose tissue is not an inert storage area but rather a tissue with a high metabolic potential that is delicately controlled by various homeostatic mechanisms.

The administration of glucose or of insulin brings about a decrease in the nonesterified fatty acid level of the plasma with inhibition of release of the fatty acids from the depots. Tolbutamide (Orinase®, see p. 340), when given to either normal or diabetic subjects, also brings about a decrease in plasma NEFA if hypoglycemia occurs. Both epinephrine and norepinephrine, on the other hand, cause a prompt and marked rise in plasma nonesterified fatty acids, an effect which occurs only slowly during fasting.

The content of nonesterified fatty acids in the plasma of diabetic patients without ketosis was found to be higher than in normal subjects and still higher when diabetic patients with ketonuria were studied[3]. In nondiabetic obese persons the NEFA levels were also found to be higher than normal. Administration of glucose and insulin to diabetic patients in coma produces a decrease in plasma nonesterified fatty acids which correlates with the fall in blood sugar; the fall in NEFA precedes the decrease in ketonemia by several hours. It is assumed that the administration of insulin brings about a prompt decrease in the release of fatty acids from the depot fats, thus removing the cause of the ketosis, but the metabolism of the ketones already accumulated requires an additional period of time.

The finding of consistently elevated levels of plasma nonesterified fatty acids in diabetic patients suggests that the normal mechanism for control of the release of fatty acids from depots is impaired in this disease, perhaps as a result of impaired utilization of glucose by adipose tissue.

MOBILIZATION AND DEPOSITION OF FAT

The liver is not an important storage organ for fat as it is for carbohydrate and protein. In fact, the accumulation of fat in the liver is a pathologic process. The body stores fat principally in three places: in the subcutaneous connective tissue (the panniculus adiposus), the abdominal cavity, and the intermuscular connective tissue. The distribution of fat in the body is exemplified by the data in the following list, which was obtained from analysis of the tissues of the female rat:

Subcutaneous tissues	- 50%
Abdominal cavity	
Genital tissue	- 20%
Perirenal tissue	- 12%
Mesenteric tissue	- 10%
Omental tissue	- 3%
Intermuscular tissue	- 5%

The degree of unsaturation of stored fat varies with the location. Liver fat is the most unsaturated (i.e., has the highest iodine number); the subcutaneous, perirenal, and omental fats follow in that order. The melting point of stored fat is higher in animals living in warm climates than in animals living in cooler areas. Fat depots of animals living in cool climates generally contain more unsaturated fat.

All energy-yielding foods - carbohydrate, protein, fat, or mixtures of all three - contribute to the body fat. Although each animal tends to deposit a fat peculiar to itself, the diet of the animal will alter this. The fat deposited tends to be similar to the fat ingested. For example, when soybean oil is fed to rats, the iodine number of the deposited fat is 132, whereas if rats are fed coconut oil under the same conditions the number is only 35. A carbohydrate-rich diet causes the deposition of hard fats (higher melting point and low iodine number); a high-protein diet decreases fat deposition but increases the cholesterol ester content of the liver.

Hormonal Effects on Metabolism of Fat[4]:

The role of the thyroid in control of the serum lipids is well established. Hyperactivity of the thyroid brings about a rapid loss of body fat; hypothyroidism is characterized by increase of fat storage; and after total thyroidectomy there is a significant increase in serum lipids.

The levels of cholesterol in the blood are notably affected by the thyroid. Hypothyroidism is characterized by elevations in total cholesterol. The diagnostic value of hypercholesterolemia, while limited in adults, is particularly useful in the diagnosis of hypothyroid states in children. In hyperthyroid states there is a tendency to a lowering of the serum lipid levels and especially to hypocholesterolemia. The diagnostic value of a decreased blood cholesterol in hyperthyroidism is not as useful as is the finding of hypercholesterolemia associated with hypothyroidism.

Pituitary growth hormone, when injected into hypophysectomized rats, causes, after an eight-hour interval, a significant rise in concentration of nonesterified fatty acids of the plasma. The fatty acids are derived from the lipid of adipose tissue. The mobilization of fatty acids by growth hormone may be necessary in order to provide a source of energy in connection with the protein anabolic effect of growth hormone.

The posterior pituitary gland either elaborates or stores a hormone which mobilizes fat (triglycerides) from mesenteric and omental fat depots to the liver. The hormone has been termed **lipid mobilizer** (LM)[5]. It has been isolated from the posterior pituitary as well as from the plasma of patients undergoing stress, e.g., as in pregnancy or surgery. LM has also been identified in the plasma of some patients with certain types of familial hyperlipemia.

The metabolic hormones of the adrenals (the 11-oxysteroids) act as a stimulus to the posterior pituitary gland for release of LM. As a result of stress, the anterior pituitary secretes ACTH (corticotropin), which in turn brings about increased production of adrenocortical steroids. The effect of stress on the production of lipid mobilizer is apparently mediated through this pituitary-adrenal axis.

In contrast to the omental and mesenteric depots, which release triglycerides, the subcutaneous and perirenal depots appear to release nonesterified fatty acids, which may then be utilized to resynthesize the fat removed by LM from the omental and mesenteric depots during a period of acute need such as occurs in stress. Epinephrine and norepinephrine are hormones which bring about such a rise in plasma nonesterified fatty acids as a result of stress.

The hormonal factors which regulate the mobilization of depot fat to the liver have been discussed by Levin and Farber[6]. These authors also conclude that mobilization of fat to the liver is dependent upon an endocrine mechanism involving both the anterior pituitary and the adrenal cortex. Although ACTH is a part of the mobilization system, a separate pituitary factor (which may be LM) is also active. The term "adipokinin" has been applied to this pituitary factor.

There is increasing evidence that the gonads have an important influence on the levels of circulating lipid and lipoproteins. Sex differences in the distribution of serum lipoproteins have frequently been reported. Normal young women have a relatively greater cholesterol concentration in the serum α lipoprotein fraction and correspondingly less in the β lipoprotein fraction than do normal men of the same age. After the menopause, this difference is not apparent. In connection with the hypothesized relationship of hyperlipemic states to a higher incidence of coronary heart disease, it is pertinent to recall that coronary disease occurs only rarely in women during the reproductive years of life.

Fat mobilization in response to fasting does not occur in the male in the absence of functioning testicular tissue. The response to the stress of cold or to the administration of pituitary extract is, however, unaffected by the testes.

A relationship of the estrogens to fat metabolism is well demonstrated in birds. During the period of egg-laying, the blood fats are about ten times higher than otherwise. The injection of estrogenic ovarian hormone reproduces this lipemia in immature female birds.

Insulin or the administration of glucose brings about a decrease in the plasma nonesterified fatty acids as well as inhibition of their release from the fat depots. This effect is directly opposed to that of fasting or of the administration of epinephrine. The effect of insulin on synthesis of fat (lipogenesis) from carbohydrate is discussed below.

Lipogenesis:

The synthesis of fat by the tissues (lipogenesis) has been studied by Winograd and Renold in rat adipose tissue in vitro[7]. For this purpose, the epididymal fat pads were used. The incorporation of a labeled carbon of glucose into the ether-soluble lipids of adipose tissue was found to be markedly stimulated by insulin added in vitro. An insulin effect on glucose uptake as well as on oxidation of glucose carbons to CO_2 within adipose tissue was also apparent. It was concluded that adipose tissue is a major site of insulin action.

In the experiments described above, lipogenesis from labeled acetate or pyruvate was not stimulated by added insulin unless glucose was also present in the medium. These observations suggest that the effect of insulin on lipogenesis is actually an indirect one. Reductive steps in the synthesis of long-chain fatty acids require reduced TPN (TPN.H) as hydrogen donor (see p. 207), and this co-factor is most efficiently obtained by the oxidation of glucose-6-phosphate through the direct oxidative pathway (see p. 186). The insulin-glucose relationship for lipogenesis in adipose tissue may therefore be explained by the need for insulin and glucose in the operation of the direct oxidative pathway to produce the necessary TPN.H for fatty acid synthesis.

Milstein[8], while studying the regulation of lipogenesis in adipose tissue, compared the rate of oxidation of glucose-1-C^{14} to that of glucose-6-C^{14} as a means of determining the relative quantities of glucose which disappeared through the direct oxidative pathway or through the Embden-Meyerhof glycolytic pathway (see p. 179). He found that the oxidative pathway was particularly active in adipose tissue from fed rats but that starvation greatly decreased its activity, as evidenced by the yield of C^{14} from glucose-1-C^{14}. However, the oxidation of glucose through the E-M glycolytic pathway was apparently not affected by starvation. Administration of cortisone had an effect on lipogenesis similar to that of starvation, as did also the production of alloxan diabetes. In contrast to these effects on adipose tissue, the rates of formation of labeled CO_2 from C-1 and C-6-labeled glucose were unchanged in liver tissue under identical experimental conditions. These experiments support the belief, expressed above, that lipogenesis is dependent upon simultaneous metabolism of glucose through the direct oxidative pathway, presumably as a means of supplying TPN.H.

Lipogenesis in the liver, as in adipose tissue, is connected with utilization of carbohydrate. There is also a requirement for insulin.

The rate of synthesis of fat from carbohydrate is more markedly influenced by the fat content than by the carbohydrate content of the antecedent diet of an experimental animal. Both lipogenesis and oxidation of glucose are diminished in adipose tissue by fasting or by feeding a diet high in fat (60% fat, 1.2% carbohydrate). Treatment with insulin for three days does not restore the depressed lipogenesis brought about by the high-fat diet. Dietary carbohydrate (glucose), on the other hand, stimulates lipogenesis and glucose utilization and, as was shown by the in vitro experiments of Winograd and Renold described above, the carbohydrate is necessary before the action of insulin on lipogenesis will be manifested.

A similar effect of diet on lipogenesis in the liver has also been reported by Hill et al.[9] Lipogenesis in the liver, as in adipose tissue, is connected with utilization of carbohydrate, and there is also a requirement for insulin. The feeding of fat depresses lipogenesis in the liver, an increase of as little as 2.5% fat in the diet exerts a measurable effect on hepatic lipogenesis as evidenced by decreased incorporation of C^{14} from labeled glucose or acetate into fatty acids.

These experimental observations on the influence of the composition of the diet on lipogenesis both in liver and in adipose tissue are of interest in relation to the reports of Kekwick and Pawan[10] on weight loss as influenced by the composition of a reducing diet. After ingesting a low-calorie diet which contained 90% of the calories in the form of fat, their patients experienced a rapid loss of weight during the seven-day period of observation. Patients receiving isocaloric diets in which 90% of the calories were derived from carbohydrate maintained their weight. It was established that the weight loss on the high-fat diets was not due to reduced absorption of fat from the intestine. On the high-fat diet or on a high-protein diet, 41-52% of the weight loss was due to loss of body water and 48-59% to loss of body fat, which suggests that the effect of dietary composition on weight loss is mediated by an effect on alteration of metabolic pathways in accord with the composition of the diet.

The Essential Fatty Acids[11]:

In 1929, Burr and Burr[12] showed that rats maintained on a diet from which fat was rigidly excluded ceased to grow and developed scaliness of the skin, kidney damage, and impaired reproductive ability. It was possible to cure or prevent these abnormalities by feeding less than 100 mg. per day of linoleic or arachidonic acids. Linolenic acid restored growth in the deficient animal but did not relieve the dermal symptoms. These three polyunsaturated acids, linoleic, linolenic, and arachidonic, are generally referred to as the essential fatty acids (EFA), but it is possible that the term should be restricted to those substances which are effective both in promoting growth and in maintaining the integrity of the skin. In this sense, only linoleic and arachidonic acids - and such acids as may be derived metabolically from them - would qualify. Because linoleic acid, by the addition of an acetate molecule, is convertible to arachidonic acid[13], it may be thought of as the most important of the essential fatty acids from a dietary standpoint.

It is well known that desaturation of fats by introduction of double bonds into the fatty acid chain occurs in the body, particularly in the liver, where the iodine number of fats is higher than is the case with the fats of other tissues. However, it appears that the introduction of additional double bonds in unsaturated fatty acids is limited to the area between the carboxyl group and the existing double bonds and that it is not possible to introduce a double bond between the methyl group at the opposite end of the molecule and the first unsaturated linkage.

Essential fatty acids are required for a number of biological functions. In the lipids associated with the structural elements of the tissues, polyunsaturated fatty acids occur in higher concentration than in depot fats, and deficiency symptoms are most easily induced during physiological conditions where rapid cell division occurs. The lipids of the gonads also contain a high concentration of polyunsaturated fatty acids, and this suggests the importance of these compounds in the reproductive function. A deficiency of essential fatty acids or exposure to hypotonic solutions have the same effect on the mitochondria. The effect is on the physical state of the mitochondrial membrane, producing a swelling of the structure which also occurs with added thyroxin (see p. 122). The result is a reduction in efficiency of oxidative phosphorylation which may be responsible for the increased heat production noted in animals deficient in the essential fatty acids.

Diets which are high in saturated fats, generally the fats of animal origin, tend to result in high levels of cholesterol in the serum, whereas fats of plant origin, the oils, in which there is a high content of unsaturated fats (and particularly of the fats which contain the essential fatty acids), tend to depress serum cholesterol levels. There is also an effect on the clotting time and on the fibrinolysin activity of the blood. A prolongation of clotting time and an increase in fibrinolytic activity follow the ingestion of fats rich in essential fatty acids, whereas diets rich in saturated fat shorten clotting time and decrease fibrinolytic activity.

Atherogenesis, the formation of lipid deposits (atheromata) within the blood vessels, is believed to be related to the metabolism of lipid and in particular to that of cholesterol because of the association of prolonged hypercholesterolemia with the development of atheromatous lesions in experimental animals. Formation of blood clots is believed to occur more readily in vessels already narrowed by atheromatous deposits. Because of the presumed relationship of the essential fatty acids to the levels of cholesterol in the blood and to clotting time and fibrinolytic activity, there has been much interest in these compounds in current studies of the problems of atherosclerosis.

The effect of essential fatty acids on cholesterol metabolism may be related to the need for these fatty acids in cholesterol transport and metabolism. Gordon et al.[14] attempted to investigate this hypothesis by analysis of the fecal steroids, particularly the bile acids (see p. 159), which are the main excretory products of cholesterol. They concluded that the effect of the unsaturated fats on serum cholesterol is due to a promotion of sterol catabolism and excretion. It was suggested that the polyunsaturated fatty acids, when esterified to cholesterol, promote its emulsification and incorporation into serum lipoproteins which is essential to its transport to the liver for oxidation and excretion via the bile. In this connection it is of interest to note that conditions which are accompanied by hypercholesterolemia (administration of cholesterol in the diet, alloxan diabetes, hypothyroidism) all accelerate essential fatty acid deficiency in rats. This may be due to a greater need for EFA for the purpose of transporting and metabolizing the additional cholesterol.

A deficiency of essential fatty acids has not yet been unequivocally demonstrated in humans. It is difficult even in animals to produce the deficiency in the adult. However, in weanling animals symptoms of EFA deficiency are readily produced, and it should be noted that a requirement for an essential fatty acid (linoleic) to cure skin lesions observed in infants on a low-fat formula has actually been reported. Although dermatitis has not been observed in adult humans on a fat-free diet, changes in the nature of dietary fat do alter the composition of the plasma lipids.

Linoleic and linolenic acids occur in high concentrations in the common vegetable oils. The more highly unsaturated acids are found only in animal glandular tissue (not to any extent in muscle meats). Fish oils are also rich in polyunsaturated fatty acids. A diet containing a variety of vegetables, grains, and meats should provide adequate quantities of essential and polyunsaturated fatty acids.

Typical Fatty Acid Analyses of Some Fats of Animal and Plant Origin*
(All values in weight percentages of component fatty acids.)

	Saturated			Unsaturated		
	Palmitic	Stearic	Other	Oleic	Linoleic	Other
Animal Fats						
Lard	29.8	12.7	1.0	47.8	3.1	5.6
Chicken	25.6	7.0	0.3	39.4	21.8	5.9
Butterfat	25.2	9.2	25.6	29.5	3.6	7.2
Beef fat	29.2	21.0	3.4	41.1	1.8	3.5
Vegetable Oils						
Corn	8.1	2.5	0.1	30.1	56.3	2.9
Peanuts	6.3	4.9	5.9	61.1	21.8	-
Cottonseed	23.4	1.1	2.7	22.9	47.8	2.1
Soybean	9.8	2.4	1.2	28.9	50.7	7.0†
Olive	10.0	3.3	0.6	77.5	8.6	-
Coconut	10.5	2.3	78.4	7.5	trace	1.3

*Reproduced, with permission, from N.R.C. Publication No. 575: The role of dietary fat in human health; a report of the Food and Nutrition Board of the National Academy of Sciences.
†Mostly linolenic acid.

THE ROLE OF THE LIVER IN FAT METABOLISM

The concept of a central role for the liver in fat metabolism has been modified somewhat by the recent finding that many other tissues can accomplish complete oxidation of fatty acids in the absence of the liver. However, in certain circumstances, increased amounts of fat accumulate in the liver as a result of excessive mobilization of fat from the depots by the action of hormones or neurogenic factors. This type of fatty liver, which occurs, for example, during starvation or in uncontrolled diabetes, is accompanied by hyperlipemia and ketonemia, indicating that increased emphasis on the metabolism of fat is the basis for the development of the fatty liver.

A fatty liver may also be produced by the administration of hepatotoxins such as carbon tetrachloride, chloroform, phosphorus, lead, or arsenic, or by the use of diets deficient in so-called lipotropic factors, as described below. In the latter type of fatty liver, hyperlipemia and ketonemia are not observed; and recent studies of the metabolic defect in these instances indicate that faulty oxidation of fat by the liver is the cause of lipid accumulation in that organ.

Fatty livers may also be induced by special diets, e.g., high-cholesterol diets or particularly in the presence of an increased intake of certain B vitamins, high-carbohydrate, high-fat, or low-protein diets. A fatty liver followed by acute, massive hepatic necrosis has been produced in rats by a number of diets all of which are characteristically low in protein and, particularly, in the sulfur-containing amino acids. A deficiency of vitamin E enhances the necrogenic action of these diets, whereas added vitamin E or a dietary factor termed "Factor 3" by Schwartz have a protective effect (see p. 330). Diets in which the protein was largely yeast have been shown to be necrogenic, although not all yeasts were equally effective.

Factor 3 has now been identified as an organic compound containing selenium[15]. It is effective in very small doses and is replaceable in the diet by relatively small amounts of inorganic selenium compounds. It is suggested that the biologically occurring active compound functions as a co-factor, possibly for respiratory enzymes in the cell. The requirement for yeast as the sole source of protein in the necrogenic diet is apparently attributable to its low content of selenium; yeasts vary in their effectiveness according to their selenium content.

Protein, essential fatty acid, and vitamin deficiencies may also contribute to fatty infiltration of the liver. Alcoholism undoubtedly predisposes an individual to some of these dietary errors. Hepatic injuries from other causes, such as toxic diseases or the various poisons mentioned heretofore, will similarly influence liver function in the direction of fatty infiltration. There is much evidence to show that accumulation of fat in the liver may progress to actual destruction of liver cells which is characterized by fibrotic changes. This chronic liver disease, cirrhosis, is treated with high-protein and high-carbohydrate diets with a view to reducing and preventing fatty infiltration of the liver.

Lipotropic Factors:

A substance which prevents the accumulation of abnormal quantities of fat in the liver is called a lipotropic factor. Choline is probably the most important of these. When lecithin is added to the diets of experimental animals, increased accumulation of fat is prevented. The choline contained in the lecithin molecule is the effective agent; in fact, the role of choline in the formation of phospholipids such as lecithin may be the basis for its lipotropic action. Chaikoff, using radioactive phosphorus as a tracer, reported that choline does accelerate the rate of formation of liver phospholipid.

Apparently pre-formed choline need not be provided in the diet, since it may be synthesized in accordance with the need. The details of choline synthesis are described on p. 100

Role of Choline in Metabolism of Fatty Acids:

It was suggested above that the lipotropic action of choline may be attributed to its effect on the synthesis of phospholipid by the liver. Evidence for an increased rate of mobilization of fatty acids from the liver to the depots under the action of choline has been obtained, but lipotropic activity cannot be explained by increased transfer of fat from the liver.

Artom[16] has investigated the role of choline in lipid metabolism by studying the rate of oxidation of C^{14}-labeled long-chain fatty acids in liver preparations. In his experiments, the liver tissue taken from choline-deficient rats on low-protein diets showed a diminished ability to oxidize the added fatty acids. When choline was added to the diet of the animal or injected shortly before the liver was removed, the tissue was able to oxidize the fatty acids at a normal rate. Addition of choline to the liver preparation in vitro had no effect; thus the action of choline was not that of a direct catalyst. It was concluded that its lipotropic effect may be due in part to its ability to enhance the rate of fatty acid oxidation in the liver. In further studies Artom found that the administration of choline to protein-deficient rats enhances fatty acid oxidation not only in liver but also in the kidney and heart. However, it now appears that the lipotropic action of choline cannot be explained merely as the result of enhanced oxidation of fatty acids in the liver. A recent suggestion[17] is that choline may promote the formation of some lecithin-containing lipoproteins in the liver which are essential in mitochondrial activity or in the oxidation of fatty acids. It is also possible that these lipoproteins synthesized in the liver may be the form in which fatty acids are transferred from the liver to the depots.

Transmethylation:

A source of **labile methyl** in the body is an important factor in fat metabolism by the liver. Substances which supply methyl groups act indirectly as lipotropic agents by contributing to the synthesis of choline (see p. 100). Methionine is the most important source of labile methyl, and it is believed that the lipotropic effect of protein in the diet is due to its methionine content. Betaine, an oxidation product of choline, is another methyl donor, although formation of methionine (by methylation of homocysteine, see p. 248) is the actual route by which betaine acts as a methyl donor. In that sense, only methionine is the primary methyl donor in transmethylation.

The formation of choline by methylation processes is a reversible reaction, and so choline may also serve as a methyl donor. However, choline must be oxidized to betaine in order to donate methyl groups. Vitamins B_{12} and C and folinic acid are necessary to the activity of the enzymes concerned with the oxidation of choline to betaine (see p. 100).

The transfer of methyl groups (**transmethylation**) is important in metabolism. Methylation to form N-methylnicotinamide in the excretion of niacin derivatives has been mentioned (see p. 81). Other examples of transmethylation will be given in connection with the metabolism of methionine (see p. 247) and of creatine (see p. 260). The diversion of methyl groups for these reactions leads to the production of a fatty liver if the dietary supply of choline and methionine is inadequate.

Labile methyl groups may be lost not only by transfer to other compounds which are then excreted; they may also be destroyed by oxidation. However, the animal organism has a considerable ability to synthesize methyl groups. Vitamin B_{12} and folic acid are both involved in the synthesis of methyl groups.

Other Lipotropic Agents:

The lipotropic action of protein has been described above in connection with the role of the amino acids methionine, glycine, and serine, all of which contribute to the synthesis of choline. However, A. E. Harper has found that the lipotropic action of dietary protein is not confined to these amino acids. From his experiments[18] he has concluded that this additional lipotropic action of protein is not a choline-sparing action but that it results from the provision of certain other amino acids in the absence of which fat accumulates in the liver. Fatty livers may be induced in rats fed a 9% casein diet supplemented with choline, methionine, and tryptophan. The addition of 0.36% DL-threonine to this diet consistently reduces liver fat to about one-half that in the group whose diets were not supplemented with threonine. The mechanism of the threonine effect is not known, but it was suggested[19] that it may be related to the synthesis of enzymes which function in the formation of phospholipid or in the oxidation of fatty acids.

A substance may be lipotropic without being a methyl donor. Arsenocholine and triethylcholine are examples. Both of these substances, however, will take the place of choline in a lecithin molecule. These observations suggest that lipotropism is not necessarily identified with the ability to transfer methyl groups.

Inositol is a lipotropic agent, but its action is limited to fatty livers produced in animals on fat-free diets[20]. The manner in which inositol functions as a lipotropic agent is not known, although it is found as a constituent of some lecithins. The inositol-containing fats, or lipositols, may also provide an explanation for the role of inositol as a lipotropic substance (see p. 22).

It has long been noted that fatty livers develop in depancreatized dogs maintained on insulin. The addition of raw pancreas to the diet of these animals prevents the occurrence of the fatty liver. Dragstedt has stated that there is in raw pancreas a lipotropic factor (which he has named **lipocaic**) which is different from choline. This lipotropic factor is said to be particularly effective in alleviation of the high-cholesterol type of fatty liver. More recent work indicates that the so-called lipotropic factor in raw pancreas is actually due to proteolytic enzymes. The fatty livers in the insulin-maintained depancreatized dogs can be prevented if crystalline trypsin, or the plant proteolytic enzymes, papain or ficin, are added to the diet. These observations suggest that the cause of the fatty livers in the depancreatized dogs is inadequate digestion of protein. As a result, methionine is unavailable to the animals and a deficiency of methyl groups occurs. This explanation is supported by the fact that the addition of methionine or choline to the diet of the depancreatized dogs will also prevent the development of fatty livers.

BIOSYNTHESIS OF FATS

Synthesis of Fatty Acids:

It has been assumed that fatty acids were synthesized by a process which was essentially the reverse of that by which the fatty acids are degraded. Such a process would merely involve repeated condensation of two-carbon (acetate) units to form the long-chain fatty acids, of which the C-16 acid, palmitic acid, is the main constituent in fatty tissue. However it has recently been found that acetaldehyde is actually a better precursor of fatty acids than is acetic acid[21] and that the addition of the three-carbon dicarboxylic acid, malonic acid, markedly enhances synthesis of fatty acids[22]. On the basis of these observations and the recent isolation of a malonic acid derivative from pigeon liver[23], a scheme for fatty acid synthesis involving aldol condensation has been proposed[24].

The first step in fatty acid synthesis requires carboxylation of one mol of acetyl-Co A, by fixation of CO_2, to form malonyl-Co A. The fixation of CO_2 probably is accomplished by ''active'' CO_2, the adenyl derivative shown on p. 239. The reaction is catalyzed by an enzyme which contains biotin (see p. 91) and which also requires manganous ion.

In the second step, acetyl Co A or the Coenzyme A-derivative of a fatty acid (''active'' fatty acid), is reduced by a TPN.H-dependent enzyme to the aldehyde. An aldol condensation and decarboxylation then occur to form the β-hydroxy derivative of a fatty acid longer by two carbon atoms than the original. These reactions are shown below.

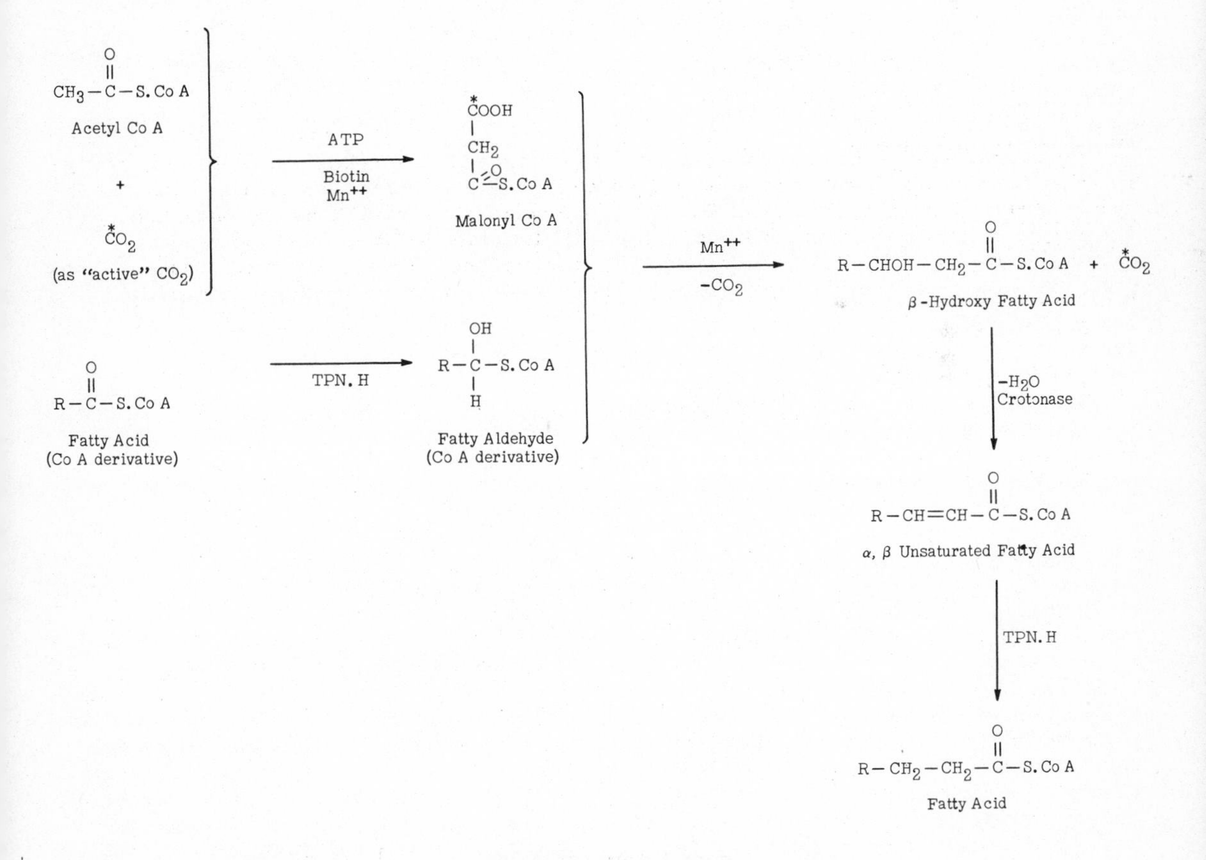

The β-hydroxy fatty acid is then dehydrated, in a reaction catalyzed by crotonase, to the α, β-unsaturated (acrylic acid) derivative. Finally, in a TPN. H-dependent reaction, the unsaturated compound is reduced to the fatty acid.

Synthesis of Triglycerides[25]:

The first step in triglyceride biosynthesis is the formation of an activated glycerol (α-glycerophosphate) by phosphorylation of glycerol with ATP as phosphate donor. This compound then reacts with activated fatty acids to form an α, β-diglyceride phosphate (1, 2-glycerophosphatidic acid), which becomes dephosphorylated to a 1, 2-diglyceride. Finally, a third mol of active fatty acid is esterified at the one free position on the diglyceride, completing the synthesis of the triglyceride.

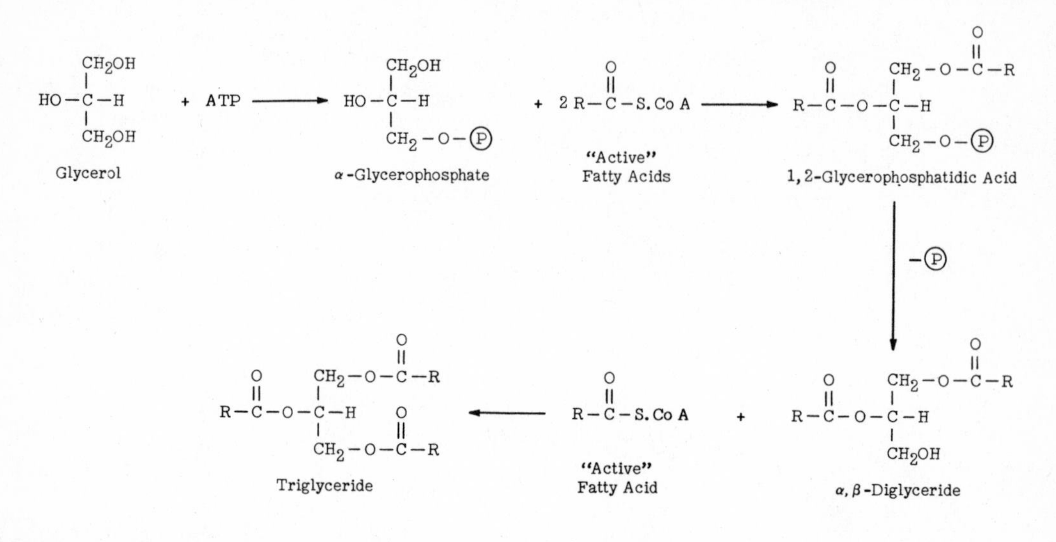

Synthesis of Phospholipids:

The biosynthesis of the phospholipids requires first the phosphorylation of choline for synthesis of lecithins, or of ethanolamines for synthesis of cephalins. An enzyme in liver catalyzes the formation of phosphoryl choline or of phosphoryl ethanolamine, with ATP as phosphate donor. The phosphorylated bases then react with cytidine triphosphate (CTP, p. 48) to form either cytidine diphosphocholine (CDP. choline) or CDP. ethanolamine. These "activated" bases will now react with an α, β-diglyceride, synthesized as shown above, to form a lecithin or a cephalin.

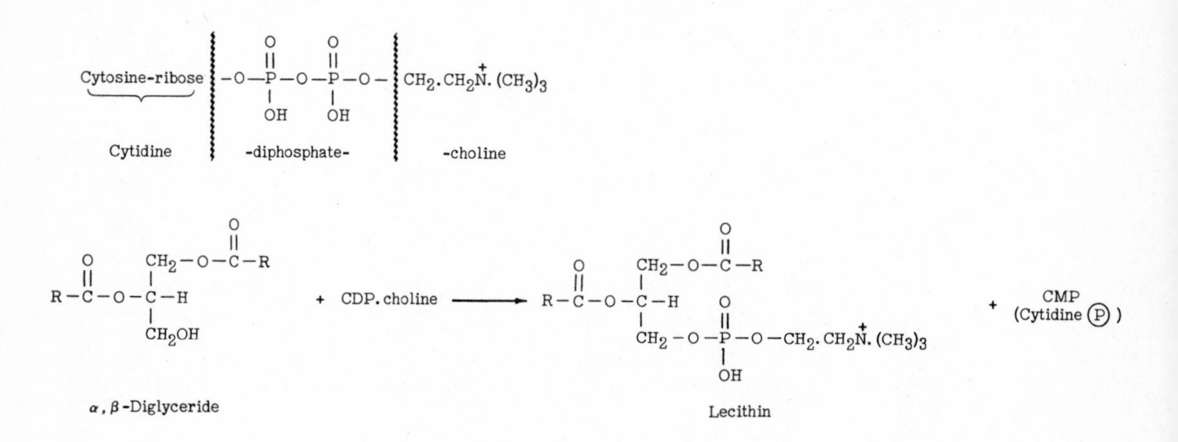

OXIDATION OF FATS

The metabolic breakdown of fat presumably requires first the hydrolysis of fat into glycerol and fatty acids. Glycerol, since it is known to be converted to glycogen in the liver, is undoubtedly metabolized as a carbohydrate. The formation of glycogen takes place by oxidation of glycerol to glyceraldehyde (a triose sugar), which is then converted to glycogen.

Oxidation of Fatty Acids:

Normally, the fatty acids are oxidized completely to carbon dioxide and water. Fatty acid oxidation probably occurs exclusively in the mitochondria.

The earliest theory on the method of the oxidative breakdown of fatty acids was the **β oxidation hypothesis** of Knoop. He postulated that the fatty acid chain is degraded in a stepwise fashion by the repeated removal of the last two carbon atoms due to oxidation at the β carbon atom. Thus, an 18-carbon chain, stearic acid, would first be oxidized to the 16-carbon palmitic acid, and this in turn to the 14-carbon acid, etc., until the entire chain disintegrates to carbon dioxide and water.

Many workers have since contributed to our knowledge of the mechanisms for the oxidation of fatty acids, but the fundamental concept of Knoop, which involved a gradual degradation of the fatty acid chain by removal of two-carbon fragments, is still valid. As a result of numerous studies on the metabolism of fatty acids, the following scheme is now proposed:

The first step is the ''activation'' of the fatty acid by the formation of the coenzyme A derivative. This requires also ATP. Following desaturation between α and β carbon atoms, the active fatty acid is then oxidized by β oxidation. The result is the formation of one mol of acetyl-Co A (active acetate) and a fatty acid which is shorter by two carbon atoms and, because of a **thiolytic** split with Co A, already activated for repetition of the oxidative cycle. These reactions are shown below.

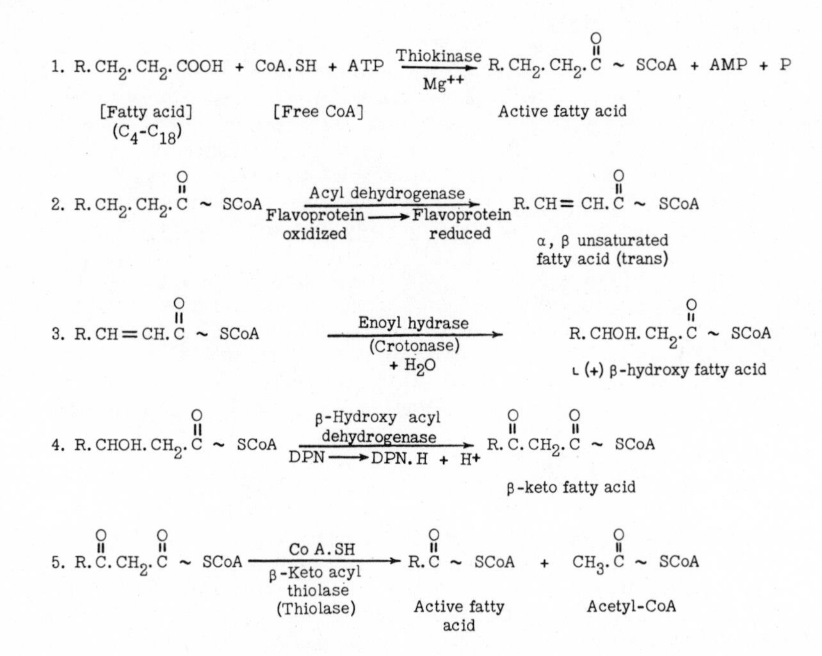

The two-carbon fragments (acetyl-Co A) derived from β oxidation may then enter the citric acid cycle for complete oxidation or they may recombine (condense) to form acetoacetyl-Co A (active acetoacetate) and other ketones as later described. The production of ketone bodies under normal conditions is minimal; rather, acetyl-Co A condenses with oxaloacetate and enters the citric acid cycle for complete oxidation.

$$\underset{\text{[2 mols of acetyl-CoA]}}{CH_3.\overset{O}{\overset{\|}{C}} \sim SCoA + CH_3.\overset{O}{\overset{\|}{C}} \sim SCoA} \longrightarrow \underset{\text{[Acetoacetyl-CoA]}}{CH_3-\overset{O}{\overset{\|}{C}}-CH_2.\overset{O}{\overset{\|}{C}} \sim SCoA} + \underset{\text{[Free CoA]}}{CoA.SH}$$

Acetoacetyl-Co A is readily converted in the liver to free acetoacetic acid because of the presence in that organ of a deacylase. The free acetoacetic acid then diffuses into the blood and is carried to the peripheral tissues, where it may then be oxidized.

$$CH_3.\overset{O}{\overset{\|}{C}}.CH_2.\overset{O}{\overset{\|}{C}} \sim SCoA \xrightarrow[\text{(liver only)}]{\text{Deacylase}} CH_3.\overset{O}{\overset{\|}{C}}.CH_2.COOH \quad + \quad CoA.SH$$

(Acetoacetyl-CoA) (Free acetoacetic acid)

In the liver, β-keto acids such as acetoacetic acid can be ''activated'' (formation of Co A derivative) by a reaction involving Co A and ATP. However, in heart muscle, kidney, and possibly in skeletal muscle, activation of some β-keto acids occurs specifically by transfer of Co A from succinyl-Co A (''active'' succinate; see p. 88), which is in the citric acid cycle by decarboxylation of ketoglutaric acid. The enzyme which catalyzes this reaction is a thiophorase (formerly termed Co A transferase[26]). It is as a result of the activity of thiophorases in the peripheral tissues that acetoacetic acid produced in the liver and deacylated there to the inactive form may be reactivated and readily oxidized via the citric acid cycle in the peripheral tissues[27].

The Enzymes for Fatty Acid Oxidation:

A system of nomenclature of the enzymes of fatty acid metabolism was proposed at the Second International Conference on Biochemical Problems of Lipids held at Ghent, Belgium, in 1955[28]. For the enzymes active in reaction 1 (see p. 209) (the fatty acid activating enzymes), the term **thiokinases** was proposed. It is possible that there may be separate thiokinases for fatty acids of separate chain lengths such as C_2 - C_3, C_4 - C_{12}, and C_{14} - C_{18}. Reaction 2 is catalyzed by an acyl dehydrogenase; reaction 3, by enoyl hydrases, also termed crotonases; reaction 4, by β-hydroxy acyl dehydrogenases; and reaction 5, by β-keto acyl thiolases (or, simply, thiolases). It is possible that different thiolases act specifically on β-keto fatty acids of different chain lengths.

Green and his co-workers have studied the characteristics of the acyl dehydrogenases, the enzymes which catalyze the oxidation of active fatty acids to α-β unsaturated derivatives (reaction 2, above). One enzyme[29], which exhibits a brilliant green color and contains flavin as flavin-adenine dinucleotide (FAD), reacts at a maximal rate with butyryl-Co A (active butyric acid). Its action seems limited to fatty acids from C_4 to C_8. The enzymatic activity can be markedly reduced with inhibitors of SH groups, such as *p*-chloromercuribenzoate, and the inhibition reversed with glutathione.

A second dehydrogenase, also a flavoprotein but with a yellow color, acts on fatty acids from C_4 to C_{16}[30]. It has therefore a much broader spectrum of activity than the green enzyme. In connection with studies of the activity of the yellow acyl dehydrogenase, it was found that the reoxidation of the flavoprotein of the yellow dehydrogenase cannot be directly accomplished but that another flavoprotein, which was termed electron-transferring-flavoprotein (ETF)[31] is required. ETF can then be reoxidized by the conventional electron acceptors such as cytochrome c or oxygen. ETF is also necessary for the reoxidation of the flavoprotein of the green enzyme, described above.

A third fatty acid dehydrogenase specific for long-chain fatty acids with maximal activity for lauryl-Co A (C_{12}) was termed palmityl-Co A dehydrogenase by Hauge et al.[32] although it attacks palmityl-Co A at only about 70% of its maximum rate. This enzyme also requires ETF.

The enzyme which catalyzes hydration of α-β unsaturated fatty acids to form the β-hydroxy derivative (reaction 3, above) is termed enoyl hydrase (or crotonase, when referring to the enzyme which hydrates the 4-carbon α-β unsaturated acid, crotonic). Stern et al.[33] have prepared and studied a crystalline crotonase from ox liver. The enzyme had a molecular weight of 210,000. Its activity at pH 9.4 was quite remarkable: one mol of crystalline crotonase catalyzed the hydration of 1.4×10^6 mols of crotonyl-Co A per minute at 25° C. This enzyme, like acyl dehydrogenase, is also inhibited by *p*-chloromercuribenzoate, which indicates that SH groups are important to its activity. Wakil and Mahler[34] have also studied a fatty acid hydrase and found that it could be inhibited by *p*-chloromercuribenzoate, iodoacetamide, or iodobenzoate and activated by glutathione or cysteine.

Wakil et al.[35] have isolated and purified from beef liver mitochondria β-hydroxyl acyl dehydrogenase, an enzyme which catalyzes the oxidation of the β-hydroxy acids (reaction 4, p. 209).

The Energy Aspect of Fatty Acid Oxidation:

The oxidation of acetyl-Co A through the citric acid cycle is believed to yield a total of 12 high-energy bonds as described on p. 181. In the process of β-oxidation as shown on p. 209 in reactions 2 and 4, flavoprotein and DPN are reduced. The transport of electrons from these coenzymes over the transfer system to cytochrome may be expected to yield additional high-energy bonds; three may be assumed for the DPN system, the number attributable to the flavoprotein system being still in doubt. To exemplify the energetics of the oxidation of a fatty acid, the breakdown of palmitic acid (C_{16}) might be considered. A total of eight C_2 units will be produced, yielding $8 \times 12 = 96$ high-energy bonds upon complete oxidation in the citric acid cycle. In addition, seven of these units will contribute to the high-energy pool, through the DPN and flavoprotein system, as they are formed in the course of β-oxidation; this will produce at least $7 \times 3 = 21$ more bonds. The total is $21 + 96 = 117$ bonds per mol of the C_{16} acid oxidized. From this must be subtracted the two bonds used in the initial activation of the fatty acid (reaction 1, p. 209), yielding a net gain of 115 bonds per mol or $115 \times 7.6 = 874$ kilocalories. The caloric value of palmitic acid is 2340 kilocalories per mol. Therefore, the coupling of phosphorylation to oxidation of the fatty acid (see p. 119) in the cell may be expected to capture about 37% ($874/2340 \times 100$) of the total available energy.

KETOSIS

Under certain conditions, the metabolic degradation of fats gives rise to considerable quantities of ketone or acetone bodies, viz., acetoacetic acid, β-hydroxybutyric acid, and acetone. These ketone bodies may be considered normal intermediates in the oxidation of fatty acids, but under the usual conditions of metabolism they are produced only in very small quantities. Normal values for ketones in the blood do not exceed 1.5 to 2.0 mg. per 100 ml.; in the urine, less than 1 mg. of ketones is excreted in 24 hours.

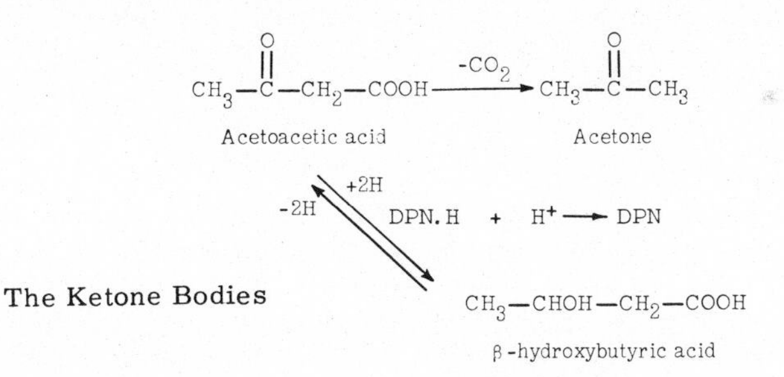

The Ketone Bodies

When carbohydrate utilization is deficient (for example, in diabetes or in starvation after carbohydrate reserves are depleted), the energy requirements of the body must be met by the use of more fat. Furthermore, a supply of carbohydrate sufficient to maintain the citric acid cycle is necessary in order to permit the smooth entry of acetic acid into the cycle. When carbohydrate metabolism is decreased, not only is the cycle impaired for lack of materials of which it is composed (pyruvate and oxaloacetate); the increased demand for energy from fat also produces more acetic acid to be oxidized. The production of the products of fatty acid catabolism at a rate which exceeds the capacity of the tissues to oxidize them causes these acetic acid residues to condense and form ketones. The ketone bodies then accumulate in the blood and are excreted in the urine. This accumulation of ketone bodies in urine or blood is termed **ketonuria** or **ketonemia**, respectively, and the over-all condition is called **ketosis**.

The liver is a particularly important source of ketone bodies when the metabolic circumstances are such as to favor ketogenesis. These circumstances include starvation and deficiencies in storage and utilization of carbohydrate, as may occur in diabetes or as a result of poisoning by carbon tetrachloride, chloroform, or phosphorus. Ketosis following starvation is more severe when preceded by the accumulation of excess fat in the liver.

The formation of acetoacetate undoubtedly takes place in the course of normal fatty acid catabolism not only in the liver but in most of the other tissues as well. In fact, the extrahepatic oxidation of ketone bodies may provide a main portion of the energy derived from the catabolism of fatty acids. However, as pointed out above, the rate of production of ketones in the liver may exceed the ability of that organ to accomplish their complete oxidation. Consequently, the ketones overflow into the blood and are excreted into the urine. At the same time, the kidney and the peripheral tissues such as muscle, including heart muscle, may not only be completely metabolizing the ketones produced within their own cells but may also be able to utilize some of the ketones which were produced in the liver and carried in the circulating blood to these tissues. An explanation of the superior ability of tissues other than liver to metabolize acetoacetate may be sought in the fact that "active" acetoacetate (acetoacetyl-Co A) is rapidly deactivated in the liver by deacylase, an enzyme which occurs in high concentration only in that organ; but reactivation of acetoacetate (a necessary preliminary to its utilization) fails to occur because of a lack of the activating enzyme thiophorase (Co A transferase, see p. 210) in the liver[36]. Other tissues have some deacylase activity, but, in contrast to the liver, their ability to activate acetoacetate through the thiophorase system is so efficient that little or no acetoacetate escapes from the cell.

Effect of Ketosis on Acid-Base Balance:

The ketone bodies are organic acids and must be neutralized with alkali in order to preserve the pH of the blood and tissue fluids. This tends to deplete the reserve of alkali and leads to a fall in pH and to an acidosis which may be fatal. The acidosis observed in uncontrolled diabetes (diabetic acidosis) is caused by ketosis.

Reduction of Ketosis:

Restoration of normal carbohydrate metabolism in order to reduce the excess utilization of fat and to restore the function of the citric acid cycle is followed by a disappearance of ketosis. This may be accomplished by carbohydrate alone, as in starvation ketosis, or by carbohydrate plus insulin, as in diabetic ketosis. Carbohydrate is therefore said to be antiketogenic, or ketolytic.

Ketosis is a manifestation of the fact that there is a limit to the **rate** at which the organism can completely utilize fat. Various races differ considerably in their ability to tolerate high-fat diets without the development of ketosis.

Metabolism of Ketone Bodies:

The disappearance of acetoacetic acid may be accomplished by a reversal of the method of its formation; this results in the formation of acetate, which is then utilized by the various metabolic pathways open to this compound. Thus the labeled carbon atoms of acetoacetic acid were found to be incorporated into cholesterol by rat liver by Brady and Gurin[37] and by Curran[38], who also proved that acetoacetic acid was oxidized to carbon dioxide, presumably via the citric acid cycle.

Another important fate of acetoacetic acid is in connection with its role as a precursor of fatty acids. Under the proper conditions the formation of fatty acids from two-carbon fragments, proceeds readily in the liver and in the peripheral tissues, including adipose tissue (see p. 202). The incorporation of labeled carbonyl ($-\overset{\overset{O}{\|}}{C}-$) carbon of acetoacetate into long-chain fatty acids by liver slices from normal **fed** rats was demonstrated by Chen, Chapman, and Chaikoff[39]. The capacity of the tissue to accomplish this conversion was lost when the liver was obtained from normal **fasted** rats or fed **diabetic** rats. Injection of insulin into the diabetic rats restored the capacity of the liver to form fatty acids from acetoacetate. From these observations it may be concluded that depressed utilization of acetoacetic acid in the liver, including a failure of lipogenesis, as well as overproduction of acetate from increased oxidation of fat, are both factors in the genesis of ketosis (see also p. 202).

The decarboxylation of acetoacetic acid produces acetone. Two pathways for the utilization of acetone are known. The first involves its conversion by oxidation to pyruvic acid, which then enters the usual glycolytic pathways.

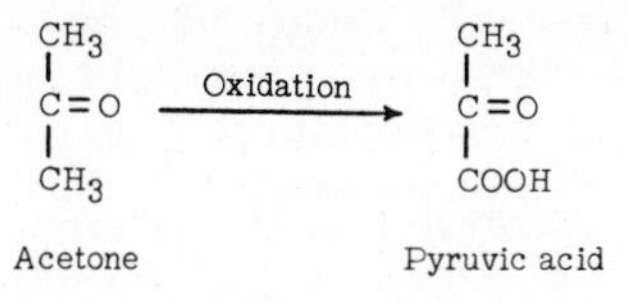

A second method for the metabolism of acetone results in its breakdown to yield a one-carbon intermediate and a two-carbon (acetate?) residue. This may be accomplished through the intermediate formation of propanediol[40]. When C^{14}-labeled propanediol was fed to rats, the β-carbon of serine and the methyl carbon of choline were found to contain the label, i.e., the carbon behaved as a typical one-carbon (formate) moiety (see p. 90). The label (C^{14}) was also found in the liver glycogen distributed in such a manner as to indicate that the propanediol might also be converted, in part, to a three-carbon glycolytic intermediate. These reactions are shown below.

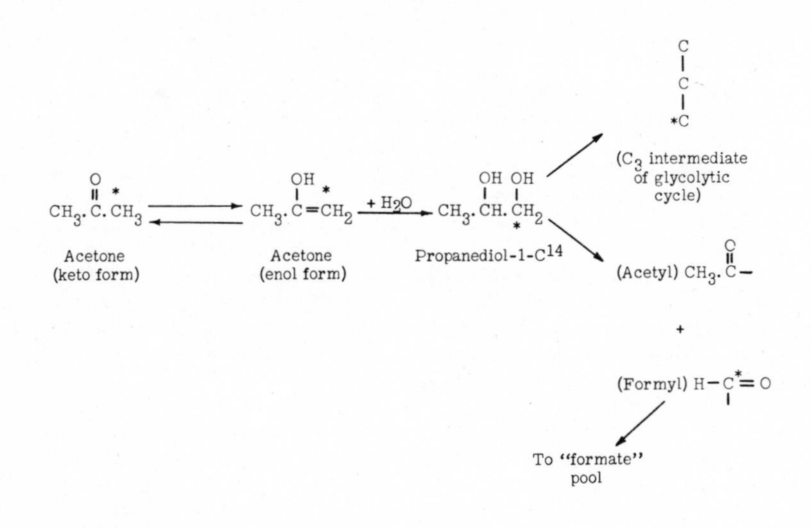

Protein as a Source of Fat:

Most amino acids are known to form glucose and glycogen. Through this pathway they could contribute to fat formation. A few amino acids increase ketone body formation (see p. 229) (i.e., they are ketogenic) and can therefore be converted directly to fat.

METABOLISM OF CHOLESTEROL

The principal sterid of the diet is cholesterol. It is found only in fats of animal origin. About 0.3 Gm. of cholesterol is ingested per day. Egg yolk is the chief source of cholesterol, although the fats of meat, liver, brain, and liver oils are good sources.

Absorption and Transport:

Cholesterol is well absorbed from the intestine. This is in contrast to the plant sterols (e.g., sitosterols), which are poorly absorbed. The absorption of cholesterol requires bile and pancreatic juice. The pancreatic juice contains an enzyme, **cholesterol esterase,** which catalyzes the formation of esters of cholesterol with fatty acids so that 80-90% of the cholesterol which is absorbed is in the ester form. Within the intestinal mucosa, virtually all of the absorbed cholesterol becomes incorporated into the chylomicrons which are being formed there. As is the case with the triglycerides, the absorbed cholesterol is transported to the blood with the chylomicrons by way of the lymphatic circulation. After the oral administration of a dose of C^{14}-labeled cholesterol, the peak of absorption into the lymph was reached in about 6-8 hours, but by 12 hours the labeled compound could no longer be detected in the lymph.

The plant sterols, such as β-sitosterol, reduce the absorption of cholesterol from the intestine possibly by competing for the enzyme, cholesterol esterase, as well as for fatty acids and bile salts.

The total concentration of cholesterol in the blood plasma is highly variable, averaging in the adult 200 mg. per 100 ml., of which 55 mg. is in the free state and about 145 mg. is esterified. The concentration of the ester fraction varies much more than does the free.

It is now well established that the levels of lipids (including cholesterol) in the serum can be altered by substitution of one dietary fat for another. The lowest levels occur when peanut, cottonseed, corn or safflower oil are added to the diet, whereas in general the addition to the diet of fats of animal origin, such as butter fat, causes an elevation in the serum lipids. Coconut oil, although a fat of vegetable origin, also raises the serum lipid levels. The rise in serum lipids varies with the type of fat used, but there is a good correlation between the iodine number (degree of unsaturation) of a fat and its ability to lower the serum lipids; those fats which have the highest iodine number produce the greatest effect on lowering of the serum lipid levels. The ability of the polyunsaturated fatty acids to accomplish a lowering of the serum cholesterol may be explained by the suggestion that cholesterol, when esterified with these fatty acids, is more readily emulsified and incorporated into lipoproteins for transport to the liver where the cholesterol is then oxidized and excreted in the bile (see p. 203).

The deposition of cholesterol and other lipids in the connective tissue of the walls of arteries is a feature of certain degenerative changes in the vessels. The disease process is called **atherosclerosis**. Considerable data from both clinical and experimental sources support the concept that atherosclerosis is a result of alteration in cholesterol and lipoprotein metabolism. Atherosclerosis is more frequent and severe among racial groups subsisting on diets rich in animal fat. It is also noteworthy that diseases characterized by prolonged elevation of blood cholesterol, such as uncontrolled diabetes, lipoid nephrosis, and hypothyroidism, are often accompanied by premature or more extensive atherosclerotic changes in the arteries.

In certain experimental animals, the ingestion of excessive quantities of cholesterol in the diet leads to the deposition of free and esterified cholesterol in the connective tissue of the walls of the arteries with lesions very similar to those of atherosclerosis in human beings. The herbivora, such as rabbits, are particularly susceptible; in the chick, atherosclerotic lesions may appear spontaneously or may be produced by cholesterol feeding. Furthermore, the administration of an estrogen (stilbestrol, p. 360) to chicks produces a hyperlipemia which even in the absence of added cholesterol leads to the production of atherosclerotic lesions in the aorta. Carnivorous animals, such as the dog, have a considerable capacity for rapid excretion of cholesterol in the bile and, in the dog, thyroid activity must be suppressed (by administration of thiouracil; see p. 335) before cholesterol administration will be effective in producing atherosclerotic changes.

Synthesis of Cholesterol[41]:

The synthesis of cholesterol has been demonstrated in many animals. The liver is the principal organ for this synthesis, but other tissues, such as the adrenal cortex, skin, intestine, testis, and even aortic tissue, have also been shown to synthesize cholesterol. Acetate serves as a direct precursor of cholesterol. After the administration of C^{14}-labeled acetate to dogs by stomach tube, C^{14}-labeled free cholesterol was detected in the plasma in a few minutes and the specific activity of the isotope reached a maximum within one hour. This indicated that the synthesis of cholesterol by the liver was very rapid. The newly-synthesized cholesterol soon appeared in the plasma and red cells, and distribution between the liver and the blood then followed. Later, an equilibrium between plasma and tissue cholesterol was accomplished. However, this was apparently not a reversible process; tissue cholesterol did not contribute to the plasma cholesterol.

The cholesterol which is directly synthesized in the extra-hepatic tissues amounts to about 0.5 Gm. per day; the liver contributes about 1 to 1.5 Gm. per day; thus, the total daily synthesis of cholesterol in the body may be as much as 1.5 to 2 Gm., which is considerably more than is present in the average diet.

When rats were fed a diet to which 1% cholesterol had been added, there was a large rise in the cholesterol of the liver but only a minimum rise in that of the serum[42]. The relatively constant level of cholesterol in the serum, despite large variations in that of the liver, suggests that there may be a homeostatic mechanism for the regulation of the serum cholesterol which depends upon the ability of the liver to suppress the synthesis of this compound when its concentration within that organ increases.

In human subjects, alteration of the dietary intake of cholesterol from as low as 200 mg. per day to as high as 1000 mg. per day does not affect the levels of cholesterol in the serum. It is now apparent that endogenous synthesis rather than dietary intake of cholesterol is the primary factor which affects the concentration of cholesterol in the blood. Efforts to control the level of cholesterol in the blood by dietary restriction of cholesterol itself are therefore of limited value. On the other hand, as mentioned above, regulation of the total quantity and of the type of dietary fat is likely to be a much more effective means of controlling the levels of lipids in the blood.

Pathways of Cholesterol Biosynthesis:

It has been noted that acetate is a direct precursor of the cholesterol which is synthesized in the body. In studies with C^{14}-labeled acetate, of the 27 carbon atoms in cholesterol, 15 were found to be derived from the methyl carbon of acetate and 12 from the carboxyl carbon[43]. In later work, the isopropyl group of isovaleric acid was shown to be a better cholesterol precursor in liver than acetate. In 1956, Wright et al.[44] isolated a new growth factor for Lactobacillus acidophilus which acted as a replacement for acetate in the growth medium, and Tavormina et al.[45] reported that the growth factor was involved in the synthesis of cholesterol and was, in fact, far superior to any other substance previously tried for this purpose. The growth factor isolated by Wright and his co-workers was identified as β-hydroxy-β-methyl-α-valerolactone. The compound is hydrolyzed to an acid which has been named mevalonic acid (MVA). It was soon apparent that mevalonic acid served as the source of five carbon atoms which form the isoprenoid units that had previously been suggested as the fundamental recurring unit in the cholesterol nucleus.

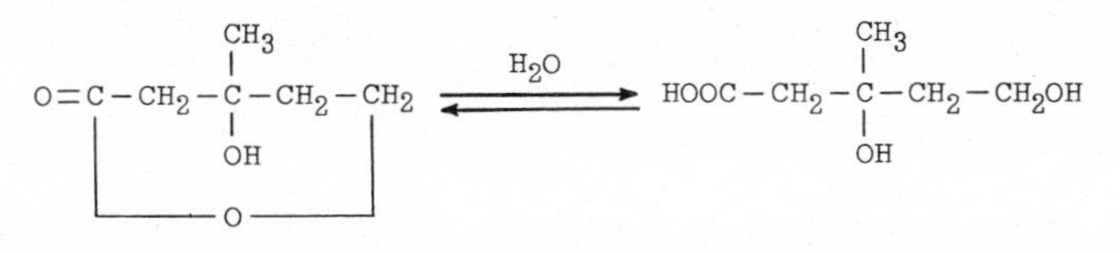

β-Hydroxy-β-methyl-δ-valerolactone Mevalonic Acid

The immediate precursor of mevalonic acid is not definitely known, but among the compounds which could serve in this role is β-hydroxy-β-methylglutaric acid. It is formed from leucine (see p. 239), and it would also be formed by condensation of acetate and acetoacetate, both of which are certainly direct precursors of cholesterol. Reduction of the glutaric acid derivative would produce mevaldic acid, which would be further reduced to mevalonic acid.

BIOSYNTHESIS OF CHOLESTEROL

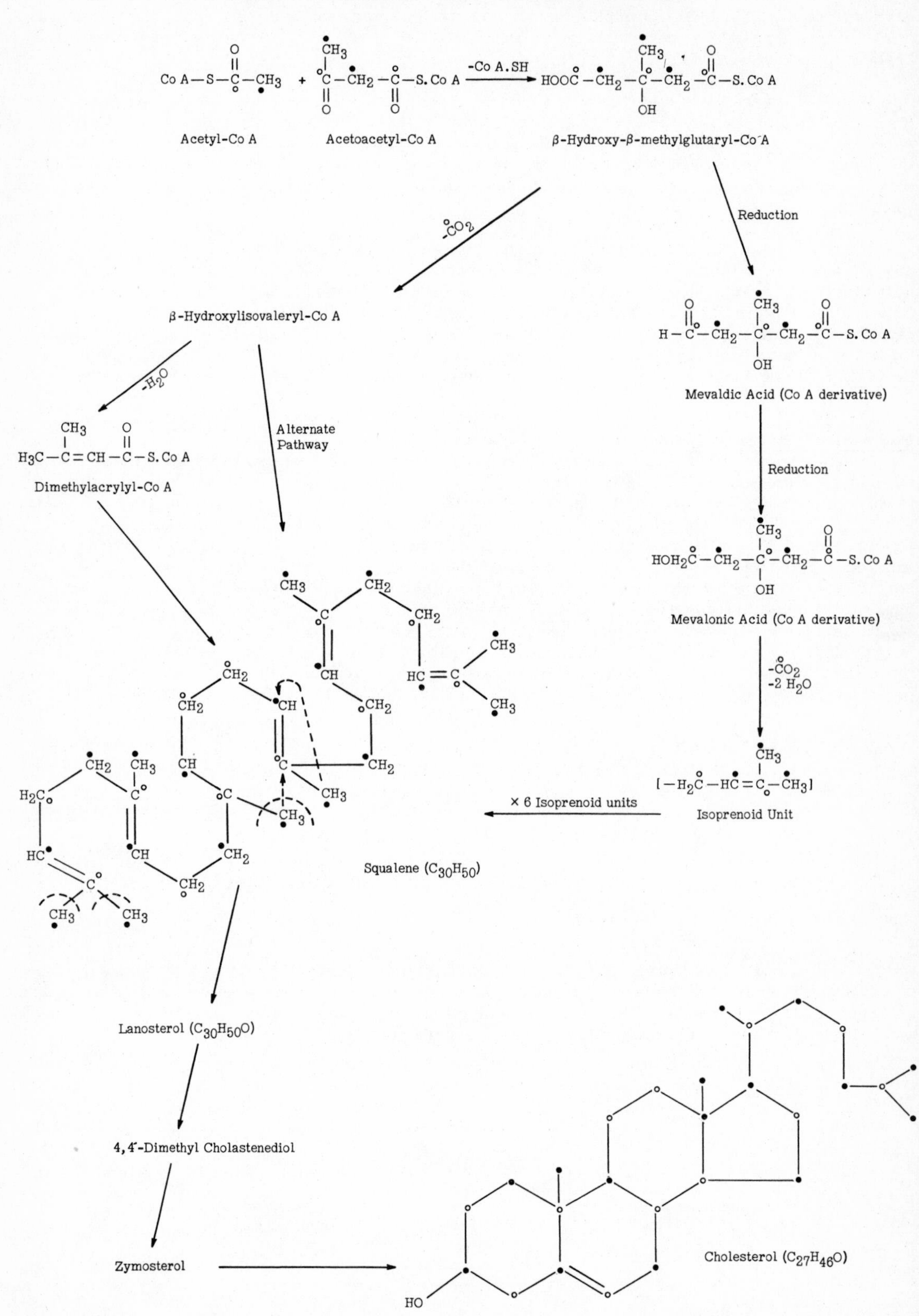

The condensation of six isoprenoid units, presumably derived from mevaldic acid, to form squalene is probably the next stage in biosynthesis of cholesterol. Squalene may then be converted to lanosterol by closure to form the ring structure of a steroid. Before this closure occurs, two methyl groups of squalene are shifted: that on C-14 to C-13, and that on C-8 to C-14. In addition, hydroxylation at C-3 occurs. Removal of a methyl group at C-14 by oxidation to CO_2 produces 4, 4'-cholastenediol; two methyl groups (at C-4) are then removed from this compound to produce zymosterol. Formation of cholesterol from either lanosterol or zymosterol requires also reduction of the double bond in the side chain and transfer of the double bond between carbons 8 and 9 to carbons 5 and 6.

This pathway for biosynthesis of cholesterol is shown on p. 216. It begins with acetyl-Co A and acetoacetyl-Co A. If both the methyl and carboxyl carbons of acetate and the corresponding carbons of acetoacetate are labeled, the isoprenoid unit which results would be labeled as shown. When such labeled intermediates, including mevalonic acid, are used in tissue preparations synthesizing cholesterol, squalene labeled as shown can be isolated. The labeling of cholesterol derived from these labeled precursors is also as predicted from the scheme shown on p. 216.

An alternate pathway to squalene from β-hydroxy-β-methylglutaryl-Co A is also indicated. This pathway may explain the superiority of the isopropyl group of isovaleric acid over acetate as a cholesterol precursor.

Excretion of Cholesterol:

In the metabolism of cholesterol, the liver occupies a central position as it does in the case of other lipids. As noted, it synthesizes cholesterol and both adds it to the blood and removes it from the blood. The liver also excretes in the bile, as cholesterol or as cholic acid, the great majority of the cholesterol lost from the body. Very little is excreted in urine; some is lost by way of the skin and intestinal secretions. The cholesterol excreted in the bile is constant in quantity; it reflects the hepatic synthesis of cholesterol. The cholesterol removed from the blood by the liver is excreted in the bile largely as cholic acid; that obtained from the diet is carried to the liver by the lymph, stored for a time in the liver, and gradually lost as cholic acid. It is probable that the blood cholesterol levels are the resultant of the rate of **synthesis** versus the rate of **destruction**; in many diseases characterized by high blood cholesterol levels, an inadequate rate of destruction seems responsible for the hypercholesterolemia.

About 20% of the cholic acid excreted in the bile is combined with taurine to form taurocholic acid; the remaining portion is combined with glycine to form glycocholic acid. These acids are neutralized by alkali and constitute the bile salts. Their role in fat digestion and absorption has been noted. In complete obstruction of the common duct, bile salts disappear from the hepatic bile. A diminished concentration of bile salts in drainage bile is indicative of severe damage to the hepatic cells, which are failing to form bile acids.

Effects of Hormones on Cholesterol Metabolism:

A. Thyroid: The thyroid is undoubtedly an important regulator of the plasma cholesterol level. The addition of desiccated thyroid to the diet of cholesterol-fed rabbits or chicks results in a significant depression of the hypercholesterolemia as well as a slowing down of the atherosclerotic process. Dogs are refractory to these effects of cholesterol feeding unless thyroid activity is inhibited. Thyroid hormone causes a shift of cholesterol from the blood into the tissues. Low blood cholesterol is therefore characteristic of hyperthyroidism.

B. Estrogens: In experiments on chicks, the administration of estrogens along with cholesterol markedly inhibited atherogenesis in the coronary arteries although atherosclerotic changes in the aorta continued to occur. Furthermore, estrogen administration reversed coronary lesions, which were previously induced by cholesterol feeding, without affecting similar lesions of the aorta.

C. Pancreas: Plasma cholesterol levels and those of other lipids are characteristically increased in uncontrolled diabetes. Experiments carried out with depancreatized chicks indicated that such animals developed much higher levels of plasma cholesterol after the addition of cholesterol and fat to the stock diet than did control animals on the same diet but with an intact pancreas.

D. Pituitary and Adrenals: An effect on fat metabolism of the anterior pituitary and the adrenal cortex has been described on p. 201. Cholesterol participates in these effects, although not in a manner different from that of other lipids.

METABOLISM OF PHOSPHOLIPIDS AND CEREBROSIDES

The phospholipids are absorbed from the intestine along with the triglycerides and cholesterol, as a component of the chylomicrons. In the blood, they are found associated in almost equal amounts with both the β and α lipoproteins, although to a slightly greater extent with the latter. The phospholipids are essential elements in every cell, but they seem to play a special role in the metabolism of lipid within the liver (see p. 205). Studies of phospholipid turnover (formation and disappearance) with radioactive phosphorus as a label show that liver is virtually the only organ involved in synthesis and removal of the phospholipids. The pathway for phospholipid synthesis is shown on p. 208.

The cephalins are phospholipids which differ from the lecithins in containing aminoethanol rather than choline. A cephalin is the prosthetic group of the protein which is found in the blood platelets and liberated when they are ruptured, i. e., a thromboplastic factor for blood coagulation.

Enzymes for the hydrolysis of lecithin (lecithinases; phospholipases) have been identified in kidney, liver, pancreas, and various other organs. Removal of one fatty acid from lecithin by the action of such enzymes produces a lysolecithin, a compound which possesses extremely potent hemolytic properties.

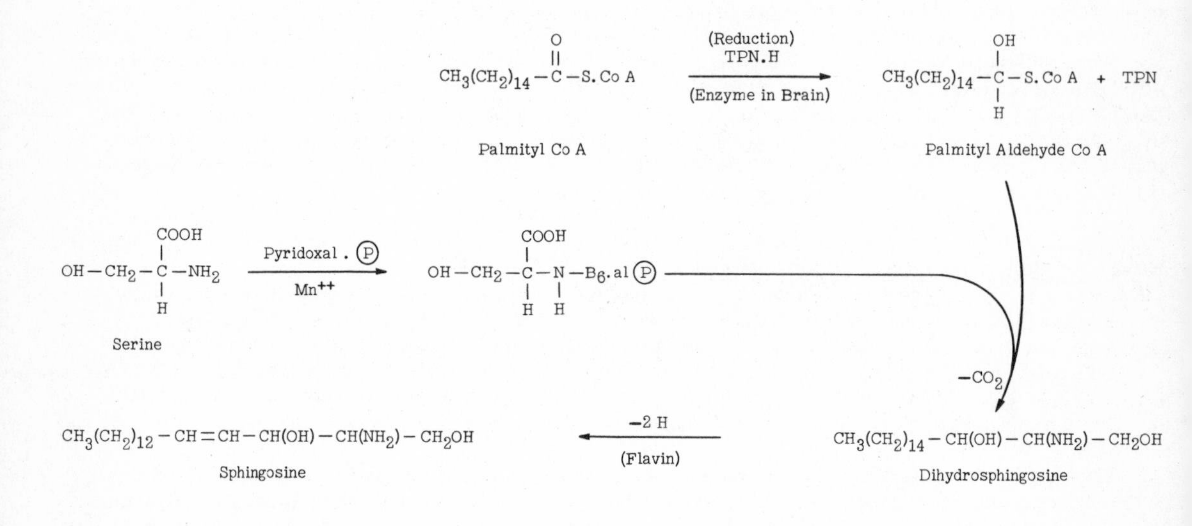

Sphingomyelins:

The sphingomyelins are phospholipids containing a fatty acid, phosphoric acid, choline, and a complex amino alcohol, sphingol (sphingosine). No glycerol is present.

The synthesis of sphingosine has been studied by Brady and Koval[46]. They have prepared an enzyme system from brain tissue which catalyzes the synthesis of sphingosine by the reactions shown on p. 218. The reduction of palmityl-Co A to the aldehyde is the first step in the synthetic pathway. (It will be recalled that reduction of fatty acids to fatty aldehydes is also involved in synthesis of long-chain fatty acids; see p. 207.) The amino acid serine, after activation by combination with vitamin B_6 (pyridoxal phosphate) and decarboxylation, then condenses with palmityl aldehyde Co A to form dihydrosphingosine, which loses 2 H atoms to form sphingosine.

The biosynthesis of sphingomyelin utilizes cytidine-diphosphate choline (CDP-choline), which is also the active form in which choline was incorporated into lecithins (see p. 208). The incorporation of CDP-choline, catalyzed by the enzyme phosphoryl choline-ceramide transferase, takes place after attachment of a fatty acid through the amino group on sphingosine (shown below as an N-acyl group). The N-substituted sphingosine is called **ceramide.** The final step in sphingomyelin synthesis may then be represented as:

$$\text{Cytidine}-\text{Ⓟ}-\text{Ⓟ}-\text{choline} + \text{N-Acyl sphingosine} \xrightleftharpoons{\text{Transferase}} \text{Sphingomyelin} + \text{Cytidine}.\text{Ⓟ} + \text{Ⓟ}$$

The sphingomyelins accumulate in the spleen, brain, and liver in Niemann-Pick disease, a disease of infancy or childhood marked also by anemia and leukocytosis (particularly of the lymphocytic fraction). The disease is also sometimes spoken of as lipoid histiocytosis because the histiocytes throughout the body become filled with phospholipid and take on a foamy appearance. The disease is congenital in origin and seems to represent an abnormality in sphingomyelin metabolism.* It usually develops rapidly and leads to death within the first two years of life.

Cerebrosides:

The cerebrosides contain the sphingosine–fatty acid combination (ceramide) found in the sphingomyelins, but a galactose-sulfate moiety is attached to the ceramide rather than choline phosphate, as is the case with sphingomyelin. The requirement for galactose in the formation of cerebrosides is the only known physiologic role of this sugar other than in formation of lactose in milk.

The transfer of sulfate to galactose is accomplished by the action of "active" sulfate and catalyzed by a sulfokinase (see p. 168). Galactose sulfate is then attached to the ceramide moiety to form the cerebroside.

The cerebrosides are found in high concentration in the myelin sheath of nerves. Phrenosin is a cerebroside which is prominent among the fats extensively deposited in the tissue in Gaucher's disease. In several cases of this disease, the cerebroside contained glucose rather than galactose.

*The lipoidoses have been reviewed by Thannhauser and Schmidt[47]. Adlersberg[48] has discussed inborn errors in lipid metabolism.

References:

1. Gofman, J.W., DeLalla, O., Glazier, F., Freeman, N.K., Lindgren, F.T., Nichols, A.V., Strisower, B., and Tamplin, A.R.: Plasma **Anno II, N. 4**:413, 1954.
2. Korn, E.D.: J. Biol. Chem. **215**:1, 1955.
3. Bierman, E.L., Dole, V.P., and Roberts, T.N.: Diabetes **6**:475, 1957.
4. Adlersberg, D.: Am. J. Med. **23**:769, 1957.
5. Zarafonetis, C.J.D., Miller, G.M., Seifter, J., Baeder, D.H., Myerson, R.M., and Steiger, W.A.: Am. J. Med. Sci. **234**:493, 1957.
6. Levin, L., and Farber, R.K.: "Recent Progress in Hormone Research," Vol. VII, p. 399. Academic Press, 1952.
7. Winograd, A.I., and Renold, A.E.: J. Biol. Chem. **233**:267, 1958.
8. Milstein, S.W.: Proc. Soc. Exp. Biol. & Med. **92**:632, 1956.
9. Hill, R., Linazasoro, J.M., Chevallier, F., and Chaikoff, I.L.: J. Biol. Chem. **233**:305, 1958.
10. Kekwick, A., and Pawan, G.L.S.: Lancet, **II**, 155, 1956.
11. Sebrell, W.H., Jr., and Harris, R.S.: "The Vitamins," Vol. II, p. 267. Academic Press, 1954.
12. Burr, G.O., and Burr, M.M.: J. Biol. Chem. **82**:345, 1929.
13. Steinberg, G., Slaton, W.H., Howton, D.R., and Mead, J.F.: J. Biol. Chem. **220**:257, 1956.
14. Gordon, H., Lewis, B., Eales, L., and Brock, J.F.: Nature, **180**:923, 1957.
15. Schwartz, K., and Foltz, C.M.: J. Biol. Chem. **233**:245, 1958.
16. Artom, C.: J. Biol. Chem. **205**:101, 1953.
17. Artom, C.: Am. J. Clin. Nutrition **6**:221, 1958.
18. Harper, A.E., Benton, D.A., Winje, M.E., and Elvehjem, C.A.: J. Biol. Chem. **209**:171, 1954.
19. Harper, A.E., Monson, W.J., Benton, D.A., Winje, M.E., and Elvehjem, C.A.: J. Biol. Chem. **206**:151, 1954.
20. Best, C.H., Ridout, J.H., Patterson, J.M., and Lucas, C.C.: Biochem. J. **48**:448, 1951.
21. Brady, R.O., and Gurin, S.: J. Biol. Chem. **189**:371, 1951.
22. Popjak, G., and Tietz, A.: Biochem. J. **60**:147, 1955.
23. Wakil, S.J.: J. Am. Chem. Soc. **80**:6465, 1958.
24. Brady, R.O.: Proc. Nat. Acad. Sci. **44**:993, 1958.
25. Kennedy, E.P.: Ann. Rev. Biochem. **26**:119. Annual Reviews, 1957.
26. Stern, J.B., Coon, M.J., and Del Campillo, A.: J. Biol. Chem. **221**:1, 1956.
27. Stern, J.B., Coon, M.J., Del Campillo, A., and Schneider, M.C.: J. Biol. Chem. **221**:15, 1956.
28. Nomenclature of Enzymes of Fatty Acid Metabolism. Science **124**:614, 1956.
29. Green, D.E., Mii, S., Mahler, H.R., and Bock, R.M.: J. Biol. Chem. **206**:1, 1954.
30. Crane, F.L., Mii, S., Hauge, J.G., Green, D.E., and Beinert, H.: J. Biol. Chem. **218**:701, 1956.
31. Crane, F.L., and Beinert, H.: J. Biol. Chem. **218**:717, 1956.
32. Hauge, J.G., Crane, F.L., and Beinert, H.: J. Biol. Chem. **219**:727, 1956.
33. Stern, J.R., Del Campillo, A., and Rau, I.: J. Biol. Chem. **218**:971, 1956.
34. Wakil, S.J., and Mahler, H.R.: J. Biol. Chem. **207**:125, 1954.
35. Wakil, S.J., Green, D.E., Mii, S., and Mahler, H.R.: J. Biol. Chem. **207**:631, 1954.
36. Green, D.E.: Clinical Chemistry **1**:53, 1955.
37. Brady, R.O., and Gurin, S.: J. Biol. Chem. **189**:371, 1951.
38. Curran, G.L.: J. Biol. Chem. **191**:775, 1951.
39. Chen, R.W., Chapman, D.D., and Chaikoff, I.L.: J. Biol. Chem. **205**:383, 1953.
40. Rudney, H.: J. Biol. Chem. **210**:361, 1954.
41. Popjak, G.: Ann. Rev. Biochem. **27**:533, Palo Alto, Ann. Rev., 1958.
42. Frantz, I.D., Schneider, H.S., and Hinkelman, B.T.: J. Biol. Chem. **206**:465, 1954.
43. Little, H.N., and Block, K.: J. Biol. Chem. **183**:33, 1950.
44. Wright, L.D., Cresson, E.L., Skeggs, H.R., MacRae, G.D.E., Hoffman, C.H., Wolf, D.E., and Folkers, K.: J. Am. Chem. Soc. **78**:5273, 1956.
45. Tavormina, P.A., Gibbs, M.H., and Huff, J.W.: J. Am. Chem. Soc. **78**:4498, 1956.
46. Brady, R.O., and Koval, G.J.: J. Biol. Chem. **233**:26, 1958.
47. Thannhauser, S.J., and Schmidt, G.: Physiol. Rev. **26**:275, 1946.
48. Adlersberg, D.: Am. J. Med. **11**:600, 1951.

15...

Protein and Amino Acid Metabolism

The complete digestion of dietary protein by the gastric and intestinal proteases liberates amino acids which are readily absorbed into the portal circulation. This can be demonstrated by measurements of the total amino nitrogen of the plasma. Between meals, the normal plasma amino nitrogen level is 4 to 6 mg. per 100 ml., but during the absorption of a protein meal the amino nitrogen increases by 2 to 4 mg. per 100 ml. The amino acids are rapidly taken up by the tissues, particularly the liver, intestine, and kidney. By the sixth or seventh hour, the plasma level returns to the base line value. The lymph is not an important route of transport for amino acids. The normal concentrations of amino acids in human plasma measured after a 12-hour fast are shown in the table on p. 225.

Fate of the Absorbed Amino Acids:

Amino acids are apparently not stored as such in the tissues to any great extent but are metabolized by incorporation into protein or by deamination and further oxidation. Reserves of protein accumulate in the liver and possibly in the muscle. These labile reserves can be called upon when the protein intake is inadequate, but are actually incorporated into the architecture of the tissue. Storage of protein thus produces cellular hypertrophy; excessive depletion of protein produces atrophy.

Protein Synthesis:

The basic structural linkage of the protein molecule is the peptide bond (see p. 35). In an attempt to clarify the synthesis of proteins from the amino acids, studies have been made on the synthesis of this bond. The bulk of information on this subject has been obtained by investigating the synthesis of glutathione (see p. 235), the naturally occurring tripeptide of glutamic acid, cysteine, and glycine, and the **transpeptidation** reactions.

The synthesis of glutathione has been studied by Snoke et al.[1] Using pigeon liver extracts as a source of the necessary enzymes, these authors found that the tripeptide is formed by the following reactions:

$$\text{Glutamic acid} + \text{Cysteine} \xrightarrow[\text{ATP} \; Mg^{++},\, K^{+}]{\text{Peptide synthetase}} \text{Glutamyl-cysteine} + \text{ADP} + \text{P}$$

$$\text{Glutamyl-cysteine} + \text{Glycine} \xrightarrow[Mg^{++},\, K^{+}]{\text{ATP}} \underset{\text{(Glutathione)}}{\text{Glutamyl-cysteinyl-glycine}} + \text{ADP} + \text{P}$$

The synthesis of the peptide bonds requires a source of high energy, which may be ATP as shown in the above reactions. It will be recalled that the formation of hippuric acid, which also involves the production of a peptide bond, requires coenzyme A as well as ATP (see p. 167).

Another aspect of protein synthesis is **transpeptidation**. Extracts of pancreas and of kidney tissue have been prepared which contain enzymes for the transfer of the glutamic acid portion of glutathione (and of other glutamic acid–containing peptides) to other amino acids, with the result that a new peptide is formed. For example, glutathione may react with leucine, valine, or phenylalanine to form glutamyl-leucine, glutamyl-valine, or glutamyl-phenylalanine, respectively, and the dipeptide cysteinyl-glycine which would remain after transfer of the glutamic residue. The enzymes catalyzing such transfers, **transpeptidases**, are undoubtedly active in peptide bond synthesis. In the presence of ATP, the glutamyl–amino acid dipeptide (e.g., glutamyl-leucine) may react with another amino acid to form a tripeptide, and the process continues to lengthen the peptide chain, using a mol of ATP for each peptide linkage formed. Ultimately the original glutamic acid group may be transferred back to a cysteinyl-glycine dipeptide, in another transpeptidation reaction, to re-form glutathione. According to this scheme (Waelsch), glutathione is essential to peptide synthesis; by transferring the glutamyl group to an amino acid, it activates the carboxyl group of the amino acid so that in the presence of ATP it can unite with a third amino acid to form a new peptide bond.

The cathepsins (see p. 109) are proteolytic enzymes which occur in particularly high concentration in the liver, kidney, and spleen. By in vitro experiments with purified preparations of certain types of cathepsins, as well as with other proteolytic enzymes such as chymotrypsin, it has been found possible to catalyze a type of **transamidation** reaction which results in the formation of peptides. This is, of course, a reversal of the protein-splitting action usually attributed to these enzymes. For example:

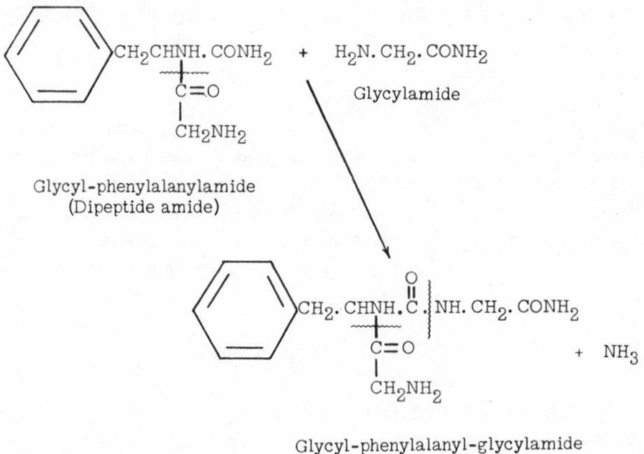

All of the above examples of peptide bond synthesis are derived from in vitro experiments. Although their precise significance in the intact organism is not known as yet, they probably indicate at least the starting points in protein synthesis. There is also evidence that not all of the amino acids in a protein are incorporated into the protein molecule at the same rate. There may occur, therefore, a **partial** breakdown of the protein with substitution of some amino acids within the peptide chains without the necessity of a complete breakdown of the entire chain.

Role of Ribonucleoprotein in Protein Synthesis:

The concentration of ribonucleic acid is always greatest in cells which are actively synthesizing protein, and there is a great deal of evidence that ribonucleoproteins are associated with the actual process of protein synthesis. It is thought that ribonucleoproteins act as a template to control arrangement of the constituent amino acids of a protein in the proper sequence.

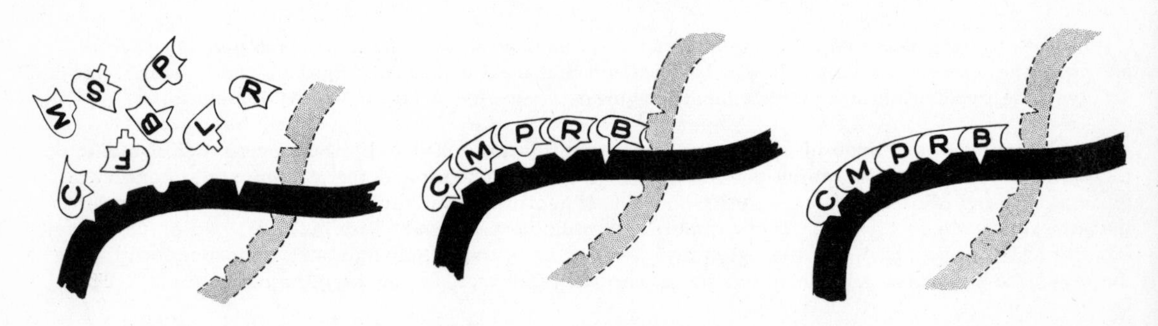

The first step in utilization of amino acids for protein synthesis is presumed to require activation of the carboxyl group of the amino acid[2]. This occurs by formation of an adenosine-monophosphate (AMP)-amino acid complex in which the 5'-phosphate group of AMP is linked as a mixed anhydride to the carboxyl group of an amino acid.

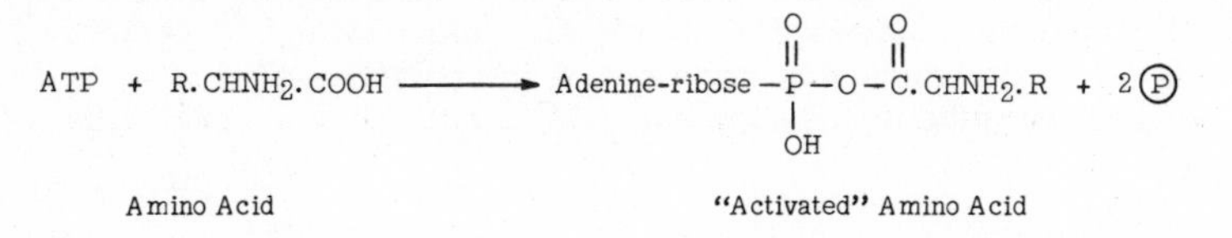

The activated amino acid is then transferred to some form of soluble ribonucleoprotein in the microsomes of the cell, which may be the template to arrange each amino acid in specific sequence as diagrammatically shown on p. 222. The activated amino acids may unite through their carboxyl groups with the phosphate radical of nucleic acids. The reaction requires guanosine triphosphate (GTP, p. 48), and is catalyzed by a specific enzyme in the soluble cell fraction.

Through the catalytic action of a second enzyme, the amino acids then become linked by peptide bonds and leave the template as polypeptides; shortly thereafter, specific coiling must occur - controlled either by amino acids or an inducer, as in the case of adaptive enzyme formation, or by an antigen, as in the case of antibody formation.

The key role of ribonucleoprotein in protein synthesis is attested by the fact that those cell fractions with the highest content of ribonucleic acid incorporate amino acids most rapidly into protein. In studies with cell preparations of Staphylococcus aureus, treatment with ribonuclease abolishes protein synthesis, and any factor which inhibits ribonucleic acid synthesis halts synthesis of protein.

Hormonal Effects on Amino Acid Utilization:

At least three hormones are known to affect protein synthesis: pituitary growth hormone, insulin, and testosterone. Both growth hormone and testosterone are protein anabolic in their action. The action of growth hormone seems to be related to a decrease in the rate of amino acid catabolism and an increase in the rate of storage of protein. The effect of insulin is probably indirect in that it favors glycolysis, which in turn provides a source of energy for the formation of peptide bonds. The 11-oxycorticosteroids tend to favor the breakdown of amino acids; in this sense they are protein anti-anabolic.

The level of amino acids in the blood can be lowered by injections of insulin and of growth hormone; a similar effect can be produced with epinephrine[3]. Luck et al. believe that epinephrine is actually the effector substance and that the effects of the other hormones may be mediated through epinephrine. The administration of thyrotropic hormone (TSH) also reduces the amino acid content of the blood, which suggests that the thyroid may also be involved in the hormonal control of these metabolites within the blood.

Tracer Studies of Protein Metabolism:

A. N^{15} Studies: The dynamic aspects of metabolism are well illustrated by the proteins. Tracer studies of amino acids or proteins labeled with isotopic nitrogen (N^{15}) have yielded valuable information on this subject. These studies have shown that synthesis and hydrolysis of proteins occurs constantly in the cells - particularly in the cells of organs like the liver and muscle which readily hypertrophy. After the administration of various amino acids labeled with N^{15}, only 10 to 12% of the administered nitrogen was retained in the body as nonprotein nitrogen (e.g., urea and unchanged amino acids). From 28% (in the case of leucine) to 52% (in the case of arginine) of the labeled nitrogen was excreted within three days, but 30% (of the arginine nitrogen) to 66% (of the lysine nitrogen) was incorporated into the body protein. The body weight of the animal did not change. Apparently the ingested amino acids constantly replaced those of the tissue proteins by a process of continuous opening and closing of the peptide linkages of the protein without alteration in the structure or quantity of the protein itself.

B. Heavy Hydrogen and S^{35} Studies: Heavy hydrogen (deuterium) has also been used as a "label" in studies of overall protein metabolism, and radiosulfur incorporated into methionine has been employed to study the uptake of this amino acid and the turnover of protein. In the deuterium studies, 10% of the liver proteins and 2.5% of the muscle proteins of animals were regenerated in three days. The "half-life" (the time required for half of the original activity to disappear) of liver and plasma proteins was seven and 14 days, respectively (N^{15} experiments). Such studies also supply information on the relative rates of turnover of protein (most rapid in liver and plasma, intermediate in visceral organs, and slowest in muscle and skin).

Protein and Nitrogen Balance:

Nitrogen losses from the body are measured by analysis of the urine and feces, and nitrogen intake is measured by analysis of the total nitrogen of the diet. When losses equal intake, the subject is in nitrogen balance. Normal adults are in nitrogen balance on an adequate diet which is constant in its protein content. When the nitrogen intake is changed, the equilibrium adjusts to a new level after a few days. This adjustment is characterized by an increase or decrease in urinary nitrogen excretion.

A. Positive nitrogen balance, which is characterized by an excess of nitrogen intake over loss, occurs during the growth of children, the repair of tissue losses as in convalescence, and during pregnancy.

B. Negative nitrogen balance, where losses exceed intake, is found in starvation, malnutrition, febrile diseases, and after burns or trauma. A protein catabolic period seems to be inevitable after surgical operations. During this period of negative nitrogen balance, the body may draw on its stores of "labile protein"; when these are exhausted, circulating plasma protein will be depleted and hypoproteinemia, most evident in the albumin fraction, results.

SPECIAL METABOLISM OF BLOOD PROTEINS

(See also p. 128.)

In experiments carried out by the technic of plasmapheresis*, the normal quantity of circulating plasma protein was replaced in one week when the experimental animal received sufficient protein in the diet. However, if the animal was on a protein-free diet (in which case the tissue reserves served as the only source of protein nitrogen), only one-tenth of this amount could be regenerated in a similar period. Regeneration of hemoglobin protein has priority over that of plasma protein. In plasmapheretic dogs receiving inadequate dietary protein, 40 to 50 Gm. of hemoglobin were regenerated in one week, whereas only 3 to 4 Gm. of plasma protein were regenerated in the same period. This observation emphasizes the necessity of blood transfusion to correct anemia before the administered dietary protein can be expected to contribute to extensive synthesis of plasma protein.

The liver regenerates the plasma fibrinogen easily but the albumins with more difficulty. After hemorrhage, a portion of the serum albumin is restored rapidly during the first day, presumably by the mobilization of the limited albumin reserve of the tissue fluids. The amount of albumin thus available is equivalent to somewhat more than the total circulating protein.

"Dynamic Equilibrium":

Plasma proteins are constantly interchanging with tissue protein. This exchange between these two groups of proteins in the body is described as the "dynamic equilibrium" (Whipple) of plasma and tissue protein. Injections of plasma protein can replace dietary protein and maintain nitrogen balance. This plasma protein can also be converted into body protein. Thus plasma protein supplies all of the nitrogenous substances necessary for the maintenance of the animal, but this is an expensive and physiologically uneconomical method of supplying nutritional protein. Tracer studies provide some evidence that the plasma albumin fraction functions as a sort of protein-in-transit for supply to tissues.

*Removal of plasma from circulation by the process of bleeding, separation of cells, and reinjection of cells suspended in Locke's or Ringer's solution.

Concentrations of Amino Acids in Plasma[4]

Amino Acid	Range (mg./100 ml.)	Mean
Glutamine	4.6 - 10.6	7.51 (±1.63)
Alanine	2.4 - 7.6	3.96 (±1.47)
Lysine	2.3 - 5.8	3.68 (±1.15)
Valine	2.5 - 4.2	3.23 (±0.53)
Cysteine/Cystine	1.8 - 5.0	3.02 (±1.16)
Glycine	0.8 - 5.4	2.91 (±1.36)
Proline	1.5 - 5.7	2.61 (±1.26)
Leucine	1.0 - 5.2	2.48 (±0.90)
Arginine	1.2 - 3.0	2.26 (±0.47)
Histidine	1.0 - 3.8	2.11 (±0.61)
Threonine	0.9 - 3.6	2.06 (±0.69)
Isoleucine	1.2 - 4.2	2.00 (±0.79)
Phenylalanine	1.1 - 4.0	1.99 (±0.82)
Tryptophan	0.9 - 3.0	1.74 (±0.51)
Serine	0.3 - 2.0	1.39 (±0.44)
Tyrosine	0.9 - 2.4	1.32 (±0.37)
Glutamic acid	0.0 - 1.3	0.89 (±0.39)
Methionine	0.25 - 1.0	0.57 (±0.27)
Aspartic acid	0.0 - 1.2	0.33 (±0.37)

A portion of the plasma protein normally penetrates the capillary walls and enters the tissue fluid. This is partially returned to the blood stream by the lymphatic system; in fact, 20 minutes after isotopically-labeled plasma protein is injected into the circulation it can be detected in the thoracic duct lymph. Furthermore, labeled plasma protein which has been injected intraperitoneally in dogs soon appears in the blood, presumably returned to the blood by way of the lymph. About 50% of N^{15}-labeled plasma protein injected into normal dogs was found to leave the circulation in 24 hours. This transudation was opposed by a balanced inflow of unlabeled plasma protein into the circulation, partly through the lymph. In shock, the mobilization of stored protein is abnormal; and while plasma protein leaves the blood at a normal rate the return flow is delayed.

GENERAL REACTIONS OF AMINO ACIDS

Amino acids are required for incorporation into protein of the blood and tissues to replace amino acids already present and to form new protein. In addition, many of the amino acids are specifically utilized in the formation of hormones, enzymes, and other constituents of the metabolic machinery. Certain amino acids in the protein molecule must be ingested as such in the diet; the remaining amino acids of the body may be synthesized from other substances. Those amino acids which cannot be synthesized from other substances are called essential amino acids. The essential amino acids for man and the tentative daily requirements for each are listed in the table on p. 377.

Synthesis and Interconversion of Amino Acids:

Examples of synthesis of amino acids have been cited in connection with carbohydrate metabolism. Certain α-keto acids formed in the course of the metabolism of carbohydrate can be converted to an amino acid by replacing the α oxygen with an amino group. Undoubtedly many of the so-called dispensable amino acids, i.e., amino acids found in the body protein but not required in the diet, are formed in this way. The reactions by which amino acids may be synthesized or interconverted are illustrated in the following sections.

Transamination:

By the use of isotopic nitrogen as a label, the transfer of nitrogen from a labile nitrogen pool in the body to other compounds has been repeatedly demonstrated. This process, **transamination,** is important for the synthesis of amino acids from carbohydrate intermediates. It is catalyzed by specific enzymes known as transaminases. In 1937, Braunstein and Kritzman[5] reported the discovery of transaminases which catalyzed the transfer of amino groups from glutamic and aspartic acids to certain α-keto acids. The reactions shown below, which are of considerable biological importance, illustrate the actions of the glutamic and aspartic systems. Pyridoxal phosphate (a vitamin B_6 derivative; see p. 82) serves as coenzyme (cotransaminase) for these systems; it probably functions as an amino group transfer mechanism through formation of pyridoxamine.

In the following reaction, glutamic acid participates in the synthesis of alanine from pyruvic acid by transamination.

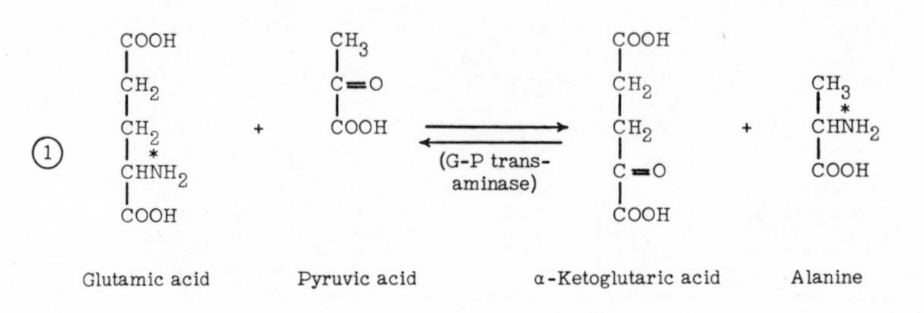

Ketoglutaric acid, one of the compounds in the citric acid cycle, by amination from an amino donor, forms glutamic acid. This is illustrated by a reversal of the above reaction, in which alanine is serving as an amino donor.

Oxaloacetic acid, another compound of the citric acid cycle, forms aspartic acid by amination of its α-keto group.

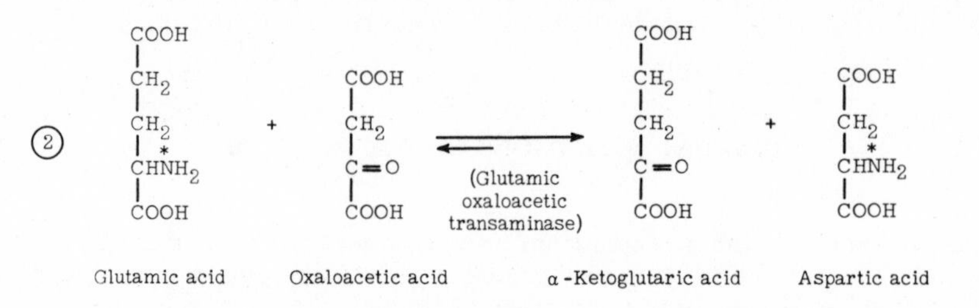

The reaction is reversible, although it proceeds to the right much faster than to the left. It also demonstrates the interconversion of amino acids (in this case, glutamic acid to aspartic acid).

The activity of transaminases in the blood is markedly increased in certain disease states (see p. 108). The reaction by which ketoglutaric acid is converted to glutamic acid (reaction 2, above) is one of those used to measure transaminase activity in tissues and in the blood. In the original assay procedure[6], whole blood or serum is added to solutions of a mixture of ketoglutaric acid and aspartic acid and the production of glutamic acid is measured under standardized conditions of pH, temperature, and time of incubation. The transaminase catalyzing this reaction is designated glutamic-oxaloacetic transaminase (G-OT). In order to apply the method to spectrophotometric assay, a second step is added. This consists of the addition of malic dehydrogenase and reduced DPN to convert the oxaloacetic acid produced in the first reaction to malic acid and thus to oxidize DPN. The reoxidation of DPN can then be followed on the spectrophotometer by the change in absorption of light in the ultraviolet. This second reaction is shown below (reaction 3). In a recent modification, serum glutamic-oxaloacetic transaminase (SGO-T) is measured as shown in reaction 2, but the oxaloacetic acid produced is directly assayed.

Activity of blood glutamic-pyruvic transaminase (G-PT) is now also measured in the clinical laboratory. The reaction on which this test is based is shown on p. 226, reaction 1. Activity of the enzyme is estimated from the amounts of pyruvic acid produced when ketoglutaric acid is transaminated.

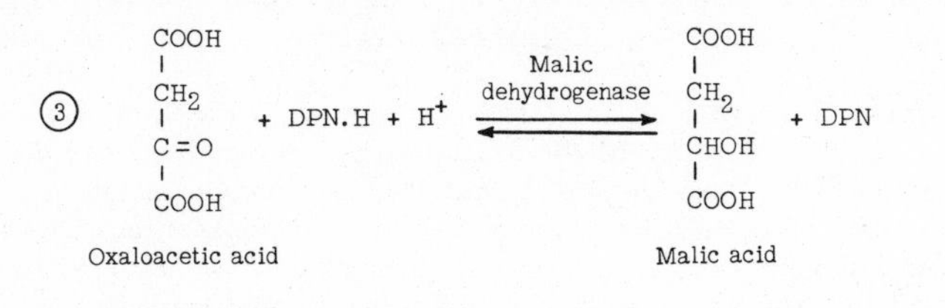

Meister and co-workers[7,8] have discovered other types of transamination reactions in which **glutamine** and **asparagine** rather than glutamic and aspartic acids serve as amino donors. The transfer of the α-amino nitrogen to certain keto acids is accompanied by the simultaneous removal of the amide nitrogen to form ammonia. This is illustrated for glutamine in the following reactions:

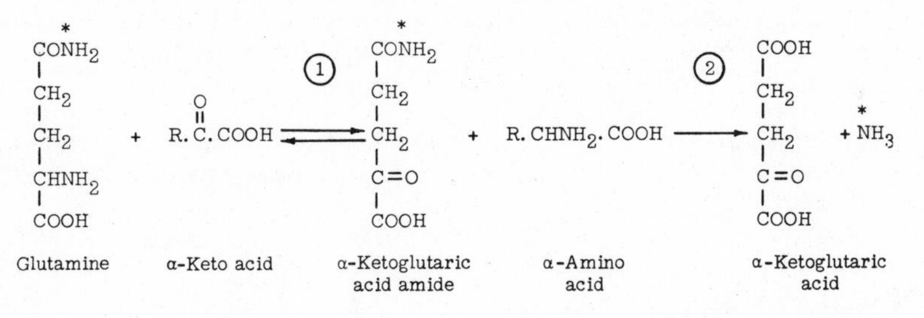

Reaction 1 (above) is reversible, with asparagine or with many other amino acids serving as amino donors.

A similar reaction occurs with asparagine, the corresponding amide of aspartic acid, although another transaminating enzyme is involved. A suitable keto acid is required for this reaction, but, with the exception of two keto acids (pyruvic and α-ketobutyric), if glutamic acid was substituted for glutamine all other keto acids which were tested were either not transaminated at all or only very slowly. Another interesting fact is that all keto acids which are active in these systems must have two H atoms on the β carbon; for example, α-ketoisovaleric acid (the keto acid of valine) is not active. An exception is glyoxylic acid, the deamination product of glycine.

Glyoxylic acid may be transaminated with glutamine, glutamic acid, asparagine, aspartic acid, or ornithine[9]. In the reaction with ornithine the δ-amino group transaminates and glutamic semialdehyde is formed. Thus an aldehyde is both a reactant (glyoxylic acid) and a product (glutamic semialdehyde).

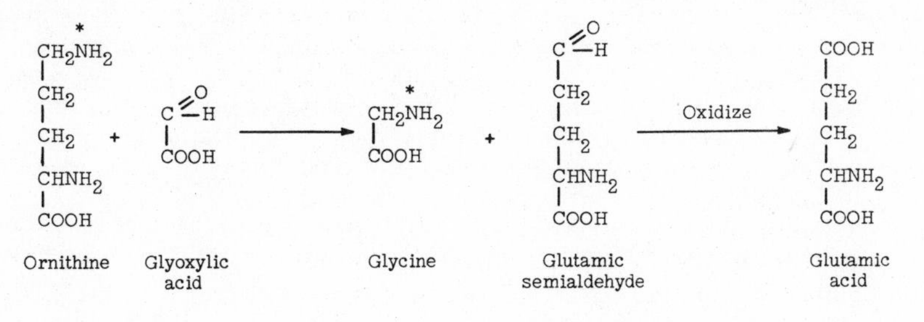

Present evidence thus indicates that both glutamine and asparagine are of considerable importance in transamination reactions - at least in the liver, which is the only organ which has yet been shown definitely to possess these systems. With the exception of three keto acids (pyruvic, α-ketobutyric, and glyoxylic), transamination between glutamine or asparagine and α-keto acids was found to be considerably more rapid than the corresponding reactions with glutamic acid. Glutamine transaminase, like other transaminases, requires a pyridoxal coenzyme[10].

In the rat, after feeding of ammonium salts, glycine, L- or D-leucine, or L-tyrosine labeled with isotopic nitrogen (N^{15}), almost all of the amino acids subsequently isolated from the proteins of the tissue contain some N^{15}. This suggests that transamination is taking place continuously and involves the large majority of the amino acids of the protein molecule. The amide nitrogen in glutamine and the amino nitrogen of glutamic and aspartic acids always have the highest concentrations of the isotope since these compounds participate in transamination more extensively than do the other amino acids. However, some other amino acids, even those considered essential, also exchange reversibly their α-amino nitrogen with that of other amino acids or even with ammonia.

Certain amino acids do not appear to participate in transamination. In rabbit and rat experiments, after the feeding of N^{15}-labeled glycine, only lysine and threonine were observed never to contain the isotopic N^{15}. It is concluded from these observations that these two amino acids do not participate in transamination; after deamination, they apparently cannot be reaminated. These two amino acids appear to be the only exceptions, however; all other amino acids are believed to be involved at least to some extent in transamination reactions.

Transamination plays an essential role in urea formation. This will be shown on p. 232 in connection with the synthesis of glutamic and aspartic acids as a means of transferring ammonia to the production of urea.

Deamination of Amino Acids:

Deamination, or removal of the α-amino group of amino acids, is the first step in the metabolic breakdown of these compounds. The liver is the chief site of deamination, although the kidney and perhaps other organs may accomplish deamination. Deamination is catalyzed by a flavin-dependent specific enzyme, amino acid oxidase, which catalyzes the oxidative removal of the amino group as follows:

$$1.\ R.CHNH_2.COOH + Flavin \rightleftharpoons R.CH{=}NH.COOH + Flavin.H_2$$

$$2.\ R.CH{=}NH.COOH + H_2O \rightleftharpoons R.\overset{\overset{O}{\|}}{C}.COOH + NH_3$$

$$3.\ Flavin.H_2 + O_2 \longrightarrow Flavin + H_2O_2 \xrightarrow{Catalase} H_2O + 1/2\ O_2$$

The Enzymes for Deamination:

The amino acid oxidases which catalyze removal of the amino group by deamination are found in liver and kidney. One oxidase, the L-amino acid oxidase, attacks only L- or natural isomers of the amino acids. It does not attack glycine, the dicarboxylic acids, the diamino acids, or the β-hydroxyamino acids. Other oxidases are required to deaminate these compounds.

An oxidase specific to the deamination of D-amino acids has been found in liver and kidney. This D-amino acid oxidase deaminates many D-amino acids, but not at equal rates. D-Alanine and D-methionine are attacked by this enzyme at a relatively rapid rate.

Both L- and D-amino acid oxidases are flavoprotein enzymes; the L- oxidase contains riboflavin phosphate as the prosthetic group; the D- oxidase contains FAD (flavin-adenine-dinucleotide) (see p. 79).

Utilization of D-Amino Acids:

The deamination of the unnatural isomer, the D- isomer of an amino acid, by a specific D-amino acid oxidase, produces an α-keto acid which is no longer optically active. This keto acid, by reamination to the α-amino acid, is then utilizable by the body. It is believed that deamination, followed by reamination to the L- isomer, is the method by which the body uses the so-called unnatural isomer of many amino acids.

In growth experiments with animals it has been demonstrated that many D-amino acids may serve as the sole source of an essential amino acid in the diet. Since utilization of the D- isomer by the animal presumably requires transamination, these experiments serve as further evidence of the exchangeability of the amino nitrogen of certain amino acids in animals.

Neither D-lysine nor D-threonine is available for growth in the rat. This would be expected because they fail to participate in transamination, as noted above.

The reabsorptive capacity of the kidney tubule for D- and L-amino acids is different. Thus, when a racemic DL-amino acid is administered, considerable quantities of the D- isomer are promptly excreted while the L- isomer is not found in the urine to any comparable extent.

Decarboxylation:

Amino acids may be converted to amines by decarboxylation. The reaction is catalyzed by specific pyridoxal-dependent decarboxylases.

$$R.CH.NH_2.COOH \xrightarrow[\text{Pyridoxal (P)}]{\text{Amino acid decarboxylase}} R.CH_2.NH_2 + CO_2$$

This reaction is characteristically brought about by bacterial attack on the amino acids. It is a common putrefactive mechanism in the spoilage of food protein, resulting in the formation of ptomaines (see p. 160).

Further Metabolism of Keto Acids:

A. The Fate of the Keto Acid After Deamination Includes:
   1. Reconversion to an amino acid - As noted above, the keto acid may be reconverted to an amino acid by reamination or transamination.
   2. Direct catabolism - The keto acid may be directly catabolized to carbon dioxide and water.
   3. Conversion to carbohydrate or fat - The conversion of the keto acid to carbohydrate is the major source of glycogen obtained by gluconeogenesis. This reaction is catalyzed by certain adrenocortical steroids, the 11-oxysteroids, such as cortisone. These steroids are sometimes spoken of as "S" hormones or sugar-forming hormones.

B. The D:N Ratio (Dextrose to Nitrogen Ratio): The D:N ratio is the ratio between the amount of dextrose and the amount of nitrogen excreted in the urine of an animal rendered completely diabetic by phlorhizin; according to Lusk, it is about 3.65:1 (in the phlorhizinized dog). The feeding of protein increases the excretion of sugar in the diabetic or phlorhizinized animal because a certain amount of carbohydrate is produced from the protein by the conversion of certain amino acids (glycogenic amino acids; see below) to glycogen. The usual D:N ratio of 3.65:1, which is obtained in these experimental animals when on a mixed diet, suggests that about 58% of the protein in such a diet is made up of glycogenic amino acids. This is an average percentage; not all pure proteins have the same ratio.*

Ketogenic Amino Acids:

The feeding of certain amino acids to experimental animals gives rise to acetoacetic acid and other ketone bodies. Such amino acids are presumably metabolized as a fat. They are said to be ketogenic.

Leucine, phenylalanine, and tyrosine are definitely ketogenic. Isoleucine is both ketogenic and glycogenic.

Summary of Metabolism of Deaminated Residues:

The fate of the deaminated residues of amino acids as it is now known is summarized below.

*Protein contains, on the average, 16 Gm. of nitrogen per 100 Gm. Since, according to the D:N ratio, 3.65 Gm. of dextrose is obtained for each gram of nitrogen in the protein, 58 Gm. (3.65 × 16) of dextrose would be produced by 100 Gm. of the protein, i.e., 58% of the protein is glycogenic.

The Fate of the Deaminated Amino Acids

Glycogenic			Ketogenic
Glycine	Glutamic acid	Citrulline	Leucine
Alanine	Aspartic acid	Cystine	Isoleucine (weak)
Serine	Histidine	Methionine	Phenylalanine
Threonine	Arginine	Proline	Tyrosine
Valine	Lysine	Hydroxyproline	
	Isoleucine (weak)		

The Fate of the Ammonia Removed in Deamination:

The ammonia removed from amino acids may be used in amination or transamination or excreted as an ammonium salt (particularly in metabolic acidosis; see p. 295). Glutamine and possibly asparagine are active as ammonia carriers to prevent the accumulation of ammonia to toxic levels. Their prominent roles in exchanging ammonia in transamination have also been mentioned (see p. 227). The synthesis of glutamine from glutamic acid and ammonia requires energy derived from ATP.

The great majority of the ammonia lost to the body is utilized in the formation of urea. This compound is therefore the principal end product of protein metabolism in man and makes up a large percentage of the nitrogen excreted in the urine.

Urea Formation; the Krebs-Henseleit Cycle:

The liver is believed to be the only organ capable of making urea. In a completely hepatectomized animal, the urea content of the blood and urine promptly declines; ammonia and amino acids concomitantly increase. Krebs and Henseleit studied urea formation in liver slices which were respiring in an oxygenated medium containing ammonium salts. As a result of these studies, they formulated a series of reactions to explain the formation of urea in the liver. Three amino acids - ornithine, citrulline, and arginine - were found to be involved; and the enzyme **arginase***, which is present only in the liver, accomplished the final hydrolysis of arginine to ornithine and urea.

It is now apparent that the synthesis of urea is somewhat more complicated than the original scheme of Krebs and Henseleit had indicated. The introduction of carbon dioxide and ammonia into the cycle is accomplished by the formation of "carrier" molecules, and these reactions require the presence of ATP as a source of energy in connection with their formation. Two steps in the original cycle require expansion: (1) the formation of citrulline from ornithine, and (2) the conversion of citrulline to arginine.

Formation of Citrulline From Ornithine:

Citrulline is formed by the addition of carbon dioxide and ammonia to ornithine. This is accomplished by the process of "transcarbamylation." The metabolically active source of the carbon dioxide and ammonia is **carbamyl phosphate**, formed by a reaction between carbon dioxide, ammonia, and ATP[11].

$$CO_2 + NH_3 + ATP \xrightarrow{Mg^{++}} H_2N-\overset{O}{\overset{\|}{C}}-O-\overset{O}{\overset{\|}{P}}=(OH)_2$$

(Carbamyl phosphate)

The carbamyl groups may then be transferred to aspartic acid to form carbamyl aspartic acid[12]. Carbamyl phosphate–aspartate transcarbamylase is the enzyme which catalyzes this reaction. The distribution of this enzyme as studied in the rat indicates that carbamyl aspartate is probably synthesized by most (if not all) tissues.

*Arginase is not found in the livers of birds, and urea is not formed in these animals. Uric acid is the principal end product of protein metabolism in all animals which lack arginase. Arginase is found in most bony fish, but in these animals ammonia rather than urea is the principal end product of the metabolism of protein.

Carbamyl aspartate readily transfers the carbamyl group to ornithine to form citrulline (which is actually carbamyl ornithine), and aspartic acid is regenerated to reenter the cycle and react again with carbamyl phosphate.

It has been shown[11] that carbamyl phosphate can also transfer the carbamyl group directly to ornithine to form citrulline.

Conversion of Citrulline to Arginine:

The formation of arginine from citrulline involves two intermediate steps.

1. Condensation of citrulline with aspartic acid, in the presence of ATP and magnesium ion, to form argininosuccinic acid[13,14]

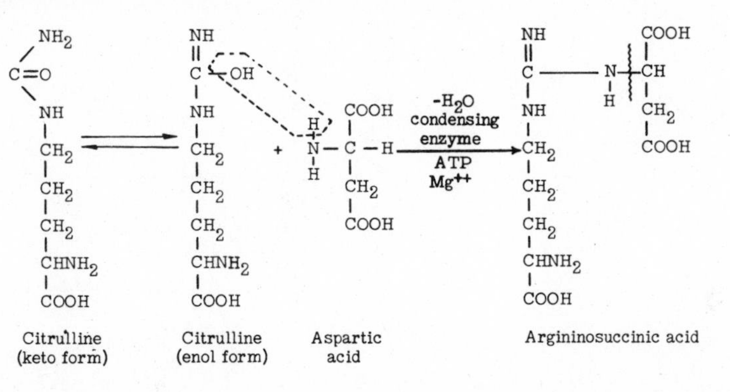

2. Formation of arginine by splitting off of fumaric acid from argininosuccinic acid[15]. The fumaric acid is later converted in the citric acid cycle to malic acid and oxaloacetic acid, which by transamination re-forms aspartic acid.

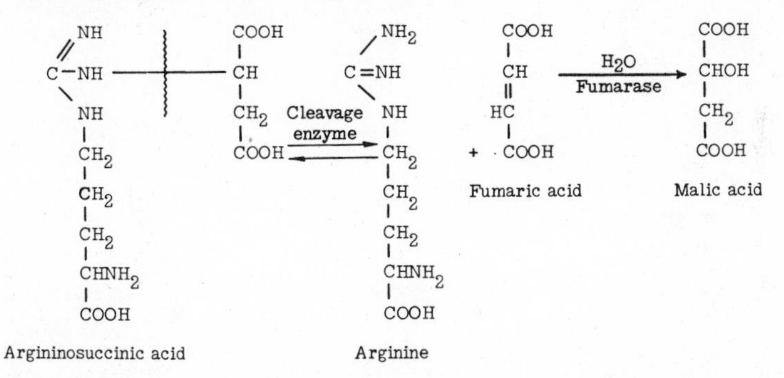

All of the reactions involved in the formation of urea, as described above, are summarized in the diagram on p. 232.

These additions to the Krebs-Henseleit cycle indicate that a number of reactions are connected with the central process for the formation of urea. Most important is the integration with the citric acid cycle to supply oxaloacetic acid, which, after transamination, produces aspartic acid. This amino acid may itself be considered to act as a catalyst in the urea cycle, functioning as a carrier molecule to transfer carbon dioxide and ammonia as a carbamyl group or ammonia as an amino group to the ultimate formation of urea.

Four amino acids are therefore concerned directly with the cycle by which urea is formed: aspartic acid, arginine, ornithine, and citrulline. Aspartic acid may be readily available by transamination of oxaloacetic acid, and ornithine and citrulline arise from arginine. This last amino acid has been shown to increase markedly the production of urea and at the same time to lower the ammonia in the blood of animals in which ammonia intoxication has been produced experimentally[16]. Arginine has also been used clinically to lower the blood ammonia in various disease states where ammonia intoxication may be a pathogenic factor[17].

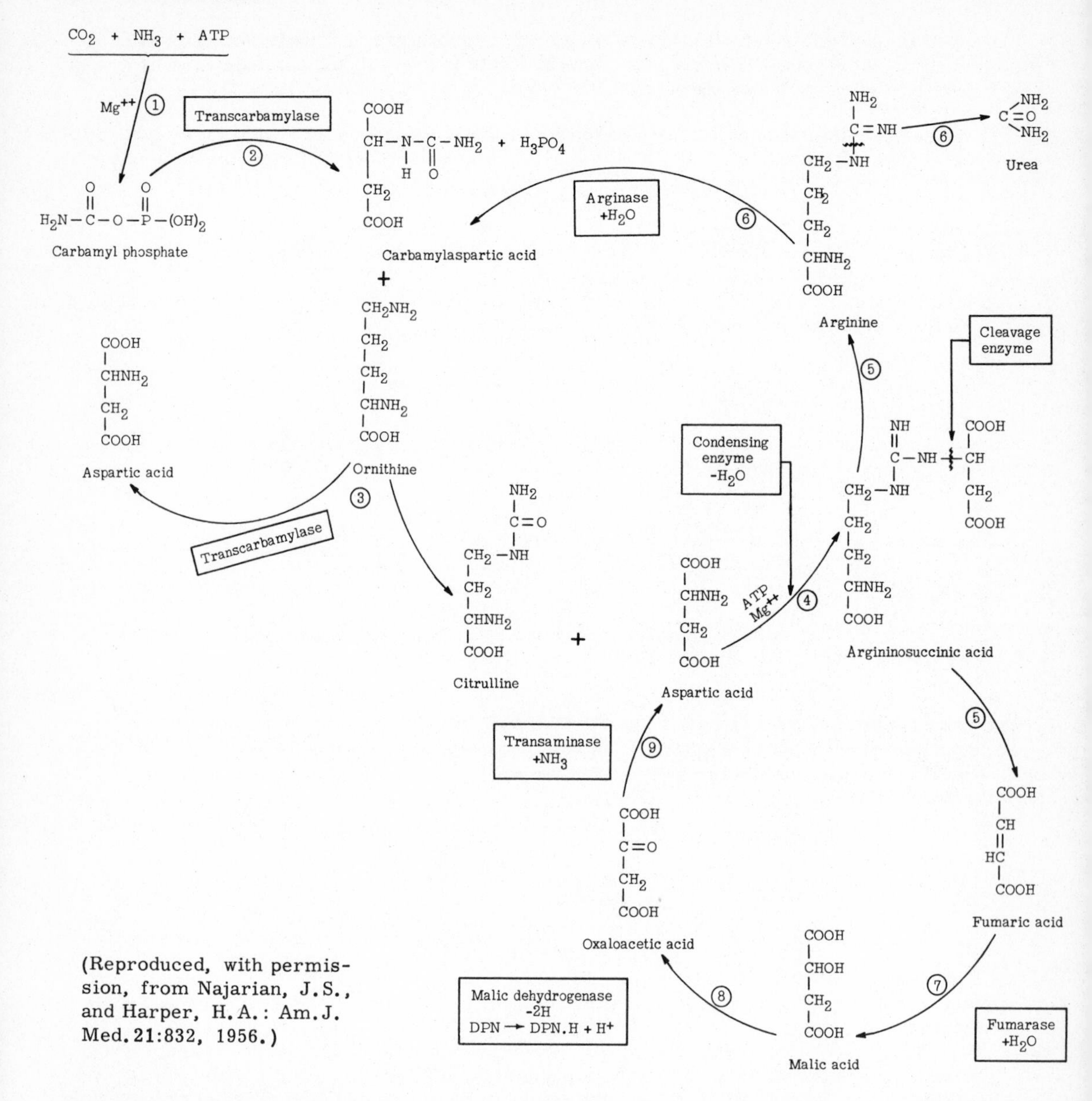

(Reproduced, with permission, from Najarian, J.S., and Harper, H.A.: Am.J. Med.21:832, 1956.)

Ammonia Formation:

Although ammonia is produced in the body in reactions involving the metabolism of proteins and amino acids, the ability of the normal liver to remove it, principally by formation of urea, is so great that ammonia is virtually absent from the peripheral blood. The production of ammonia within the gastrointestinal tract by the action of the intestinal bacteria on nitrogenous substrates accounts for the high ammonia content of the portal blood. In the presence of impaired hepatic function and/or collateral communications between the portal and systemic veins that may develop in cirrhosis, the portal blood ammonia may bypass the liver and the ammonia content of the peripheral blood may then increase to toxic levels. Surgically produced shunting procedures are also conducive to this method of production of ammonia intoxication. Elevated levels of ammonia in the blood are particularly harmful to the central nervous system, and the symptoms of ammonia intoxication include a peculiar flapping tremor, slurring of speech, blurring of vision, and, in severe cases, coma and death.

The ammonia content of the blood leaving the kidney by way of the renal vein is always higher than that of the renal artery. This indicates that the kidney produces ammonia and adds it to the blood. However, the excretion of the ammonia produced by the kidney tubule cells into the urine is much more important. This reaction is an important mechanism for conservation of fixed base (see p. 294). Renal production of ammonia is markedly increased in metabolic acidosis and depressed in alkalosis. Ammonia produced by the renal tubular cells is not derived from urea, as was formerly believed. This is demonstrated by the fact that isotopically labeled ammonia cannot be detected in the urine after the administration of labeled urea. The amino acids are now known to serve as the source of urinary ammonia, and glutamine is the most important amino acid for this purpose. A special enzyme, glutaminase, is present in the kidney to catalyze this reaction.

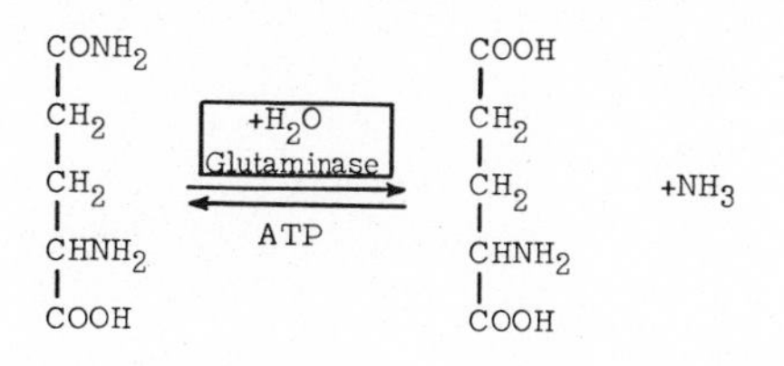

It is of interest to note that glutamine accounts for a larger fraction of the total α-amino nitrogen of the plasma than any one other amino acid (see p. 225).

METABOLISM OF INDIVIDUAL AMINO ACIDS

In addition to their general functions, many individual amino acids participate in special metabolic reactions. The fate of these amino acids will now be discussed.

Glycine:

This amino acid is considered dispensable since it can be synthesized by many animals (but not the chick). However, it is used in many important physiologic functions.

A. Deamination and Transamination: Liver and kidney possess a special enzyme, glycine oxidase, a flavoprotein, which is capable of oxidatively deaminating glycine to glyoxylic acid.

$$\underset{\text{Glycine}}{CH_2.NH_2.COOH} \xrightarrow{-2H} \underset{\text{Imino acid}}{CH{=}NHCOOH} \xrightarrow{+H_2O} \underset{\text{Glyoxylic acid}}{H.\overset{O}{\overset{\|}{C}}.COOH} + NH_3$$

The deamination is readily reversible. Transamination of glyoxylic acid, using either glutamine, glutamic acid, asparagine, aspartic acid, or ornithine (see p. 227) as an amino donor produces glycine.

Glycine is glycogenic and antiketogenic. When glycine labeled with isotopic carbon in the carboxyl group is fed, the carbon is found in the liver glycogen. The amino nitrogen of glycine is exchanged readily with other amino acids. It has therefore been added in considerable quantities to mixtures of the indispensable amino acids for use in parenteral nutrition to serve as a source of nitrogen in the synthesis of the dispensable amino acids by amination and transamination. The labeled nitrogen of glycine is transferred readily to urea and to the amidine group of arginine on its way to urea.

B. Special Functions of Glycine:

1. Hemoglobin synthesis - By the use of labeled glycine it has been found that the α-carbon and nitrogen atoms are used in the synthesis of the hemin of hemoglobin; but the carboxyl carbon is found only in the protein, i.e., the globin moiety. The detailed steps involved in the synthesis of the porphyrins are shown on page 55. The nitrogen in each pyrrole ring is derived from the α-nitrogen and an adjoining carbon from the α-carbon of glycine. The α-carbon is also the source of the methylene bridges (α, β, γ, δ) of the porphyrin structure. It is thus apparent that for every four glycine nitrogen atoms utilized, eight α-carbon atoms enter the porphyrin molecule.
2. Formation of glycine from choline by way of betaine - The oxidation of choline forms betaine[18] as follows:

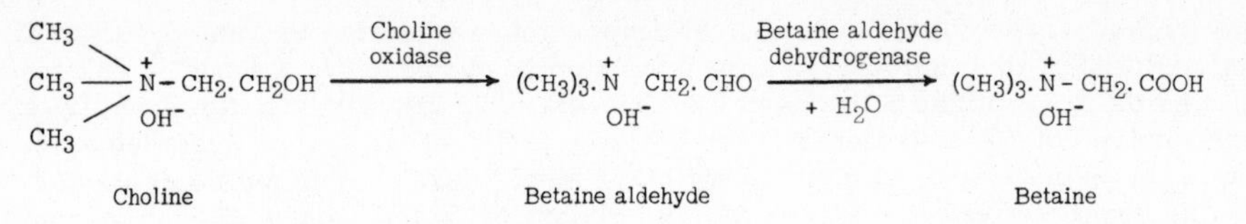

Betaine is demethylated to glycine by the loss of the first methyl group in transmethylation (see pp. 100 and 206); the last two, by oxidation to formate.

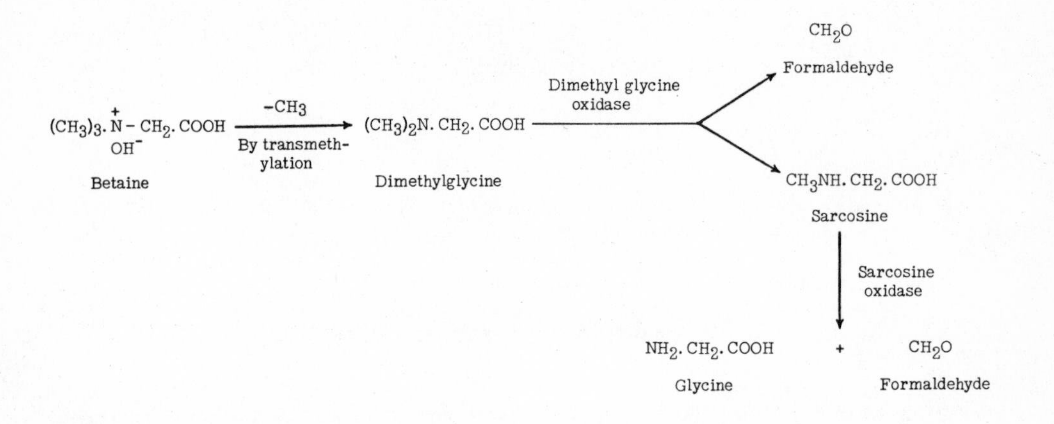

3. Conversion of glycine to serine - Glycine is readily converted to serine. Tracer studies suggest the following reaction as the mechanism; note also how the final product of serine decarboxylation, ethanolamine, may be re-converted to glycine:

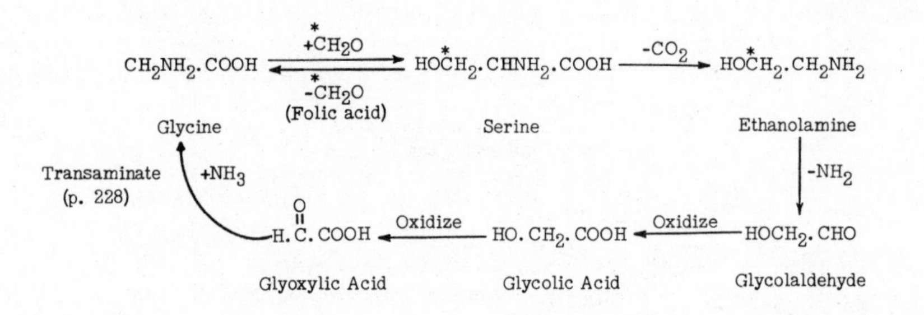

The source of the formate carbon which is added to glycine to form the β carbon of serine has been studied by Mackenzie[19]. He has shown that a 50% yield of serine occurs in the metabolism of sarcosine by liver mitochondria. Tracer studies reveal that the β carbon of the serine is derived from the methyl carbon of sarcosine. It is suggested that oxidation of this methyl group produces an ''active'' one-carbon moiety (active formaldehyde) which either condenses with glycine to form serine or accumulates as formaldehyde. Labeled formaldehyde does not participate in the synthesis of serine in this mitochondrial system. The reaction is as follows:

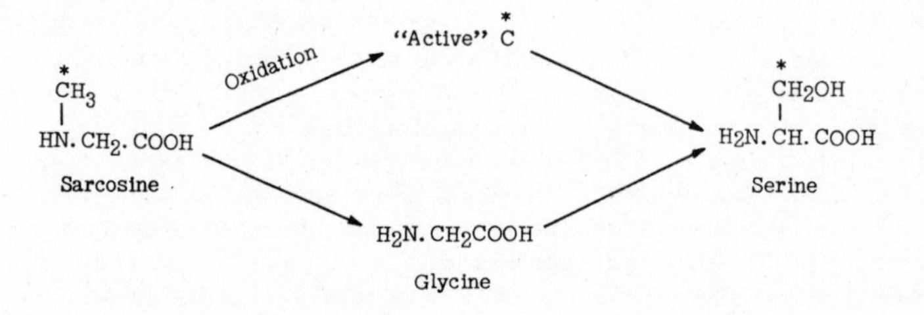

It is known that the β carbon of serine serves as a source of methyl groups for choline or thymine. From a consideration of the above reaction it is apparent that this represents transfer of methyl groups by way of a one-carbon (formate) moiety utilizing glycine and serine as carrier molecules. This type of methyl carbon transfer has been called "transformalation" by Mackenzie.

4. Synthesis of purines - The entire glycine molecule is utilized to form positions 4, 5, and 7 of the purine skeleton as shown on p. 264.*
5. Synthesis of creatine - The sarcosine (N-methyl glycine) component of creatine (see p. 260) is derived from glycine.
6. A constituent of glutathione - The tripeptide **glutathione**, a combination of glutamic acid, cysteine, and glycine, accepts glycine nitrogen much faster than does the protein in the tissues. The nitrogen in glutathione is not available for transamination.

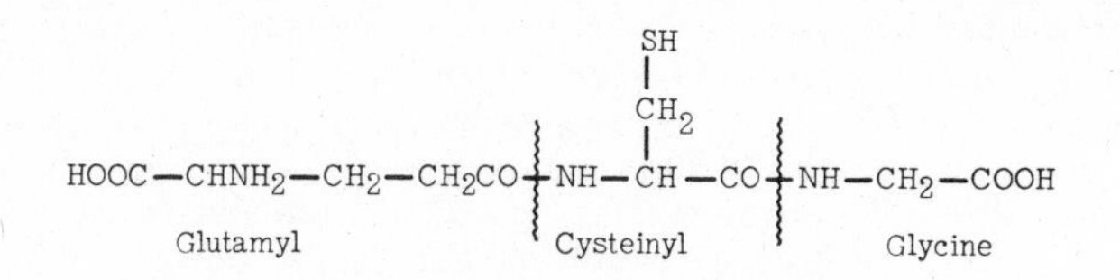

Glutathione (Reduced Form)

7. Conjugation with glycine - Glycine is an important amino acid in conjugation reactions. Conjugated with cholic acid (see p. 159), it forms the bile acid, glycocholic acid. With benzoic acid, it forms hippuric acid. This reaction, shown on p. 235, is used as a test of liver function (see p. 279).

$$\underset{\text{Benzoic Acid}}{C_6H_5.COOH} + CoA.SH + ATP \longrightarrow \underset{\text{Benzoyl-CoA}}{C_6H_5.\overset{O}{\overset{\|}{C}} \sim S.CoA} + ADP + P$$

$$\underset{\text{Benzoyl-CoA}}{C_6H_5-\overset{O}{\overset{\|}{C}} \sim S.CoA} + \underset{\text{Glycine}}{H_2NCH_2.COOH} \longrightarrow \underset{\text{Hippuric Acid}}{C_6H_5.CO.NHCH_2.COOH} + CoA.SH$$

8. Synthesis of ribose - When glycine labeled on the carboxyl carbon is fed to rats, the level of the isotope in the ribose of nucleic acid is even higher than in the proteins. This indicates that glycine contributes to the synthesis of pentose sugars, possibly by contributing a two-carbon fragment to a three-carbon compound.
9. Oxidation of glycine - Two pathways for glycine breakdown have been shown. One involves conversion to serine (para. 3, above); the other, formation of glyoxylic acid after deamination by glycine oxidase. The further breakdown of glyoxylic acid has been studied in rat liver and kidney[20], where formate results from oxidative decarboxylation of glyoxylic acid as shown at right:

   Glyoxylic acid may also be oxidized to form oxalic acid, as shown at right:

   Formate is oxidized readily by many tissues, probably using hydrogen peroxide and catalase (see p. 90).

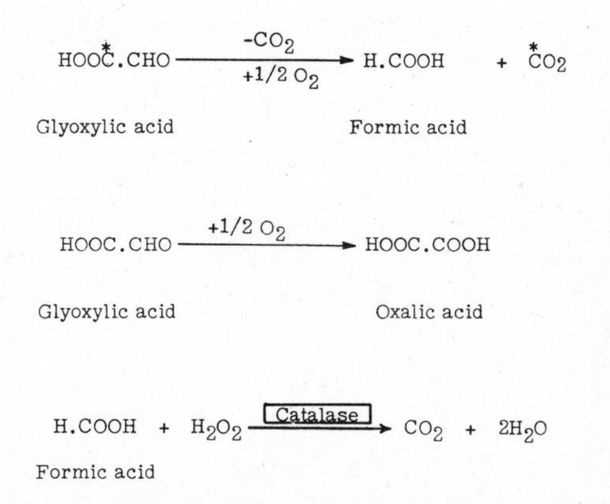

*The δ-carbon atom of δ-amino-levulinic acid (see p. 55) is derived from the α-carbon of glycine. This δ-carbon was found to be incorporated into positions 2 and 8 (the ureido carbons) of the purine nucleus to an even greater extent than α-carbons of glycine or β-carbons of serine. This suggests that δ-aminolevulinic acid is actually the carrier molecule for the transfer of this carbon atom to purines (see p. 236).

In summary, the pathway of glycine oxidation in rat liver is as follows:

$$\text{Glycine} \longrightarrow \text{Glyoxylic acid} \longrightarrow CO_2 + \text{Formic acid} \longrightarrow CO_2$$

10. The succinate-glycine cycle - The important role of glycine in porphyrin (see p. 55) and in purine (see p. 264) synthesis, as described in paragraphs 1 and 4 above, suggested to Shemin and Russell[21] a relation of the metabolism of glycine to the citric acid cycle. These pathways for the metabolism of glycine they have summarized in their "**succinate-glycine cycle.**" Succinate, as "active" (Co A) succinate (see p. 88), condenses on the α-carbon atom of glycine to form α-amino-β-ketoadipic acid; it is at this point that the metabolism of glycine is linked to the citric acid cycle, which serves as the source of active succinate (derived from the decarboxylation of α-ketoglutaric acid). α-Amino-β-ketoadipic acid is then converted by loss of carbon dioxide to δ-aminolevulinic acid. This compound serves as a common precursor for porphyrin synthesis as well as, after deamination, a carrier molecule for the introduction of the ureido carbons (2 and 8) into the purine ring (see also p. 264). Succinic acid and ketoglutaric acid, which may return to the citric acid cycle, are also formed. These integrated reactions are shown below.

Alanine:

This amino acid is not known to have any specific function, but together with glycine it makes up a considerable fraction of the amino nitrogen in human plasma. It is dispensable and can be synthesized easily by amination of pyruvic acid with the aid of glutamic acid as ammonia donor (see p. 226). Alanine is glycogenic and antiketogenic, probably more so than glycine. Deamination of alanine produces pyruvic acid, which is oxidized in the citric acid cycle. Both D- and L-alanine appear to be utilized by the tissues, but at differing rates.

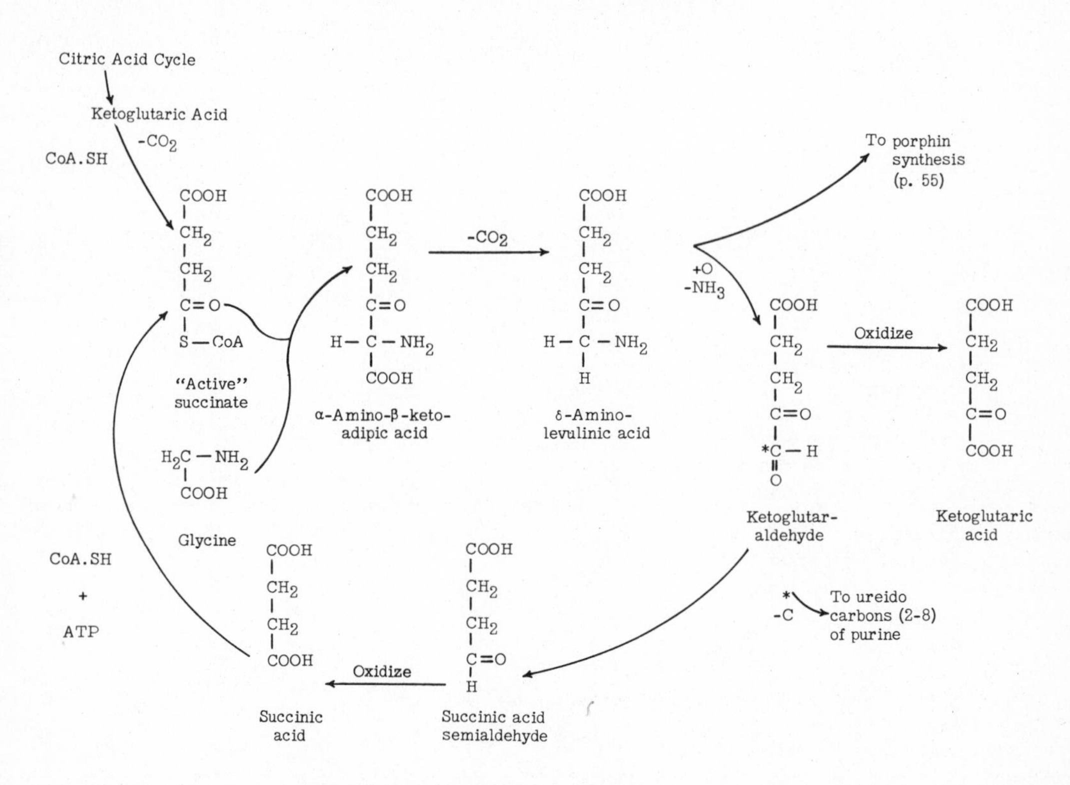

β-Alanine is a constituent of pantothenic acid and an end product in the catabolism of certain pyrimidines (cytosine and uracil; see p. 268). Studies on the catabolism of β-alanine in the rat[23] indicate that this amino acid is degraded to acetic acid as follows:

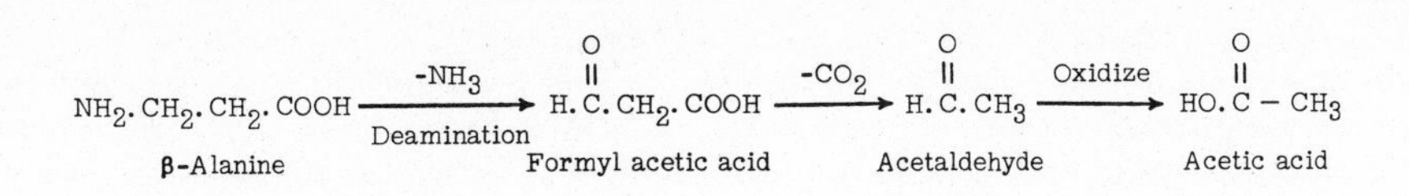

Serine:

Serine can be readily synthesized in the tissues, so it is not considered among the essential amino acids. Serine can be formed by addition of a β carbon to glycine through transfer of a hydroxymethyl group by way of a formylated tetrahydrofolic acid (see p. 92). Sarcosine (N-methyl glycine) may be considered to serve as the source of the hydroxymethyl group which forms the β carbon of serine. The reaction may be visualized as an intramolecular transformation (catalyzed by tetrahydrofolic acid) as follows:

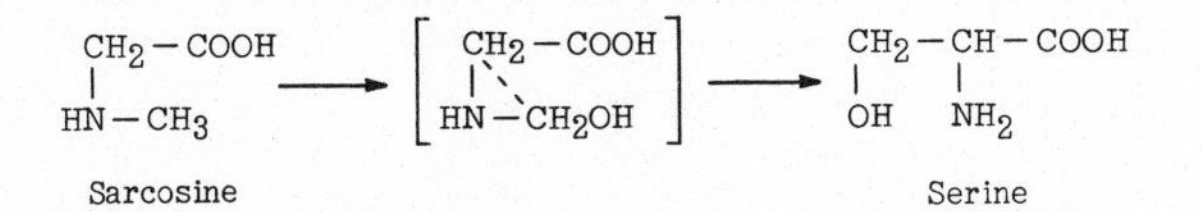

Phosphoglyceric acid formed in glycolysis may also serve as a precursor of serine by oxidation at carbon 2 to the keto derivative followed by amination of the keto group[22].

COOH–CHOH–CH_2–O–(P) (3-Phosphoglyceric Acid) —(DPN → DPN.H, −2 H)→ COOH–C=O–CH_2–O–(P) (3-Phosphohydroxypyruvic Acid) —($+NH_3$ (From Alanine))→ COOH–$CHNH_2$–CH_2–O–(P) (*O*-Phosphoserine) → COOH–$CHNH_2$–CH_2OH (Serine) + (P)

Much of the serine in phosphoproteins appears to be present in the form of *O*-phosphoserine.

The conversion of serine to glycine (see p. 234) involves the loss of a single carbon. This may be the β carbon, which thus yields an active one-carbon moiety for methylation or for purine synthesis; or it may be the carboxyl carbon, in which case ethanolamine is formed. Since this latter compound, on methylation, forms choline, serine may be considered to participate in the formation of lipotropic agents (see p. 205).

A cephalin fraction containing serine has been isolated from the brain. The production of this compound, phosphatidyl serine (see p. 23), may be another lipotropic function of serine.

Serine is involved directly in the synthesis of sphingol and therefore in the formation of sphingomyelins of brain. The details of this reaction are shown on p. 219.

Serine participates in purine and pyrimidine synthesis. The β-carbon is a source of the methyl groups of thymine (and of choline) and of the carbon in positions 2 and 8 of the purine nucleus.

Serine is glycogenic through the formation of pyruvic acid, which has been shown to arise from serine incubated with cell-free liver extracts. The amino acid is deaminated by a special **serine dehydrase** which is activated by biotin. Pyridoxal phosphate is a prosthetic group on the enzyme. The reaction proceeds as follows:

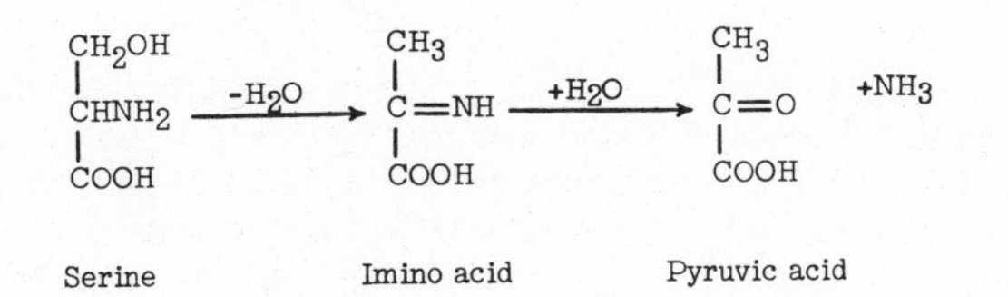

Threonine:

This amino acid is a constituent of most proteins and is essential for growth and maintenance of animals, including man. It can be deaminated to form α-ketobutyric acid, and while this keto acid could be transaminated by the glutamic acid and the glutamine systems described on pp. 226 and 227, the resulting compound, α-amino butyric acid, is apparently not convertible to threonine. This essential amino acid does not therefore participate in transamination reactions. The D-isomer is not utilized by the body, probably because of the nonconvertibility of the keto acid to the amino acid. A specific function for threonine other than as a constituent of body proteins has not been discovered. It may be related to the utilization of fat in the liver (see p. 206).

Threonine, like serine, serves as a carrier for phosphate in the phosphoproteins. Threonine-*O*-phosphoric acid has been isolated from casein.

The deamination of threonine is catalyzed by a threonine dehydrase. Pyridoxal phosphate may be a prosthetic group for this enzyme. The reactions, which produce α-ketobutyric acid, are similar to those for the deamination of serine:

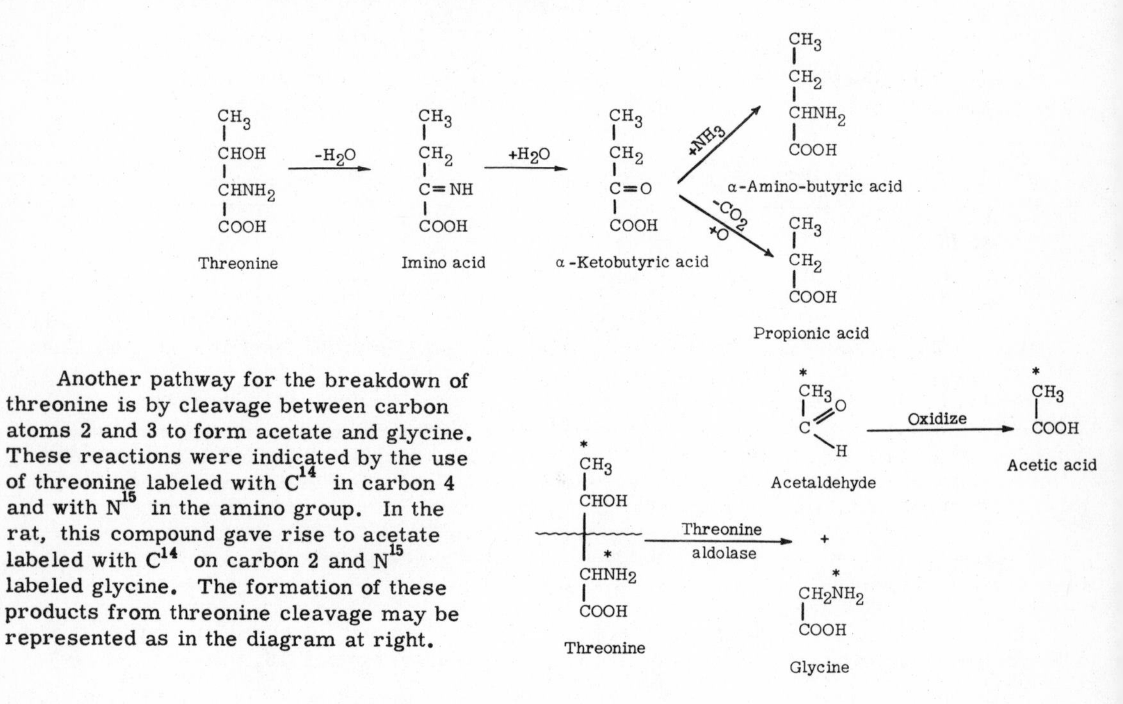

Another pathway for the breakdown of threonine is by cleavage between carbon atoms 2 and 3 to form acetate and glycine. These reactions were indicated by the use of threonine labeled with C^{14} in carbon 4 and with N^{15} in the amino group. In the rat, this compound gave rise to acetate labeled with C^{14} on carbon 2 and N^{15} labeled glycine. The formation of these products from threonine cleavage may be represented as in the diagram at right.

Leucines:

Both leucine and isoleucine are essential. Their D- isomers are not available for growth in the rat and mouse, but their keto acids are utilized (the chick can utilize D-leucine). N^{15}-labeled leucine exchanges its nitrogen with other amino acids, and D-leucine is inverted to L-leucine in the rat. The negative results in growth experiments with the unnatural isomer can probably be explained on the basis of a rate of deamination which is too slow to support growth.

Rose[24] has concluded from nitrogen balance studies that neither D-leucine nor isoleucine is utilized by humans.

Leucine is the most strongly ketogenic of all the amino acids. Bachhawat et al.[25] proposed the following scheme for the metabolism of leucine. It involves activation with Co A and decarboxylation as a preliminary to oxidative steps which in some aspects are reminiscent of the oxidation of fatty acids but which include carbon dioxide fixation by "active" carbon dioxide. Fixation of carbon dioxide is catalyzed by a specific 'carboxylase" which requires biotin as cofactor.

β-Hydroxy-β-methylglutaryl Co A, the product of CO_2 fixation on β-hydroxy-isovaleryl Co A, is a key intermediate in the synthesis of cholesterol. As shown on p. 216, it is a direct precursor of mevaldic and mevalonic acids.

The formation of free β-hydroxy-β-methylglutaric acid occurs through the action of a specific deacylating enzyme (deacylase) followed by enzymatic cleavage to form acetoacetic acid and acetic acid. The cleavage enzyme has been found in liver, kidney, and heart tissues. It requires magnesium or manganese ions as activators and a source of thiol (SH) groups such as cysteine or glutathione[25].

The deacylase may couple with the cleavage enzyme to accomplish the following reversal of the above reactions, which leads to synthesis of cholesterol precursors[26]:

$$\text{Acetoacetic Acid} + \text{Acetyl Co A} + H_2O \rightleftharpoons \beta\text{-Hydroxy-}\beta\text{-methylglutaric Acid} + \text{Co A}$$

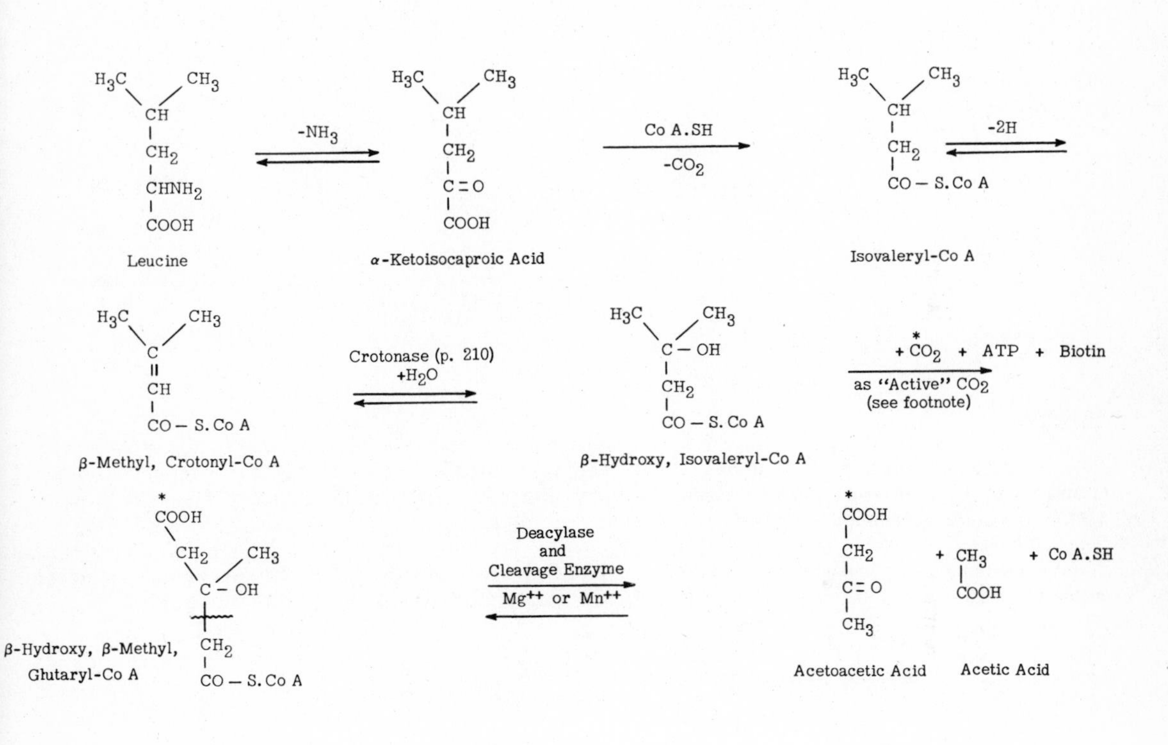

* "Active" $\overset{*}{CO_2}$ is a nucleotide based on adenosine monophosphate (AMP). Its structure is shown at right.

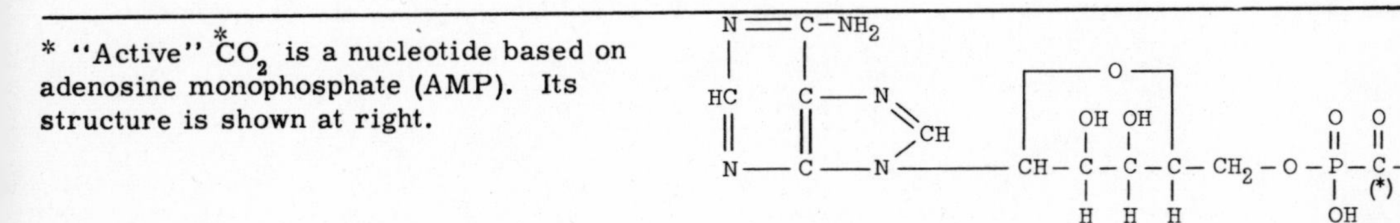

In contrast to leucine, isoleucine is only weakly ketogenic and also has slight but definite glycogenic properties. The metabolism of isoleucine as proposed by Robinson et al.[27] is shown below.

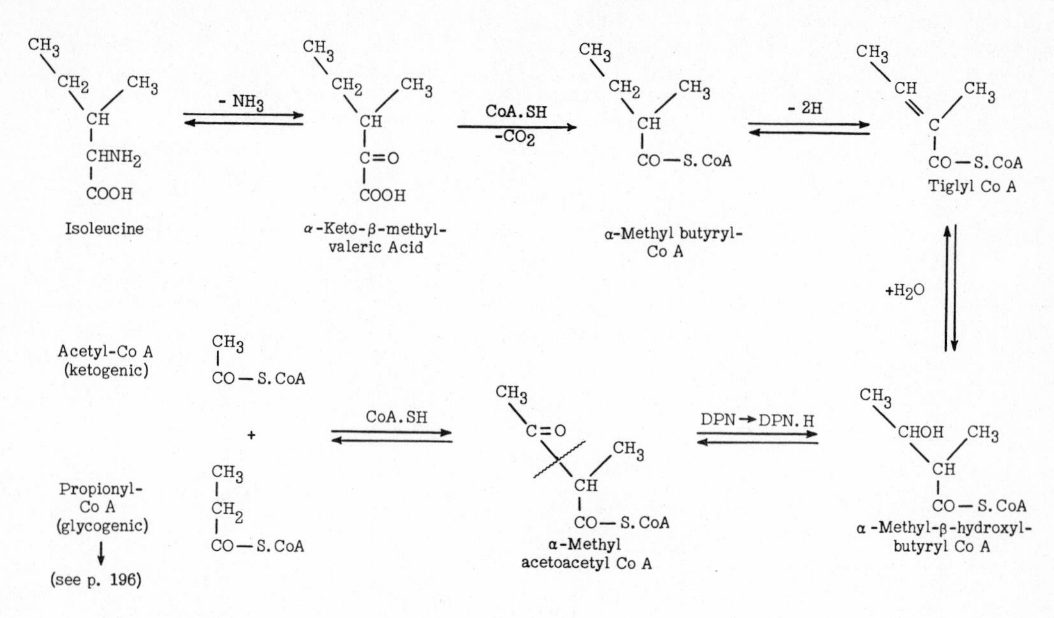

Valine:

This amino acid is also indispensable in animals and man, but its specific function is not known. It is deaminated in the body. After feeding it to rats, the keto acid was recovered in the urine. The D- isomer supports growth in the rat at a rate about one-half that of the L- isomer. The keto acid also supports growth on a valine-free diet.

In the metabolic breakdown of valine, three of its five carbon atoms are converted to glucose. In the following scheme the source of the glucose derived from valine is propionic acid.

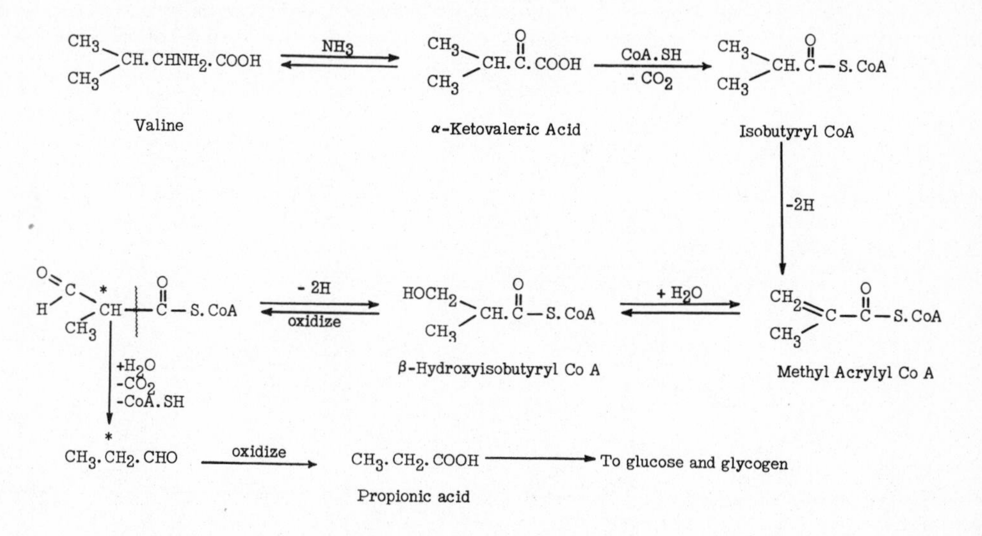

Lysine:

Lysine is an essential amino acid. The absence of lysine from many cereal proteins is a conspicuous reason for the failure of such protein to support growth when fed as the sole source of dietary protein.

In the proteins of one collagen group, 5-hydroxylysine is found. Because labeled hydroxylysine is not incorporated into collagen, it is concluded that hydroxylation of lysine must occur only after lysine has been incorporated into the collagen.

Lysine, like threonine, does not participate in reversible transamination. The D-isomer is therefore not utilized. If the failure of the D-isomer of valine and leucine adequately to support growth be excepted, threonine and lysine are the only two indispensable amino acids whose D-isomer cannot be used under any experimental conditions.

However, Meister[28] has found that a transaminase in rat liver will catalyze the transfer of amino groups from glutamine to an α-keto derivative of lysine **if the nitrogen on the ε-carbon has a substituent group** but not if it is unsubstituted (as would be the case after simple deamination of lysine). Such ε-N-substituted lysines can also replace lysine in the diet. Examples are shown below.

CH_3 | (ε-) CH_2—NH | CH_2 |

ε-N-Methyl Lysine

CH_3 | C=O | (ε-) CH_2—NH | CH_2 |

ε-N-Acetyl Lysine

The L-amino acid oxidases of animal origin do not attack lysine, but if lysine labeled with N^{15} in the α-amino group is fed, labeled nitrogen appears in other proteins and in urea. On the basis of these findings and the experiments of Meister mentioned above, it has been suggested that deamination of lysine in the body may first require protection of the ε-amino group such as occurs by N-methylation or N-acetylation, shown above.

The metabolic breakdown of lysine is very peculiar. The following scheme, based on studies in the rate, has been proposed[29 a, b]. Note that the ε-nitrogen is incorporated into the cyclic intermediate, **pipecolic acid**, the α-nitrogen having been removed in the first step. Apparently the α-amino nitrogen is excreted because it does not appear in any other amino acids. The pathway is not reversible in the rat since dietary lysine cannot be replaced by either pipecolic acid or aminoadipic acid. The first step in the breakdown of lysine which results in the oxidative deamination of the amino acid to the keto acid has not been proved by isolation of lysine keto acid (α-keto-ε-aminocaproic acid), probably because it is rapidly converted by ring closure to dehydropipecolic acid. This may also be a reason for the failure of lysine to participate in transamination, since the keto acid is rapidly and irreversibly converted to the cyclic structure. However, substitution on the ε-carbon would prevent cyclization, so that deamination and reamination at the α-carbon could then take place. This has actually been observed, as noted above.

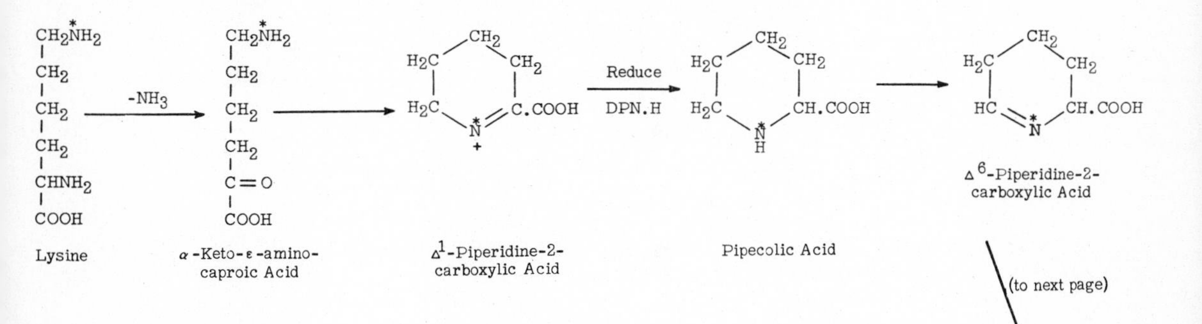

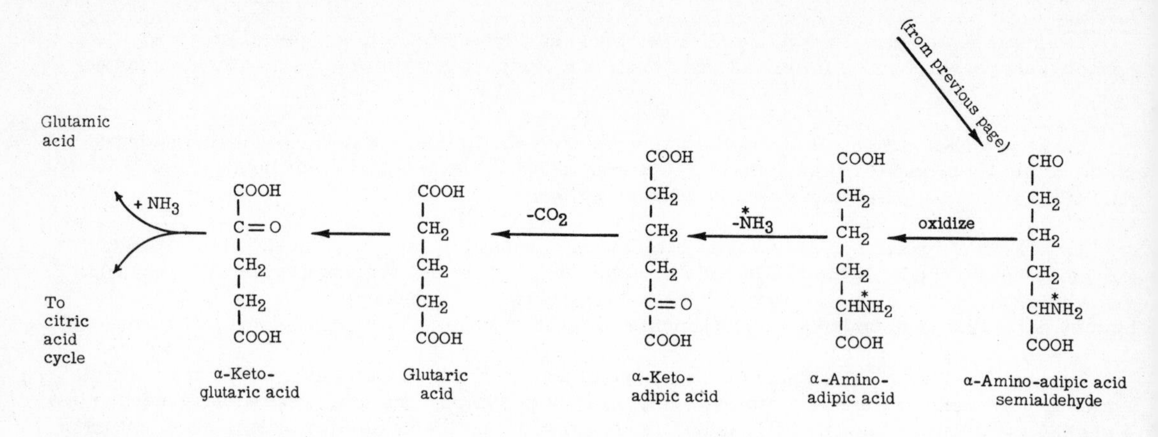

Arginine and Ornithine:

The formation of arginine in the urea cycle has been described on p. 232. As a result of the action of **arginase,** this amino acid is broken down to urea and ornithine. The further metabolism of arginine may therefore be considered to proceed through ornithine. Studies of the metabolism of ornithine have been made by the use of the N^{15}-labeled amino acid (M. R. Stetten). These have revealed that the α-amino nitrogen of ornithine is an important source of the ring nitrogen of proline and hydroxyproline, whereas the ε-nitrogen is found in glutamic acid. It is known that the ε-amino nitrogen transaminates to ketoglutaric acid[30]; this would form glutamic acid and leave as the product glutamic γ-semialdehyde, which would then either be oxidized to glutamic acid or converted to proline. All of these reactions are shown below.

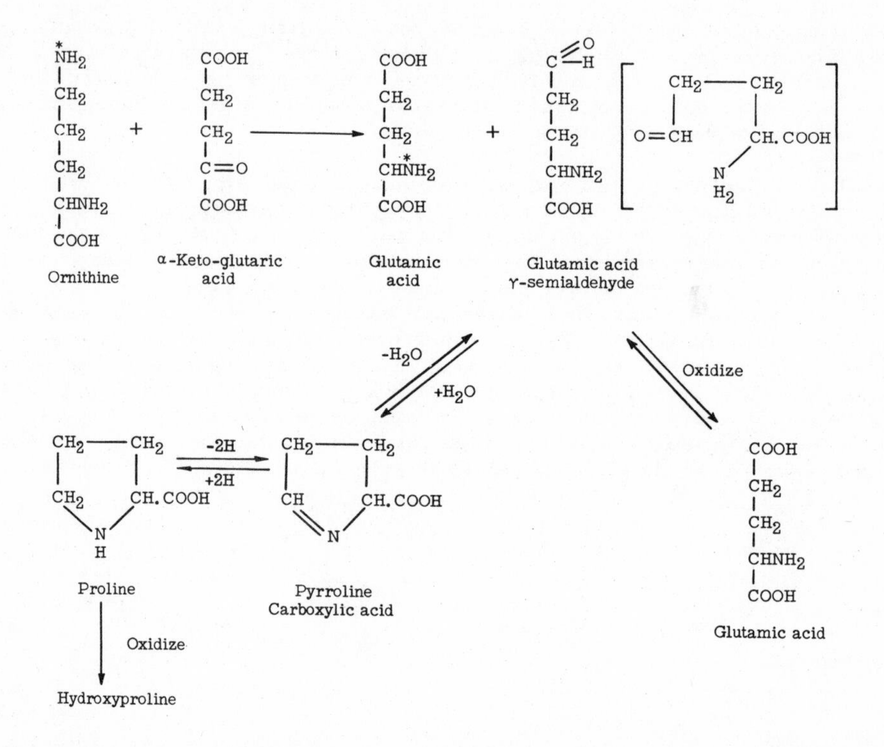

The amidine group of arginine is utilized in the synthesis of creatine. In the kidney, transamidination of glycine forms glycocyamine or guanidoacetic acid (see Creatine Synthesis, p. 260).

Arginine is not considered a completely essential amino acid for man; but it is not entirely indispensable in that although synthesized by the body, the rate of synthesis may be inadequate to provide for all of the requirements of the tissues. In experimental animals on an arginine-free diet, growth is depressed by about 25%. This indicates that additional arginine is required by the young growing animal.

The Dicarboxylic Acids, Aspartic and Glutamic Acids:

These amino acids are not essential from a nutritional standpoint since they can be synthesized with great ease; but no amino acids are more active in deamination, reamination, transamination (see p. 226), inversion of D- to L- forms, or in ammonia and urea formation. After the feeding of isotopic nitrogen, it is always noted that the amide nitrogen of aspartic and glutamic acids (asparagine and glutamine), as well as their amino nitrogen, have the largest concentration of the isotope.

A. Glycogenic Functions: Both of these amino acids are glycogenic. Deamination produces oxaloacetic acid from aspartic acid, and ketoglutaric acid from glutamic acid. These are important components of the citric acid cycle. The reactions are reversible. This provides a mechanism for the synthesis of glutamic and aspartic acids.

B. Transamination: The role of glutamic acid and glutamine as sources of ammonia in transamination, and of glutamine in ammonia formation by the kidney, has been mentioned. Asparagine may act similarly as an ammonia donor. It is particularly important in plants, where it occurs in considerable quantities along with glutamine. Both dicarboxylic acids are particularly important as nitrogen donors in synthesis of purines (see p. 266).

C. Citric Acid Cycle Function: It is postulated that the glutamic acid–glutamine system, by serving as a source of ketoglutaric acid, may aid in regulating the concentration of metabolites entering the citric acid cycle. In this connection, it is interesting that glutamic acid restored hypoglycemic patients in insulin coma to consciousness at a lower blood sugar level than when glucose alone was used.

D. Urea Cycle Function: This is described on p. 231 in connection with the formation of carbamyl aspartic acid and argininosuccinic acid.

E. Functions in Central Nervous System: There is evidence that glutamic acid plays a special role in brain metabolism (see Chapter 22). For example, the transport of potassium to the brain is accomplished with glutamate.

Decarboxylation of glutamic acid produces γ-aminobutyric acid. An enzyme which catalyzes its formation from glutamic acid by alpha decarboxylation is found only in the tissues of the central nervous system, principally in the gray matter. This enzyme requires pyridoxal phosphate (B_6 . al Ⓟ) as a coenzyme.

$$\underset{\text{Glutamic acid}}{HOOC.H_2N.CH.CH_2.CH_2.COOH} \xrightarrow[(B_6.\text{al}\ Ⓟ)]{-CO_2} \underset{\gamma\text{-Amino butyric acid}}{H_2N.CH_2.CH_2.CH_2.COOH}$$

γ-Aminobutyric acid is now known to serve as a normal regulator of neuronal activity, being active as an inhibitor when studied in various reflex preparations. It is further metabolized by deamination to succinic semialdehyde. The deamination is accomplished by a pyridoxal-dependent enzyme and the ammonia removed is transaminated to ketoglutaric acid, thus forming more glutamic acid. Succinic semialdehyde is then oxidized to succinic acid. These reactions, which are shown below, provide a "by-pass" (shown in dotted lines, p. 244) around the citric acid cycle in the brain from ketoglutaric acid, which is aminated to form glutamic acid, to succinic acid.

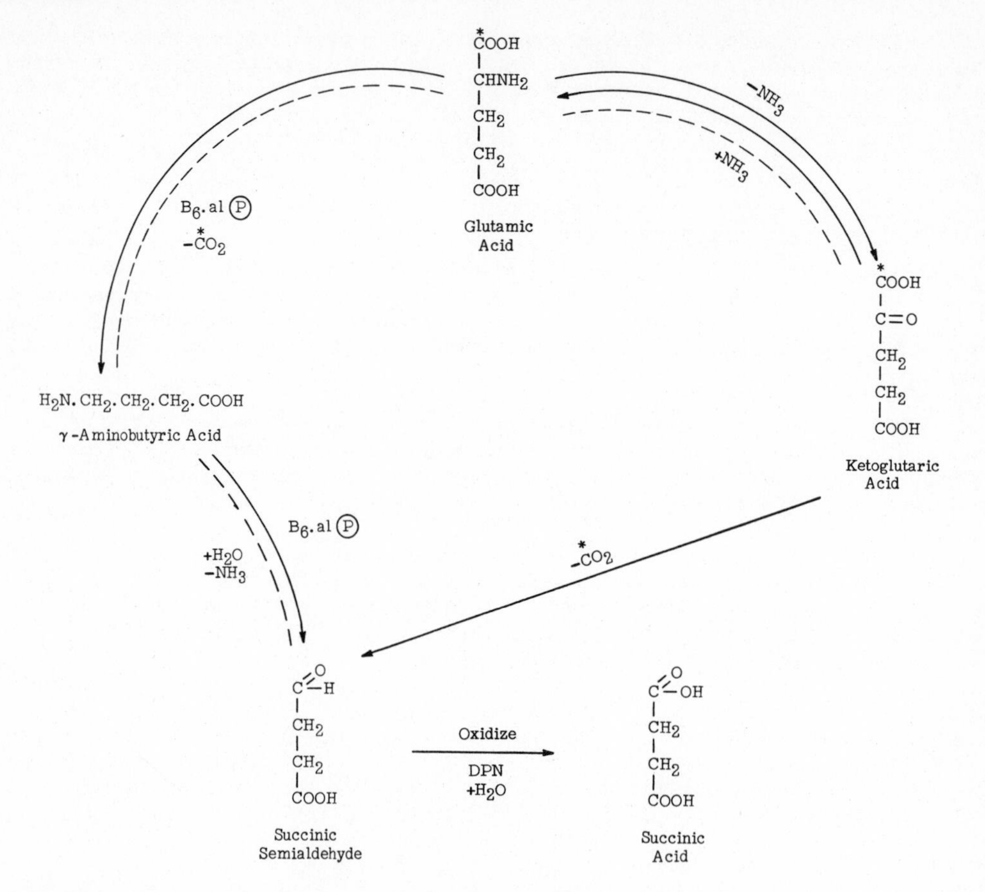

Pyridoxal phosphate dissociates easily from the glutamic acid decarboxylase apoenzyme. As a result this enzyme is very susceptible to pyridoxal deficiencies or to the action of pyridoxal antagonists.

In both rat and cat brain, N-acetylaspartic acid has been found in concentrations as high as 100 mg. per 100 Gm.[31] Very little of this N-acetylated amino acid is present in liver, kidney, muscle, or in the urine. Its significance in the brain is not known.

Proline and Hydroxyproline:

Both of these amino acids are dispensable. Their synthesis is accomplished by the reactions shown on p. 242. The conversion of proline to hydroxyproline is not reversible.

Hydroxyproline occurs only in collagen. As is true also of hydroxylysine, hydroxylation of proline takes place only after incorporation of proline into collagen or a collagen precursor. The hydroxylation may require ascorbic acid as a co-factor.

Proline is glycogenic, presumably through the formation of glutamic acid and α-ketoglutaric acid. Proline also gives rise to ornithine. All of these reactions take place by a reversal of the reactions by which proline is synthesized (see p. 242). The metabolic breakdown of hydroxyproline is probably analogous to that of proline.

Histidine:

This amino acid is classed like arginine as relatively dispensable. Adult human beings and adult rats have been maintained in nitrogen balance for short periods in the absence of histidine. The growing animal does, however, require histidine. If studies were to be carried on for longer periods, it is probable that a requirement for histidine in adult human subjects would also be elicited. In N^{15} experiments, histidine exchanges α-amino nitrogen. In experimental animals, there is some evidence that the D- isomer will substitute for the natural L- isomer, but it has been said that this does not occur in man.

Histidine is glycogenic, although it forms glycogen very slowly.

A. Metabolism: The enzyme, **histidase**, which is present in the liver, produces glutamic acid, formic acid, and ammonia when incubated with histidine. This glutamic acid may be the source of glycogen derived from histidine.

Recent studies on the metabolic breakdown of histidine may be summarized as follows:

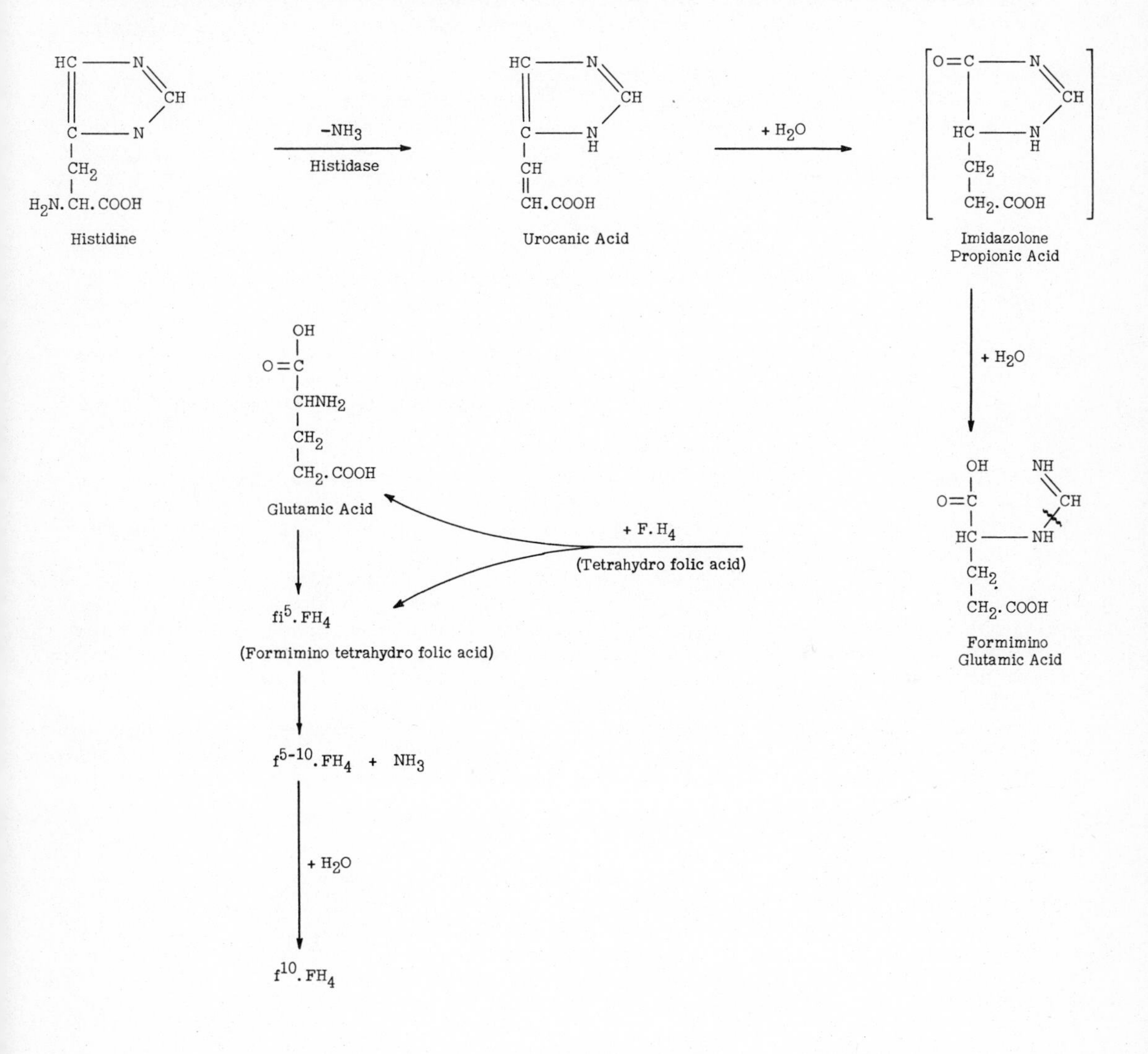

The formation of free glutamic acid occurs by transfer of a formimino group (fi) to tetrahydro folic acid (FH_4) (see p. 93), the group being attached at position 5 on the tetrahydro folic acid. Ammonia is then split off and the formyl folic acid ($f^{10}.FH_4$) derivative is thus produced. It will be recalled that this latter compound is the source of the one-carbon (formyl) moiety, as shown on p. 90. Histidine is therefore an important source of the one-carbon groups for positions 2 and 8 of the purine nucleus, as well as for synthesis of methyl groups.

The imidazole nucleus of histidine apparently cannot be synthesized by mammalian tissues. This suggests that the amino acid should be essential, although a specific function for histidine has not yet been found. Isotopic nitrogen is not found in the nucleus of histidine. However, as noted heretofore, it does enter the α-amino group when made available to the organism.

B. Histidinuria in Pregnancy: The quantity of histidine found in normal urine is relatively large (see p. 302). For this reason it may be more readily detected than most other amino acids. It has been reported that a conspicuous increase in histidine excretion is a characteristic finding in normal pregnancy but does not occur in toxemic states associated with pregnancy. The existence of a metabolic defect in the metabolism of histidine has been invoked to explain the apparent alterations in histidine excretion of pregnant women. However, Page et al. [32a,b] have shown by renal clearance studies that the phenomenon may be largely explained on the basis of the changes in renal function which are characteristic of normal pregnancy as well as the pregnancy toxemias. Furthermore, the alterations in amino acid excretion during pregnancy are not confined to histidine.

C. Histamine Formation: Histamine is derived from histidine by decarboxylation. Intestinal bacteria can bring about this reaction, but a histidine decarboxylase which catalyzes this reaction is also found in kidney and intestine. A histamine-destroying **histaminase** is present in the tissues.

D. Histidine Compounds: Three histidine compounds are found in the body: **ergothioneine,** in red blood cells and liver; **carnosine,** a dipeptide of histidine and β-alanine; and **anserine,** 1-methylcarnosine. The latter two compounds occur in muscle. The functions of these histidine compounds are not known.

Ergothioneine was found to be widely distributed in the tissues of the rat, particularly in the liver, where its concentration was even higher than in the red blood cells. It was associated with the cytoplasmic fraction, apparently in an unbound form; none was found in the blood plasma, testes, or brain. When rats were placed on a purified diet with casein as the sole source of protein, the ergothioneine content of the blood and tissues was reduced to very low levels; this suggests that the diet affects the amounts of ergothioneine in the blood.

Using radioactive sulfur (S^{35}), Melville et al. [33a,b] could find no evidence in the pig or the rat that synthesis of ergothioneine occurs. In a search for the dietary precursors which are therefore presumed to serve as the sole source of ergothioneine, it was found that both corn and oats contained this compound.

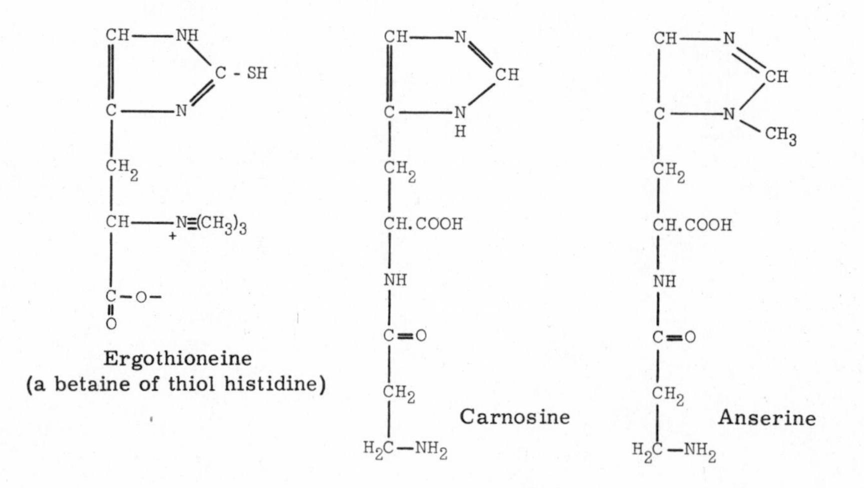

Ergothioneine
(a betaine of thiol histidine)

Carnosine

Anserine

The presence of 1-methylhistidine in human urine has been reported. This is undoubtedly derived from anserine, and larger amounts were found in the urine after the ingestion of rabbit muscle, which is particularly high in anserine. 3-Methylhistidine has been identified in human urine in amounts of about 50 mg. per day[34]. The origin of this compound, an isomer of 1-methylhistidine, which is a component of anserine, is not known. There is no evidence that it occurs in muscle as a constituent of a peptide similar to anserine. It is of interest that 3-methylhistidine is unusually low in the urine of patients with Wilson's disease (see p. 327).

Carnosine can be used to replace histidine in the diet. When injected into animals, it has a circulatory depressor action similar to but not as potent as that of histamine, which in large doses may cause vascular collapse.

Sulfur-containing Amino Acids, Methionine and Cystine:

These amino acids are the principal sources of organic sulfur for body processes; but methionine (or homocysteine plus a source of methyl groups; see next page) is essential whereas cystine is not, presumably because it can be synthesized.

A. Metabolism of Methionine: This amino acid readily donates its terminal methyl group for methylation of various compounds (see p. 206). This role of methionine in methylation reactions is an important function of this amino acid, because methionine is the principal methyl donor in the body. The methyl group may be transferred to other compounds for the synthesis of choline or of creatine, for example, or for use in detoxication processes, such as the methylation of pyridine derivatives like nicotinic acid.

In the **transmethylation** reactions which utilize methionine as a methyl donor, it is first required that methionine be "activated." This requires adenosine triphosphate (ATP) and a methionine-activating enzyme of liver. The reaction is believed to proceed as follows:[35]

$$\text{L-Methionine} + \text{ATP} \xrightarrow[\text{Methionine-activating enzyme}]{(Mg^{++})\ \text{Glutathione}} \text{Active methionine} + 3\ \textcircled{P}$$

Active methionine is believed to be an adenine-containing nucleoside, i.e., adenosine attached through the sulfur of methionine (S-adenosyl methionine):

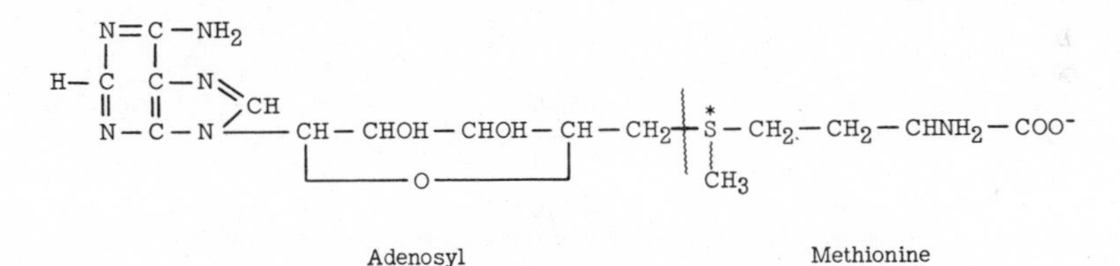

Methyl-(5-desoxy-ribosyladenine)(2-aminobutyro-)thetin

Structure of "Active" Methionine

The S-methyl bond is "high-energy"; this is a reason for the lability of the methyl group acting as a source of methyl for transmethylations.

In addition to utilization of the methyl group in the intact form, there is evidence in experiments with methionine containing a labeled methyl carbon that this methyl group is also oxidized. In the rat, one-fourth of the labeled methyl carbon appears in the expired carbon dioxide during the first day; and about one-half of the labeled carbon is excreted in the urine, feces, or respiratory carbon dioxide in two days. It has already been noted that the methyl carbon may also be used to produce the one-carbon moiety which conjugates with glycine in the synthesis of serine (see p. 234).

The demethylation of methionine, either for transmethylation or oxidation of the methyl group, produces homocysteine, as shown in the reactions below.

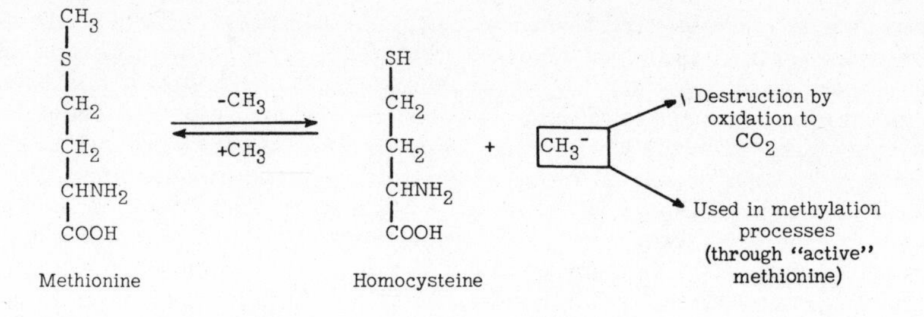

Homocysteine together with a source of labile methyl (e.g., betaine, or choline by way of betaine; see p. 206) can be used to replace methionine in the diet. This observation suggests that the demethylation process is reversible.

The animal organism has considerable ability to synthesize methyl groups, and vitamin B_{12} and folic acid are involved in the synthesis of these labile methyl groups.

There is evidence that homocysteine is involved in the utilization of the one-carbon (formate) moiety (see also p. 90). This may occur through the formation of an intermediary compound of homocysteine with a one-carbon moiety derived from formic acid or formaldehyde. The single carbon could then be transferred to the synthesis of purines, the formation of the β-carbon of serine or the methyl of methionine. A scheme outlining this postulated role for homocysteine in utilization of formate is shown below[36].

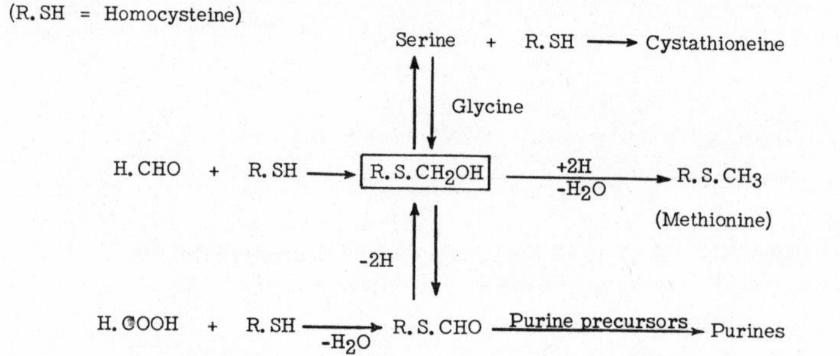

Methionine may undergo oxidative deamination to form the corresponding keto acid. This reaction is reversible and thus inversion of the D- form is possible.

B. Synthesis of Cystine; Transulfuration: In the synthesis of cystine (through cysteine) the -SH (sulfhydryl) group of homocysteine derived from demethylated methionine is transferred to serine. This is an example of transulfuration, a process which is catalyzed by enzymes designated transulfurases; pyridoxal (Vitamin B_6) phosphate is required as coenzyme. It is thus apparent that the carbon chain of serine is used for the structure of cysteine, and that methionine serves only as a source of sulfur. The reaction probably proceeds through the intermediate formation of cystathioneine, as follows:

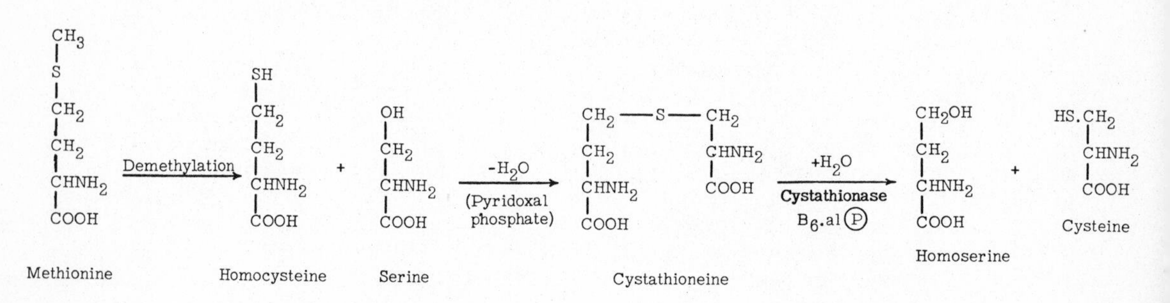

C. Metabolism of Homoserine: As shown above, after demethylation and transfer of sulfur, methionine is converted to homoserine. Recent studies[37] on the metabolism of methionine and of homocysteine indicate that the pyridoxal-dependent enzyme which cleaves cystathioneine (cystathionase) acts also as a homoserine deaminase, so that homoserine once formed is quickly deaminated to ketobutyric acid. This latter compound may then be aminated to form aminobutyric acid or it may be decarboxylated to yield propionic acid, a precursor of glucose and glycogen.

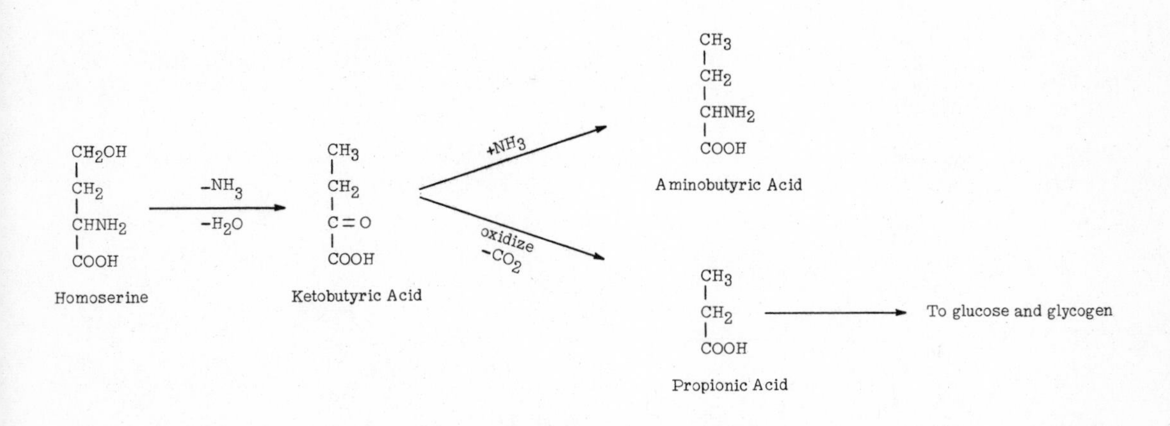

α-Aminobutyric acid has been identified in the urine of human subjects and the excretion of this amino acid is increased after the administration of methionine.

The glycogenic ability of methionine is explicable by the intermediate formation of propionic acid.

D. Oxidation of Cysteine: The oxidation of cysteine to cystine proceeds readily. In fact, this conversion of two -SH radicals to S-S is probably an important oxidation-reduction system in the body (see Glutathione, p. 235).

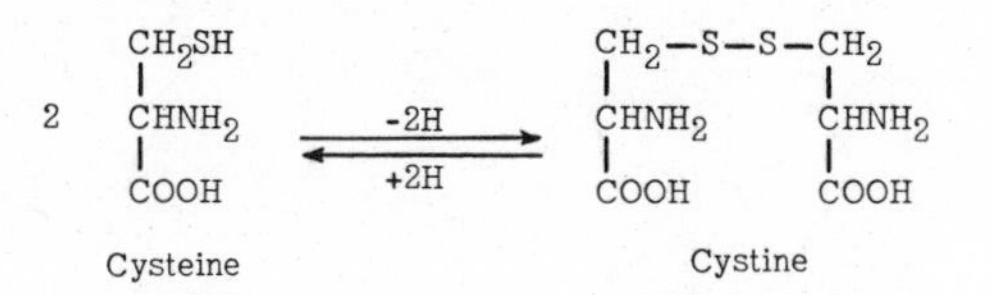

E. Function of Cysteine in Enzyme Proteins: Many enzymes depend on a free -SH group for maintenance of their activity. The importance of the -SH groups on coenzyme A has also been noted (see p. 86). In the case of enzymes, mild oxidation, converting the -SH group to the S-S linkage, will inactivate them. They may be reactivated by reduction of the S-S group, as with glutathione. A relationship of glutathione to the protection of S-S groups of insulin has also been postulated. Heavy metals like mercury or arsenic also combine with -SH linkages and cause inactivation of enzyme systems (see p. 104).

F. Cystine in Conjugation Reactions: The functions of glutathione are attributable to its cysteine content. In addition to the function of cysteine/cystine in glutathione synthesis, this amino acid is important in conjugation with aromatic halogens to form mercapturic acids (see p. 167.

Taurine, the cholic acid conjugate in bile which forms the bile acid, **taurocholic** acid, is derived from cysteine. Its origin is shown in the reactions outlined on p. 250.

G. Further Metabolism of Cysteine and Cystine: The sulfur occurring in the urine originates almost entirely from the oxidation of cystine. Methionine is believed to contribute its sulfur to cystine rather than allowing it to be oxidized directly. The pathways by which cysteine and cystine are metabolized are shown on p. 250. It will be noted that the major pathway is via pyruvic acid.

METABOLISM OF CYSTINE AND CYSTEINE

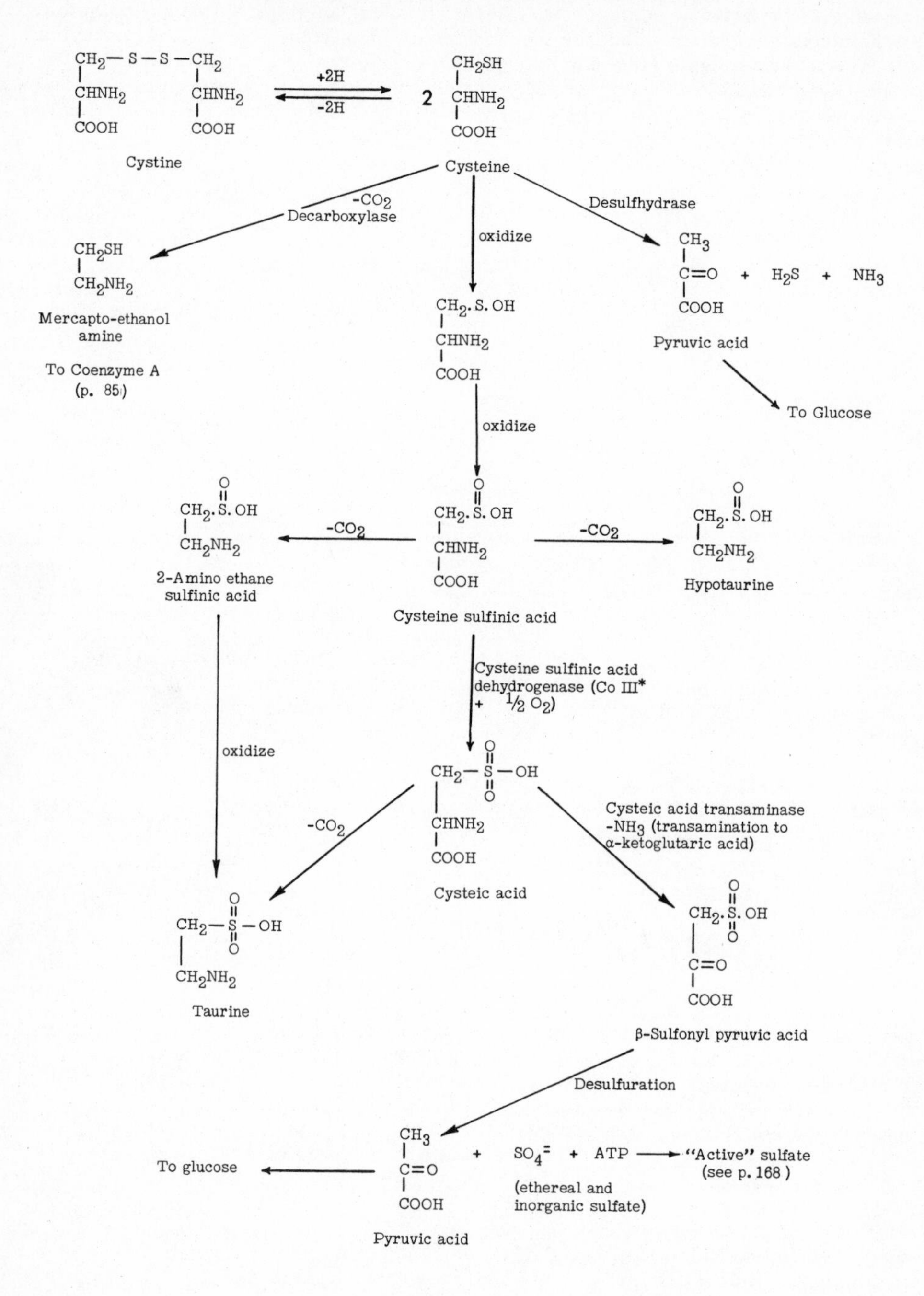

*The oxidation of cysteine sulfinic acid to cysteic acid is catalyzed by an enzyme, **cysteine sulfinic acid dehydrogenase,** which has been found in liver, kidney, and heart. This dehydrogenase requires as coenzyme a pyridine nucleotide similar in structure to DPN or TPN but not identical with these coenzymes (although they will react to a slight extent with the dehydrogenase). The new co-factor has been named Coenzyme III[39]. Co III has been found in every animal tissue examined, the highest concentration being in liver and kidney. It also occurs in microorganisms, extensive purification of the coenzyme from yeast having been achieved.

The D- isomer of cystine is not utilized for growth of animals. It is, however, slowly oxidized since it increases the urinary sulfate. Cystine is converted into glucose in the phlorhizinized dog.

Cystine is a particularly prominent amino acid in the protein of hair, hoofs, and the keratin of the skin. The S-S linkage may be important in the mechanical properties of the hair (see also p. 35).

H. Cystinuria: In this rare congenital metabolic disease excretion of cystine in the urine is greatly increased to 20 to 30 times normal. However, it has been shown that in cystinuric patients the excretion of lysine, arginine, and ornithine is also markedly increased. The content of all of the remaining amino acids in the urine is normal. Cystinuria is now considered to be due entirely to a renal defect. The greatly increased excretion of lysine, arginine, and ornithine as well as cystine in the urine of cystinuric patients suggests that there exists in these individuals a defect in the renal reabsorptive mechanisms for these four amino acids. It is possible that a single reabsorptive site is involved. Thus, as far as renal mechanisms are concerned cystinuria is not an uncomplicated defect which affects only cystine; the term "cystinuria" is therefore actually a misnomer.

Because cystine is a relatively insoluble amino acid, in cystinuric patients it may precipitate in the kidney tubules and form cystine calculi. This may be a major complication of the disease; were it not for this possibility, cystinuria would be an entirely benign anomaly and probably would escape recognition in many cases.

I. Cystinosis (Cystine Storage Disease): Cystinuria should be differentiated from cystinosis. In this disease, which is also congenital, cystine crystals are deposited in many tissues and organs (particularly the reticuloendothelial system) throughout the body. It is usually accompanied by a generalized aminoaciduria in which **all** amino acids are considerably increased in the urine[38]. Various other renal functions are also seriously impaired, and these patients usually die at an early age with all of the manifestations of acute renal failure. On the other hand, except for the likelihood of the formation of cystine calculi, cystinuria is compatible with a normal existence.

Phenylalanine and Tyrosine:

Phenylalanine is an essential amino acid, but tyrosine is not. Phenylalanine is readily converted to tyrosine, but the reaction is not reversible. Both amino acids participate in transamination reactions. Thus it is expected that their keto acids would support growth and that the D-isomer would also be utilized. However, D-tyrosine cannot be utilized by human subjects.

A. Abnormalities of Metabolism: Inability to metabolize completely phenylalanine and tyrosine occurs in the rare congenital diseases designated as **alcaptonuria** and **tyrosinosis**.

1. Alcaptonuria - In alcaptonuria, the urine darkens on standing exposed to air. This is due to the formation of an oxidized product of **homogentisic acid**, an intermediary in phenylalanine and tyrosine metabolism. The cartilage of the body may also darken in alcaptonuria; this is presumably due to deposition of oxidized homogentisic acid. The condition is known as **ochronosis**.

   A study of the enzymatic basis for the metabolic block in alcaptonuria has been made with hepatic tissue obtained by biopsy from an alcaptonuric patient who required surgical exploration of the abdomen for other reasons[40]. In this study it was found that the metabolic defect is limited to an essentially complete lack of the enzyme, homogentisate oxidase, the enzyme which converts homogentisic acid to maleyl-acetoacetic acid (see p. 256). It may therefore be assumed that alcaptonuria is another example of a heritable disease characterized by complete inability to synthesize an enzyme.

2. Tyrosinosis - In tyrosinosis, tyrosine and/or *p*-hydroxyphenylpyruvic acid, the corresponding keto acid, are excreted in considerable amounts. Homogentisic acid, on the other hand, can be completely utilized by the patient with tyrosinosis. Only one case of tyrosinosis has been reported.

3. Phenylketonuria - Another congenital anomaly of phenylalanine metabolism is phenylketonuria. In children exhibiting this metabolic defect, retarded mental development frequently also occurs. As a result, the condition has been designated phenylpyruvic oligophrenia.

   Phenylketonuric patients are unable to metabolize phenylalanine in a normal manner. Consequently, phenylalanine accumulates in the blood; serum levels as high as 15 to 63 mg. per 100 ml. have been reported. (The normal range of phenylalanine in plasma is 1 to 4 mg. per 100 ml.) In addition, considerable quantities of phenylalanine and its metabolites are excreted in the urine.

Deamination of phenylalanine produces the corresponding keto acid, phenylpyruvic acid. This metabolite usually appears in the urine of phenylketonuric patients in sufficient quantity to be readily detected by simple colorimetric tests, and it is by this test that the disease is usually first detected biochemically. Small amounts of phenylpyruvic acid also appear in the serum. The prominence of phenylpyruvic acid in the urine is the reason for designating the disease as phenylketonuria. However, the colorimetric tests may not detect phenylpyruvic acid in the urine when the serum phenylalanine levels fall below 15 mg. per 100 ml. In such circumstances the diagnosis would be missed unless serum phenylalanine levels were measured.

The other metabolites of phenylalanine which may appear in the urine of a phenylketonuric patient are shown in the table below. In man phenylacetic acid is mainly conjugated with glutamine in the liver to form phenylacetyl glutamine, so that this compound rather than free phenylacetic acid is found in the urine. It is this conjugate which is said to be responsible for the so-called "mousy" odor of the urine of phenylketonuric patients.

Metabolites of Phenylalanine in Plasma and Urine of Normal and Phenylketonuric Subjects

Metabolite	Plasma (mg./100 ml.)		Urine (mg./day)	
	Normal	Phenylketonuric	Normal	Phenylketonuric
Phenylalanine	1-2	15-63	30	300-1000
Phenylpyruvic acid		0.3-1.8		300-2000
Phenyllactic acid				290-550
Phenylacetic acid				Increased
Phenylacetyl glutamine			200-300	2400

It has recently been reported[41] that abnormal indole derivatives (see p. 257) are also excreted in phenylketonuria and that some phenolic acids which are present in normal urine in small amounts occur in phenylketonuric urine in greatly increased amounts. The most prominent of these is *o*-hydroxyphenylacetic acid, which is excreted in quantities of 100 to 400 mg. per Gm. of creatinine (normal is 1 mg. per Gm.). The amounts of hydroxyphenylacetic acid excreted in the urine are directly related to the levels of phenylalanine in the blood.

The most specific biochemical diagnostic test for phenylketonuria is measurement of serum phenylalanine. This is also the most useful means of assessing the progress of therapy with low phenylalanine diets, as mentioned below.

It would be expected that the administration of phenylalanine to a phenylketonuric subject would result in prolonged elevation of the level of this amino acid in the blood, i.e., diminished tolerance to phenylalanine. However, it has been found that an abnormally low tolerance to injected phenylalanine and a high fasting level of phenylalanine are also characteristic of the parents of the phenylketonuric individual[42 a,b]. Evidently the recessive gene responsible for phenylketonuria can be detected biochemically in the phenotypically normal parents.

The conversion of phenylalanine to tyrosine as described below and illustrated on p. 254 involves hydroxylation of phenylalanine at the para position. The reaction requires molecular oxygen and reduced TPN; it is presumably catalyzed by an enzyme of the nature of the oxygenases described on p. 114. This enzyme is termed phenylalanine hydroxylase. It occurs in the liver, where it catalyzes the following reaction:

$$C_6H_5-CH_2.CHNH_2COOH + TPN.H + H^+ \xrightarrow{+O_2} HO-C_6H_4-CH_2.CHNH_2.COOH + TPN + H_2O$$

Phenylalanine — Tyrosine

The biochemical defect in phenylketonuria has been proved to be due to a virtually complete absence of the phenylalanine hydroxylase in liver. This has been determined by in vitro studies with liver tissue obtained from phenylketonuric patients during surgical exploration of the abdomen for other purposes[43 a,b]. Although other co-factors are also involved in the system by which phenylalanine is hydroxylated, none of these are missing in phenylketonuric liver tissue[44].

A biochemical basis for the profound impairment of mental function which occurs in phenylketonurics has not yet been discovered. Meister[45] has suggested that phenylalanine in excess may act as an antimetabolite to tyrosine. In this connection it is of interest that defects in skin and hair pigmentation are not infrequent occurrences in phenylketonurics. This might be a reflection of an interference with the metabolism of some derivatives of tyrosine in the pathway to formation of melanin.

The mental performance of phenylketonuric children can be improved as long as they can be maintained on a diet very low in phenylalanine. Concurrently, the blood phenylalanine levels return to normal and there is a reduction in the excretion of phenylalanine and its metabolites. The provision of such diets is unfortunately very difficult and impractical for long periods.

B. **Metabolism of Phenylalanine and Tyrosine:** Studies on the abnormalities in the metabolism of phenylalanine and tyrosine which have been described above have served to suggest the intermediate pathways for the catabolism of these amino acids in the normal individual. Phenylalanine is converted very early in its metabolism to phenolic derivatives in which a hydroxy group is added to the aromatic ring, either in the para- or in the ortho- position. The para-derivative is normally favored, so that tyrosine (*p*-hydroxyphenylalanine) is the major product of the oxidation of phenylalanine. However, the finding of small amounts of ortho-hydroxy derivatives in the urine of normal individuals and of very large amounts in phenylketonuria suggested to Armstrong et al.[46] that both types of tyrosine could be formed in normal subjects. The phenylketonuric patient, however, was unable to produce the para-derivative, which shows that the basic defect in phenylketonuria is the lack of an enzyme for the para- oxidation of phenylalanine. Since this major pathway for the utilization of phenylalanine is blocked, phenylalanine and some of its metabolites accumulate and more phenylalanine is diverted to the ortho-hydroxy pathway, normally a minor route. Phenylketonuric patients do not exhibit impaired ability to metabolize *o*-hydroxyphenyl compounds; the excess amounts of these compounds in the urine are therefore attributable to greater production because of inability to produce para-hydroxy derivatives.

Both phenylalanine and tyrosine are ketogenic as well as glycogenic when added to liver slices. The production of acetoacetic acid and fumaric acid by the oxidation of homogentisic acid (which is itself derived from tyrosine) has been studied by Edwards and Knox[47]. The first product of homogentisate oxidation is maleyl-acetoacetate, produced by the action of a liver enzyme, homogentisate oxidase. This enzyme is completely lacking in alcaptonuric patients. The reaction is aerobic, and the enzyme requires SH groups and ferrous iron as well as ascorbic acid to maintain the iron in the reduced state[48]

A cis-trans isomerization of maleyl-acetoacetate to form fumaryl acetoacetate then occurs. This is catalyzed by maleyl-acetoacetate isomerase, an enzyme which has a specific requirement for reduced glutathione as a coenzyme. Finally, fumaryl acetoacetate is split into fumaric and acetoacetic acids by the action of fumaryl acetoacetate hydrolase. These reactions are shown on p. 256.

The metabolism of phenylalanine and tyrosine prior to the formation of homogentisic acid is shown on p. 254. It will be noted that tyrosine is also a direct precursor of the adrenal medullary hormones, norepinephrine (arterenol) and epinephrine (adrenalin) as well as of the thyroid hormones, which are iodotyrosines. The decarboxylation of tyrosine produces tyramine, a vasopressor substance.

Melanin, the pigment of the skin and hair, is derived from tyrosine by way of 3, 4-dihydroxyphenylalanine (DOPA) and its oxidation product, 3, 4-dioxyphenylalanine (dopaquinone), which progresses to further melanin precursors.

The phenols which occur in the blood and urine are derived from tyrosine. In the urine, the phenols are largely conjugated with sulfate (see p. 303), and this comprises a portion of the so-called ethereal sulfate fraction of the total urinary sulfur. Tyrosine itself is also excreted in the urine, not only in the free state but also as a sulfate in which the sulfate moiety is conjugated through the para-hydroxy group. The excretion of tyrosine-*O*-sulfate averaged 28 mg. per day in the five adult males studied by Tallan et al.[49]. This accounted for about one-half of the bound tyrosine and 3 to 8% of the ethereal sulfate sulfur in the urine.

METABOLISM OF PHENYLALANINE AND TYROSINE

The metabolism of phenylalanine and tyrosine and the points at which metabolic impairment occurs in phenylketonuria, tyrosinosis, and alcaptonuria are summarized below.

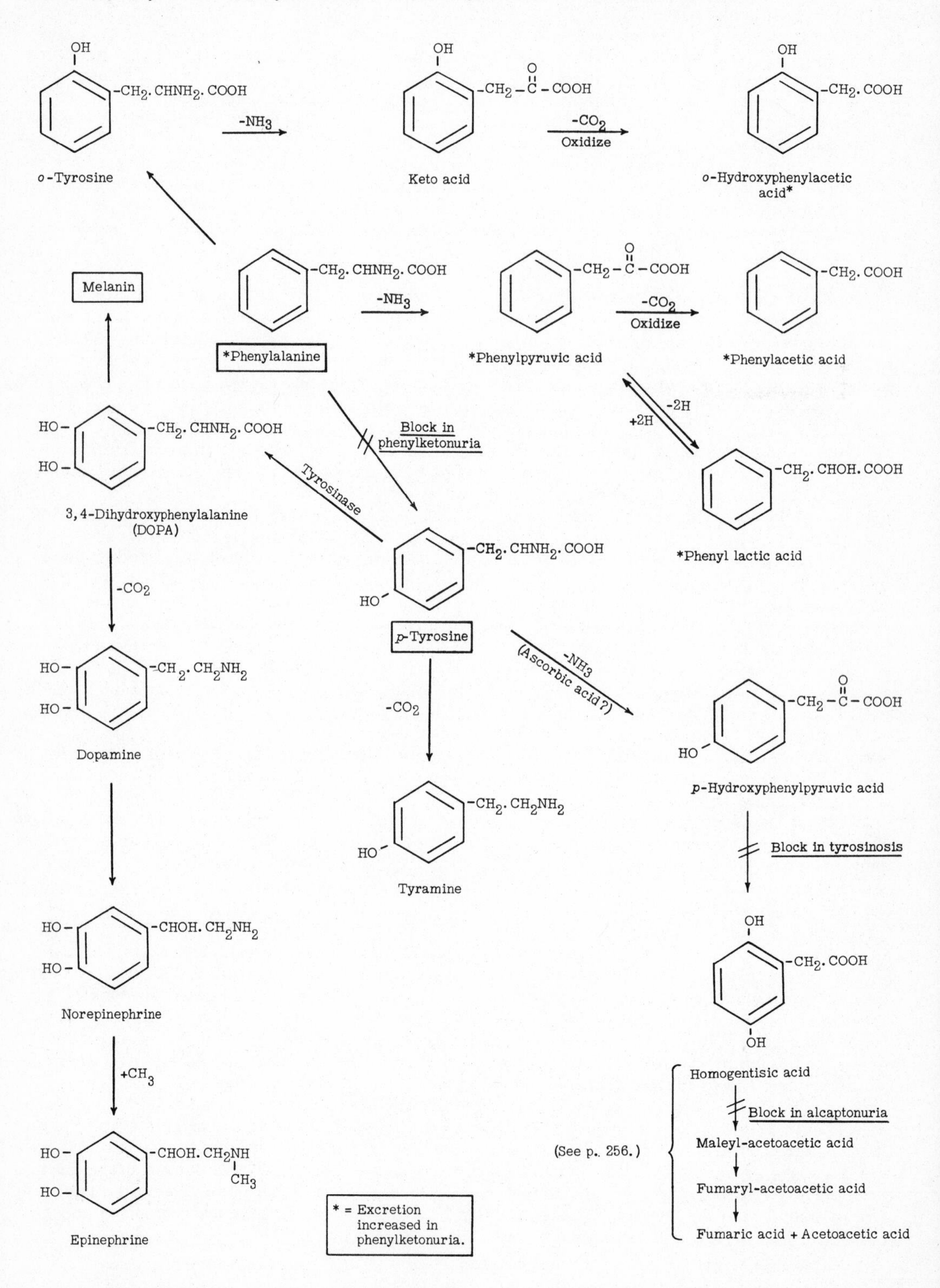

METABOLISM OF TRYPTOPHAN

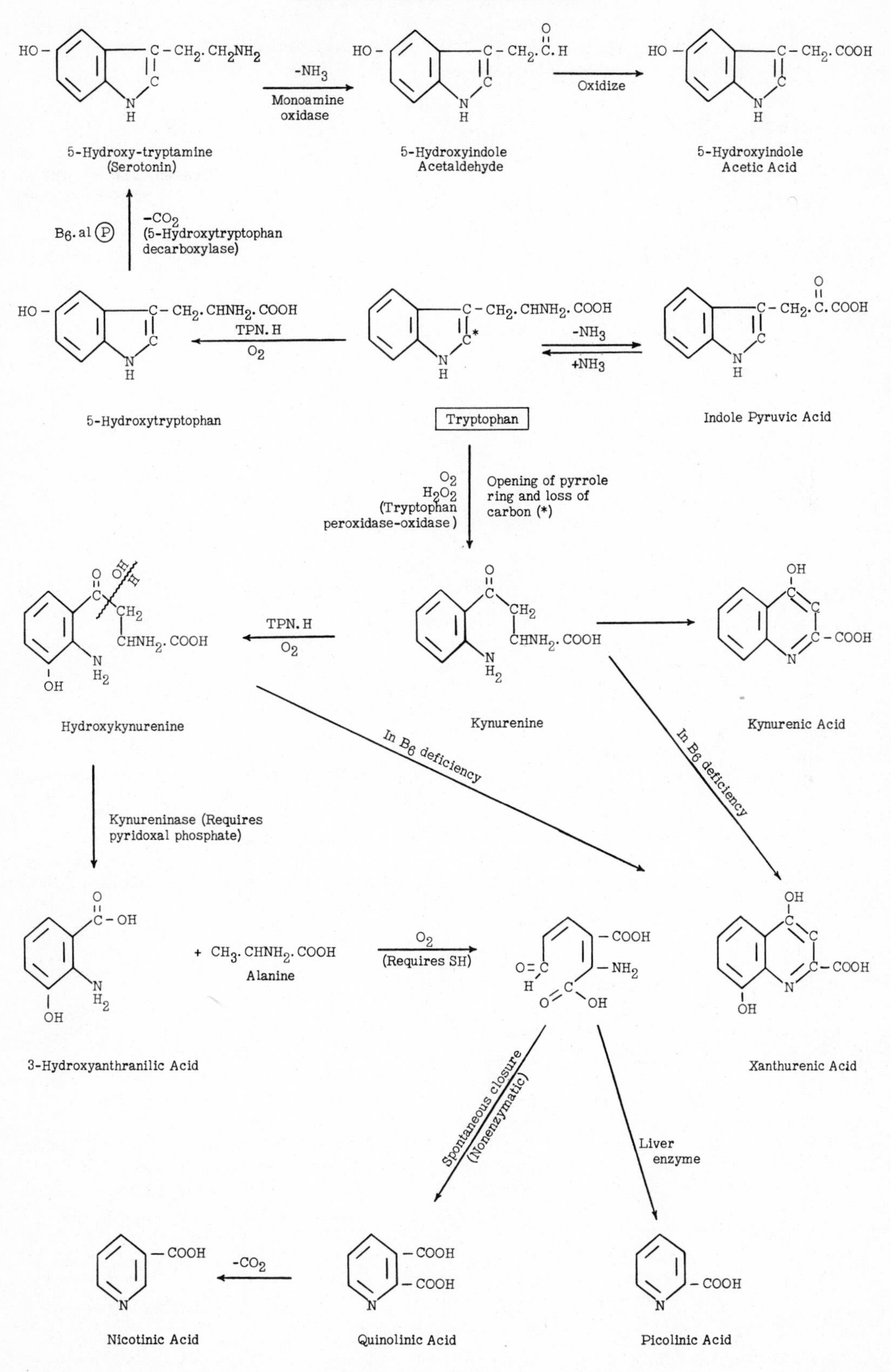

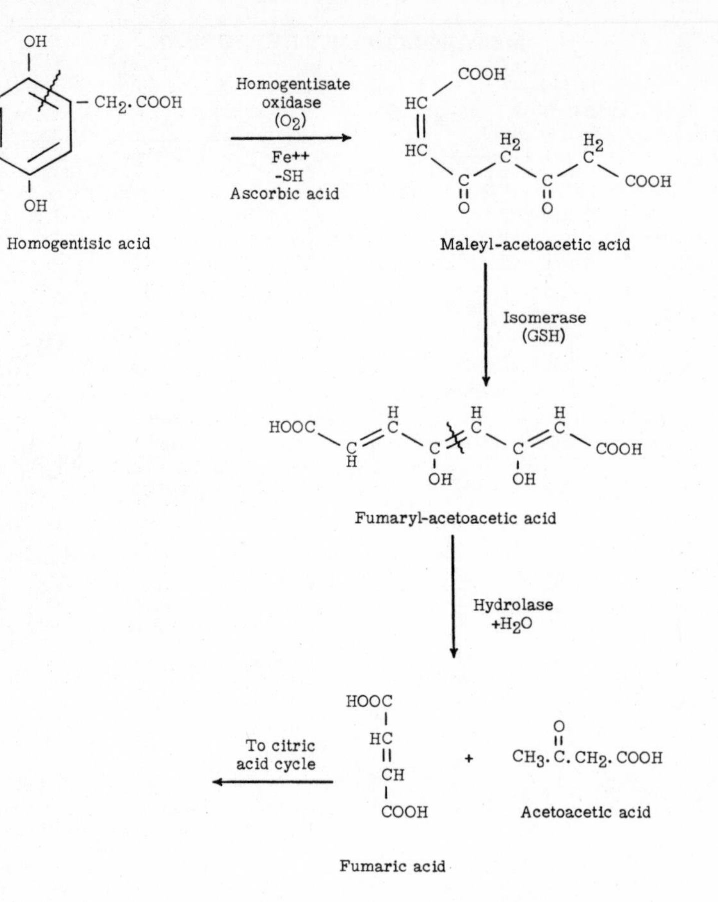

Vitamins Involved in Tyrosine Metabolism:

Ascorbic acid and folic acid are both involved in tyrosine metabolism. Both vitamins prevent the defect in tyrosine oxidation observed in guinea pigs maintained on diets deficient in these substances. Alcaptonuria is observed not only in scorbutic guinea pigs but also in premature infants deprived of vitamin C. When the vitamin is supplied the alcaptonuria promptly disappears. Vitamin C is not effective, however, in alcaptonuria of congenital origin. A direct association of ascorbic acid with tyrosine oxidation at the level of formation of *p*-hydroxyphenylpyruvic acid as well as in the oxidation of homogentisic acid has been demonstrated.

Tryptophan:

Tryptophan is the only amino acid containing an indole nucleus. It is an essential amino acid but can be transaminated so that the keto acid, indole pyruvic acid, can replace tryptophan in a tryptophan-deficient diet. The D- isomer will also support growth, but there is evidence that it is not metabolized in the same manner as the L- isomer.

In dogs and rabbits, it has long been known that **kynurenic acid**, a quinoline derivative, is derived from tryptophan. This compound can be identified in the urine of these animals. A precursor of kynurenic acid is **kynurenine**; this has also been found in the urine and is produced when tryptophan is added to liver slices. Kynurenine is formed from tryptophan by opening of the pyrrole ring of indole to form N-formyl kynurenine, which then loses the formyl carbon immediately adjacent to the indole nitrogen (the starred carbon on p. 255). Kynurenine is a major metabolite formed in the breakdown of tryptophan. It may be deaminated, the amino group being transaminated to ketoglutaric acid, and the resulting keto derivative, *o*-aminobenzoyl pyruvic acid, loses water and then undergoes spontaneous ring closure to form kynurenic acid[50]. However, kynurenic acid is actually a by-product of kynurenine; as shown on p. 255, it is not in the main pathway of tryptophan degradation.

The further metabolism of kynurenine involves its conversion to hydroxykynurenine and this, in turn, goes to 3-hydroxyanthranilic acid. The hydroxylation occurs with molecular oxygen in a TPN.H-catalyzed reaction similar to that in hydroxylation of phenylalanine to tyrosine (see p. 254). The reaction by which kynurenine is converted to hydroxyanthranilic acid is catalyzed by an enzyme, **kynureninase**, which requires vitamin B_6 (pyridoxal phosphate) as coenzyme. A deficiency of vitamin B_6 results in failure to convert kynurenine. This compound then reaches various extrahepatic tissues, where it is converted to **xanthurenic acid**. This abnormal metabolite has been identified in the urine of man, monkeys, and rats in the presence of a B_6 deficiency. If a B_6 deficiency exists, the excretion of xanthurenic acid is increased when extra tryptophan is fed. The kidney is one organ which has been shown to produce xanthurenic acid derivatives from kynurenine.

The metabolism of 3-hydroxyanthranilic acid is shown on p. 255. It is oxidized to an intermediate compound (1-amino-4-formyl-1, 3-butadiene-1, 2-dicarboxylic acid) which involves opening of the ring structure of anthranilic acid between C-3 and C-4. Spontaneous closure then occurs, with formation of quinolinic acid. However, Mehler has shown[51] that in the liver there is an enzyme which can also catalyze the formation of picolinic acid from the intermediate compound, and this isomer of nicotinic acid must therefore be considered another end product of tryptophan metabolism in liver. In the formation of quinolinic or picolinic acid it will be noted that the nitrogen which was originally a part of the indole nucleus of tryptophan has become the nitrogen of a pyridine ring. Tracer studies with labeled indole nitrogen prove this to be so.

It is likely that the formation of quinolinic acid is quantitatively the more important pathway for anthranilic acid metabolism. Decarboxylation of quinolinic acid produces nicotinic acid as the end product of this pathway of tryptophan metabolism. All of the above reactions for the metabolism of tryptophan are shown on p. 255.

In many animals, the conversion of tryptophan to nicotinic acid makes a supply of the vitamin in the diet unnecessary. In the rat, rabbit, dog, and pig, tryptophan can completely replace the vitamin in the diet; in man and other animals tryptophan increases the urinary excretion of nicotinic acid derivatives (e.g., N-methylnicotinamide). In vitamin B_6 deficiency it has been noted that the synthesis of pyridine nucleotides (DPN and TPN) in the tissues may be impaired. This is a result of the inadequate conversion of tryptophan to nicotinic acid which occurs in B_6 deficiency. Vitamin B_6 has no direct effect on the utilization of nicotinic acid for nucleotide synthesis; if an adequate supplement of nicotinic acid is supplied, nucleotide synthesis proceeds normally even in the presence of the B_6 deficiency.

It is likely that in many diets, tryptophan normally provides a considerable amount of the nicotinic acid requirement. In man, approximately 60 mg. of tryptophan produces 1 mg. of nicotinic acid. Nutritional deficiency states such as pellagra must therefore be considered combined protein (tryptophan) as well as vitamin (nicotinic acid) deficiencies.

Indole Derivatives in Urine:

In addition to the excretion of various products derived from the metabolism of phenylalanine as described on p. 252, patients with phenylketonuria also excrete increased quantities of indoleacetic and indolelactic acids as well as 5-hydroxyindoleacetic acid and traces of many other indole acids. Hartnup disease[52] is a hereditary abnormality in the metabolism of tryptophan, characterized by a pellagra-like skin rash, intermittent cerebellar ataxia, and mental deterioration. The urine of patients with Hartnup disease contains greatly increased amounts of indole acetic acid, α-N(indole-3-acetyl)glutamine, as well as tryptophan. Armstrong et al.[53] have studied the indole acids of human urine by paper chromatography. A total of 38 different indole acids were chromatographed. The most strikingly "abnormal" patterns of indole acid excretion were found in the urine of severely mentally retarded patients and in urine from the mentally ill. These authors questioned the significance of these findings insofar as the causes of mental disease were concerned, particularly in view of the fact that the urinary excretion patterns tended to revert to normal after administration of broad-spectrum antibiotics.

Serotonin:

Another pathway for the metabolism of tryptophan involves its hydroxylation to 5-hydroxytryptophan. The hydroxylation step is probably carried out by a system similar to that involved in formation of hydroxykynurenine. There are several examples among natural substances of formation of 5-hydroxyindoles[54]. The oxidation of tryptophan to the hydroxy derivative is analogous to the conversion of phenylalanine to tyrosine (see p. 254).

Decarboxylation of 5-hydroxytryptophan produces 5-hydroxytryptamine. This compound, also known as serotonin, enteramin, or thrombocytin (see p. 126), is a potent vasoconstrictor and stimulator of smooth muscle. In these systemic effects it is probably equal in importance to epinephrine, norepinephrine, and histamine as one of the regulatory amines of the body[55]. However, serotonin has also a potent effect in the metabolism of the brain. The serotonin produced in the rest of the body does not pass the blood-brain barrier. Therefore, serotonin must be produced within the brain itself from precursors which do gain access to the brain. While its functions in brain are not yet entirely clear, it seems reasonable to assume that an excess of serotonin brings about stimulation of cerebral activity and that a deficiency produces a depressant effect.

Serotonin is metabolized by oxidative deamination to form 5-hydroxyindoleacetic acid (5-HIAA). The enzyme which catalyzes this reaction is a monoamine oxidase. A number of inhibitors of this enzyme have been found. Among them is iproniazid (Marsilid®). It is hypothesized that the psychic stimulation which follows the administration of this drug is attributable to its ability to prolong the stimulating action of serotonin through inhibition of monoamine oxidase.

There is evidence that serotonin when first produced in the brain exists in a bound form which is not susceptible to the action of monoamine oxidase. The depressant drugs such as reserpine may effect a rapid release of the bound serotonin, thus subjecting it to rapid destruction by monoamine oxidase. The resultant depletion of serotonin would then bring about the calming effect which follows administration of reserpine.

The preparation and properties of the 5-hydroxytryptophan decarboxylase which forms serotonin from hydroxytryptophan have been described[56]. The enzyme is highly specific. It has been obtained from hog and guinea pig kidney, and these tissues will form serotonin from hydroxytryptophan. The enzyme also occurs in liver and stomach, and the gastric mucosa is high in its content of tryptamine. The blood platelets also contain a considerable amount of serotonin (see p. 126), but it is believed that it is merely concentrated there since the platelets do not contain the decarboxylase, indicating that they do not manufacture the hydroxytryptamine.

The further metabolism of serotonin by deamination and oxidation results in the production of 5-hydroxyindoleacetic acid (5-HIAA), and this end product is excreted in the urine. In normal human urine, 2 to 8 mg. of 5-HIAA are excreted per day, which indicates that the 5-hydroxyindole route is a significant pathway for the metabolism of tryptophan. Other metabolites of serotonin have been identified in the urine of patients with carcinoid[57]. These include 5-hydroxyindoleaceturic acid (the glycine conjugate of 5-hydroxyindoleacetic acid), N-acetyl serotonin, conjugated with glucuronic acid, some unchanged serotonin, and very small amounts of oxidation products of the nature of indican.

Greatly increased production of serotonin occurs in malignant carcinoid (argentaffinoma), a disease characterized by the widespread development of serotonin-producing tumor cells in the argentaffin tissue throughout the abdominal cavity. Patients exhibit cutaneous vasomotor episodes (flushing) and occasionally a cyanotic appearance. There may also be a chronic diarrhea. These symptoms are attributed to the effects of serotonin on the smooth muscle of the blood vessels and digestive tract. In over half of the patients observed there is also respiratory distress with bronchospasm. Cardiac involvement may occur late in the disease[58]. The serotonin in the blood of carcinoid patients, all of which occurs in the platelets, is 0.5 to 2.7 mcg. per ml. (normal is 0.1 to 0.3 mcg. per ml.). The most useful biochemical indication of increased production of serotonin, such as may occur in metastasizing carcinoid tumors, is the measurement of the urinary hydroxyindoleacetic acid. In the carcinoid patient, excretion of 5-HIAA has been reported as 76 to 580 mg. in 24 hours (normal is 2 to 8 mg.). Several assay methods for 5-HIAA have been described[59-62]. For diagnostic purposes, it is recommended that a quantitative measurement of the 5-hydroxyindoleacetic excretion be made on a urine specimen collected over a 24-hour period. Random specimens of urine are useful for qualitative screening tests, but confirmation of the diagnosis should be made only on the 24 hour specimen. It has been reported[63] that bananas have a relatively high content of 5-hydroxy indoles. The ingestion by normal adults of 12 Gm. of banana per Kg. over a 24-hour period produced an average doubling of the excretion of 5-hydroxy indoles in the urine. The ingestion of bananas could therefore cause erroneous diagnosis of carcinoid if the urinary findings only were considered.

From a biochemical point of view, carcinoid has been considered to be an example of an abnormality in tryptophan metabolism in which a much greater proportion of tryptophan than normal is metabolized by way of the hydroxy indole pathway. One per cent of tryptophan is normally converted to serotonin, but in the carcinoid patient as much as 60% may follow this pathway. This metabolic diversion markedly reduces the production of nicotinic acid; consequently, symptoms of pellagra as well as negative nitrogen balance may occur.

Anthranilic Acid:

Altman and Miller[64] have described a so-called "inborn error of metabolism" characterized by a failure to metabolize anthranilic acid. The metabolic defect is associated with a hypoplastic anemia, which is normocytic and normochromic in type and totally unresponsive to treatment with liver extract, iron, folic acid, or vitamin B_{12}. There is no reduction in leukocytes or platelets; the hematologic defect seems to be one affecting erythrogenesis, with a failure to produce mature reticulocytes or erythrocytes. The syndrome was first described by Diamond and Blackfan[65] and by Cathie[66], but the association with the excretion of anthranilic acid in the urine was made by Altman and Miller, who found anthranilic acid in eight cases of the disease. Anthranilic acid is not present in the urine of normal individuals, but in the patients investigated the administration of 1.6 Gm. of L-tryptophan increased the excretion of anthranilic acid. This metabolite has also been found in the urine of rats subsisting on riboflavin-deficient diets, but riboflavin administration (100 to 200 mg. per day) to the patients, although it caused a reduction in anthranilic acid output, did not improve their hematologic status.

THE ORIGIN OF CREATINE AND CREATININE

Creatine is present in muscle, brain, and blood, both phosphorylated as phosphocreatine and in the free state (see p. 175). Traces of creatine are normally present in the urine. **Creatinine** is the anhydride of creatine. It is found in blood and urine. The conversion of creatine to creatinine is not reversible in the body. It is apparently a preliminary to the excretion of most of the creatine.

The 24-hour excretion of creatinine in the urine of a given subject is remarkably constant from day to day. The creatinine coefficient is the 24-hour urinary creatinine expressed in terms of body size. When expressed in this manner, the creatinine excretion of different individuals of the same age and sex is also quite constant.

The origin of creatine shown in the reactions below has been established by metabolic studies and confirmed by isotope technics. Three amino acids - glycine, arginine, and methionine - are directly involved. The first reaction is between glycine and arginine to form glycocyamine (guanidoacetic acid). This has been shown by **in vitro** experiments to occur in the kidney but not in the liver or in heart muscle. The synthesis of creatine is completed by the methylation of glycocyamine in the liver. In this reaction, methionine is the methyl donor. Other methyl donors, such as betaine or choline after oxidation to betaine, may also serve indirectly by producing methionine through the methylation of homocysteine (see p. 248). The methylation of glycocyamine is not reversible. Neither creatine nor creatinine can methylate homocysteine to methionine. ATP and oxygen are required in the methylation of creatine.

The enzymatic mechanisms for the synthesis of creatine have been studied by Cantoni and Vignos[67]. They were found to be similar to those required for the formation of N-methyl nicotinamide (see p. 81). The first step is the formation of active methionine (S-adenosyl methionine; see p. 247), which requires ATP, magnesium ions, and glutathione (GSH), and a methionine-activating enzyme. The second step involves the methylation of guanidoacetic acid (glycocyamine) by active methionine, a reaction which is catalyzed by a soluble enzyme, **guanidoacetate methylpherase**, found in cell-free extracts of guinea pig, rabbit, beef, and pig liver. Glutathione or other reducing substances are required for the optimal activity of this enzyme; there is as yet no evidence for the need of metal ions or other cofactors.

Synthesis of Creatine and Creatinine

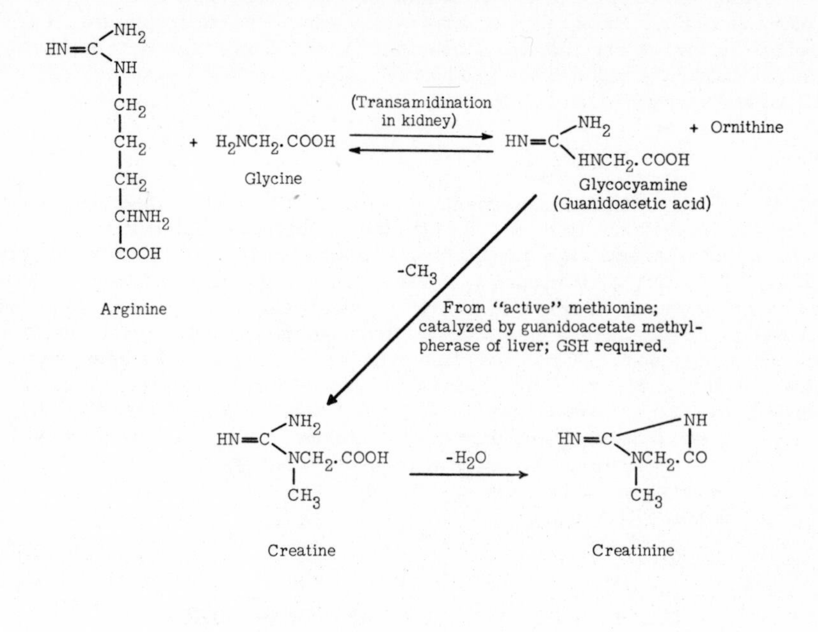

• • •

References:

1. Snoke, J. E., Yanari, S., and Bloch, K.: J. Biol. Chem. **201**:573, 1953.
2. Borsook, H.: J. Cellular and Comparative Physiology **47**, Suppl. 1, 1956.
3. a. Luck, J. M., Griffin, A. C., Boer, G., and Wilson, M.: J. Biol. Chem. **206**:767, 1954.
   b. Griffin, A. C., Luck, J. M., Kulakoff, V., and Mills, M.: J. Biol. Chem. **209**:387, 1954.
4. Harper, H. A., Hutchin, M. E., and Kimmel, J. R.: Proc. Soc. Exper. Biol. and Med. **80**: 768, 1952.
5. Braunstein, A. E., and Kritzmann, M. G.: Enzymologia **2**:129, 1937-38.
6. Karmen, A., Wroblewski, F., and LaDue, J. S.: J. Clin. Invest. **34**:126, 1955.
7. Meister, A., and Tice, S. V.: J. Biol. Chem. **187**:173, 1950.
8. Meister, A., Sober, H. A., Tice, S. V., and Fraser, P. E.: J. Biol. Chem. **197**:319, 1952.
9. Meister, A.: J. Biol. Chem. **206**:587, 1954.
10. Meister, A., and Downey, P. F.: Proc. Soc. Exper. Biol. and Med. **91**:49, 1956.
11. Jones, M. E., Spector, L., and Lipmann, F.: J. Am. Chem. Soc. **77**:819, 1955.
12. Lowenstein, J. M., and Cohen, P. P.: J. Biol. Chem. **220**:57, 1956.
13. Ratner, S., and Petrack, B.: J. Biol. Chem. **200**:161, 1952.
14. Ratner, S., Petrack, B., and Rochovansky, O.: J. Biol. Chem. **204**:95, 1953.
15. Ratner, S., Anslow, W. P., Jr., and Petrack, B.: J. Biol. Chem. **204**:115. 1953.
16. Najarian, J. S., and Harper, H. A.: Proc. Soc. Exper. Biol. and Med. **92**:560, 1956.
17. Najarian, J. S., and Harper, H. A.: Am. J. Med. **21**:832, 1956.
18. Soloway, S., and Stetten, D. W., Jr.: J. Biol. Chem. **204**:207, 1953.
19. Mackenzie, C. G., and Abeles, R. H.: J. Biol. Chem. **222**:145, 1956.
20. Nakada, H. L., and Weinhouse, S.: Arch. Biochem. and Biophys. **42**:257, 1953.
21. Shemin, D., Russel, C. S., and Abramsky, T.: J. Biol. Chem. **215**:613, 1955.
22. Sallach, H. J.: J. Biol. Chem. **223**:1101, 1956.
23. Pihl, A., and Fritzson, P.: J. Biol. Chem. **215**:345, 1955.
24. Rose, W. C., Eades, C. H., Jr., and Coon, M. J.: J. Biol. Chem. **216**:225, 1955.
25. Bachhawat, B. K., Robinson, W. G., and Coon, M. J.: J. Biol. Chem. **216**:727, 1955.
26. Dekker, E. E., Schlesinger, M. J., and Coon, M. J.: J. Biol. Chem. **233**:437, 1958.
27. Robinson, W. G., Bachhawat, B. K., and Coon, M. J.: J. Biol. Chem. **218**:391, 1956.
28. Meister, A.: J. Biol. Chem. **206**:587, 1954.
29. a. Rothstein, M., and Miller, L. L.: J. Biol. Chem. **211**:851, 1954.
    b. Rothstein, M., and Miller, L. L.: Fed. Proc. **13**:285, 1954.
30. Meister, A.: J. Biol. Chem. **206**:587, 1954.

31. Tallan, H.H., Moore, S., and Stein, W.H.: J.Biol.Chem. **219**:257, 1956.
32. a. Page, E.W., Glendening, M.B., Dignam, W., and Harper, H.A.: Am.J.Obst. and Gyn. **68**:110, 1954.
    b. Page, E.W., Glendening, M.B., Dignam, W., and Harper, H.A.: Am.J.Obst. and Gyn. **70**:766, 1955.
33. a. Melville, D.B., Otken, C.C., and Kovalenko, V.: J.Biol.Chem. **216**:325, 1955.
    b. Melville, D.B., and Eich, S.: J.Biol.Chem. **218**:647, 1956.
34. Stein, W.H.: J.Biol.Chem. **201**:45, 1953.
35. Cantoni, G.L.: J.Biol.Chem. **204**:403, 1953.
36. Berg, P.: J.Biol.Chem. **205**:145, 1953.
37. Matsuo, Y., and Greenberg, D.M.: J.Biol.Chem. **230**:545, 1958.
38. Harper, H.A., Grossman, M., Henderson, P., and Steinbach, H.: Am.J.Dis.Child. **84**: 327, 1952.
39. Singer, T.P., and Kearney, E.B.: Biochim. et Biophys. Acta **8**:700, 1952.
40. La Du, B.N., Zannoni, V.G., Laster, L., and Seegmiller, I.E.: J.Biol.Chem. **230**:251, 1958.
41. Armstrong, M.D., Shaw, K.N.F., and Robinson, K.S.: J.Biol.Chem. **213**:797, 1955.
42. a. Hsia, D.Y., Driscoll, K.W., Troll, W., and Knox, W.E.: Nature **178**:1239, 1956.
    b. Knox, W.E., and Messinger, E.C.: Am.J.Human Genetics **10**:53, 1958.
43. a. Moldave, K., and Meister, A.: Proc.Soc.Exp.Biol. and Med. **94**:632, 1957.
    b. Mitoma, C., Auld, R.M., and Udenfriend, S.: Ibid. p. 634.
44. Kaufman, S.: Science **128**:1506, 1958.
45. Meister, A.: Pediatrics **21**:102, 1958.
46. Armstrong, M.D., and Shaw, K.N.F.: J.Biol.Chem. **213**:805, 1955.
47. Edwards, S.W., and Knox, W.E.: J.Biol.Chem. **220**:79, 1956.
48. Schepartz, B.: J.Biol.Chem. **205**:185, 1953.
49. Tallan, H.H., Bella, S.T., Stein, W.H., and Moore, S.: J.Biol.Chem. **217**:703, 1955.
50. Miller, I.L., and Adelberg, E.A.: J.Biol.Chem. **205**:691, 1953.
51. Mehler, A.H.: J.Biol.Chem. **218**:241, 1956.
52. Baron, D.M., Dent, C.E., Harris, H., Hart, E.W., and Jepson, J.B.: Lancet **2**:421, 1956.
53. Armstrong, M.D., Shaw, K.N.F., Gortatowski, J., and Singer, H.: J.Biol.Chem. **232**:17, 1958.
54. Udenfriend, S., Titus, E., Weissbach, H., and Peterson, R.E.: J.Biol.Chem. **219**:335, 1956.
55. Page, I.: Physiol.Rev. **34**:563, 1954.
56. Clark, C.T., Weissbach, H., and Udenfriend, S.: J.Biol.Chem. **210**:139, 1954.
57. McIsaac, W.M., and Page, I.H.: Science **128**:537, 1958.
58. Sjoerdsma, A., Weissbach, H., and Udenfriend, S.: Am.J.Med. **20**:520, 1956.
59. Udenfriend, S., Titus, E., and Weissbach, H.: J.Biol.Chem. **216**:499, 1955.
60. Curzon, G.: Lancet **269**:1361, 1955.
61. Hanson, A., and Serin, F.: Lancet **269**:1359, 1955.
62. Sjoerdsma, A., Weissbach, H., and Udenfriend, S.: J.Am.Med.Assoc. **159**:397, 1955.
63. Puente-Duany, G.A., Riemer, W.E., and Miale, J.B.: Proc.Soc.Exp.Biol. and Med. **98**:499, 1958.
64. Altman, K.I., and Miller, G.: Nature **172**:868, 1953.
65. Diamond, L.K., and Blackfan, K.D.: Am.J.Dis.Child. **56**:464, 1938.
66. Cathie, I.A.B.: Arch.Dis.Child. **25**:313, 1950.
67. Cantoni, G., and Vignos, P.J., Jr.: J.Biol.Chem. **209**:647, 1954.

Bibliography:

Greenberg, D.M., Ed.: Chemical Pathways of Metabolism. 2 Vols. Academic Press, 1954.

McElroy, W.D., and Glass, H.B., Eds.: Symposium on Amino Acid Metabolism. Johns Hopkins Press, 1955.

Annual Review of Biochemistry. Annual Reviews, Inc.

Meister, A.: Biochemistry of the Amino Acids. Academic Press, 1957.

16...

Metabolism of Nucleic Acids and Their Derivatives

The chemistry of the nucleic acids derived from nucleoproteins, as well as of the purine and pyrimidines contained in the nucleic acids has been described in Chapter 4 (see p. 43). A summary of these derivatives is given in the table on p. 263. In this chapter, the metabolism of these compounds will be discussed.

Digestion:

The pancreatic juice contains enzymes (nucleinases) which degrade nucleic acids into nucleotides. These include ribonuclease and deoxyribonuclease; each acts on the type of nucleic acid for which it is specific. In the intestinal juices there are enzymes which supplement the action of the pancreatic nucleinases in producing mononucleotides from nucleic acids. These intestinal enzymes are called polynucleotidases, or, more recently, phosphodiesterases. Certain intestinal phosphatases (specifically termed mononucleotidases) may then remove phosphate from the mononucleotides to produce nucleosides.

The final step in this degradation is an attack on the nucleosides by the action of the nucleosidases of the intestinal secretions. One nucleosidase reacts with purine nucleosides liberating adenine and guanine, and another enzyme causes the breakdown of the pyrimidine nucleosides liberating uracil, cytosine, or thymine. The action of these nucleosidases is peculiar; the process actually involves a transfer of phosphate so that the end products are the free purine or pyrimidine and a phosphorylated pentose.

$$\underset{\text{(Nucleoside)}}{\text{Ribose-purine}} + \text{Phosphate} \xrightarrow[\text{(Nucleoside phosphorylase)}]{\text{Nucleosidase}} \text{Ribose-phosphate} + \text{Purine}$$

Fate of the Absorbed Products:

A. Free Purines and Pyrimidines: The metabolic fate of purine and pyrimidine derivatives has been studied by the oral or parenteral administration of isotopically labeled (N^{15}) compounds. When labeled guanine was fed to rats or pigeons, the tagged nitrogen appeared in large amounts in the urinary allantoin of the rat or the uric acid of the pigeon but did not appear to any appreciable extent in the tissue nucleoproteins. This indicates that ingested guanine is largely catabolized after absorption. A similar fate was indicated for uracil, cytosine, and thymine after the oral administration of the free purines or pyrimidines to rats. On the other hand, the oral administration of labeled adenine resulted in its incorporation into the tissue nucleoprotein and some of the labeled nitrogen was also found in guanine in the body. These experiments suggest that dietary adenine is a precursor of guanine. However, all of the guanine is probably not derived from adenine. After the feeding of labeled glycine, more of the isotope appeared in the guanine than in the adenine, indicating that all of the guanine had not been produced through adenine. In summary, it appears from these experiments that, with the exception of adenine, none of the free purines or pyrimidines of the diet serves as a direct precursor of the purines or pyrimidines of the tissue nucleic acids. It is also of interest that the naturally produced breakdown products of the nucleic acids in the tissues are not re-utilized to any extent as is demonstrated by the fact that C^{14} labeled adenine incorporated into nucleoprotein of one test animal does not appear in that of another connected in a parabiotic experiment.

B. Nucleosides and Nucleotides: Somewhat different results were obtained when the purines or pyrimidines were administered not as the free bases but as nucleosides or nucleotides. Thus the subcutaneous injection of the labeled pyrimidine nucleoside, cytidine, into rats resulted in the incorporation of this compound, as well as the labeled nucleoside, uridine, into the ribonucleic acids (RNA), and, to a smaller extent, into the cytosine and thymine of the deoxyribonucleic acids (DNA). There was even some incorporation of the label into the purines of the DNA although not those of the RNA. This is noteworthy since it indicates that pyrimidine nitrogen has been transferred to purines. The injection of uridine resulted in considerably less incorporation of the N^{15} label, whereas injected thymidine remained unaltered. Apparently demethylation of thymidine did not take place since no labeled cytosine was detected.

The utilization of **purine** nucleotides when given by mouth to rats was, however, not as effective as that of free adenine. Intraperitoneal injection of these nucleotides resulted in a somewhat improved utilization for the synthesis of tissue nucleic acids, but it was still not equivalent to that observed after the administration of adenine itself.

The Purine and Pyrimidine Bases and Their Related Nucleosides and Nucleotides

Base	Nucleoside (Base + Sugar)	Nucleotide (Base + Sugar + Phosphoric Acid)	Source
PURINES			
Adenine (6-aminopurine)	Adenosine	Adenylic acid	From RNA and DNA
Guanine (2-amino-6-oxypurine)	Guanosine	Guanylic acid	From RNA and DNA
Hypoxanthine (6-oxypurine)	Inosine (hypoxanthine riboside), hypoxanthine deoxyriboside	Inosinic acid (hypoxanthine ribotide), hypoxanthine deoxyribotide	From adenine by oxidative deamination
Xanthine (2, 6-dioxypurine)	Xanthine riboside (or deoxyriboside)	Xanthine ribotide (or deoxyribotide)	From guanine by oxidative deamination
PYRIMIDINES			
Cytosine (2-oxy-6-amino-purine)	Cytidine	Cytidylic acid	From RNA and DNA
Thymine (2, 6-dioxy-5-methylpyrimidine)	Thymidine	Thymidylic acid	From DNA
Uracil (2, 6-dioxypyrimidine)	Uridine	Uridylic acid	From RNA

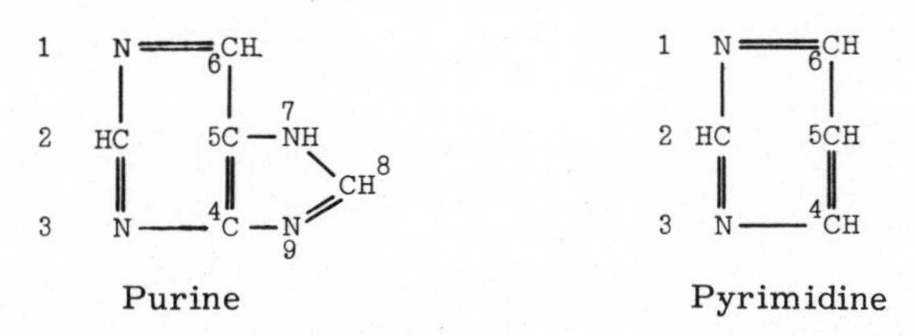

SYNTHESIS OF PURINES AND PYRIMIDINES IN THE BODY

Preformed purines or pyrimidines are not required in the diet since the animal organism is able to synthesize these compounds. In fact, from the evidence cited above, it is likely that the tissue nucleic acids are very largely derived from synthetic or endogenous sources rather than from preformed, exogenous sources of the diet. The reactions by which these compounds are synthesized have been studied in both plant and animal tissues. Such studies have yielded considerable evidence as to possible pathways for the synthesis of the purine and pyrimidine derivatives, although it must be remembered that the evidence is derived from various experimental preparations, notably microorganisms. Therefore it does not necessarily represent the established pathways in any one species.

Biosynthesis of Purines:

Information on the sources of the various atoms of the purine nucleus has been obtained by tracer studies in birds, rats, and man. The amino acid glycine is utilized in the intact form to form the carbons in positions 4 and 5 while its α-nitrogen forms position 7. The nitrogen at position 1 is derived from the amino nitrogen of aspartic acid; those at positions 3 and 9, from the amide nitrogen of glutamine. The carbon atom in position 6 is derived from respiratory carbon dioxide while the carbons in positions 2 and 8 come from a one-carbon compound such as formate or from the β carbon of serine, which is itself derived from a one-carbon (formate) moiety, when glycine is converted to serine (see p. 90). The δ-carbon of δ-aminolevulinic acid, which is itself derived from the α-carbon of glycine, may actually serve as a carrier molecule for the transfer of the α-carbons of glycine to the purine ring (see p. 235 note). A similar role has been proposed for homocysteine (see p. 248).

Tetrahydro folic acid (FH_4) derivatives, acting as formyl carriers, are required for the incorporation of the one-carbon units into positions 2 and 8 (see p. 90).

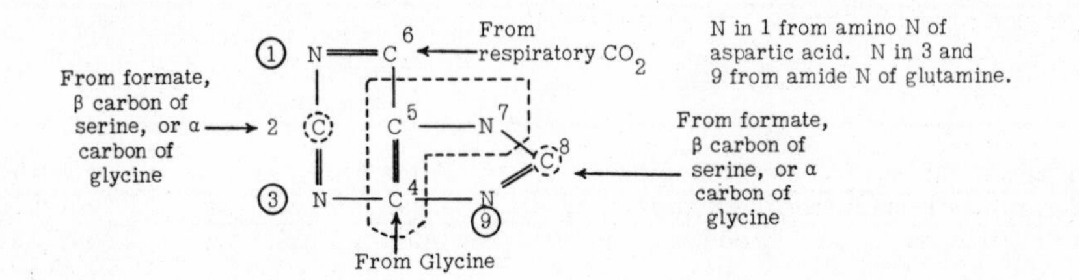

Diagram Illustrating Sources of Carbon and Nitrogen in the Purine Nucleus

The biosynthetic pathway for the synthesis of purines is shown on p. 266[1]. The initial step includes the formation of a nucleotide structure with glycine.* For this purpose, 5-phosphoribosyl-pyrophosphate (PRPP) serves as phosphate donor. PRPP is frequently used in nucleotide synthesis, e.g., in the formation of DPN. This sugar-phosphate is formed in the liver[3] according to the following reaction:

$$\text{Ribose-5-phosphate} + \text{ATP} \xrightarrow{Mg^{++}} \underset{\text{(Adenylic acid, p. 47)}}{\text{Adenosine-5-P}} + \underset{\text{(PRPP)}}{\text{5-phosphoribosylpyrophosphate}}$$

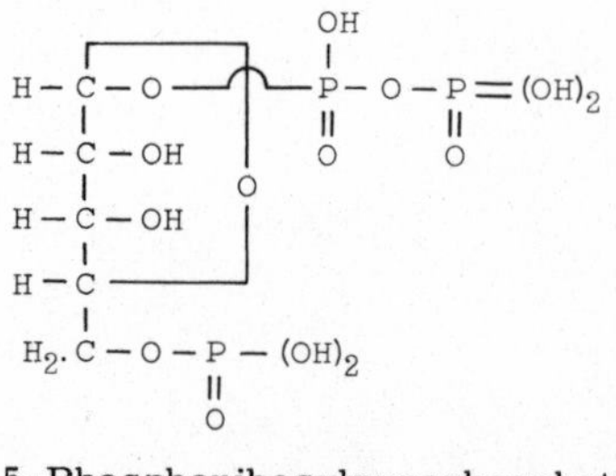

5-Phosphoribosylpyrophosphate
(PRPP)

PRPP then reacts (reaction 1) with glutamine in a reaction catalyzed by the enzyme PRPP-amidotransferase to form 5-phosphoribosylamine, which (reaction 2) reacts in turn with glycine, resulting in the production of glycinamide-ribosyl-phosphate (GAR). This compound is the source of positions 4, 5, 7, and 9 of the purine nucleus. The enzyme catalyzing reaction 2 is designated GAR-kinosynthase.

---

*There is considerable evidence that the ribosides or ribotides are biologically much more active than the free bases. This has previously been noted (see p. 262) in connection with the administration of pyrimidines. None is utilized for the synthesis of nucleic acids if fed as the free base. In the synthesis of purines and pyrimidines all of the intermediate compounds are also first converted into ribotides or deoxyribotides.

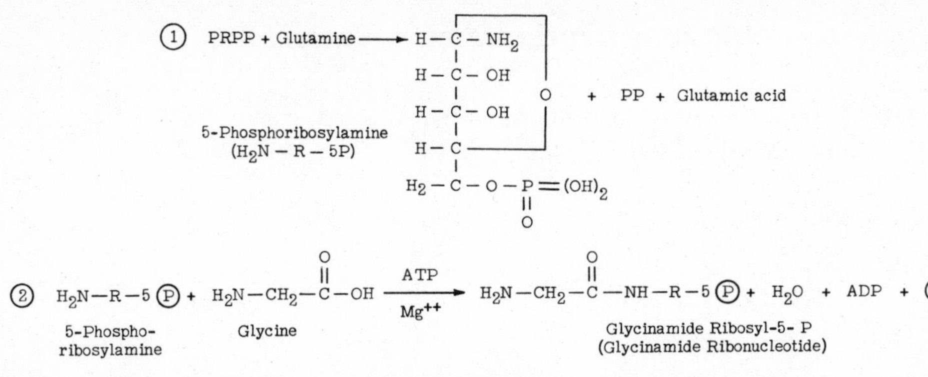

Glycinamide-ribosyl-phosphate is then formylated (reaction 3), in a reaction which requires FH_4 and the enzyme transformylase, to transfer the one-carbon moiety which will become position 8 of the purine nucleus. In reaction 4, with glutamine as the amino donor, amination occurs at carbon 4 of the formylated glycinamide. The added nitrogen will be position 3 in the purine. Ring closure (reaction 5) forms an amino imidazole which progresses (reaction 6) to 5-amino-4-imidazole-N-succinyl carboxamide ribotide. This is formed by addition of a carbamyl group to the precursor compound. The source of the carbon is respiratory CO_2, and the source of the nitrogen is the amino nitrogen of aspartic acid. The utilization of CO_2 as in other reactions of CO_2-fixation, apparently requires biotin; in fact, the precursor substance, amino-imidazole-ribosyl phosphate, has been found to accumulate in biotin-deficient animals. Next (reaction 7), fumaric acid is split off from the amino-imidazole-succinyl carboxamide and 5-amino-4-imidazole carboxamide ribotide remains. This latter compound is then formylated (reaction 8) to form 5-formamido-4-imidazole carboxamide ribotide. The formyl carbon is transferred from a tetrahydro folic acid derivative catalyzed by a transformylase. This newly-added carbon, which, like carbon 8 of the purine nucleus, is derived from the one-carbon pool, will become carbon 2 of the purine nucleus. Ring closure now occurs (reaction 9), and the first purine to be synthesized is formed.

It is clear from the above scheme that hypoxanthine nucleotide (inosinic acid) is apparently the first purine to be synthesized and that adenine (reaction 10) and guanine nucleotides (reactions 11 and 12) are then derived from it by amination from the nitrogen pool. Glutamine and aspartic acid serve as nitrogen donors in these reactions.

The biosynthesis of guanosine nucleotide has been studied by Lagerkvist[2]. Using preparations of pigeon liver, he found that xanthine nucleotide (xanthosine) is first formed from inosine nucleotide (reaction 11). Guanosine is then produced by amination of xanthosine at position 2, using the amide nitrogen of glutamine as nitrogen donor (reaction 12).

Several antimetabolites are effective at various points in purine biosynthesis, as follows: Azaserine blocks reactions 1, 3, and 12. Deoxynorleucine (DON) blocks reaction 3. Folic acid antagonists (e.g., amethopterin) block reaction 8. Purinethol® (6-mercaptopurine) is presumed to block reaction 10.

Biosynthesis of Pyrimidines:

The pathway for the biosynthesis of pyrimidines is shown on p. 267[3]. It may be considered to begin with the formation of carbamyl aspartic acid (ureidosuccinic acid). Carbamyl aspartic acid is also involved in the urea cycle; its synthesis is described on p. 232. By ring closure, dihydroorotic acid is produced which is then oxidized to orotic acid. The nucleotide structure orotidine-5-phosphate is next formed by a reaction with PRPP (see p. 264). Decarboxylation of orotidine phosphate produces the primary pyrimidine, uridine-5-phosphate (uridylic acid). This compound may be aminated to form cytidylic acid (cytidine phosphate) or methylated to produce thymidylic acid (thymidine phosphate).

The carbon required in the methylation of uridylic acid to form thymidylic acid is derived either from formate, the β-carbon of serine, which is a major source, or the α-carbon of glycine.

THE BIOSYNTHETIC PATHWAY FOR PURINES

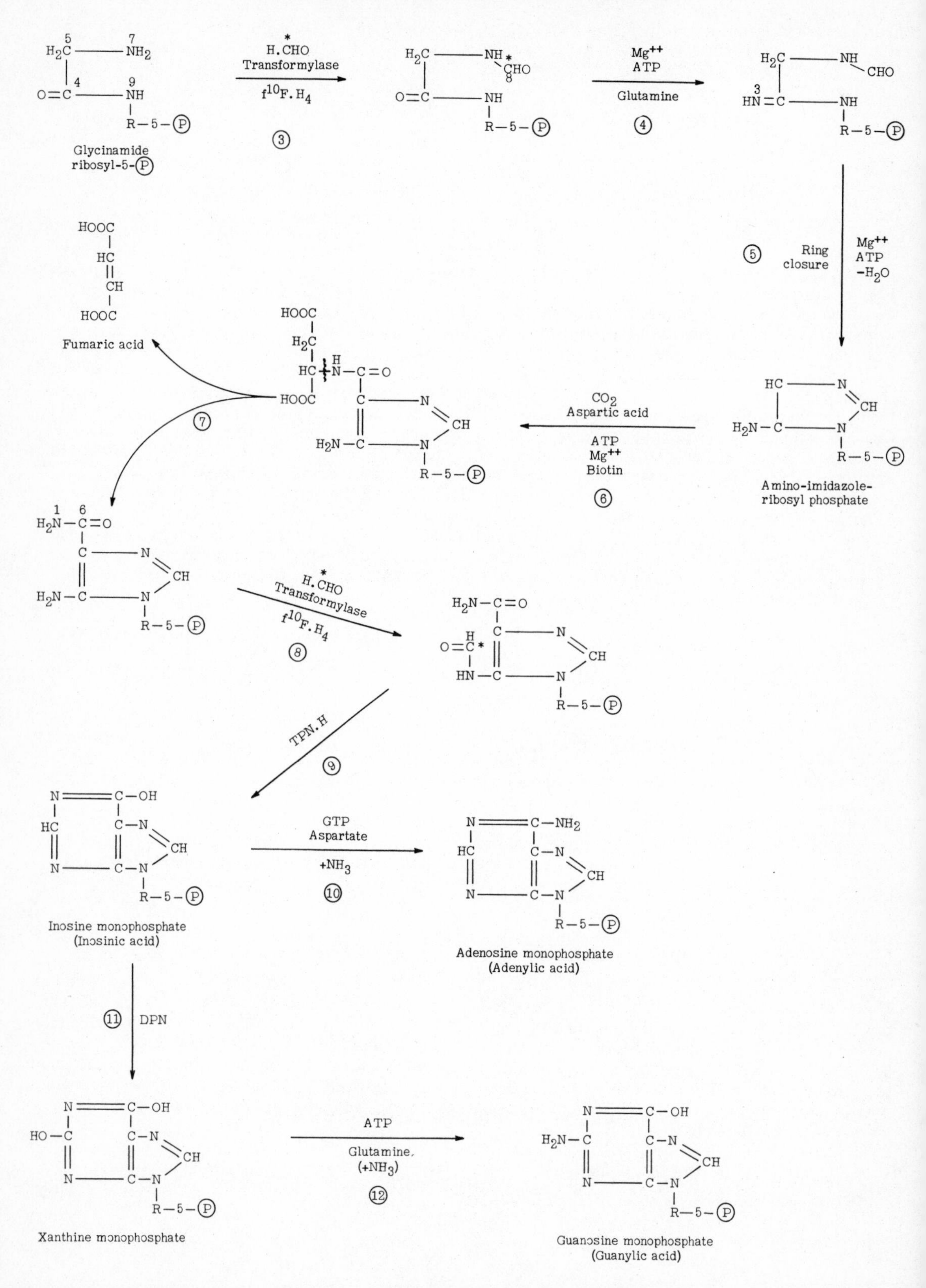

Tetrahydro folic acid (FH_4) derivatives are involved in the transfer of these carbon atoms, and the methylation reaction itself requires vitamin B_{12}. These vitamins, and especially vitamin B_{12}, have a marked effect not only on the synthesis of the purine and pyrimidine nucleus but possibly also on the formation of their nucleosides and nucleotides, which, as has been pointed out, is probably a necessary preliminary step in their synthesis.

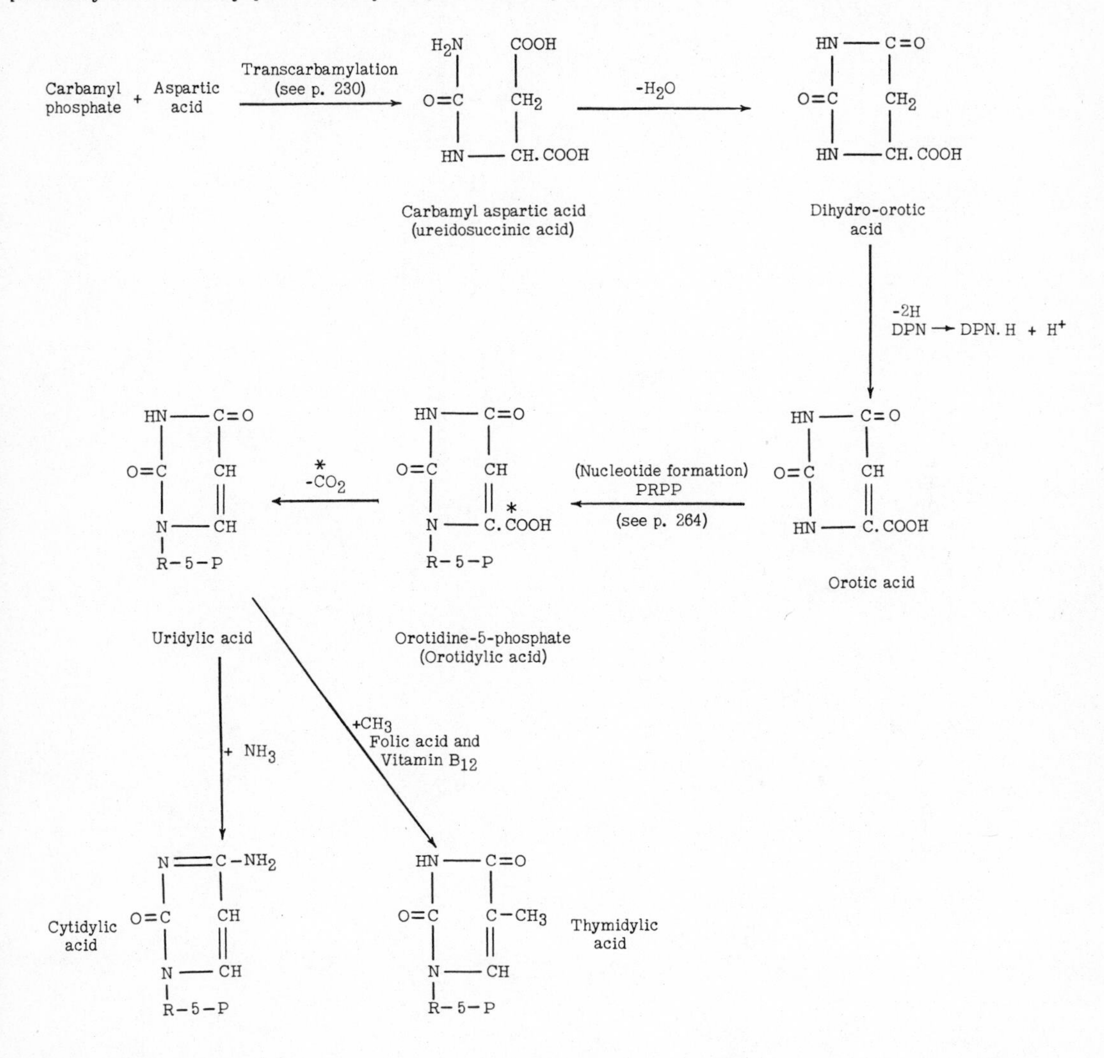

Synthesis of Pentoses:

The pentoses, ribose and deoxyribose, required for the synthesis of nucleosides and nucleotides are presumably readily manufactured from glucose. The HMP shunt suggested as an alternate pathway of glycolysis (see p. 186) would provide a method for the synthesis of these sugars. It is probable that the pentoses are also formed from the combination of three-carbon and two-carbon intermediates. They are also formed by decarboxylation of uronic acids (see p. 193).

CATABOLISM OF PURINES AND PYRIMIDINES

Various tissues, particularly the liver, contain enzyme systems similar to those described in the intestine for the digestion of nucleic acids.

CATABOLISM OF PYRIMIDINES

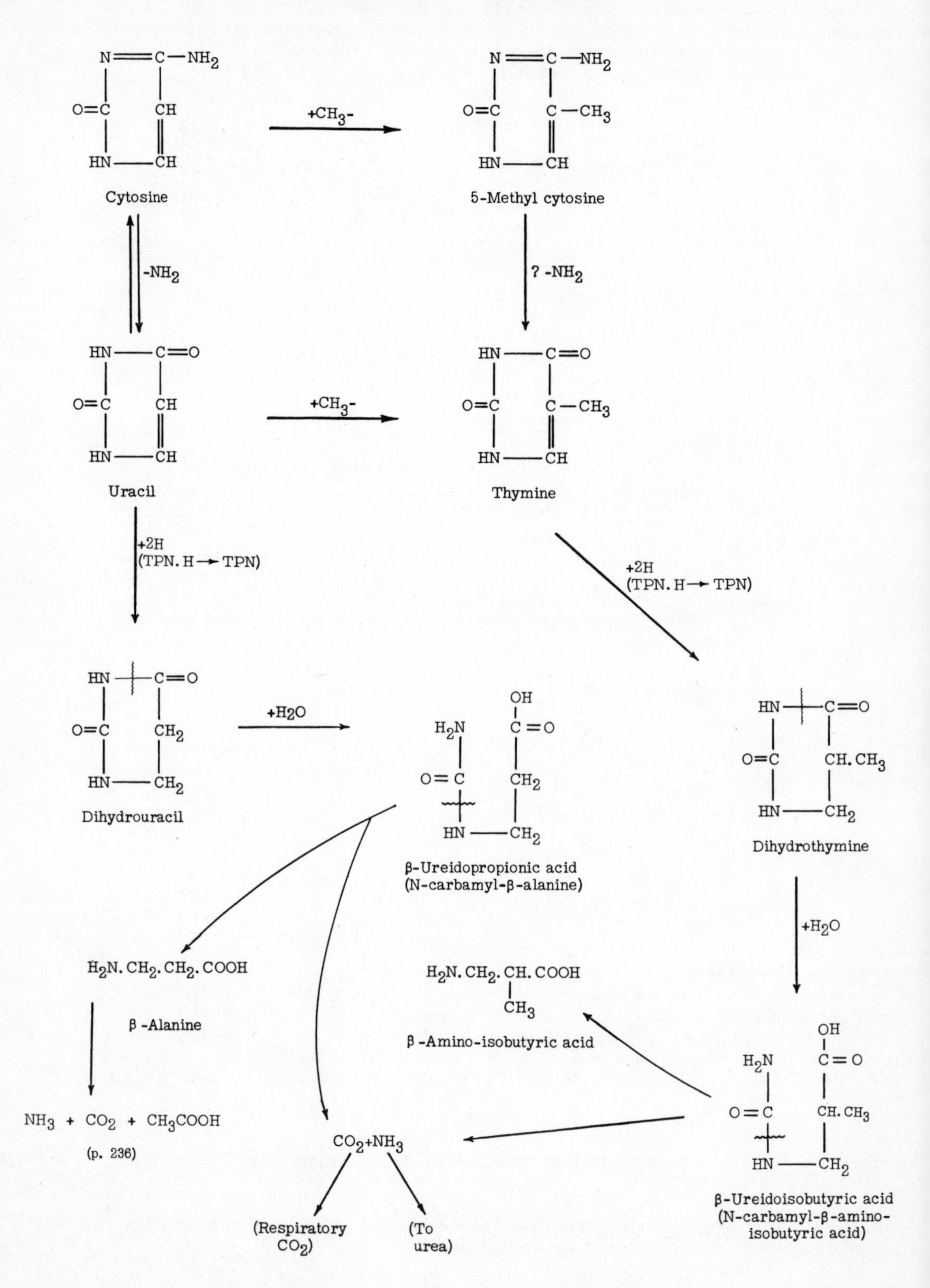

Purine Catabolism:

The further metabolism in the tissues of the purine nucleosides may proceed according to the following scheme:

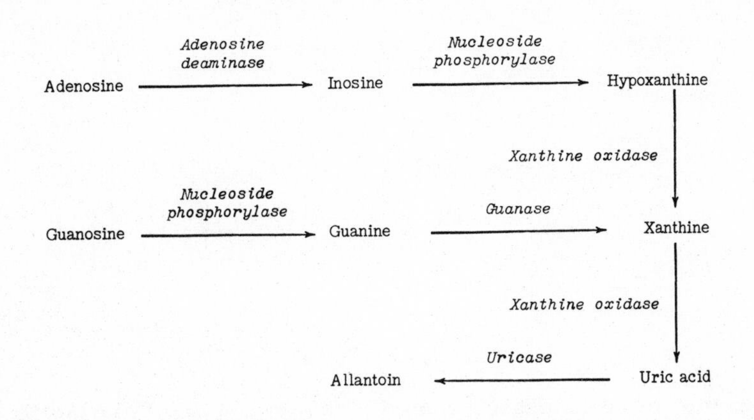

Adenase is an enzyme corresponding in its action to guanase. These enzymes deaminate the free amino purines. Adenase is present in only very low concentrations in animal tissues, whereas an adenosine deaminase as well as the other enzymes listed above are more generally distributed. It is therefore likely that adenine is deaminated while still in the nucleoside form, as shown above.

Xanthine oxidase, the enzyme which oxidizes hypoxanthine and xanthine, has been isolated from liver and from milk. Recent studies on the activity of this enzyme indicate that the enzyme complex contains riboflavin as a prosthetic group and that trace amounts of iron and molybdenum are also part of the enzyme molecule. Both of these minerals are required in the diet to provide for the deposition and maintenance of normal levels of xanthine oxidase in rat liver and intestine.

Abnormalities of xanthine metabolism have been reported in several patients who excreted only minute quantities of uric acid, although the amounts of xanthine excreted were about equal to the amount of uric acid expected in a normal individual. The disease was considered to be a congenital metabolic defect possibly caused by a lack of xanthine oxidase. The extremely low levels of blood uric acid are believed to be due to the rapid glomerular filtration of xanthine together with a failure of reabsorption by the renal tubules, so that little or no xanthine remains to form uric acid. In the xanthinuric patient, xanthine stones may be formed in the kidney. The xanthine calculus is not opaque to x-ray; hence this condition might be suspected in any patient presenting symptoms of urinary calculi without x-ray evidence of their presence.

Several purine derivatives in addition to uric acid have been identified in the urine of normal subjects. On a purine-free diet, a total of more than 30 mg. per day of purine derivatives other than uric acid are excreted. In order of diminishing abundance they are: hypoxanthine, xanthine, 7-methylguanine, adenine, 1-methylguanine, and guanine.

Pyrimidine Catabolism:

The catabolism of pyrimidines occurs mainly in the liver. The release of the ureido carbon (carbon No. 2) of the pyrimidine nucleus as respiratory carbon dioxide represents a major pathway for the catabolism of uracil. Diets rich in thymine or in desoxyribonucleic acid (DNA) have been found to evoke in rats increased excretion of **β-amino isobutyric acid**. Dihydrothymine is even more effective as a precursor of this amino acid than is thymine. The feeding of ribonucleic acid (RNA), which contains no thymine, does not induce the excretion of the amino acid. The hypothetical pathways for the breakdown of the pyrimidines, based on the fragmentary evidence given above, are shown on p. 268. β-Alanine and β-amino isobutyric acids, according to this scheme, are the major end-products of cytosine-uracil or thymine catabolism, respectively.

Uric Acid Metabolism:[4]

In man, the end product of purine metabolism is mainly uric acid, but in subprimate mammals the uric acid is further oxidized to **allantoin** by the action of **uricase**. Allantoin is, therefore, the principal end-product of purine metabolism in such animals.*

In birds and reptiles uric acid is synthesized, and this corresponds in its function to urea in man since it is the principal end-product of nitrogen metabolism. Such animals are said to be **uricotelic**. Man is **ureotelic**.

The metabolism of uric acid in man has been studied by the use of isotopically labeled uric acid (N^{15} in carbons 1 and 3). Single doses of the labeled uric acid were injected intravenously into a normal human subject and into patients suffering from gout, a disease which is characterized by a disturbance in purine metabolism. The dilution of the injected labeled compound was used to calculate the quantity of uric acid which is present in the body water. This quantity, the so-called "miscible pool," contained in the normal subjects an average of 1131 mg. of uric acid. The plasma was found to be considerably higher in uric acid than other portions of the body water. In gouty subjects, the miscible pool was much larger; for example, 4742 mg. in a mild case having a serum uric acid of 6.9 mg. per 100 ml. and no symptoms of the disease; up to 31,000 mg. in a patient with severe symptoms.

From the rate at which N^{15} declined in the uric acid, it was possible to estimate the "turnover" of this compound, i.e., the rate at which uric acid is synthesized and lost to the body. The uric acid formed in the normal subject was 500 to 580 mg. per day. It was also noted that the quantities of uric acid entering the miscible pool exceeded those which were lost by the urinary route by 100 to 250 mg. per day. This suggested that about 20% of the uric acid lost to the body is not excreted as such but is chemically broken down. This conclusion is supported by the finding of N^{15} in urea and ammonia. It is known that some uric acid is excreted in the bile. This would then be subject to degradation by intestinal bacteria, and the liberated N^{15} would be absorbed and reappear in various nitrogenous metabolites. More recent experiments indicate that the breakdown of uric acid in man is, however, independent of the bacteria of the intestinal tract. When administered intravenously to a normal human subject, 18% of labeled uric acid was degraded to other nitrogenous products and 6% of the label was recovered in the feces over a period of two weeks. However, 78% of the uric acid injected was recovered unchanged from the urine. The experiment was repeated in the same subject, but sulfonamides were given by mouth to induce a reduction in the activity of the intestinal bacteria. The results were the same as had been obtained in the original experiment. It was concluded that some uricolysis does occur in normal human beings and that the intestinal flora does not make a major contribution to the process.

Excretion of Uric Acid:

Uric acid in the plasma is filtered by the glomeruli but is later partially reabsorbed by the renal tubules. Glycine is believed to compete with uric acid for tubular reabsorption. Certain uricosuric drugs block reabsorption of uric acid, e.g., salicylates, cincophen, neocincophen, caronamide (4'-carboxyphenylmethanesulfonanilide), probenecid [Benemid®, *p*-(di-N-propylsulfamyl)-benzoic acid], and zoxazolamine (Flexin®). The urinary excretion of urate by human

*In the Dalmatian dog more uric acid is excreted in comparison to body size than in man, although the plasma levels are only 10 to 20% as high. This is not due to a deficiency of uricase but is apparently a renal phenomenon caused by a failure in reabsorption of uric acid by the kidney tubules. In fact, there is some evidence for tubular secretion as well as filtration of uric acid in this animal.

subjects can also be increased by the administration of hormones of the adrenal cortex (the 11-oxysteroids) as well as by corticotropin (ACTH).

Uric acid is very slightly soluble, so that in acid urines it tends to precipitate on standing. This factor may also increase the tendency to form renal calculi in the gouty patient. The urates, alkali salts of uric acid, are much more soluble.

Gout:

Gout is characterized by elevated levels of uric acid in blood and urine. Deposits of urates (tophi) in the joints are also common, so that the disease in this form is actually a type of arthritis.

It has been noted above that the miscible pool of uric acid in the gouty patient is much larger than in the normal subject. This excess is largely stored in the tophi. The uric acid in the body fluids is in some degree equilibrated with that in the tophi since the miscible pool can be decreased by the administration of uricosuric drugs even though these changes may not be reflected in the serum uric acid levels.

The administration of N^{15}-labeled glycine has been used to study the rate of synthesis of uric acid in the gouty patient. In some patients there was evidence not only for an increased rate of incorporation of dietary glycine nitrogen into uric acid but also of a cumulative incorporation of N^{15} into uric acid which was three times greater than in normal subjects. It was suggested, because of the rapidity and extent of the isotope incorporation, that in some gouty subjects there may be a metabolic shunt whereby glycine nitrogen can enter the purine nucleus of uric acid more promptly than in normal individuals[5]. A diet rich in protein would be expected to increase the formation of uric acid, and it has in fact been found that uric acid synthesis from glycine is accelerated both in normal and in gouty individuals when the protein content of the diet is increased.

Elevated blood levels of uric acid may occur in other diseases in which there is abnormally great turnover of nucleic acids. Examples are polycythemia, myeloid metaplasia, chronic leukemia, and other hematopoietic diseases. Occasionally in such cases there may be attacks of gout. Such cases are referred to as secondary gout. This form of gout is thought to be due to increased catabolism of nucleic acids and a consequent flooding of the organism with the products of nucleic acid breakdown. The etiology of secondary gout is therefore different from that of primary gout discussed above, which is presumably due to an inherited metabolic error.

An elevation in the levels of uric acid in the blood (hyperuricemia) which is not secondary to increased destruction of nucleic acids is the most consistent biochemical criterion of primary gout. Hyperuricemia may be the only manifestation of the genetic trait for gout, occurring in as many as 25% of a group of asymptomatic individuals in families of gouty patients. In men with the genetic trait for gout, hyperuricemia occurring after puberty may be the first sign of the disease. There may never be any further evidence of the abnormality, or gouty arthritis may develop in middle age. However, women carrying the trait usually do not develop hyperuricemia until after the menopause, and clinical gout, if it occurs at all, does not appear until some years later.

There is no evidence that a renal defect in excretion of uric acid is a factor in the causation of gout. However, renal impairment with a decline in glomerular filtration is a common complication of gout in its later stages. The result may be an accumulation of uric acid as well as of other catabolites normally removed by glomerular filtration.

The cause of acute gouty arthritis is still unknown. It is agreed that uric acid itself is not the offending substance. Acute gout cannot be produced in normal or gouty subjects by the administration of uric acid orally, intravenously, or subcutaneously when injected around the joints. Uricosuric drugs (see above) lower the levels of uric acid in the blood by increasing excretion of uric acid, but there is no correlation between the uricosuric effect of a drug and the clinical response.

• • •

References:

1. Buchanan, J.M.: Texas Rep. Biol. and Med. 15:148, 1957.
2. Lagerkvist, U.: J. Biol. Chem. 233:138; 143, 1958.
3. Carter, C.E.: Ann. Rev. Biochem. 25:123, 1956.
4. Wyngaarden, J.B.: Metabolism 6:245, 1957.
5. Stetten, D., Jr.: Geriatrics 9:163, 1954.

17...

The Functions and Tests of the Liver

The liver is the largest and, from a metabolic standpoint, the most complex internal organ in the body. Metabolic disturbances in hepatic disease are therefore quite characteristic and may serve as diagnostic aids. However, because the liver performs so many diverse metabolic functions a great many tests have been devised, some of which are not clinically practicable. Furthermore, these tests differ widely in sensitivity in various pathologic processes. This is particularly notable in assessing the extent of liver damage, since the less sensitive tests may be normal even when only about 15% of the liver parenchyma is functioning.

Anatomic Considerations: (See illustration below.)

The basic structure of the liver is the lobule (see below), consisting of cords of cells extending out from the portal triad, which contains the intralobular bile duct and the final small branches of the circulatory vessels: portal vein, hepatic artery, and lymphatics. A system of capillaries and open spaces (sinusoids) containing blood spreads from the lobules to surround individual liver cords and to conduct blood to the central hepatic vein, by which blood leaves the liver. Every lobule is well supplied with a capillary network originating from the portal vein. In this manner, each hepatic cell is provided with an adequate amount of blood. Furthermore, in each lobule there are five afferent veins to each efferent vein. Such an arrangement slows down blood flow through the liver and so facilitates the exchange of materials between the blood and the liver tissue.

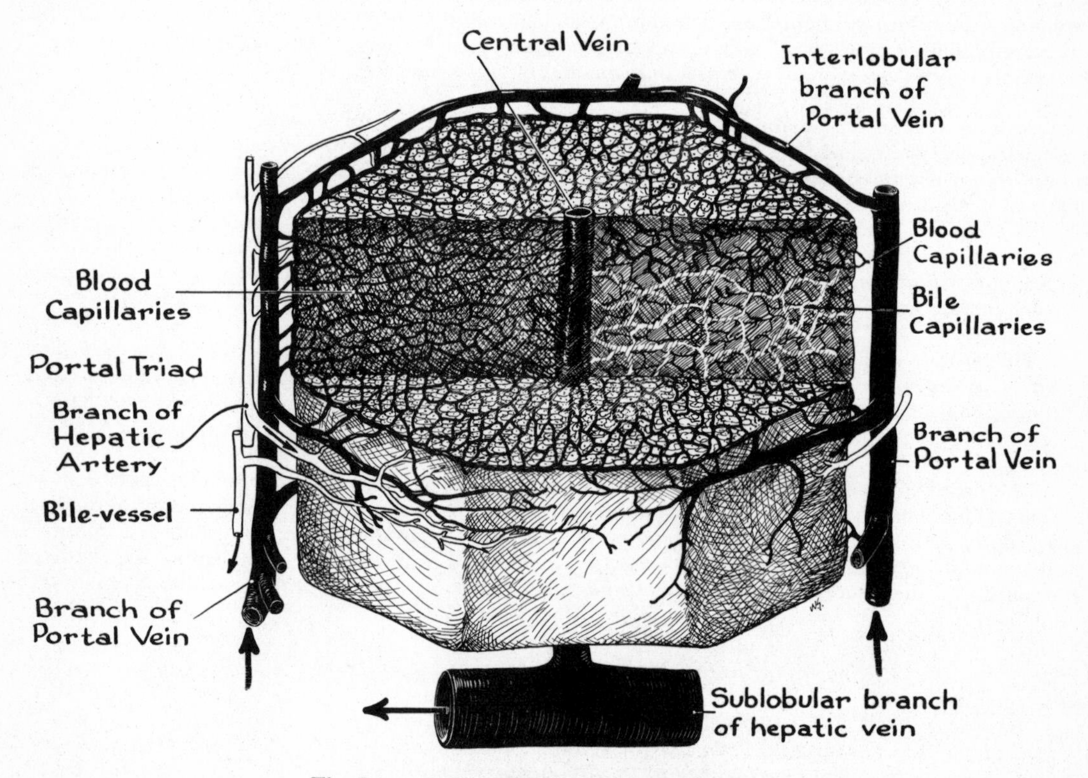

The Liver Lobule (Redrawn from Piersol)

The liver is supplied with arterial blood by the hepatic arteries. Within the liver, the arterial and portal venous circulation anastomose. In general, about 70% of the blood supply to the liver is delivered by the portal veins, and only 30% by the hepatic arteries.

The Functions of the Liver:

The functions of the liver may be classified in five major groups:

A. Circulatory Functions:
1. Transfer of blood from portal to systemic circulation.
2. Activity of its reticuloendothelial system (Kupffer's cells) in immune mechanisms.
3. Blood storage; regulation of blood volume.

B. Excretory Functions:
1. Bile formation and excretion of bile into the intestine.
2. Secretion in the bile of products emanating from the liver parenchymal cells, e.g., bilirubin conjugates, cholesterol, cholic acid as bile salts.
3. Excretion of substances withdrawn from the blood by hepatic activity, e.g., heavy metals, dyes such as bromsulfalein, and alkaline phosphatase.

C. Metabolic Functions:
1. Carbohydrate metabolism.
2. Protein metabolism.
3. Lipid metabolism.
4. Mineral metabolism.
5. Vitamin metabolism.
6. Heat production.

D. Protective Functions and Detoxication:
1. Kupffer cell activity in removing foreign bodies from the blood; phagocytosis.
2. Detoxication by conjugation, methylation, oxidation, and reduction.
3. Removal of ammonia from blood, particularly that absorbed from the intestine by way of the portal vein.

E. Hematologic Functions (Hemopoiesis and Coagulation):
1. Blood formation in the embryo and, in some abnormal states, in the adult.
2. Production of fibrinogen, prothrombin, and heparin.
3. Erythrocyte destruction.

PHYSIOLOGIC AND CHEMICAL BASIS FOR TESTS OF LIVER FUNCTION

Many functions of the liver have been discussed in other portions of this book. Certain other functions and their applications in tests of liver function will now be described.

Bile Pigment Metabolism: (See also pp. 63 to 66.)

The two principal bile pigments are bilirubin and biliverdin. Bilirubin is the chief pigment in the bile of carnivora, including man. Biliverdin is present in only small amounts in human bile, although it is the principal pigment of avian bile.

Bile pigments originate in the reticuloendothelial cells of the liver, including the Kupffer cells, or in other reticuloendothelial cells where the erythrocytes are destroyed. In the course of the destruction of erythrocytes, the protoporphyrin ring of heme derived from hemoglobin is opened to form the bile pigment, biliverdin (see p. 63).

Normally, 0.1 to 1.5 mg. of bilirubin loosely associated with protein (mainly albumin) is present in 100 ml. of human serum. It has been estimated that 1 Gm. of hemoglobin yields 35 mg. of bilirubin.

Bilirubin produced in the reticuloendothelial tissue from the catabolism of heme pigments is carried to the liver, where it is conjugated with glucuronic acid[1,2]. The bilirubin glucuronide is much more soluble in an aqueous medium than is the free (unconjugated) bilirubin. For this reason, the bilirubin conjugate is readily excreted into the intestine with the bile. Indeed, the formation of the conjugated bilirubin within the liver seems to be a necessary preliminary to its excretion in the bile (see below).

Within the intestine, bilirubin ($C_{33}H_{36}O_6N_4$) is successively reduced ultimately to stercobilinogen ($C_{33}H_{48}O_6N_4$), also called L-urobilinogen, the reductions being accomplished by the metabolic activity of the intestinal bacteria. A portion of the urobilinogen is then absorbed from the intestine into the blood. Some of this is excreted in the urine (1 to 4 mg. per day); the remainder is re-excreted in the bile. The unabsorbed urobilinogen is excreted in the stool as fecal urobilinogen (40 to 280 mg. per day). On exposure to air, urobilinogen is oxidized to urobilin. This is the cause of the darkening of the stools on exposure to air.

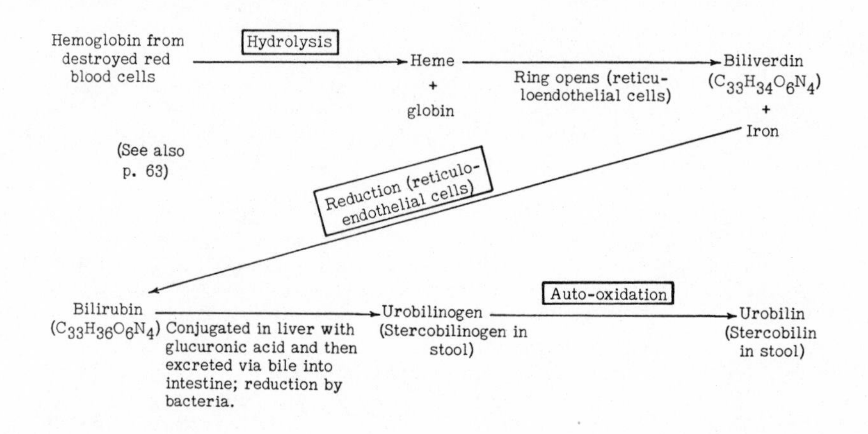

Jaundice:

When bile pigment in the blood is excessive, it escapes into the tissues, which then become yellow. This condition is known as jaundice or icterus.

Jaundice may be due to the production of more bile pigment than the normal liver can excrete, or it may result from the failure of a damaged liver to excrete the bilirubin produced in normal amounts. In the absence of hepatic damage, obstruction of the excretory ducts of the liver by preventing the excretion of bilirubin will also cause jaundice. In all of these situations, bile pigment accumulates in the blood; and when it reaches a certain concentration, it diffuses into the tissues. Jaundice is frequently due to a combination of factors.

A. Types of Jaundice: According to its mode of production, jaundice is sometimes subdivided into three main groups: hemolytic, hepatic, or obstructive.
   1. Hemolytic jaundice - Any condition which increases erythrocyte destruction also increases the formation of bile pigment. If erythrocytes are destroyed faster than their products, including bilirubin, can be excreted by the liver, the concentration of bilirubin in the serum rises above normal; hemolytic jaundice is the result.
   2. Hepatic jaundice - This type of jaundice is caused by liver dysfunction resulting from damage to the parenchymal cells. Examples are the jaundice caused by various liver poisons (chloroform, phosphorus, arsphenamine, carbon tetrachloride), toxins, hepatitis virus, engorgement of hepatic vessels in cardiac failure, and cirrhosis.
   3. Obstructive (regurgitation) jaundice results from blockage of the hepatic or common bile ducts. The bile pigment is believed to pass from the blood into the liver cells as usual; however, failing to be excreted by the bile capillaries, it is absorbed into the hepatic veins and lymphatics.

B. Summary of the Types of Jaundice:
   1. Hemolytic.
   2. Hepatic.
      a. Toxic or infective.
      b. Obstructive.

(1, 2, 2a: } Nonobstructive)

Constitutional Nonhemolytic Hyperbilirubinemia:

In addition to the occurrence of elevated levels of bilirubin in the serum as a result of the various pathologic states described above, hyperbilirubinemia has been detected in individuals who are otherwise free of any symptoms of hepatic disease or of hemolysis as a cause of hyperbilirubinemia. The abnormality is familial and is therefore believed to be genetically transmitted[3,4]. An impairment of excretion of bilirubin has been suggested as the cause of this syndrome and the results of the bilirubin tolerance test (see p. 278) in these subjects support this suggestion. The increased serum bilirubin is entirely of the indirect-reacting type (i. e., unconjugated bilirubin; see p. 276). This could be the result of excessive red blood cell destruction, i. e., a hemolytic jaundice; but, as noted above, the metabolism of the red cells is not abnormal. An alternate explanation is that hyperbilirubinemic subjects have an impaired ability to conjugate bilirubin and thus to excrete it in the bile. This latter explanation has proved to be correct, the defect having been found to reside in an impairment in the mechanism for conjugation of bilirubin due to a deficiency of the hepatic enzyme, glucuronyl transferase, that catalyzes the transfer of the glucuronide moiety from uridine diphosphoglucuronic acid (UDPGluc, p. 192) to bilirubin[5]. Because the bile of an individual with constitutional nonhemolytic hyperbilirubinemia does contain some direct-reacting bilirubin, it must be assumed that the conjugative defect is not complete. However, there are also reports of rare instances of what appears to be a complete absence of conjugation of bilirubin. These individuals develop a very intense jaundice (serum bilirubin as high as 80 mg. /100 ml.). In one such case the bile contained no bilirubin ("white" bile), and there was also complete inability to form glucuronide conjugates with substances such as menthol and tetrahydrocortisone that are normally excreted in the urine as glucuronides. This defect resembles that which has also been found in a mutant strain of rats wherein there seems to be a complete absence of the mechanism for detoxification by conjugation with glucuronic acid[6]. It is possible that the mild disease in humans (so-called Gilbert's disease) represents the "trait" or heterozygous form of inheritance of the genetic defect and that the severe form represents the homozygous inheritance of the defect.

The occurrence of some degree of jaundice in the newborn (icterus neonatorum) is not infrequent, particularly in premature infants. This is believed to be a reflection of a temporary inadequacy in the function of the hepatic system for conjugating bilirubin. If, however, hemolysis (as in Rh incompatibility) occurs during this period of decreased ability to excrete bilirubin, it is likely that very high levels of bilirubin will be found in the serum. As a result bilirubin may accumulate in the tissues, producing a generalized jaundice. In the brain, the localized deep pigmentation of basal ganglia which occurs is termed "kernicterus." Such "brain jaundice" is associated with objective signs of disturbances in nervous system function, and permanent damage to the central nervous system will occur if death does not supervene.

TESTS BASED ON SECRETORY AND EXCRETORY FUNCTIONS

Tests of liver function based on the secretory and excretory functions of the liver and on bile pigment metabolism are of major importance.

Estimation of Serum Bilirubin; Van den Bergh Test:

A method of quantitatively assaying the bilirubin content of the serum was first devised by van den Bergh[7] by application of Ehrlich's test for bilirubin in urine[8]. The Ehrlich reaction is based on the coupling of diazotized sulfanilic acid (Ehrlich's diazo reagent) and bilirubin to produce a reddish-purple azo compound. In the original procedure as described by Ehrlich, alcohol was used to provide a solution in which both bilirubin and the diazo reagent were soluble. Van den Bergh[9] inadvertently omitted the alcohol on an occasion when assay of bile pigment in human bile was being attempted. To his surprise, normal development of the color occurred "directly." This form of bilirubin which would react without the addition of alcohol was thus termed "direct-reacting." It was then found that this same direct reaction would also occur in serum from cases of jaundice due to obstruction. However, it was still necessary to add alcohol to detect biliburin in normal serum or that which was present in excess in serum from cases of hemolytic jaundice where no evidence of obstruction was to be found. To that form of bilirubin which could be measured only after the addition of alcohol, the term "indirect-reacting" was applied.

In the years intervening since van den Bergh first described the two types of bilirubin, a number of theories have been proposed in an attempt to explain the chemical and clinical significance of his observations. It has now been demonstrated that the indirect bilirubin is "free" (unconjugated) bilirubin en route to the liver from the reticuloendothelial tissues where the bilirubin was originally produced by the breakdown of heme porphyrins. Since this bilirubin is not water-soluble, it requires extraction into alcohol to initiate coupling with the diazo reagent. In the liver, the free bilirubin becomes conjugated with glucuronic acid[1,2] and the conjugate, bilurubin glucuronide, can then be excreted into the bile. Furthermore, conjugated bilirubin, being water-soluble, can react directly with the diazo reagent so that the "direct bilirubin" of van den Bergh is actually a bilirubin conjugate (bilirubin glucuronide). In jaundice due to obstruction (regurgitation jaundice), the conjugated bilirubin may return to the blood, which accounts for the presence of direct as well as indirect bilirubin in such cases. In hemolytic jaundice there is no obstruction and the increase in bilirubin is confined to the indirect type, since the cause of the bilirubinemia is increased production and not an abnormality in hepatic conjugation or excretion.

When bilirubin appears in the urine, it is almost entirely of the direct type. It is thus probable that bilirubin is excreted into the urine only when conjugated and so made water-soluble. It also follows that bilirubinuria will be expected to occur only when there is an increase in the direct bilirubin content of the serum.

"Direct-reacting" bilirubin is a diglucuronide of bilirubin in which the two glucuronyl groups transferred from "active" glucuronide (UDPGluc) by the catalytic action of glucuronyl transferase are attached through ester linkages to the propionic acid carboxyl groups of bilirubin[10]. In the conjugation reaction, the monoglucuronide is formed first; this is followed by attachment of a second glucuronyl group to form the diglucuronide. Some monoglucuronide is normally found in the bile, but in hepatic disease, because of impairment of the conjugating mechanism, greater quantities of the monoglucuronide may be formed. Because the monoglucuronide does not react as rapidly with the van den Bergh reagent as does the diglucuronide, it is possible that the so-called "delayed direct-reacting" bilirubin may reflect the presence of the monoglucuronide.

The determination of serum bilirubin is frequently carried out by the colorimetric method of Malloy and Evelyn[11a] as modified by Ducci and Watson[12]. In this method, the color developed within one minute after addition of diluted serum (usually 1:10 with distilled water) to the diazo reagent is a measure of the "direct" bilirubin. This is therefore referred to as the "one-minute bilirubin." Another sample of diluted serum is then added to a mixture of the diazo reagent and methyl alcohol and the color developed after thirty minutes is measured. This "thirty-minute bilirubin" represents the **total** bilirubin content of the serum, i.e., both conjugated (direct) and unconjugated (indirect). The indirect bilirubin concentration is then obtained simply as the difference between the one-minute and the 30-minute readings. In summary:

Total serum bilirubin	=	30 minute bilirubin
Direct (conjugated) bilirubin	=	one-minute bilirubin
Indirect (unconjugated) bilirubin	=	30 minute minus one-minute bilirubin

The normal concentrations of these various fractions of the serum bilirubin are given by Watson as follows:

	Total (mg./100 ml.)	Direct (mg./100 ml.)	Indirect (mg./100 ml.)
Mean	0.62 ± 0.25	0.11 ± 0.05	0.51 ± 0.20
Upper limit of normal	1.50	0.25	1.25

A quantitative measurement of the total serum bilirubin is of considerable value in the detection of latent jaundice, which is represented by concentrations between 1.5 and 2 mg./100 ml. Hyperbilirubinemia may be noted not only in diseases of the liver or biliary tract but also in disease states involving hemolysis such as occur in infectious diseases, pernicious anemia, or hemorrhage.

The progress of a case of manifest jaundice may be followed by repeated serum bilirubin determinations. A rising concentration is an unfavorable sign; a progressive decline in serum bilirubin signifies improvement in the course of liver disease or of biliary obstructions.

The Icterus Index (Meulengracht Test):

In this test, the intensity of the yellow color of the serum is compared with a 1:10,000 solution of potassium dichromate as a standard. One ml. of serum is diluted to 10 ml. with 5% sodium citrate solution, and the density of the color of the diluted serum is measured in a photoelectric colorimeter at a wave length of 420 mμ. The colorimetric density of the standard has an icterus index of 10; the ratio of the colorimetric density of the unknown to that of the standard multiplied by 10 equals the icterus index. Normal values for the icterus index are 4 to 6 units. In latent jaundice (increased bilirubin in the blood without clinical signs of jaundice), the index is between 6 and 15. Above this value symptoms of icterus are usually noted. Since the yellow color is chiefly due to bilirubin, an icterus index of 5 corresponds to about 0.1 to 0.2 mg. of bilirubin per 100 ml. of serum. Certain other pigments which resemble the yellow color due to bilirubin (e.g., the presence of carotene in the blood following the ingestion of carrots) lead to apparent high icterus indices which are, of course, not due to increased blood bilirubin; carrots should not be eaten the day before the test. Blood should be drawn before breakfast in order to obtain a clear serum, and hemolysis must be carefully avoided.

This test is no longer frequently used because a direct determination of serum bilirubin is more accurate and specific and, with modern analytical methods, almost as simple to perform.

Fecal Urobilinogen and Urobilin:

Fecal urobilinogen usually varies directly with the rate of breakdown of red blood cells. It is therefore increased in hemolytic jaundice provided the quantity of hemoglobin available is normal. A decrease in fecal urobilinogen occurs in obstruction of the biliary tract or in extreme cases of diseases affecting the hepatic parenchyma. It is unusual to find a complete absence of fecal urobilinogen. When it does occur, a malignant obstructive disease is strongly suggested. The normal quantity of urobilinogen excreted in the feces per day is from 50 to 250 mg. Fecal urobilinogen is determined by conversion of urobilin into urobilinogen and determination of urobilinogen by methods similar to those used in the urine analyses (see below).

Test for Bilirubin in the Urine:

The presence of bilirubin in the urine suggests that the direct bilirubin concentration of the blood is elevated due to hepatic parenchymatous or duct disease. Bilirubin in the urine may be detected even before clinical levels of jaundice are noted. Usually only direct-reacting bilirubin is found in the urine. The finding of bilirubin in the urine will therefore accompany a direct van den Bergh reaction in blood (see p. 276).

Bilirubin may be detected in the urine by the Harrison test, the Gmelin test, or the Huppert-Cole test. This first test is the most sensitive.

A. Harrison Test: Mix 5 ml. of urine and 5 ml. of a 10% barium chloride solution in a test tube. Collect the precipitate on filter paper and spread it to dry on another filter paper. When dry, add 1 or 2 drops of Fouchet's reagent (25 Gm. trichloracetic acid and 0.9 Gm. ferric chloride in 150 ml. water) to the precipitate. A green color appears in the presence of bilirubin. Disregard other colors. Strips of filter paper impregnated with barium chloride may also be used. The strip of barium chloride paper is moistened with urine, and a drop of Fouchet's reagent is then added to the wet area.

B. The Gmelin Test: Two or 3 ml. of urine are carefully layered over about 5 ml. of concentrated nitric acid in a test tube. If bilirubin is present in the urine, rings of various colors - green, blue, violet, red, and reddish-yellow - will appear at the zone of contact.

C. The Huppert-Cole Test: Five ml. of a suspension of calcium hydroxide in water are added to 10 ml. of urine; the tube is well shaken and the mixture filtered. The bile pigment is removed with the calcium hydroxide. The residue on the filter is dissolved with about 10 drops of concentrated hydrochloric acid; the pigment set free is dissolved by adding 10 ml. of alcohol to the filter; and the alcoholic solution of the bile pigments is then caught in a clean tube. The filtrate in the test tube is warmed on a water bath; the development of a green color indicates the presence of bilirubin.

Urine Urobilinogen:

Normally there are mere traces of urobilinogen in the urine (average, 0.64 mg.; normal, up to 4.0 mg. in 24 hours). In complete obstruction of the bile duct, no urobilinogen is found in the urine since bilirubin is unable to get to the intestine to form it. In this case, the presence of bilirubin in the urine without urobilinogen suggests obstructive jaundice, either intrahepatic or posthepatic. In hemolytic jaundice, the increased production of bilirubin leads to increased production of urobilinogen, which appears in the urine in large amounts. Bilirubin is not usually found in the urine in hemolytic jaundice, so that the combination of increased urobilinogen and absence of bilirubin is suggestive of hemolytic jaundice. Increased blood destruction from any cause (e.g., pernicious anemia) will, of course, also bring about an increase in urine urobilinogen. Further, infection of the biliary passages may increase the urobilinogen in the absence of any reduction in liver function because of the reducing activity of the infecting bacteria.

Urine urobilinogen may also be increased in damage to the hepatic parenchyma because of inability of the liver to re-excrete into the stool by way of the bile the urobilinogen absorbed from the intestine.

The **Wallace-Diamond test** is used for detection of urobilinogen in the urine. One ml. of the aldehyde reagent of Ehrlich (a solution of *p*-dimethylaminobenzaldehyde acidified with hydrochloride) is added to 10 ml. of undiluted urine and allowed to stand one to three minutes. The quantity of urobilinogen in the urine is estimated by noting the rapidity and intensity of color development. If the color remains a light red, which is normal, further test with diluted urine is not required. When urobilinogen is present in larger concentrations than normal, additional tests with various dilutions from 1:10 to 1:200 or higher must be carried out. The highest dilution which shows a faint pink discoloration is sought, and the results are expressed in terms of this dilution. The appearance of color in dilutions up to 1:20 is considered normal, whereas persistence of color in higher dilutions is indicative of abnormal concentrations of urine urobilinogen.

For quantitative measurement of urinary urobilinogen the method of Schwartz et al[13] as improved by Balikov[14] may be used.

Excretion Tests:

A. The Bromsulfalein Excretion Test (B.S.P.): In this test the ability of the liver to remove a dye from the blood is determined, and this is considered to be indicative of its efficiency in removing other substances from the blood which are normally excreted in the bile. In normal persons, a constant proportion (10 to 15% of the dye present in the blood stream) is removed per minute. In hepatic insufficiency, B.S.P. removal is impaired by cellular failure or decreased blood flow; in fact, the test gives more useful information than does the concentration of serum bilirubin if impairment of function is still marginal.

Technic. The B.S.P. test is carried out by the intravenous injection of 5 mg. of the dye per Kg. body weight, withdrawal of a blood sample 30 or 45 minutes later, and estimation of the concentration of the dye in the plasma.

Interpretation. The normal retention of this dye is 5% at 30 minutes and no dye at 45 minutes. The B.S.P. excretion test is a useful index of liver damage, particularly when the damage is diffuse and extensive, as in portal or biliary cirrhosis. The test is of no value if obstruction of the biliary tree exists.

The rate of removal of substances other than B.S.P. has also been studied in liver disease.

B. Rose Bengal Dye Test: Ten ml. of a 1% solution of the dye are injected intravenously. Normally, 50% or more of the injected dye disappears within eight minutes.

C. Bilirubin Tolerance Test: One mg. per Kg. body weight of bilirubin is injected intravenously. If more than 5% of the injected bilirubin is retained after four hours, the excretory (or bilirubin conjugating ?) function of the liver is considered abnormal.

The bilirubin excretion test has been recommended by some authorities as a better test of excretory function of the liver than the dye tests mentioned above because bilirubin is a physiologic substance whereas these dyes are foreign substances. However, the test is not used extensively because of the high cost of bilirubin and the difficulty of measuring small differences in blood bilirubin by the analytical technics in common use.

The three substances listed above, with the exception of B.S.P., are excreted almost entirely by the liver. No significant amounts are taken up by the reticuloendothelial cells.

Plasma Alkaline Phosphatase: (See p. 108.)

The normal values for plasma alkaline phosphatase are 2 to 4.5 Bodansky units per 100 ml. of plasma for adults, and 3.5 to 11 for children. Since alkaline phosphatase is normally excreted by the liver, these values are increased in obstructive jaundice. In a purely hemolytic jaundice there is no rise. Unfortunately, various other factors may affect phosphatase activity so that the results of this test must be correlated with clinical findings and with other tests.

DETOXICATION TESTS

Hippuric Acid Test: (See also p. 167.)

The best-known and most frequently used test of the protective functions of the liver is based on a conjugation reaction: the detoxication of benzoic acid with glycine to form hippuric acid, which is excreted in the urine. In a sense, the test also measures a metabolic function of the liver since the rate of formation of hippuric acid depends also on the concentration and amount of glycine available.

A. Technic: The test may be carried out by either of two methods.
   1. Intravenous injection of 1.77 Gm. of sodium benzoate in 20 ml. of water over a five to ten minute period with collection of all of the urine secreted during the hour after injection.
   2. Oral administration of 6 Gm. of sodium benzoate with a four-hour collection of urine.

B. Normal Values: The hippuric acid in the urine is then isolated, hydrolyzed, and the benzoic acid produced by hydrolysis measured by titration. Using the intravenous test, more than 0.7 Gm. of hippuric acid should be excreted in one hour by normal persons; in the oral test, excretion of 3 Gm. in the four-hour period is normal.

C. Interpretation of Results: This test is a valuable method of determining the presence or absence of intrinsic liver disease as distinguished from extrahepatic involvement as in obstruction. It is in no way affected by jaundice and may therefore be used to distinguish between intrahepatic and extrahepatic jaundice. It may also be used to determine liver function and reserve in jaundiced patients when biliary operation is contemplated. However, there is difficulty in the interpretation of the hippuric acid test when the liver damage is secondary to extrahepatic obstruction. Therefore it is useful in distinguishing between intrahepatic and extrahepatic jaundice only early in the course of obstruction.

   The output of hippuric acid is low in intrahepatic disease such as hepatitis or cirrhosis. Results are normal in cholecystitis, cholelithiasis, and biliary obstruction from stones in the common duct if the condition is of short duration. Later in the disease, liver damage secondary to the obstruction or inflammatory process leads to reduced hippuric acid output.

   The hippuric acid test is valueless if renal function is impaired, and in the presence of normal renal function a minimum output of 100 ml. of urine is necessary for a valid result.

TESTS BASED ON METABOLIC FUNCTIONS

Carbohydrate Metabolism:

The most important tests of this type are based on tolerance to various sugars since the liver is involved in removal of these sugars by glycogenesis or in the conversion of other sugars to glucose in addition to glycogenesis.

A. Glucose Tolerance (See p. 173): This test may be used to evaluate hepatic function in the absence of other abnormalities in glucose metabolism.

B. Galactose Tolerance: This test is used primarily to detect liver cell injury. It is applicable in the presence of jaundice. The normal liver is able to convert galactose into glucose; but this function is impaired in intrahepatic disease, and the amount of galactose in the blood and urine is excessive. Since the test measures an intrinsic hepatic function, it may be used to distinguish obstructive and nonobstructive jaundice, although in the former case, after prolonged obstruction, secondary involvement of the liver leads to abnormality in the galactose tolerance.

1. In the intravenous test, 0.5 Gm. of galactose per Kg. body weight is given after a 12-hour fast. Blood galactose is then measured at various intervals. Normally, none is found in the 75-minute sample; but in intrahepatic jaundice, the 75-minute blood value is greater than 20 mg. per 100 ml. In obstructive jaundice without liver cell damage, galactose is present at 75 minutes; but it is less than 20 mg. per 100 ml.
2. In the oral test, 30 Gm. of galactose in 500 ml. of water are taken by mouth after a 12-hour fast. Normally, or in obstructive jaundice, 3 Gm. or less of galactose are excreted in the urine within three to five hours; and the blood sugar returns to normal in one hour. In intrahepatic jaundice, the excretion amounts to from 4 to 5 Gm. or more during the first five hours.

The intravenous galactose tolerance test is a good liver function test, whereas the oral test is a poor one since in addition to liver function other factors, such as rate of absorption of the sugar from the intestine, affect the outcome of the test.

C. Glycogen Storage (Adrenaline Tolerance): The response to adrenaline as evidenced by elevation of the blood sugar is a manifestation of hepatic glycogenolysis. However, this is directly influenced by hepatic glycogen stores. This test is therefore designed to measure the glycogen storage capacity of the liver.

The subject is placed on a high-carbohydrate diet for three days before the test. After an overnight fast, the blood sugar is determined and 0.01 ml. of a 1:1000 solution of epinephrine per Kg. body weight is injected. The blood sugar is then determined at 15-minute intervals up to one hour. Normally, in the course of an hour, the rise in blood sugar over the fasting level exceeds 40 mg. per 100 ml. In hepatic disease, the rise is less.

This test may also be used for diagnosis of glycogen storage disease (von Gierke's disease; see p. 179).

Protein Metabolism:

A. Hypoalbuminemia: The importance of the liver in the manufacture of albumin and other blood proteins has been noted on p. 132. In acute and chronic liver disease, such as cirrhosis, there is a general tendency to hypoproteinemia which is manifested mainly in the albumin fraction. Parenchymal liver damage may also elicit a rise in globulin, mainly gamma globulins, although the α_2 and β globulins are notably changed in patients with relapsing hepatitis, between the 14th and 30th days.

The severity of hypoalbuminemia in chronic liver disease is of diagnostic importance and may serve as a criterion of the degree of damage. Furthermore, a low serum albumin which fails to increase during treatment is usually a poor prognostic sign.

B. Prothrombin Time and Response to Vitamin K: Measurement of the plasma prothrombin time (see p. 127) and of the response of the liver to administered vitamin K is a useful liver function test. In moderate impairment of liver function, the prothrombin time is reduced to 60 to 90% of the normal control. The persistence of a prothrombin deficit or only a slight increase after the administration of vitamin K is evidence of severe liver damage. In conducting this liver function test, vitamin K must be given parenterally in order to insure absorption. The test must be properly interpreted since many other diseases as well as drugs may reduce the prothrombin content of the plasma. However, the measurement of prothrombin before and after the administration of vitamin K, although not very sensitive, is considered a valuable test in the diagnosis of liver disease, particularly in the distinction between early obstructive and hepatic jaundice. It is also a valuable prognostic sign. It should be noted that the test cannot be used unless the prothrombin time is at least 25% below normal before administration of vitamin K; furthermore, a rise in prothrombin time of at least 20% must occur if the results are to be considered diagnostically significant. The test dose of vitamin K should be no more than 4 mg.

C. Amino acid tolerance tests based on the rates of disappearance from the blood of intravenously or orally administered amino acids may also measure a protein metabolic function of the liver. Methionine and tyrosine have been so used. In liver disease the rates of removal of these amino acids are retarded.

D. Tests of Altered Protein Fraction Production: In liver disease there is evidence that altered protein fractions are produced, and these can be detected in the blood. A group of tests to detect these proteins has been devised for clinical use. Electrophoretic analysis, however (see p. 129), is much more specific. In all of these tests an attempt is made to precipitate the altered proteins of the plasma by the use of various ''antigens.''

1. The Takata-Ara test was the pioneer in this category; but due to its lack of specificity, it is not used today.
2. The cephalin-cholesterol flocculation (Hanger) test is said to be positive if (1) γ globulin is increased, (2) albumin of the plasma is decreased below the concentrations which can inhibit the reaction, or (3) the inhibiting ability of plasma albumin is decreased. The test is conducted by mixing 4 ml. of saline with 1 ml. of the cephalin-cholesterol emulsion and adding to this mixture 0.2 ml. of serum. This is then kept in the dark at room temperature and read at 24 and 48 hours.

   In normal serum the emulsion remains stable, and flocculation does not occur. With pathologic serum, varying degrees of both precipitation and flocculation occur. The test is reported in terms of plus signs, 4+ indicating complete precipitation and flocculation. Equivocal ± to 2+ results are considered negative. It is not a quantitative test but is intended to aid in the diagnosis of an active pathologic process in the liver such as in acute hepatitis. Thus, it is possible that the cephalin-flocculation test may be negative and the results of the hippuric acid test abnormal, e.g., in a case of currently inactive chronic liver disease with permanent liver damage. It is also valuable in infectious hepatic disease in measuring the progress of the infection. Its value in distinguishing between intrahepatic and surgical obstructive jaundice is questionable.
3. The colloidal gold test is similar to the cephalin-cholesterol test. It is based on alteration in the protective power of the serum proteins against the precipitation of colloidal gold. Although it is highly sensitive, being positive in over 90% of cases with hepatic damage and almost uniformly negative in obstructive jaundice, it is a difficult procedure for the routine laboratory.
4. The colloidal red test was designed to simplify the colloidal gold test. The test reagent, a colloidal suspension of the dye, scarlet red, is easier to prepare and more stable than the colloidal gold reagent. In this test 0.05 ml. of serum is mixed with 0.5 ml. of a barbiturate buffer, and 2.5 ml. of the scarlet red reagent are then added. The tubes are left undisturbed at room temperature overnight. The results are graded as follows: 0 = no change in the scarlet red suspension; 1 = a faint precipitate, visible only after shaking; 2 and 3 are intermediate grades of flocculation judged mainly by the color of the supernatant fluid; 4 = a supernatant fluid which is slightly colored and a heavy precipitate.

   This test is said to measure primarily the antibody globulin, and so is particularly valuable in detecting infectious processes involving the liver. It is therefore useful in differentiating infectious hepatitis, in which a 4+ reaction occurs, from toxic hepatitis (e.g., hepatitis incident to poisoning due to carbon tetrachloride or chloroform), which gives a 1+ reaction, or from obstructive jaundice, which gives 1+ or negative reactions.
5. The thymol turbidity test - 0.1 ml. of serum is added to 6 ml. of a thymol solution and allowed to stand for 30 minutes. Turbidity is then read in a colorimeter against a barium sulfate standard. Normal is 0 to 4 units.

   Sera with high β and γ globulin fractions give positive tests. The thymol turbidity test correlates well with the findings of the chephalin-cholesterol test, although it is more sensitive. It measures only an acute process in the liver, but the degree of turbidity is not proportional to the severity of the disease. Positive results are obtained in the thymol turbidity test for a longer period in the course of the disease than in the case of the cephalin-cholesterol flocculation test. A negative thymol turbidity test in the presence of jaundice is very useful for distinguishing between hepatic and extrahepatic jaundice.

   The mechanisms operating in the thymol turbidity test are not identical with those of cephalin-cholesterol flocculation. The thymol test requires lipids, which are not necessary in the cephalin-cholesterol test; and, conversely, γ globulin, so important in cephalin flocculation, is not necessarily involved in thymol turbidity. The flocculate in the thymol test is a complex lipothymoprotein. The thymol seems to decrease the dispersion and solubility of the lipids; and the protein is mainly β globulin, although some γ globulin is also precipitated.

   Thymol flocculation may occur in a thymol turbidity test some time after maximal turbidity has developed. The occurrence of thymol flocculation may also be used as a diagnostic test in liver disease. Flocculation greater than 1+ in 18 hours is considered abnormal. Early in hepatitis, thymol flocculation may be abnormal before turbidity.
6. The zinc turbidity test - When serum with an abnormally high content of γ globulin is diluted with a solution containing a small amount of zinc sulfate, a turbid precipitate forms; the amount of precipitation is proportional to the concentration of γ globulin. In conducting this test, 0.05 ml. of serum is mixed with 3 ml. of the buffered zinc reagent and the mixture allowed to stand for 30 minutes. The tubes are then shaken and the turbidity is read in a spectrophotometer at 650 mμ against barium sulfate standards similar to those used in

the thymol turbidity test. Normal values are 4 to 12 units, with an average normal of 8 units.

The reaction in the zinc turbidity test is specific for γ globulin and not for liver disease in itself; thus it may reflect only antibody formation. The principal advantage of the zinc turbidity test over the cephalin-cholesterol flocculation or thymol turbidity tests is that a single alteration in the serum is measured, i.e., an elevation in the γ globulin fraction. The test is useful, when serial determinations are made, in following the course of infectious hepatitis.

7. Turbidimetric measurement of γ globulin - A direct turbidimetric measurement of γ globulin by precipitation with ammonium sulfate–sodium chloride solutions has been described[15].
8. Measurement of serum mucoprotein - The level of mucoproteins in the blood (see p. 132). is increased in many diseases characterized by cell proliferation or destruction. In normal male subjects, the mucoprotein levels in the serum range from 48 to 75 mg. per 100 ml.; in females, between 40 and 70 mg. In approximately 90% of 91 patients with infectious or homologous serum hepatitis, and in about 70% of 89 patients with portal cirrhosis, Greenspan[16] found that the mucoprotein levels were below normal. However, only three of 125 patients with obstructive biliary tract diseases (inflammatory or neoplastic) had lowered levels of mucoprotein in the serum. This test may be of value in differentiating jaundice due to obstruction of the biliary tract from that due to hepatocellular disease.

Lipid Metabolism:

A. Cholesterol-Cholesterol Ester Ratio: There is much evidence that cholesterol metabolism is controlled largely by reticuloendothelial cells. The liver, with its large reticuloendothelial content, thus plays an active part in the intermediary metabolism of cholesterol. Cholesterol is esterified in the parenchymal cells. Normal total blood cholesterol ranges between 150 and 250 mg. per 100 ml., and about 60 to 70% of this is esterified.

In obstructive jaundice an increase in total blood cholesterol is common, but the ester fraction is also raised so that the percentage esterified does not change. In parenchymatous liver disease there is either no rise or even a decrease in total cholesterol, and the ester fraction is always definitely reduced. The degree of reduction roughly parallels the degree of liver damage.

B. Phenol Turbidity: A measure of the total lipids in the serum can be obtained by a turbidimetric reaction with a phenol reagent. The method was described by Kunkel et al.[17]

Other aspects of lipid metabolism in liver disease have been discussed in Chapter 14 (see p. 204). They are not used in chemical tests of liver function, however.

Enzymes in Liver Disease:

The activity of a number of enzymes in the blood has been studied in liver disease. The increase in serum alkaline phosphatase which may occur in obstructive hepatic disease (see p. 279) is considered to be due to the fact that this enzyme is normally excreted by the liver. However, alterations in the serum concentrations of a number of other enzymes may reflect other aspects of hepatic dysfunction. An example is the appearance in the serum of certain enzymes which are liberated from the liver as a result of the tissue breakdown which occurs in hepatocellular disease. Thus, in acute hepatitis during its clinical peak, high levels of serum lactic dehydrogenase (LDH) have occasionally been found[18]. The content of this enzyme is also reported to be elevated in the serum of patients with obstructive jaundice or with metastatic disease in the liver[19].

Glutamic-pyruvic transaminase (G-PT; see p. 226) as compared to glutamic-oxaloacetic transaminase (G-OT) activity is relatively greater in liver than in other tissues. It is not surprising, therefore, that measurement of serum G-PT has been found to be useful in the diagnosis and study of acute hepatic disease[20]. G-PT is not significantly altered by acute cardiac necrosis as is G-OT.

Aldolase[21,22] and phosphohexose isomerase[23,24] are both markedly increased in the serum of patients with acute hepatitis. No increase is found in cirrhosis, latent hepatitis, or biliary obstruction.

The serum lipase and amylase levels may be low in liver disease. The activity of cholinesterase may also be lowered when disease involves the hepatic parenchyma[25].

The activity in the serum of the TPN-dependent enzyme, isocitric dehydrogenase (ICD), is increased in the early stages of hepatitis of viral origin. There is a lesser elevation of the activity of this enzyme in serum in some malignancies with metastases to the liver. The ICD levels are well within the normal range in cirrhosis of the liver and in extrahepatic obstructive jaundice. There is also no change in the activity of this enzyme in a number of other diseases, including myocardial infarction. The present evidence indicates that determination of serum ICD is much more specific to diseases of the liver than are determinations of the transaminases or of lactic dehydrogenase. Serum ICD activity is of value in recognizing the presence of acute hepatic injury, in following the course of viral hepatitis, in detecting metastatic malignancy to the liver, and in differentiating between intrahepatic and extrahepatic obstructive jaundice[26].

Excretion of Porphyrins:

Coproporphyrin excretion in the urine of patients with liver disease often rises markedly [27,28]. In viral hepatitis the increase is mainly in Type I coproporphyrin; in "alcoholic" cirrhosis, it is mainly Type III. Uroporphyrin and porphobilinogen may also be detected in the urine of patients with liver disease[29].

Vitamin Metabolism:

The liver converts carotene to vitamin A and stores both this vitamin and vitamin D. Since the absorption of all of the fat-soluble vitamins is dependent upon a supply of bile to the intestine, some abnormalities in fat-soluble vitamin metabolism accompany obstructive jaundice. The concentration of vitamin A in the blood is reduced, but it rises after a test dose of the vitamin. It is possible that the lowered serum calcium observed in obstructive jaundice is traceable to a vitamin D deficiency occasioned by the absence of bile in the intestine. Failure to absorb vitamin K in obstructive jaundice produces a prothrombin deficiency. It must also be noted that a prothrombin deficiency may result from hepatic insufficiency even though vitamin K is present in adequate amounts. This is further discussed on p. 280.

Iron Metabolism:

The concentration of iron in the serum is significantly altered by liver disease. In hepatitis, two- to three-fold increases may occur; in cirrhosis, serum iron tends to be lower than normal, whereas in biliary obstruction normal values for iron are the rule.

Hepatic disease histologically suggestive of cirrhosis is a feature of hemochromatosis. The serum iron in these cases is unusually high and the unsaturated iron-binding capacity extremely low, with, therefore, a high percentage saturation (see p. 325) (e.g., 74 to 99% saturation in hemochromatosis as compared to 14 to 69% in cirrhosis and 28 to 58% in normal subjects[30]). These changes in serum iron-binding capacity are not confined to hemochromatosis, so that histologic studies by liver biopsy are also required to confirm the diagnosis.

MOST USEFUL TESTS OF LIVER FUNCTION

Opinions differ concerning which tests are most useful. The relative value of various liver function tests in jaundice and liver disease is tabulated below. Obviously, in the diagnosis of liver disease, a number of tests must be used and the final diagnosis made by consideration of all of the results. The choice of which test is to be used is determined by the differential diagnosis which must be made and by the status of the disease process. All things considered, it is possible that the most useful panel of tests might be as follows (not necessarily in this order):

1. Cholesterol-cholesterol ester ratio.
2. Thymol turbidity and flocculation.
3. Cephalin flocculation.
4. Bromsulfalein excretion.
5. Serum bilirubin.

(1–5: Most sensitive)

6. Urine urobilinogen; urine bilirubin (in early subclinical cases).
7. Prothrombin and response to vitamin K.
8. Serum alkaline phosphatase.
9. Serum isocitric dehydrogenase.

RELATIVE VALUE OF VARIOUS LIVER FUNCTION TESTS IN JAUNDICE AND LIVER DISEASE †

Physiological Basis for Test	TEST		ACUTE 'HEPATIC' DISEASE					Latent or Subclinical Hepatic Disease	Chronic Hepatic Disease	Normal Range of Values
				Icteric Phase						
			Preicteric Phase	Obstructive	Intrahepatic	Hemolytic	Convalescent Phase			
Bile Pigment Metabolism	Urine Bilirubin	Presence	A (incr.)	B	B	-	B (decr.)	-	-	None
		Absence	-	-	-	A	-	-	-	
	Urine Urobilinogen	Increase	A (incr.)	-	A	A	B (decr.)	B	B	1-4 mg./24 H.
		Absence	-	B	-	-	-	-	-	
	Feces Urobilinogen	Increase	-	-	-	A	B (decr.)	-	-	50-250 mg./24 H.
		Absence	-	A	-	-	B (incr.)	-	-	
	Serum Bilirubin (Icterus Index)		A (incr.)	-	-	-	A (decr.)	B	B	0.05-0.50 mg. % 3-8 I.I. Units
Enzyme Activity	Serum Alkaline Phosphatase		-	A (incr.)	-	-	-	-	-	2-4.5 Units (Bodansky)
	Serum ICD‡		-	A	A	±	±	±	B	47-264 mμM./hr.
Cholesterol Metabolism	Plasma Cholesterol	Esterified	-	-	B (decr.)	-	-	-	-	60-75% of Total
		Total	-	B (incr.)	-	-	-	-	-	100-250 mg. %*
Protein Synthesis	Plasma Prothrombin	Before Vit. K	-	-	-	-	-	-	-	90-100%
		After Vit. K	-	A (response)	A (**)	-	-	-	B (**)	15% incr. in 48 H.
	Serum Proteins	Albumin	-	-	-	-	-	B	B	3.4-6.5 Gm. %
		Globulin	-	-	-	-	-	B	B	2.0-3.5 Gm. %
		Total	-	-	-	-	-	-	B	5.7-8.2 Gm. %
	Serum Thymol Turbidity		-	-	-	-	B	B	A	0-4 Units
	Cephalin Flocculation		B	-	B	-	-	B	A	0-1 plus
Dye Excretion	Bromsulfalein (5 mg./Kg.)		-	-	-	-	B	A	A	5%-(30 minutes) 0%-(45 minutes)

Letters indicate relative diagnostic value of the various tests in each of the clinical phases mentioned in the table:
A - Excellent B - Good (-) - Limited or no value

*Varies widely with various labs. Serial determination important.
**(No response).
***With jaundice values > 20 significant.
Without jaundice values > 10 significant.

†Modified, with permission, from Chatton et al.: Handbook of Medical Treatment, 6th Ed., 1958. Lange Medical Publications, Los Altos, Calif.
‡ICD = Serum isocitric dehydrogenase (see p. 283).

References:

1. Billing, B.H., and Lathe, G.H.: Biochem.J.**63**:VI, 1956.
2. Schmid, R.: Science **124**:76, 1956.
3. Gilbert, A., and Lereboullet, P.: Gas.hebd.d.sc.méd.de Bordeaux **49**:889, 1902.
4. Baroody, W.G., and Shugart, R.T.: Am.J.Med.**20**:314, 1956.
5. Arias, I.M., and London, I.M.: Science **126**:563, 1957.
6. Carbone, J.V., and Grodsky, G.M.: Proc.Soc.Exper.Biol. and Med.**94**:461, 1957.
7. Van den Bergh, A.A.H., and Snapper, I.: Dtsch.Arch.f.klin.Med.**110**:540, 1913.
8. Ehrlich, P.: Centralb.f.klin.Med.**45**:721, 1883.
9. Van den Bergh, A.A.H., and Müller, P.: Biochem.Ztschr.**77**:90, 1916.
10. Schachter, D.: Science **126**:507, 1957.
11. Malloy, H.T., and Evelyn, K.A.: J.Biol.Chem.**119**:481, 1937.
12. Ducci, H., and Watson, C.J.: J.Lab. and Clin.Med.**30**:293, 1945.
13. Schwartz, S., Sborov, Y., and Watson, C.J.: Am.J.Clin.Path.**14**:598, 1944.
14. Balikov, B.: Clin.Chem.**1**:264, 1955.
15. De La huerga, J., and Popper, H.: J.Lab. and Clin.Med.**35**:459, 1950.
16. Greenspan, E.M., and Dreiling, P.A.: Arch.Int.Med.**91**:474, 1953.
17. Kunkel, H.G., Ahrens, E.H., and Eisenmenger, W.J.: Gastroenterology **11**:499, 1948.
18. Wróblewski, F., Ruegsegger, P., and La Due, J.S.: Science **123**:1122, 1956.
19. Hsieh, K.M., and Blumenthal, H.T.: Proc.Soc.Exper.Biol. and Med. **91**:626, 1956.
20. Wróblewski, F., and La Due, J.S.: Proc.Soc.Exper.Biol. and Med.**91**:569, 1956.
21. Bruns, F., and Puls, W.: Klin.Wchnschr. **32**:656, 1954.
22. Cook, J.L., and Dounce, A.L.: Proc.Soc.Exper.Biol. and Med.**87**:349, 1954.
23. Bruns, F., and Hinsberg, K.: Biochem.Ztschr.**325**:532, 1954.
24. Bruns, F., and Jacob, W.: Klin.Wchnschr.**32**:1041, 1954.
25. Mann, J.D., Mandel, W.I., Eichmann, P.L., Knowlton, M.A., and Sborov, V.M.: J.Lab. and Clin.Med.**39**:543, 1952.
26. Sterkel, R.L., Spencer, J.A., Wolfson, S.K., Jr., and Williams-Ashman, H.G.: J.Lab. and Clin.Med.**52**:176, 1958.
27. Watson, C.J., Hawkinson, V., Capps, R.B., and Rappaport, A.M.: J.Clin.Invest. **28**:621, 1949.
28. Watson, C.J., Sutherland, D., and Hawkinson, V.: J.Lab. and Clin.Med.**37**:8, 1951.
29. Watson, C.J., Lowry, P., Collins, S., Graham, A., and Ziegler, N.R.: Tr.A.Am. Physicians **67**:242, 1954.
30. Gitlow, S.E., Beyers, M.R., and Colmore, J.P.: J.Lab. and Clin.Med.**40**:541, 1952.

Bibliography:

Lathe, G.H.: The Chemical Pathology of Animal Pigments. Biochemical Society Symposium No. 12, pg. 34, Cambridge, England. The University Press, 1954.

Reinhold, J.G.: Chemical Evaluation of the Functions of the Liver. Clinical Chemistry **1**:351, 1955.

18...

The Kidney and the Urine

The internal environment of the cells of the body is the extracellular fluid, which comprises about 15 to 17% of the body weight. This is the medium in which the cells carry out their vital activities. Since changes in extracellular fluid necessarily are reflected in changes in the fluid within the cells and thus also in cell functions, it is essential to the normal function of the cells that this fluid be maintained relatively constant in composition.

This internal environment (the ''milieu interieur'' of Claude Bernard) is regulated mainly by two pairs of organs: the lungs, which control the levels of oxygen and carbon dioxide; and the kidneys, which maintain the chemical composition of the body fluids at the proper concentrations. Thus, the kidney is an organ which not merely removes metabolic wastes but actually performs highly important homeostatic functions. It also has a considerable metabolic capacity.

Role of Kidney in Homeostasis:

The regulation of the internal environment by the kidneys is a composite of four processes:

1. Filtration of the blood plasma by the glomeruli.
2. Selective reabsorption by the tubules of materials required in maintaining the internal environment.
3. Secretion by the tubules of certain substances from the blood into the tubular lumen for addition to the urine.
4. Exchange of hydrogen ions and production of ammonia for conservation of base.

Urine is the result of these four processes. The anatomic unit which carries out these functions is called a nephron. There are about one million nephrons in each kidney. The urine is collected by the collecting tubules and carried to the renal pelvis; from the renal pelvis it is carried by way of the ureter to the bladder.

Anatomy of the Nephron:

Urine formation begins as the blood enters the glomeruli, which are tufts or networks of arteriolar capillaries. Each glomerulus is surrounded by Bowman's capsule, a double-walled epithelial sac (like a rounded funnel) which leads to the tubule[1].

The tubule includes Bowman's capsule, with which it begins, and the following components: a proximal convolution, the descending limb of the loop of Henle, the loop of Henle itself, an ascending limb, and a distal convolution. The latter joins a collecting tubule or duct which will carry the urine to the remainder of the renal drainage system. The tubule ends and urine formation ceases at the junction of the distal tubule with the collecting duct. The anatomy of the nephron is illustrated on p. 290.

FORMATION OF URINE

Filtration:

The first step in urine formation is filtration of the blood. A large volume of blood, approximately one liter per minute (or 25% of the entire cardiac output at rest), flows through the kidneys. Thus in four to five minutes a volume of blood equal to the total blood volume passes through the renal circulation. This is made possible by a very extensive circulatory system in these organs. By the same token, the kidneys are particularly susceptible to damage by diffuse vascular disease.

The energy for filtration is derived from the hydrostatic pressure of the blood. About 70% of the mean pressure in the aorta is actually exerted on the glomerular capillaries - about 75 mm. Hg. Thus, these capillaries are able to sustain a considerably higher pressure than that of the other capillaries of the body, in which the pressure is only 25 to 30 mm. Hg. The osmotic pressure of the plasma proteins (about 30 mm. Hg) opposes this outflow pressure exerted by the aortic blood. The interstitial pressure on the capillaries themselves, together with the resistance to flow in the tubular system, also contributes about 20 mm. Hg. This reduces the pressure supplying the energy for filtration to a net of 25 mm. Hg. The control of filtration through the glomeruli is illustrated below. It is apparent that the quantity of filtrate produced by the glomerulus is governed by the net filtration pressure and by the amount of blood flowing through the kidneys. Normally, these factors are maintained relatively constant by compensatory adjustments.

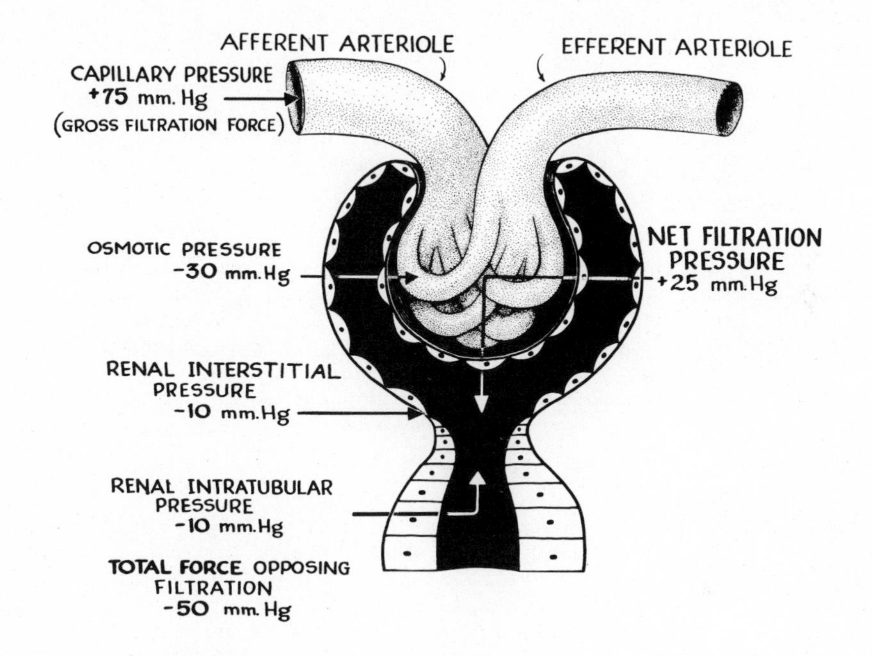

Factors Affecting the Net Filtration Pressure. (Reproduced, with permission, from Merck Sharp and Dohme, "Seminar," Vol. 9, No. 3, 1947.)

A fall in blood pressure too severe for compensation would of course reduce renal filtration. For example, a decline in the aortic systolic pressure to 70 mm. Hg results in a pressure of about 50 mm. Hg in the glomerular capillaries. This would reduce the net glomerular hydrostatic pressure, after correction for resistance to flow within the system, to zero, and filtration would cease. Urine would not be formed (anuria) until the blood pressure was restored.

Other factors which affect filtration include obstruction of the arterial pathway of the glomerulus; increased interstitial pressure on the glomerulus, as in renal edema or in an interstitial inflammatory process; and increased resistance to flow in the tubular systems of the excretory system, such as obstruction in the collecting tubules, ureters, or urethra.

The glomerular membrane may also be so injured by disease that it fails to function as a filter for the blood. Ultimately the capillary may be completely occluded and thus removed from the active circulation. During the progress of such a disease, blood cells and plasma protein will leak through the injured capillary and will be excreted in the urine. Such a pathologic process is illustrated by the syndrome of glomerulonephritis.

Glomerular Filtration Rate (GFR):

In the normal adult, one liter of blood is filtered each minute by the two million nephrons of both kidneys. At a net filtration pressure of 25 mm. Hg, 120 ml. of glomerular filtrate are formed at Bowman's capsule. The **glomerular filtration rate** (GFR) in adults is therefore about **120 ml. per minute.** Chemically, glomerular filtrate is essentially a protein-free extracellular fluid or a protein- and cell-free filtrate of whole blood.

The Action of the Tubule:

The composition of the urine is quite different from that of glomerular filtrate. There is also a vast difference in the volume of fluid formed at the glomerulus per minute and the amount which arrives during the same period at the collecting tubule. The glomeruli act only as a filter; the composition of the glomerular filtrate is thus determined solely by permeability of the capillary membrane to the constituents of the blood. As a result, the glomerular filtrate contains many substances necessary for normal metabolism, such as water, glucose, amino acids, and chlorides, as well as substances to be rejected, such as urea, creatinine, and uric acid. Furthermore, under various conditions, greater or lesser amounts of essential substances are retained in accordance with the need to maintain constancy in the internal environment.

This highly selective function of the kidney is the task of the tubule. By reabsorption and secretion it modifies the glomerular filtrate and thus produces the urine.

Threshold Substances:

Certain substances are reabsorbed almost completely by the tubule when their concentrations in the plasma are within the normal range but appear in the urine (i.e., are not completely reabsorbed) when their normal plasma levels are exceeded. These are spoken of as threshold substances. A substance reabsorbed only slightly or not at all is a low-threshold substance (e.g., creatinine, urea, and uric acid). Those materials necessary to the body which are reabsorbed very efficiently are high-threshold substances (e.g., amino acids and glucose). The reabsorption of glucose is typical of this mechanism of the renal tubule.

Reabsorption of Glucose:

At an arterial plasma level of 100 mg. per 100 ml. and a glomerular filtration rate of 120 ml. per minute, 120 mg. of glucose are delivered into the glomerular filtrate each minute. Normally, all of this glucose is reabsorbed into the blood in the proximal convoluted tubule. This results from an enzymatic mechanism involving coupling of the glucose with phosphate. It is catalyzed by a hexokinase possibly identical with that in the intestine which is similarly utilized in that organ for the absorption of glucose (see p. 162). The phosphorylated glucose is then transferred through the tubule cell to the capillary, where the complex is broken and free glucose is added to the blood. Inhibition of phosphorylation by phlorhizin produces loss of sugar in the urine which is termed ''phlorhizin glycosuria'' or phlorhizin diabetes (see pp. 173 and 341).

The capacity of this enzymatic phosphorylating system is limited. If the arterial plasma level of glucose rises, say to 200 mg. per 100 ml., and the glomerular filtration rate remains the same, twice as much glucose (240 mg. per minute) is presented for reabsorption as before. All of this additional glucose will be reabsorbed until the full capacity of the tubular transfer system is reached; the excess, which is filtered but cannot be reabsorbed, will remain in the tubular fluid and pass on into the urine. This excess glucose will also carry water with it, and this results in the characteristic diuresis of glycosuria.

The maximum rate at which glucose can be reabsorbed has been determined to be about 350 mg. per minute. This is designated the Tm_G, or tubular maximum for glucose. If the glomerular filtration rate decreases, less glucose is presented per minute to the reabsorbing cells. This would permit the concentration of glucose in the blood to rise above those levels at which glucose would normally spill into the urine, without glycosuria actually being observed. The reabsorptive capacity of the tubule for glucose probably does not change under normal conditions, although it may be above normal in hyperthyroidism and in diabetes mellitus. Thyroid extract increases Tm_G; in diabetic patients, insulin reduces it.

Reabsorption of Water: (See p. 291.)

As the glomerular filtrate passes through the tubule most of the water is reabsorbed; under normal conditions, for each 100 ml. of glomerular filtrate formed only about 1 ml. eventually reaches the collecting tubule.

A. Obligatory Reabsorption: The water is first reabsorbed as a solvent for other reabsorbed substances such as glucose and sodium. As these solutes are removed, the tubular fluid tends to become hypotonic to the interstitial fluid of the tubular cells so that more fluid is lost from the urine as adjustment to iso-osmotic conditions proceeds. This is termed **obligatory reabsorption** since it occurs without regard to the water requirement of the body. It takes place in the proximal tubule and, by the time the tubule fluid has reached the distal convoluted tubule, obligatory reabsorption has reduced its volume by about 80%, i.e., each 100 ml. of glomerular fluid is reduced to 20 ml., with a specific gravity of 1.010.

It is possible, of course, as in glycosuria, that the glomerular filtrate may contain such a high concentration of solutes that osmotic factors would operate in the opposite direction. This accounts for the diuresis of glycosuria.

Mercurial diuretics temporarily stop the reabsorption of sodium, possibly in the distal convoluted tubule, which results in an osmotic diuresis of water.

B. Facultative Reabsorption: The second phase of water reabsorption is termed **facultative reabsorption.** The urine reaching the distal tubules is believed to be isotonic with that of the glomerular filtrate (specific gravity 1.010). As a result of recent investigations by micropuncture studies on the rat nephron[2] there is a suggestion that the final concentration or dilution of the urine, in accordance with the needs of the body for water (facultative water reabsorption), takes place not only in the distal tubules but in the collecting tubules as well.

The mechanism which brings about facultative reabsorption of water is the action of the antidiuretic hormone of the posterior pituitary ADH (see p. 367). Some studies have suggested that the adrenal cortex may secrete a hormone antagonistic to pituitary ADH, thus stimulating diuresis. Normal water reabsorption would result from the interaction of these two hormones.

According to Verney[3], the mechanism for facultative reabsorption of water which operates through pituitary ADH is controlled by the activity of osmoreceptors located in the anterior hypothalamic region of the brain. When the blood is diluted, as by the ingestion of large amounts of water, the osmoreceptors detect the resultant decrease in the osmolarity of the blood brought to that region via the internal carotid artery. As a result, the osmoreceptors transmit, over nervous connections to the posterior pituitary, impulses which produce inhibition of pituitary secretion of antidiuretic hormone. The resultant suppression of ADH then permits excretion of more water in the distal and collecting tubules by impairment of facultative reabsorption, and a large volume of dilute urine results. In contrast, after deprivation of water, the blood becomes more hypertonic and the osmoreceptors then act to stimulate ADH secretion, which returns more water to the blood in an effort to compensate for the hypertonicity. Under these circumstances, a concentrated urine of small volume results. The tubular reaction to ADH is one of the most sensitive of the homeostatic mechanisms of the kidney. For this reason, determination of maximal concentrating power of the tubule (see p. 298) is one of the most useful clinical tests of renal function.

A number of drugs, including ingested alcohol, act to suppress ADH secretion and thus to increase urine flow. Certain stresses such as that of surgery or severe trauma as well as some drugs used in anesthesia all cause excessive production of ADH. In the immediate postoperative period these effects on ADH secretion contribute to excessive retention of water by the kidney and thus to oliguria[4]. Failure to recognize this has led to serious overhydration of patients in an effort to correct the oliguria which, under the circumstances, is actually a normal physiologic response.

At that point in the tubule where facultative reabsorption of water begins, the tubular cells must extract materials from a tubular fluid with a specific gravity of 1.010 - the same as that of the peritubular fluid. This they can do only by the expenditure of energy, which requires osmotic work since it is to be accomplished against an osmotic gradient. Under ordinary circumstances, the urine volume will be reduced from 100 ml. per minute at the glomeruli to a final volume of about 1 ml. per minute, and the ultimate specific gravity will thereby be raised to an average of 1.015. This **concentrating power** of the kidney is impaired when renal disease interferes with the ability of the tubular cells to do osmotic work. If the concentrating power of the kidney is at its maximum, the minimal volume of urine necessary to remove all of the body wastes formed in a day is about 700 ml. In tubular failure this minimal volume increases, and the ability to form a urine of variable solute content is impaired. In severe tubular failure the kidney is unable to vary the urine volume, and the specific gravity becomes fixed at 1.010. At this level, the damaged cells are doing practically no osmotic work since, as we have noted, obligatory reabsorption alone may provide a urine of that specific gravity.

Defects in Reabsorption by the Renal Tubules:

The mechanisms for selective reabsorption in the kidney tubules may exhibit various defects,

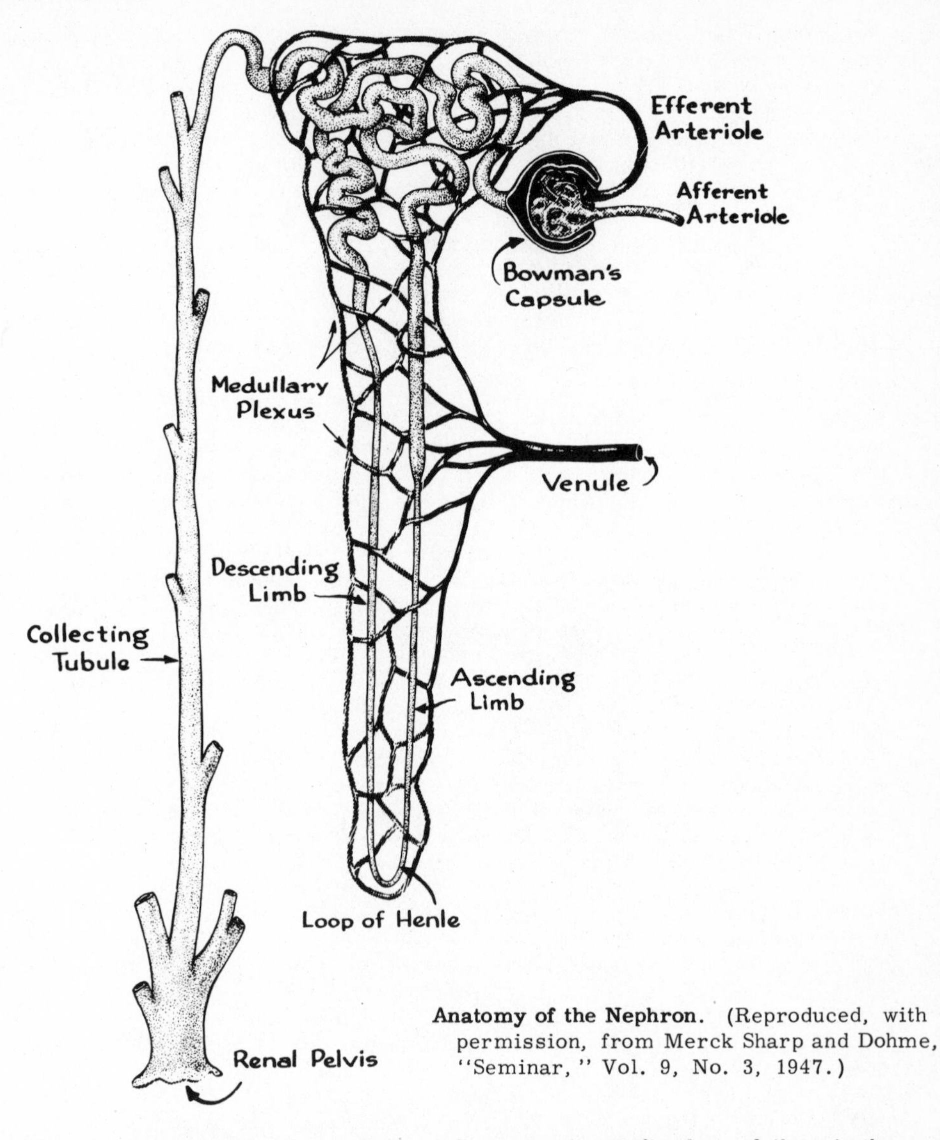

Anatomy of the Nephron. (Reproduced, with permission, from Merck Sharp and Dohme, "Seminar," Vol. 9, No. 3, 1947.)

which are frequently genetic in origin. Such tubular defects may be confined to a failure in the reabsorption of a single substance, such as phosphates or glucose, or a combination of several reabsorptive defects may be found. The presence of glucose in the urine when the level of glucose in the blood is normal is indicative of a tubular defect in the reabsorption of glucose. Such a reduction in the "renal threshold" for glucose is called renal diabetes or renal glycosuria.

Those individuals who have a decreased capacity for the reabsorption of phosphates exhibit changes in the metabolism of bone as a result of excessive losses of phosphorus from the body. Evidence for such changes can be found in the blood, where the phosphorus is low and alkaline phosphatase activity is high. Such cases are often designated as vitamin D–resistant rickets, idiopathic osteomalacia, or Milkman's syndrome. The existence of hyperphosphaturia of renal tubular origin may be detected by measurement of the serum phosphate level and of the amounts of phosphate excreted in a 24-hour urine sample. These data are then used to calculate the phosphate clearance (see p. 297). In normal individuals, the clearance of phosphate does not exceed 12% of the glomerular filtration rate. It should be emphasized that the metabolic effects of a renal tubular defect may be masked by a decline in glomerular filtration. Thus a marked reduction in filtration would compensate for a decreased tubular reabsorptive capacity for phosphate so that the serum levels would rise and healing of the bones could then occur. The urinary excretion of phosphate is therefore unreliable as a diagnostic tool in the presence of reduced glomerular filtration.

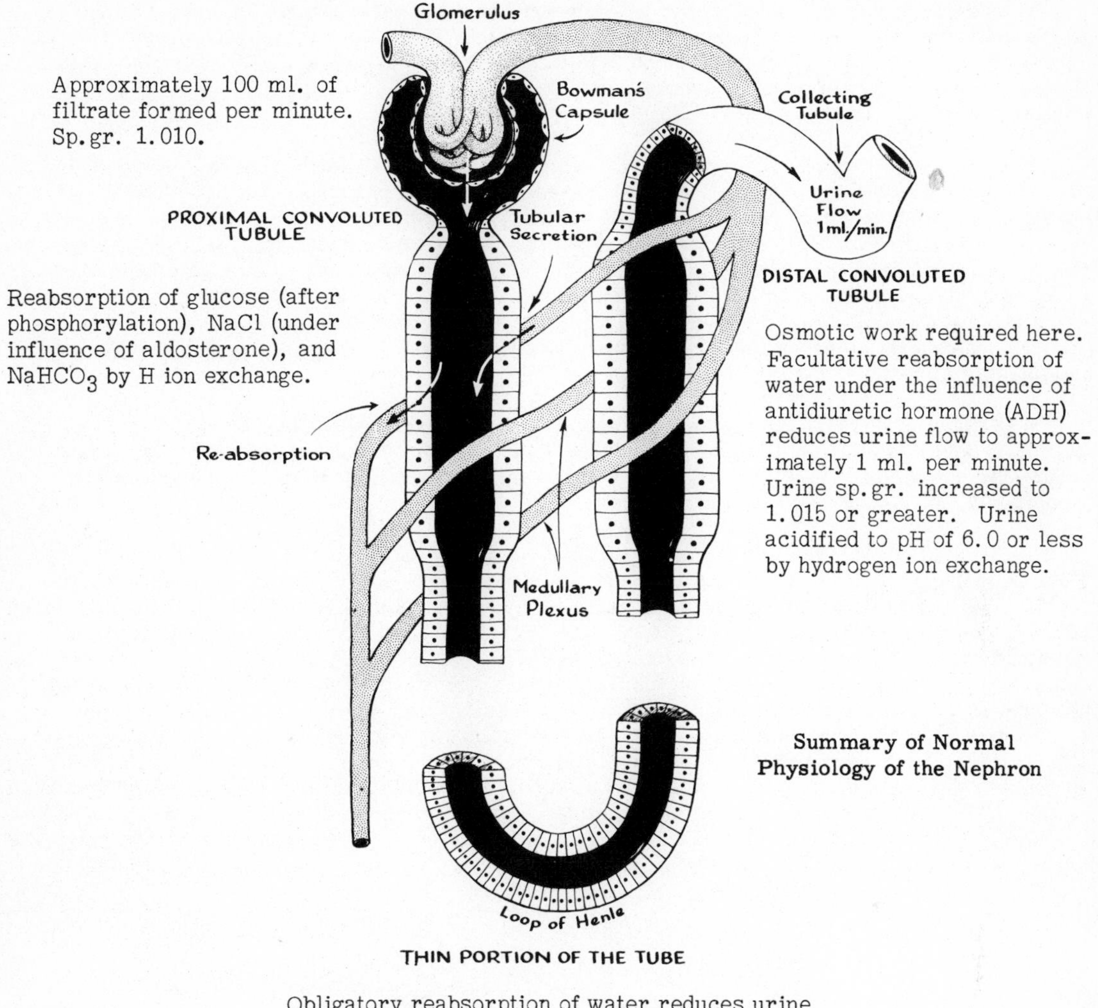

Summary of Normal Physiology of the Nephron

A third tubular defect which often occurs in combination with those mentioned above is characterized by excessive excretion of amino acids. In normal subjects, the content of amino acids in the urine is very low; but in individuals having this renal tubular deficiency a generalized aminoaciduria occurs even though the levels of amino acids in the blood are normal.

The occurrence of all three tubular defects was first described in children by Fanconi and termed by him "hypophosphatemic-glucosuric rickets." It is now referred to as the de Toni-Fanconi syndrome. Similar cases have since been reported in adults. A peculiar feature of the disease in children is the accumulation of cystine crystals in the reticuloendothelial system and other tissues, including the cornea. Such cystine storage (cystinosis) without any of the other features of the Fanconi syndrome has also been found in an otherwise normal individual[5]. This suggests that the genetic defect responsible for cystine storage is not directly related to that involved in the causation ot the renal tubular abnormalities of the Fanconi syndrome, as might otherwise have been inferred from the almost inevitable coexistence of cystinosis and Fanconi disease in children.

In severe cases of the de Toni-Fanconi syndrome there is also a failure properly to reabsorb water and potassium as well as to acidify the urine. Such patients, if untreated, suffer from a metabolic acidosis and dehydration as well as a severe potassium deficiency, and this combination of tubular defects is usually fatal at an early age.

As was pointed out above in connection with tubular hyperphosphaturia, a decline in the glomerular filtration rate may obscure the aminoaciduria which would otherwise occur in the de Toni-Fanconi syndrome. For this reason it is important to measure the GFR (e.g., creatinine clearance; see p. 297) in order properly to interpret the extent of an aminoaciduria in cases where this tubular defect is suspected.

Another group of renal tubular defects, which, unlike those described above, is never inherited, is characterized by increased urinary losses of calcium, phosphate, and potassium and a failure to acidify the urine. There may also be a lack of ammonia formation by the kidneys (nephrocalcinosis, hyperchloremic nephrocalcinosis, Butler-Albright syndrome) (see also p. 295). The primary defect in these cases is the inability to form an acid urine. A chronic acidosis results, leading to increased urinary excretion of alkali cations (sodium, potassium, and calcium), which are used to neutralize the excess acid. The losses of calcium and phosphorus in the urine produce changes in the bones (osteomalacia). Correction of the acidosis by administration of alkaline salts produces healing of the bones.

In contrast to the generalized aminoaciduria of the de Toni-Fanconi syndrome, a **limited** excretion of only a few amino acids without evidence of any other renal tubular defect may also occur. An example is the excessive excretion of cystine, lysine, arginine, and ornithine in "cystinuria" (see p. 251).

Tubular Secretion:

It is believed that some creatinine is secreted by the tubules when the blood levels rise above normal. There may be minimal tubular secretion of uric acid. Potassium is also secreted from the blood by the renal tubules. The exchange of hydrogen ions described on p. 293 might also be considered a secretory function of the tubules. These are the only **normal** constituents of the urine thus added to the tubular filtrate. It is also possible that additional secretory activity may be resorted to in glomerulonephritis when filtration is impaired. However, many foreign substances are readily secreted by the tubules. An example is the dye, phenol red, which is secreted by the proximal tubule utilizing a carrier system similar to that for glucose but acting, of course, in reverse. The same carrier system is mainly responsible for the secretion of iodopyracet (Diodrast®; 3, 5-diiodo-4-pyridone-N-acetic acid which is loosely combined with ethanolamine), iodohippurate (Hippuran®; *o*-iodohippuric acid), penicillin, and para-aminohippurate (PAH).

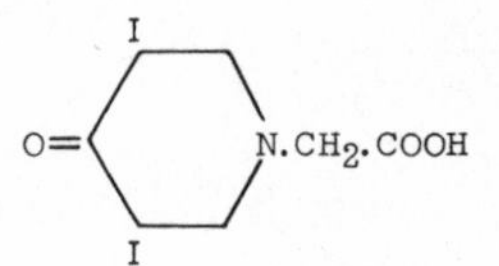

3, 5-Diiodo-4-pyridone-N-acetic Acid

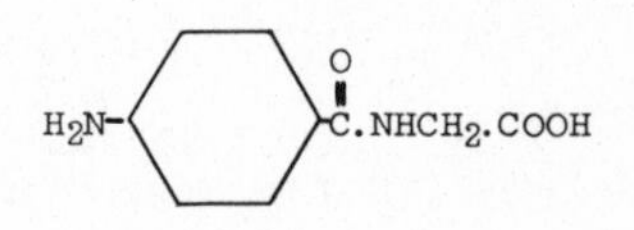

p-Aminohippurate (PAH)

Investigations of some of the energy sources for renal tubular transport indicate that energy from oxidation of succinate and other citric acid cycle oxidations is directly involved in tubular secretory processes. Various inhibitors of the succinic oxidase system are known to alter the secretion of PAH and of the dye phenolsulfonphthalein (PSP).

Maximal Tubular Secretory Capacity (Secretory Tm). The maximal rate at which the tubule can accomplish secretion is determined by the capacity of these carrier systems. This is measured and expressed as the maximal tubular secretory capacity, or secretory Tm. For PAH this is about 80 mg. per minute in normal persons. This value is useful in measuring the amount of functioning tubular tissue since it represents the total effect of all the tubules working together. Iodopyracet, which is opaque to x-ray, is used for roentgenologic examinations of the urinary tract. It may also be used like PAH to measure tubular activity. The normal secretory Tm for iodopyracet is about 50 mg. per minute.

THE ROLE OF THE KIDNEY IN ACID-BASE BALANCE

The kidney plays a very important role in regulating the electrolyte content of body fluids and thus in maintaining pH of the body.

Reabsorption of Electrolytes:

Sodium, the principal cation, and chloride and bicarbonate, the principal anions of the extracellular fluid and of the glomerular filtrate, are selectively reabsorbed, mainly in the proximal tubule. In general, chloride reabsorption and excretion roughly parallel that of sodium.

Potassium is also present in small but important quantities in the extracellular fluid and therefore in the glomerular filtrate. When a large excess of potassium is excreted the clearance of potassium may exceed the glomerular filtration rate. This indicates that potassium is excreted not only by filtration but also by tubular secretion. However, at normal rates of excretion, virtually all of the potassium which is filtered is later reabsorbed in the proximal tubule. That which appears in the urine is therefore added to the urine by tubular secretion in the distal tubule. The secretion of potassium by the renal tubule is closely associated with hydrogen ion exchange and with acid-base equilibrium. This will be discussed further on p. 296.

The fact that potassium can be excreted by tubular secretion as well as by filtration explains why serum potassium is usually normal or only slightly elevated in chronic renal failure when reduced filtration produces a marked rise in urea and other substances which depend entirely on filtration for their excretion. Tubular secretion of potassium does not usually fail until late in the course of chronic renal disease.

The adrenal corticoids, particularly aldosterone (see p. 346), have an important effect on the handling of sodium and potassium by the renal tubules. These hormones favor the reabsorption of sodium and the excretion of potassium. In adrenocortical insufficiencies, such as Addison's disease, there is, therefore, excess loss of sodium, and potassium tends to be retained. Conversely, hyperactivity of the adrenal cortex or the administration of corticoid hormones produces excess reabsorption of sodium and urinary loss of potassium. Hyperactivity of the adrenal may be caused by the presence of adrenal tumors. An adrenal tumor (aldosteroma) which is characterized by the secretion of large amounts of aldosterone (aldosteronism) leads to marked increases in sodium retention and to potassium loss. Thus, whereas the clearance of potassium in normal individuals is about 5 to 10% of the GFR, in aldosteronism it may be as much as 40%.

Acute loss of sodium, as may occur in Addison's disease or in chronic renal disease characterized by inadequate tubular reabsorption of sodium (salt-losing nephritis), leads to severe dehydration, a decline in plasma volume, and shock.

Acid-Base Regulatory Mechanisms:[6, 7]

The kidney also affects acid-base equilibrium by providing for the elimination of nonvolatile acids, such as lactic acid, the ketone bodies, sulfuric acid produced in the metabolism of protein, and phosphoric acid produced in the metabolism of phospholipids. These acids buffered with cations (principally sodium) are first removed by glomerular filtration. The cation is then recovered in the renal tubule by reabsorption in exchange for hydrogen ions which are secreted. Mobilization of hydrogen ions for tubular secretion is accomplished by ionization of carbonic acid, which is itself formed from metabolic carbon dioxide and water. In the proximal tubule, the exchange of hydrogen ions proceeds against sodium bicarbonate since this is the strongest base of the three important salts of sodium occurring in the glomerular filtrate (sodium phosphate and sodium chloride are the other two salts). The process as it occurs in the proximal tubule is diagrammed on p. 294. The formation of carbonic acid is catalyzed by **carbonic anhydrase.** The result of the reaction is not only to provide for the complete reabsorption of all of the sodium bicarbonate filtered from the plasma but also to effect a reduction in the hydrogen ion load of the plasma with little change in the pH of the urine. It will be noted that the bicarbonate moiety filtered is not that which is reabsorbed into the blood and that the carbon dioxide resulting from the decomposition of the carbonic acid formed in the tubule may diffuse back into the tubule cells for reutilization in the hydrogen ion secretory system.

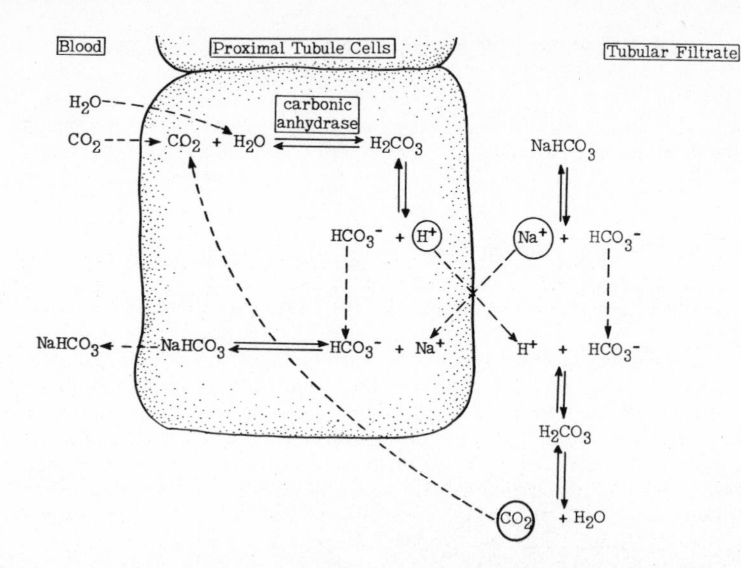

At normal levels of bicarbonate in the plasma, about half of that filtered at the glomerulus is reabsorbed in the proximal tubule and the remainder in the distal tubule. After all of the bicarbonate has been reabsorbed, hydrogen ion secretion then proceeds against Na_2HPO_4. The exchange of a sodium ion for the secreted hydrogen ion changes Na_2HPO_4 to NaH_2PO_4, with a consequent increase in the acidity of the urine and a decrease in the urinary pH. The secretion of hydrogen ions in the distal tubule may be shown as follows:

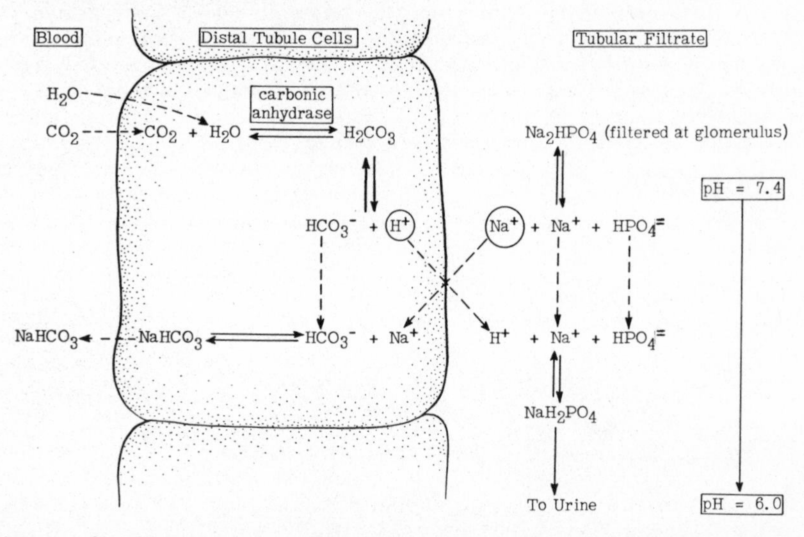

A third mechanism for the elimination of hydrogen ions and the conservation of fixed base is the production of ammonia by the distal renal tubule cells. The ammonia is obtained by deamination of amino acids. Deamination of glutamine by glutaminase (see p. 233) serves as the principal source of the urinary ammonia. The ammonia formed within the renal tubule cell may react directly with hydrogen ions so that ammonium ions rather than hydrogen ions are secreted; or ammonia may diffuse into the tubular filtrate and there form ammonium ion as shown. Such a mechanism operating against sodium chloride might be illustrated as on p. 295.

The lower the pH of the urine, i.e., the greater the concentration of hydrogen ions, the faster will ammonia diffuse into the urine. Thus ammonia production is greatly increased in metabolic acidosis and negligible in alkalosis. It has also been proposed that the activity of renal glutaminase is enhanced by an acidosis[8]. The ammonia mechanism is a valuable device for the conservation of fixed base. Under normal conditions 30 to 50 mEq. of hydrogen ions are eliminated per day by combination with ammonia and about 10 to 30 mEq. as titratable acid, i.e., buffered with phosphate.

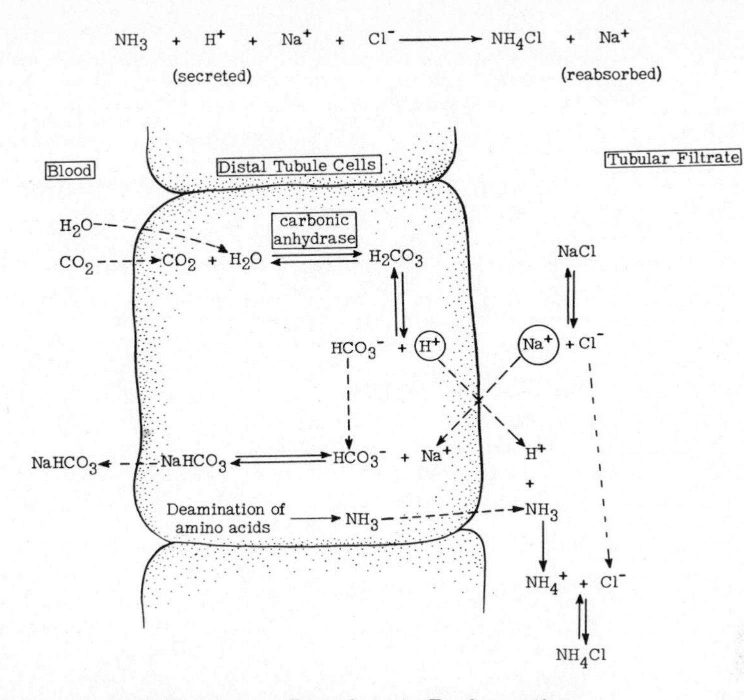

Influence of Carbon Dioxide Tension on Bicarbonate Reabsorption:

An increase in the carbon dioxide tension of the body fluids will accelerate the formation of carbonic acid and thus of hydrogen ions for secretion by the renal tubule cells. This would be expected to facilitate the reabsorption of bicarbonate. Decreased carbon dioxide tensions would act in reverse, i.e., cause a reduction in the reabsorption of bicarbonate. This response of the renal tubule to the carbon dioxide tension of the body fluids provides an explanation for the renal response to states of respiratory acidosis or alkalosis (see p. 150). In respiratory acidosis, compensation is achieved by an increase in the bicarbonate levels of the blood in an attempt to restore the normal 1:20 ratio of carbonic acid to bicarbonate. The increased bicarbonate may be obtained by the response of the kidney to the high carbon dioxide tension which prevails in respiratory acidosis. The situation is reversed in respiratory alkalosis. Here compensation is achieved by **elimination** of bicarbonate, and the lowered carbon dioxide tension which prevails in this condition provokes a renal response which reduces the rate of reabsorption of bicarbonate until the normal ratio of carbonic acid to bicarbonate is restored.

Action of a Carbonic Anhydrase Inhibitor on Excretion of Bicarbonate:

The enzyme, carbonic anhydrase, which catalyzes the important reaction in the renal tubules by which hydrogen ions are produced for secretion, may be inhibited by sulfonamide derivatives. The most potent inhibitor yet discovered is acetazolamide (Diamox®; 2-acetylamino-1, 3, 4-thiadiazole-5-sulfonamide). When this drug is administered, the urine becomes alkaline, and increased amounts of sodium bicarbonate appear in the urine. There is also a reduction in the titratable acidity and in the ammonia of the urine together with an increase in potassium excretion. All of these effects of acetazoleamide may be explained by its action on hydrogen ion secretion. The drug is used clinically to induce a loss of sodium and water in patients who will benefit from such a regimen (e.g., patients with congestive heart failure or hypertensive disease).

Failure of Acid-Base Regulation by the Kidneys in Disease:

Metabolic acidosis is a characteristic complication of renal disease. It is most frequently caused by a decrease in the glomerular filtration rate, which causes retention of fixed acid catabolites such as phosphates and sulfates, accompanied by a rise in nonprotein nitrogen in the blood. There may also be a failure of the renal tubular mechanisms for secretion of hydrogen ions and for the manufacture of ammonia. The decrease in glomerular filtration can be expected to contribute to acidosis because the retained acidic anions, such as phosphates and sulfates, are buffered by cations which cannot be recaptured and returned to the blood in exchange for hydrogen ions by the tubule until these buffered compounds have been delivered to the tubular filtrate. If, even after filtration, tubular secretion of hydrogen ions and concomitant reabsorption of sodium are also defective, there will necessarily occur a considerable loss of fixed base (cation) into the urine. Under these circumstances reduction in sodium content of the plasma may become so severe that dehydration secondary to the electrolyte depletion will result. The consequent fall in extracellular fluid volume is a factor in reducing blood flow to the kidney, which further impairs its function.

The so-called "renal tubular acidosis" mentioned on p. 292 is relatively rare. This defect is confined to failure of the renal tubular mechanisms for secretion of hydrogen ions and for manufacture of ammonia. It may be differentiated from the classical acidosis of renal insufficiency, described above, by the fact that the glomerular filtration rate (e.g., creatinine clearance) is normal and there is therefore no elevation in the plasma of urea, phosphate, or sulfate. (In fact, phosphate may actually be low.) The pH of the urine is close to neutral (6.0 to 7.0). Sodium depletion does not occur, probably because calcium and potassium are substituted as sources of fixed base for neutralization of acid. However, such substitution results in depletion of calcium from bone (osteomalacia) and increased calcium in urine (hypercalciuria), with a resultant tendency to formation of renal calculi. Furthermore, depletion of potassium may be so severe as to produce paralysis of muscle. All of the metabolic abnormalities resulting from a renal tubular acidosis will be corrected when sufficient quantities of sodium bicarbonate are administered to maintain the plasma bicarbonate at normal levels.

Relation of Potassium Excretion to Acid-Base Equilibrium:

The administration of potassium salts in excess produces a decline in hydrogen ion concentration within the cells and an increase (acidosis) in the extracellular fluid accompanied by the excretion of an alkaline urine. Conversely, potassium depletion is associated with the development of an alkalosis in the extracellular fluid and an increase in hydrogen ion concentration within the cell (intracellular acidosis) followed by the excretion of a highly acid urine despite the high bicarbonate content of the plasma (paradoxic aciduria). These latter circumstances may occur in patients treated with cortisone or corticotropin (ACTH) or in those with the hypercorticism of Cushing's syndrome. It is also a common occurrence in postoperative patients maintained largely on potassium-free fluids in whom depletion of potassium may result because of continued excretion of this cation in the urine and in the gastrointestinal fluids. Although the alkalosis in these surgical cases is usually accompanied by depletion of chloride (hypochloremic alkalosis), correction cannot be attained by the administration of sodium chloride alone but only by the administration of potassium salts as well. When adequate repletion of potassium has been accomplished, a fall in serum bicarbonate and a rise in serum chloride together with elevation of the urine pH to normal levels will then occur.

As noted on p. 293, potassium is secreted by the distal tubule. Furthermore, the same mechanisms which provide for the secretion of hydrogen ions in exchange for sodium are similarly utilized for the secretion of potassium, also in exchange for sodium. The capacity of this system is limited. Ordinarily it is represented by the sum of the secretion of hydrogen and potassium ions. When intracellular potassium levels are low and intracellular hydrogen ions are elevated, as occurs in potassium-deficient states, more hydrogen ions can be secreted by the distal tubule cells. Under these circumstances bicarbonate reabsorption proceeds together with excess hydrogen ion secretion, and a highly acid urine is formed.

The effect of acid-base balance (in this instance, alkalosis) on the excretion of potassium in the urine can be further demonstrated by administration of a quantity of sodium bicarbonate sufficient to lower hydrogen ion concentration both intracellularly and extracellularly. Within the renal tubular cells, the decline in hydrogen ion concentration permits an increase in secretion of potassium because of the normal competition for secretion which exists between hydrogen and potassium ions in the distal tubule. Marked enhancement of potassium excretion will thus be produced in response to the alkalosis following bicarbonate administration. In fact, in man, the excretion of potassium is so closely related to alkalosis that simple hyperventilation will raise the rate of excretion of potassium from 87 to 266 μEq. per minute.

The ability of acetazolamide (Diamox®) to increase excretion of potassium into the urine is attributable to its inhibiting effect on carbonic anhydrase activity, which thus brings about a reduction in hydrogen ion concentration within the renal tubule cell. Consequently, secretion of potassium can be increased.

Summary of the Mechanisms for Sodium Reabsorption:

The relative importance of each of the mechanisms for the conservation of sodium by the kidney may be assessed by a consideration of the following data[9].

A. Sodium reabsorbed with fixed anion (mostly chloride) - 12, 000 μEq. per minute.

B. Na^+-H^+ exchange.
1. $NaHCO_3$ — 3200 μEq. per minute.
2. NH_4 — 20 μEq. per minute.
3. Titratable acid (decrease in urine pH) — 30 μEq. per minute.
4. Free acid (1 ml. urine at pH 5.0) — 0. 01 μEq. per minute.

C. Na^+-K^+ exchange (K^+ excreted) — 50 μEq. per minute.

TESTS OF RENAL FUNCTION

Clearance:

As a means of expressing quantitatively the rate of excretion of a given substance by the kidney, its "clearance" is frequently measured. This is a volume of blood or plasma which contains the amount of the substance which is excreted in the urine in one minute. Alternatively, the clearance of a substance may be defined as that volume of blood or plasma cleared of the amount of the substance found in one minute's excretion of urine. The calculation of clearance can be illustrated by measurement of the clearance of inulin.

A. Inulin Clearance: The polysaccharide inulin is filtered at the glomerulus but neither secreted nor reabsorbed by the tubule. The clearance of inulin is therefore a measure of glomerular filtration rate (GFR). **Mannitol** can also be used for the same purpose. Inulin clearance, like many other physiologic phenomena, varies with body size. It is therefore expressed on the basis of a given size: normal inulin clearance (GFR) is **120 ml. per 1.73 sq. M. body surface area.** To facilitate interpretation, the results of an actual clearance study are usually calculated ("corrected clearance") on the basis of ml. per 1.73 sq. M.

In measuring inulin clearance it is desirable to maintain a constant plasma level of the test substance during the period of urine collections. Simultaneous measurement of the plasma inulin level and the quantity excreted in a given time supplies the data necessary to calculate the clearance according to the following formula:

$$C_{in.} = \frac{U \times V}{B}$$

Where $C_{in.}$ = Clearance of inulin (ml. per minute)
U = Urinary inulin (mg. per 100 ml.)
B = Blood inulin (mg. per 100 ml.)
V = Volume of urine (ml. per minute)

B. Endogenous Creatinine Clearance: At normal levels of creatinine in the blood, this metabolite is filtered at the glomerulus but not secreted nor reabsorbed by the tubule. Consequently, its clearance may also be measured to obtain the GFR. This is a convenient clinical method for estimation of the GFR since it does not require the intravenous administration of a test substance, as is the case with an exogenous clearance study using inulin. Normal values for creatinine clearance are 95 to 105 ml. per minute.

C. Urea Clearance: This test is used clinically to appraise renal function. The subject is given two glasses of water to insure adequate urine flow. He then voids, and the urine is discarded. One hour later, the patient voids a specimen of urine and a blood sample is taken. One hour later, a second urine specimen is voided. The blood and urine specimens are analyzed for urea.

In calculating the results, the volume of urine excreted per minute is noted. If it is less than 2 ml. per minute, the test is not valid. The formula above is used to calculate the urea clearance substituting urea for inulin.

The normal value for urea clearance is 75 ml. per minute. This means that the maximal amount of urea removed in one minute by the kidney is equivalent to that normally contained in 75 ml. of blood. It will be noted that urea clearance is less than that of inulin, which indicates that some of the filtered urea is subsequently passively reabsorbed by the tubules.

The clearances of many dyes or other substances like PAH which are secreted as well as filtered may therefore exceed that of inulin. It is customary to report urea clearances in percent of normal (75 ml. per minute). The normal range is 75 to 120%.

With severe renal damage, the clearance may fall to 5% or less. It may fall to 50% or lower in early acute nephritis, and to 60% or lower in early glomerulonephritis. Impaired urea clearance is due principally to a reduction of glomerular filtration.

Measurement of Renal Plasma Flow (RPF):

Para-aminohippurate (PAH) is filtered at the glomeruli and secreted by the tubules. At low blood concentrations (2 mg. or less per 100 ml. of plasma), PAH is removed completely during a single circulation of the blood through the kidneys. Thus the amount of PAH in the urine becomes a measure of the volume of **plasma** cleared of PAH in a unit of time. In other words, PAH clearance at low blood levels measures renal **plasma flow.** This is about 574 ml. per minute for a surface area of 1.73 sq. M.

Filtration Fraction (FF):

The filtration fraction, i. e., the fraction of plasma passing through the kidney which is filtered at the glomerulus, is obtained by dividing the inulin clearance by the PAH clearance (GFR/RPF = FF). For a GFR of 125 and RPF of 574, the FF would then be 125/574 = 0.217 (21.7%). The filtration fraction tends to be normal in early essential hypertension, but as the disease progresses the decrease in renal plasma flow is greater than the decrease in glomerular filtration. This produces an increase in the FF. In the malignant phase of hypertension, these changes are much greater; consequently the filtration fraction rises considerably. The reverse situation prevails in glomerulonephritis. In all stages of this disease, a decrease in the filtration fraction is characteristic because of the much greater decline in glomerular filtration than in renal plasma flow.

Measurement of Tubular Secretory Mass:

This is accomplished by measuring the Tm for PAH, i. e., the maximal secretory capacity of the tubule for PAH. In this case, PAH must be raised to relatively high levels in the blood, e. g., 50 mg. per 100 ml. At these levels, the tubular secretory carriers are working at maximal capacity. By correcting the urine PAH for that filtered as calculated from the glomerular filtration rate, the quantity of PAH secreted is obtained. The normal maximum is about 80 mg. per minute per 1.73 sq. M. Iodopyracet (Diodrast®) clearance may be similarly used to measure tubular excretion. The Tm for PAH can be used to gage the extent of tubular damage in renal disease because as tubule cells cease to function or are destroyed, excretion of PAH is proportionately diminished.

Phenolsulfonphthalein Test (PSP):

Dyes are widely used for excretion tests. An example is phenol red or phenolsulfonphthalein (PSP). The test is conducted by measuring the rate of excretion of the dye following intramuscular or intravenous administration. The intravenous test is the more valid since it eliminates the uncertainties of absorption which exist in the intramuscular test. Urine specimens may be collected at 15, 30, 60, and 120 minutes after the injection of the dye. If the 15-minute urine contains 25% or more of the injected PSP, the test is normal. Forty to 60% of the dye is normally excreted in the first hour and 20 to 25% in the second. The most useful information is obtained from the original 15-minute specimen since by the end of two hours the amount of dye excreted, although originally delayed, may now appear normal. The dye is readily excreted by the tubules, and therefore the result is not abnormal until impairment of renal function is extreme.

Concentration Tests:

Impairment of the capacity of the tubule to perform osmotic work is an early feature of renal disease. The determination of the specific gravity of the urine after a period of water deprivation becomes, therefore, a valuable and sensitive indicator of renal function. If the kidneys do absolutely no work, a fluid is excreted with a specific gravity the same as that of the glomerular filtrate, 1.010. As has been pointed out, any deviation from this specific gravity, i. e., dilution or concentration, requires osmotic work by the renal tubule.

In the **Addis test,** fluids are withheld for 24 hours (from 8:00 a. m. on one day to 8:00 a. m. on the next). This must never be done in cases of obviously impaired renal function or in hot weather. Other contraindications to the test include diabetes with polyuria, and adrenal insufficiency. The urine excreted up to 8:00 p. m. of the first day is discarded, but that excreted from 8:00 p. m. of the first day to 8:00 a. m. of the next day is collected and its specific gravity determined. Normally, this urine should have a specific gravity of more than 1.025 (up to 1.034). If the concentrating power of the kidney is such that after this period of water deprivation the specific gravity of the urine is still less than 1.025, renal damage is indicated (except during pregnancy, receding edema, or on diets inadequate in protein or salt).

The **Mosenthal test** is somewhat less rigorous than the Addis test. The patient is not restricted with respect to fluid intake. The bladder should be emptied at 8:00 a. m. on the first day and the urine discarded. Urine collections are made at two-hour intervals from 8:00 a. m. to 8:00 p. m., and all urine excreted during the 12-hour period from 8:00 p. m. to 8:00 a. m. is collected as one specimen. The specific gravity of each two-hour specimen and the volume and specific gravity of the 12-hour specimen are noted. Normally, the specific gravity of one or more two-hour specimens should be 1.018 or more, with a difference of not less than 0.009 between the highest and lowest readings. The volume should be less than 725 ml. for the 12-hour night specimen, with a specific gravity of 1.018 or above.

COMPOSITION OF URINE

Characteristics of Urine:

A. Volume: In the normal adult, 600 to 2500 ml. of urine are formed daily. The quantity normally depends on the water intake, the external temperature, the diet, and the mental and physical state. Urine volume is less in summer or in warm climates, for it is more or less inversely related to the extent of perspiration. Nitrogenous end products and coffee, tea, and alcoholic beverages have a diuretic effect. About half as much urine is formed during sleep as during activity.

B. Specific Gravity: This normally ranges from 1.003 to 1.030, varying according to concentration of solutes in the urine. The figures in the second and third decimal places, multiplied by 2.66 (Long's coefficient), give roughly the total solids in the urine in grams per liter; 50 Gm. of solids in 1200 ml. are an average normal for the day.

C. Reaction: The urine is normally acid, with a pH of about 6.0 (range: 4.7-8.0). Ordinarily, over 250 ml. of 0.1 N acid (titratable acidity) is excreted daily. When the protein intake is high, the urine is acid because excess phosphate and sulfate are produced in the catabolism of protein. Acidity is also increased in acidosis and in fevers.

   The urine becomes alkaline on standing because of conversion of urea to ammonia. It may also be alkaline in alkalosis such as that after excessive vomiting, at least in the early stages (see also p. 151).

D. Color: Normal urine is pale yellow or amber. The color varies with the quantity and concentration of urine voided. The chief pigment is urochrome, but small quantities of urobilin and hematoporphyrin are also present.

   In fever, because of concentration, the urine may be dark yellow or brownish. In liver disease, bile pigments may color the urine green, brown, or deep yellow. Blood or hemoglobin give the urine a smoky to red color. Methemoglobin and homogentisic acid color it dark brown. Drugs may color the urine. For example, methylene blue gives the urine a green appearance; and cascara and some other cathartics give it a brown color.

   The urine is usually transparent, but in alkaline urine a turbidity may develop by precipitation of calcium phosphate. Strongly acid urine precipitates uric acid salts, which have a pink color.

E. Odor: Fresh urine is normally aromatic but the odor modified by substances in the diet such as asparagus (methyl mercaptan odor?). In ketosis, the odor of excreted acetone may be detected.

Composition of Normal Urine*

Specific gravity = 1.003 to 1.030
Reaction (pH) = 4.7 to 8.0 (avg. 6.0)
Volume: Normal range = 600 to 2500 ml. per 24 hours (avg. 1200 ml.). Night/day ratio of volume = 1:2 to 1:4 if 8:00 a. m. and 8:00 p. m. are the divisions. Night urine usually does not exceed 500 to 700 ml. and usually has a specific gravity of more than 1.018.
Titratable acidity of 1000 ml. (depending on pH) = 250 to 700 ml. of 0.1 N NaOH for acid urine.
Total solids = 30 to 70 Gm. per liter (avg. 50 Gm.). Long's coefficient to estimate total solids per liter: Multiply last two figures of specific gravity by 2.66.

Inorganic Constituents (per 24 hours):

Chlorides (as NaCl)	10 (9-16) Gm. on usual diet	Sulfur (total) (as SO_3)	2 (0.7-3.5) Gm.
		Calcium	0.2 (0.1-0.2) Gm.
Sodium (varies with intake)	4 Gm. on usual diet	Magnesium	0.15 (0.05-0.2) Gm.
		Iodine	50-250 μg.
Phosphorus	2.2 (2-2.5) Gm.	Arsenic	0.05 mg. or less
Potassium (varies with intake)	2 Gm.	Lead	50 μg. or less

Organic Constituents (per 24 hours):		Nitrogen Equivalent
Nitrogenous (total)	25-35 Gm.	10-14 Gm.
Urea (half of total urine solids; varies with diet)	25-30 Gm.	10-12 Gm.
Creatinine	1.4 (1-1.8) Gm.	0.5 Gm.
Ammonia	0.7 (0.3-1) Gm.	0.4 Gm.
Uric acid	0.7 (0.5-0.8) Gm.	0.2 Gm.
Undetermined N (amino acid, etc.)		0.5 Gm.
Protein, as such ("albumin")	0-0.2 Gm.	
Creatine	60-150 mg. (increased in liver or muscle diseases or thyrotoxicosis)	

Traces of Other Organic Constituents (per 24 hours):
Hippuric acid 0.1-1 Gm. Oxalic acid 15-20 mg. Indican 4-20 mg. Coproporphyrins 60-280 μg. Purine bases 10 mg. Ketone bodies 3-15 mg. Allantoin 30 mg. Phenols (total) 0.2-0.5 Gm.

Sugar:
50% of people have 2-3 mg. per 100 ml. after a heavy meal. A diabetic can lose up to 100 Gm. per day.

Ascorbic Acid:
Adults excrete 15-50 mg. per 24 hours; in scurvy, less than 15 mg. per 24 hours.

*Modified from Krupp et al.: Physician's Handbook, 10th Ed., Lange Medical Publications, 1958.

Variations in Some Urinary Constituents With Different Protein Levels in the Diet*†

	Usual Protein Intake		Protein-Rich Diet		Protein-Poor Diet	
	Gm.	%N	Gm.	%N	Gm.	%N
Total urinary nitrogen	13.20		23.28		4.20	
Protein represented by above N	82.50		145.50		26.25	
Urea nitrogen	11.36	86.1	20.45	87.9	2.90	69.0
Ammonia nitrogen	0.40	3.0	0.82	3.5	0.17	4.0
Creatinine nitrogen	0.61	4.6	0.64	2.7	0.60	14.3
Uric acid nitrogen	0.21	1.6	0.30	1.3	0.11	2.6
Undetermined nitrogen	0.62	4.7	1.07	4.6	0.52	12.4
Titratable acidity (ml. 0.1 N)	284.0 ml.		655.0 ml.		160.0 ml.	
Volume of urine	1260.0 ml.		1550.0 ml.		960.0 ml.	
Total sulfur (as SO_3)	2.65 Gm.		3.55 Gm.		0.86 Gm.	
Inorganic sulfate (as SO_3)	2.16		2.82		0.64	
Ethereal sulfate (as SO_3)	0.18		0.36		0.11	
Neutral sulfate (as SO_3)	0.31		0.37		0.11	
Total inorganic phosphate (as P_2O_5)	2.59		4.07		1.06	
Chloride (as NaCl)	12.10		15.10		9.80	

*Reprinted, with permission, from Bodansky: Introduction to Physiological Chemistry, 4th Ed. John Wiley and Sons, Inc., 1938.
†A balance between intake of protein nitrogen and the excretion of nitrogen (nitrogen balance) is presumed to exist in these experiments.

Normal Constituents of the Urine:

Urea constitutes about one-half (25 Gm.) of the urine solids. Sodium chloride constitutes about one-fourth (9 to 16 Gm.).

A. Urea: This is the principal end product of protein metabolism in mammals. Its excretion is directly related to the protein intake. Normally it comprises 80 to 90% of the total urinary nitrogen; but on a low-protein diet this is less because certain other nitrogenous constituents tend to remain relatively unaffected by diet.

Urea excretion is increased whenever protein catabolism is increased, as in fever, diabetes, or excess adrenocortical activity. In the last stages of fatal liver disease, decreased urea production may lead to decreased excretion. There is also a decrease in urine urea in acidosis since some of the nitrogen which would have been converted to urea is diverted to ammonia formation. The urea does not, however, give rise to the ammonia directly.

Urea is measured either by conversion to ammonia by urease or by measuring gasometrically the nitrogen evolved in the hypobromite reaction:

$$CO(NH_2)_2 + 3NaOBr \longrightarrow 3NaBr + N_2 + CO_2 + H_2O$$

B. Ammonia: Normally there is very little ammonia in freshly voided urine. Its formation by the kidney in acidosis has been described on p. 294. In acidosis of renal origin, this mechanism may fail. Therefore, such acidosis is accompanied by a low concentration of ammonia in the urine. On the other hand, the ketosis and resultant acidosis of uncontrolled diabetes mellitus, in which renal function is unimpaired, will cause a high ammonia output in the urine.

The urinary ammonia, either preformed or derived from urea after treatment with urease, can be determined by liberating it from its ammonia salt by adding saturated potassium carbonate and then aerating the liberated ammonia into a measured excess of standard acid. The acid remaining unneutralized is then measured by titration, and from this the quantity of ammonia can be calculated.

C. Creatinine and Creatine: Creatinine is the product of the breakdown of creatine (see p. 260). In a given subject it is excreted in relatively constant amounts regardless of diet. The creatinine coefficient is the ratio between the amounts of creatinine excreted in 24 hours to the body weight in kilograms. It is usually 20 to 26 mg. per Kg. per day in normal men and 14 to 22 mg. per Kg. per day in normal women. Because this rate is so constant in a given individual, the creatinine coefficient may serve as a reliable index of the adequacy of a 24-hour urine collection. The excretion of creatinine is decreased in many pathologic states.

Creatine is present in the urine of children and, in much smaller amounts, in the urine of adults as well. In men the creatine excretion is about 6% of the total creatinine output (probably 60 to 150 mg. per day). In women creatinuria is much more variable (usually two to two and one-half times that of normal men), although in about one-fifth of the normal women studied the creatinuria did not exceed that found in men. In pregnancy, creatine excretion is increased. Creatinuria is also found in pathologic states such as starvation, impaired carbohydrate metabolism, hyperthyroidism, and certain myopathies and infections. Excretion of creatine is decreased in hypothyroidism.

Creatinine is measured colorimetrically by adding alkaline picrate to the urine. In the presence of creatinine, the mixture develops an amber color (Jaffé reaction). The color is read against a creatinine standard similarly treated with alkaline picrate solution.

Creatine, when heated in acid solution, is converted to creatinine, which can be measured as described. The difference in the creatinine content of the urine before and after boiling with acid gives the creatine content.

D. Uric Acid: This is the most important end product of the oxidation of purines in the body. It is derived not only from dietary nucleoprotein but also from the breakdown of cellular nucleoprotein in the body. Uric acid may also be synthesized in the body.

Uric acid is the principal end product of protein metabolism in birds and certain reptiles.

Uric acid is very slightly soluble in water but forms soluble salts with alkali. For this reason it precipitates readily from acid urine on standing.

The output of uric acid is increased in leukemia, severe liver disease, and in various stages of gout.

The blue color which uric acid gives in the presence of arsenophosphotungstic acid and sodium cyanide is the basis of the Folin colorimetric test. This is not a specific reaction, however. Salicylates raise the color value due to the excretion of gentisic acid and other similar metabolites. This may raise the apparent excretion of uric acid as much as 25% in 24 hours after a 5 Gm. dose of acetylsalicylic acid.

The specificity of the analysis for uric acid may be increased by treatment of the sample with uricase, the enzyme (from hog kidney) which causes the conversion of uric acid to allantoin. The decline in apparent uric acid concentration after uricase treatment is taken as a measure of the true uric acid content of the sample.

E. Amino Acids: In adults, only about 150 to 200 mg. of amino acid nitrogen are excreted in the urine in 24 hours. The full-term infant at birth excretes about 3 mg. amino acid nitrogen per pound body weight; this excretion declines gradually up to the age of six months, when it reaches a value of 1 mg. per lb. that is maintained throughout childhood. Premature infants may excrete as much as ten times as much amino acid nitrogen as the full-term infant.

The reason such very small amounts of amino acids are lost into the urine is that the renal thresholds for these substances are quite high. However, all of the naturally-occurring amino acids have been found in the urine, some in relatively large quantities when compared to the trace quantities characteristic of most. It is also of interest that a high percentage of some excreted amino acids is in combined forms and can be liberated by acid hydrolysis. Diet alters the pattern of amino acid excretion to a slight extent[10].

The amount of free amino acids found in the urine of normal male human subjects, after a 12-hour fast, is shown in the following table:

**Free Amino Acids in Urine*
(Normal Male Subjects After a 12-Hour Fast)**

	Range (mg. per hour)	Mean
Glycine	3.0-36.0	11.7 ± 9.45
Histidine	2.7-23.0	7.7 ± 1.33
Glutamine	1.6- 4.6	2.9 ± 0.96
Cystine	0.4- 4.3	2.3 ± 1.22
Tryptophan, tyrosine serine, proline, threonine, phenylalanine, lysine, arginine, alanine	1.0- 3.4 (for any of these amino acids)	
Leucine, isoleucine, valine, aspartic acid, glutamic acid, methionine	Less than 1 (for any of these amino acids)	

*Reproduced, with permission, from Harper, Hutchin, and Kimmel: Proc. Soc. Exper. Biol. and Med. 80:768, 1952.

In terminal liver disease and in certain types of poisoning (chloroform, carbon tetrachloride), the quantity of amino acids excreted is increased. This "overflow" type of aminoaciduria is to be distinguished from renal aminoaciduria due to a congenital tubular defect in reabsorption (see p. 291). In "cystinuria" (see p. 251) a considerable increase in excretion of only four amino acids occurs: arginine, cystine, lysine, and ornithine. The amounts of all other amino acids excreted remain normal.

F. Allantoin: This is derived from partial oxidation of uric acid (see p. 270). There are very small quantities in human urine, but in other mammals allantoin is the principal end product of purine metabolism, replacing uric acid.

G. Chlorides: These are mainly excreted as sodium chloride. Because most of the chlorides are of dietary origin, the output varies considerably with the intake. The excretion of chlorides is reduced in certain stages of nephritis and in fevers.

H. Sulfates: The urine sulfur is derived mainly from protein because of the presence of the sulfur-containing amino acids, methionine and cystine, in the protein molecule. Its output therefore varies with the protein intake. The total urine sulfur is usually partitioned into three forms. It is customary to express all sulfur concentrations in the urine as SO_3.

1. Inorganic (sulfate) sulfur - This is the completely oxidized sulfur precipitated from urine when barium chloride is added. It is roughly proportional to the ingested protein with a ratio of 5:1 between urine nitrogen and inorganic sulfate (expressed as SO_3). Together with the total urinary nitrogen, this fraction of urine sulfur is an index of protein catabolism.
2. Ethereal sulfur (conjugated sulfates) - This fraction (about 10% of the total sulfur) includes the organic combinations of sulfur excreted in the urine. Examples are the phenol and cresol sulfuric acids, indoxyl and skatoxyl sulfuric acids, and other sulfur conjugates formed in detoxication.

   The ethereal sulfate fraction is in part derived from protein metabolism; but in indican (see p. 165) and some of the phenols, putrefactive activity in the intestine is also represented.

   After hydrolysis with hot hydrochloric acid, the ethereal sulfates may be precipitated with barium chloride.
3. Neutral sulfur - This fraction is the sulfur which is incompletely oxidized, such as that which is contained in cystine, taurine, thiocyanate, or sulfides. It does not vary with the diet to the same extent that the other fractions do.

   Neutral sulfur is determined as the difference between the total sulfur and the sum of the inorganic and ethereal sulfur.

I. Phosphates: The urine phosphates are combinations of sodium and potassium phosphate (the alkaline phosphates) as well as of calcium and magnesium (so-called earthy phosphates). These latter forms are precipitated in alkaline urines.

The diet, particularly the protein content, influences phosphate excretion. Some is also derived from cellular breakdown.

In certain bone diseases, such as osteomalacia and renal tubular rickets, the output of phosphorus in the urine is increased. In hyperparathyroidism the excretion of phosphorus is also markedly increased. A decrease is sometimes noted in infectious diseases, in hypoparathyroidism, and renal disease.

Urine phosphate may be measured colorimetrically as phosphomolybdate after addition of stannous chloride, or a gravimetric method may be used. The concentration of phosphate is expressed as P_2O_5.

J. Minerals: Sodium, potassium, calcium, and magnesium - the four cations of the extracellular fluid - are present in the urine. The sodium content varies considerably with intake and physiologic requirements. Urine potassium rises when the intake is increased or in the presence of excessive tissue catabolism, in which case it is derived from intracellular materials. The excretion of potassium is also affected by acid-base equilibrium (see p. 296). Sodium and potassium excretions are also controlled by the activity of the adrenal cortex.

Most of the calcium and magnesium is excreted by the bowel; the content of these elements in the urine is, therefore, relatively low. However, this will vary in certain pathologic states, particularly those involving bone metabolism.

K. Vitamins, hormones, and enzymes can be detected in small quantities in normal urine. The urinary content of these substances is often of diagnostic importance. (See pp. 108 and 353.)

Abnormal Constituents of the Urine:

A. Proteins: Proteinuria (albuminuria) is the presence of albumin and globulin in the urine in abnormal concentrations. Normally not more than 30 to 200 mg. of protein are excreted daily in the urine.

1. Physiologic proteinuria, in which less than 0.5% protein is present, may occur after severe exercise, after a high-protein meal, or as a result of some temporary impairment in renal circulation when a person stands erect (orthostatic or postural proteinuria).

   In 30 to 35% of the cases, pregnancy is accompanied by proteinuria.
2. Pathologic proteinurias are sometimes classified as prerenal, when the primary causes are factors operating before the kidney is reached, although the kidney may also be involved; renal, when the lesion is intrinsic to the kidney; and postrenal, when the proteinuria is due to inflammation in the lower urinary tract.

a. In glomerulonephritis albuminuria is marked during the degenerative phase. The lowest excretion of albumin is during the latent phase and may increase terminally.
b. In nephrosis a marked proteinuria occurs. This is accompanied by edema and low concentrations of serum albumin.
c. Nephrosclerosis, a vascular form of renal disease, is related to arterial hypertension. The proteinuria observed in this disease increases with the increasing severity of the renal lesion. The loss of protein in nephrosclerosis is generally less than that in glomerulonephritis.
d. Eclampsia - Proteinuria occurs during eclampsia in pregnancy, not only during the convulsive stage but also in pre-eclampsia.
e. Poisoning - Proteinuria is also observed in poisoning of the renal tubules by heavy metals like mercury, arsenic, or bismuth unless the poisoning is severe enough to cause anuria.

3. Albumin may be detected by heating the urine, preferably after centrifuging to remove the sediment, then adding a little dilute acetic acid. A white cloud or precipitate which persists after addition of the acid indicates that protein is present. In quantitative measurement of urine protein, the protein is precipitated with trichloroacetic acid and then separated for analysis, either colorimetrically (biuret) or by Kjeldahl analysis.
4. Bence Jones proteins (see also p. 135) - These peculiar proteins are globulins which occur in the urine most commonly in multiple myeloma and rarely in leukemia, Hodgkin's disease, and lymphosarcoma. They may be identified in the urine by their ability to precipitate when the urine is warmed to 50 to 60° C. and to redissolve almost completely at 100° C. The precipitate re-forms on cooling.

B. Glucose: Normally not more than 1 Gm. of sugar is excreted per day. Glycosuria is indicated when more than this quantity is found. The various causes of glycosuria have been discussed. Transient glycosuria may be noted after emotional stress, such as an exciting athletic contest. Fifteen percent of cases of glycosuria are not due to diabetes. Usually, however, glycosuria suggests diabetes; this must be confirmed by blood studies to eliminate the possibility of renal glycosuria.

C. Other Sugars:

1. Fructosuria is a rare anomaly in which the metabolism of fructose but not that of other carbohydrates is disturbed.
2. Galactosuria and lactosuria may occur occasionally in infants and in the mother during pregnancy, lactation, and the weaning period. In congenital galactosemia, the inherited disease which is characterized by impaired ability to convert galactose to glucose (see p. 195), the blood levels of galactose are much elevated and galactose spills over into the urine.
3. Pentosuria may occur transiently after ingestion of foods containing large quantities of pentoses, such as plums, cherries, grapes, and prunes. Congenital pentosuria is a genetic defect characterized by inability to metabolize L-xylulose, a constituent of the uronic acid pathway (see p. 193).

All of the above sugars reduce Benedict's solution. When it is suspected that sugars other than glucose are present, it has been customary to perform a fermentation test with baker's yeast. If all of the reducing action is removed by the yeast, this suggests that only glucose is present. However, more specific tests are preferred. The recent introduction of a specific analytical test for glucose by the use of the enzyme, glucose oxidase, is one such test. A comparison of the apparent glucose content of the urine, as determined by total reducing action with the absolute glucose content (as determined by glucose oxidase), would indicate more definitely whether sugars other than glucose were present. If so, these other sugars can be identified readily by paper chromatography or in some cases by preparation of specific osazones (see p. 8).

D. Ketone Bodies: Normally, only 3 to 15 mg. of ketones are excreted in a day. The quantity is increased in starvation, impaired carbohydrate metabolism (e.g., diabetes), pregnancy, ether anesthesia, and some types of alkalosis. In many animals, excess fat metabolism will also induce a ketonuria (see p. 211). The acidosis accompanying ketosis will cause increased ammonia excretion as the result of the body's effort to conserve fixed base.

The qualitative tests for ketone bodies in the urine are usually tests for the presence of acetone. Acetoacetic acid is readily converted to acetone so that it too may be tested by applying a test for acetone after boiling. In the iodoform test for acetone, when urine containing acetone is heated with NaOH and iodine, iodoform precipitates from the solution. Alcohol will also form iodoform in this test, but the reaction is slower.

The **Gerhardt ferric chloride test**: Ferric chloride solution is added dropwise to 5 ml. of urine until a precipitate is no longer formed. In the presence of acetoacetic acid, the mixture becomes red. On boiling the urine no reaction can be obtained if it was originally due to acetoacetic acid, since the heating converts acetoacetic acid to acetone, which does not give this reaction.

Quantitative methods for determining total acetone bodies or the individual substances themselves are based on their conversion to acetone and its precipitation as a basic acetone-mercury complex in the presence of oxidizing agents, sulfuric acid, and potassium dichromate.

E. Bilirubin: The presence of bilirubin in the urine and its relationship to jaundice are discussed in Chapter 17, pp. 273 and 277.

F. Blood: In addition to its occurrence in nephritis, blood in the urine (hematuria) may be the result of a lesion in the kidney or urinary tract (e.g., after trauma to the urinary tract). However, free hemoglobin (hemoglobinuria) may also be found in the urine after rapid hemolysis, e.g., in blackwater fever (a complication of malaria) or after severe burns.

G. Porphyrins (see p. 62): The urine coproporphyrin excretion of the normal human adult is 60 to 280 μg. per day. The occurrence of uroporphyrins as well as increased amounts of coproporphyrins in the urine is termed porphyria.

THE RENAL PRESSOR SYSTEM

In addition to the excretory functions of the kidney, the tubule cells have what resembles an endocrine function in elaborating a chemical substance which is added to the blood. This chemical substance contributes to a reaction affecting tissues throughout the body. The effect is best demonstrated by experiments in which hypertension is induced after the supply of blood to the kidney is impaired. This can be accomplished by compressing the renal artery (Goldblatt) or by inducing contraction around the kidney. Perinephritis, which results from contraction around the kidney, has been brought about experimentally by enclosure of the kidney in silk cloth (Page). Clinically, it is also observed from contraction around the kidney of the scar caused by injury due to a foreign body. In all such cases, i.e., reduction of the blood supply to the kidney (renal ischemia) or the induction of perinephritis, a persistent hypertensive state develops.

The Mechanism of Renal Hypertension. The renal cortex forms a protein called renin, actually a proteolytic enzyme; this is liberated into the blood. Here it meets a specific substrate, another protein (an α_2 globulin) normally present in plasma and formed in the liver; this globulin is termed renin substrate or angiotensinogen. The renin splits off and removes from angiotensinogen a polypeptide (recently shown to be an octapeptide[11]) designated angiotensin. This latter is the pressor substance acting directly on the cardiovascular system. It increases the force of the heart beat and constricts the arterioles, often resulting in diminished renal blood flow even though peripheral blood flow is unchanged.

The normal kidneys, and other tissues to a lesser extent, contain proteolytic enzymes called **angiotensinases** which are capable of destroying angiotensin.

The reactions of this renal pressor system are as follows:

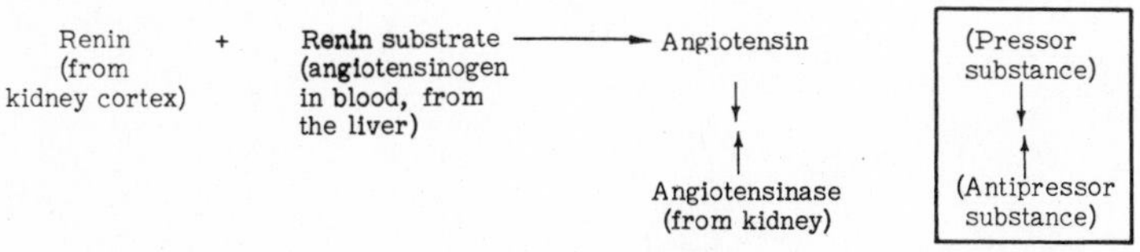

Renin (from kidney cortex) + Renin substrate (angiotensinogen in blood, from the liver) ⟶ Angiotensin ↓↑ Angiotensinase (from kidney)

(Pressor substance) ↓↑ (Antipressor substance)

It is still uncertain whether the renal pressor system is involved in blood pressure regulation under normal circumstances. When renal circulation is impaired, the pressor system is more active than in the presence of normal renal circulation. Renal hypertension may be due to a relative or absolute lack of angiotensinase.

The discovery of the renal pressor system has clarified the long-recognized association between renal and cardiovascular disease, but the part which it plays in human hypertension is still controversial. Some authorities have suggested renal participation in both essential and malignant hypertension - if not in the early, at least in the late phases of the diseases. But as far as present knowledge is concerned, this is still unconfirmed. The hypertension of acute glomerulonephritis may be caused in part by renal pressor activity.

• • •

References:

1. Mueller, C.B., Mason, A.D., Jr., and Stout, D.G.: Am.J.Med. **18**:267, 1955.
2. Wirz, H.: Helv. Physiol. et Pharmacol. Acta **14**:353, 1956.
3. Verney, E.B.: Surg., Gynecol. and Obs. **106**:441, 1958.
4. Dudley, H.F., Boling, E.A., Le Quesne, L.P., and Moore, F.D.: Ann.Surg. **140**:354, 1954.
5. Cogan, D.G., Kuwabara, T., Kinoshita, J., Sheehan, L., and Merola, L.: J.Am.Med. Assoc. **164**:394, 1957.
6. Pitts, R.F.: Am.J.Med. **9**:356, 1950.
7. Gilman, A., and Brazeau, P.: Am.J.Med. **15**:765, 1953.
8. Rector, F.C., Jr., Seldin, D.W., and Copenhaver, J.H.: J.Clin.Invest. **34**:20, 1955.
9. Mudge, G.H.: Am.J.Med. **20**:448, 1956.
10. Fowler, D.I., Norton, P.M., Cheung, M.W., and Pratt, E.L.: Arch.Biochem.Biophys. **68**:452, 1957.
11. Braun-Menendez, E., and Page, I.H.: Science **127**:242, 1958.

Bibliography:

Smith, H.: The Kidney. Oxford University Press, 1952.
Smith, H.: Principles of Renal Physiology. Oxford University Press, 1956.

19...

Water and Mineral Metabolism

WATER METABOLISM

Body Water and Its Distribution:

The total body water is equal to about 40 to 65% of the body weight, with an average of 55% for adult men and 47% for adult women. The body water is distributed throughout two main compartments: the extracellular compartment (plasma and interstitial fluid), which contains about one-fourth of the total body water; and the intracellular compartment, which contains about three-fourths of the total body water.

Distribution of Body Water in the Male

Fluid Compartment	Percent of Body Weight	ml. of Water in 154 lb. (70 Kg.) Man
Extracellular Water	15-17	10,500
Plasma	4-5	3200
Interstitial Fluid	11-12	7300
Intracellular Water	40-45	31,500

Measurement of Distribution of Body Water:

Total body water has been determined in animals by desiccation. More recently the distribution of heavy water, deuterium oxide, or of tritium oxide has been used in the living animal and in human subjects as a method of measuring total body water. Antipyrine is also used.

The values for body water given above were obtained by measuring the volume of distribution of antipyrine. Deuterium oxide or tritium oxide measurements have given somewhat higher average values for total body water (62% in males). However, in all studies of the proportion of the body which is water, considerable variation is to be expected when different subjects are compared even by the same analytical method. This is due mainly to variation in the amount of fat in the body. The higher the fat content of the subject, the smaller the percent of water that subject will contain in his body. If a correction for the fat content of the subject is made, the total body water in various subjects is relatively constant when expressed as percent of the "lean body mass," i.e., the sum of the fat-free tissue.

The composition of the adult human body has been determined by direct chemical analysis[1]. The whole body contains 19.44% ether-extractable material (lipid), 55.13% moisture, 18.62% protein, and 5.43% ash, including 1.907% calcium and 0.925% phosphorus. When these data are recalculated to the fat-free basis, the moisture content is then 69.38%, which is in agreement with data obtained by indirect methods as described above.

Specific gravity of the body may also serve as a basis for the calculation of total body water. The body is considered as a mixture of fat, which is of relatively low density, and fat-free tissue, which is of relatively high density. By measuring the specific gravity of the body (weighing the subject in air and under water), it is possible to calculate the proportion of the body which is fat tissue and that which is fat-free tissue. This technic has been used to arrive at an estimate of the lean body mass, described above.

A. Plasma Volume: Plasma volume may be measured by the Evans blue dye technic. In this procedure a carefully measured quantity of the dye is injected intravenously; after a lapse of time to allow for mixing, a blood sample is withdrawn and the concentration of the dye in the plasma is determined colorimetrically. The normal figures for plasma volume thus determined are 47 to 50 ml. per Kg. body weight.

Other methods of plasma (or blood) volume measurements are based on the intravenous injection of radiophosphorus-labeled red cells (P^{32}) or radioiodine-labeled human serum albumin. These substances distribute themselves in the blood stream and after a mixing period of ten or more minutes their volume of distribution may be calculated from their concentration in an aliquot of blood or plasma.

B. Interstitial Fluid Volume: Interstitial fluid volume is measured with some substance which does not penetrate into the cells. The test substance must be injected intravenously at a constant rate for a reasonably long period in order to provide even distribution between the plasma and interstitial spaces. When equal distribution has been accomplished, the volume of distribution is calculated by comparison of the amount injected with the plasma concentration (after correction for that excreted in the urine during the period). This volume of distribution is a measure of the entire extracellular fluid volume. A separate determination of the plasma volume is made, and this is subtracted from the total extracellular volume to obtain the interstitial volume.

Various substances such as inulin, mannitol, potassium thiocyanate, and radiosodium have been employed to measure extracellular fluid volume. Each substance yields a somewhat different value, indicating that they do not all occupy exactly the same space; in fact, thiocyanate and sodium have been discarded as unsuitable for the determination of extracellular volume because they penetrate into the cells. The values obtained with inulin are probably the most reliable. It is considered desirable, when referring to the extracellular fluid volume, to speak of the inulin space, mannitol space, etc., in accordance with the substance used to measure it.

C. Intracellular water is the difference between total body water and the extracellular water.

The Availability of Water (Water Intake):

The two main sources of water (about 2500 ml. per day) are:

A. Preformed Water: Liquids imbibed as such, 1200 ml.; water in foods, 1000 ml.

B. Water of Oxidation - 300 m.: The water of oxidation (sometimes termed "metabolic water") is derived from the combustion of foodstuffs in the body. The oxidation of 100 Gm. of fat yields 107 Gm. of water; oxidation of 100 Gm. of carbohydrate, 55 Gm. of water; oxidation of 100 Gm. of protein, 41 Gm. of water.

Losses of Water: (See the table below.)

Water is lost to the body by four routes: the skin, as sensible and insensible perspiration; the lungs, as water vapor in the expired air; the kidneys, as urine; and the intestines, in the feces. It is customary to refer to the sum of the dermal loss (exclusive of visible perspiration) and the pulmonary loss as the insensible losses.

Daily Water Losses and Water Allowances for Normal Individuals Who Are Not Working or Sweating[2]

Size	Losses				Allowances	
	Urine (ml.)	Stool (ml.)	Insensible (ml.)	Total (ml.)	ml./person	ml./Kg.
Infant (2-10 Kg.)	200-500	25-40	75-300 (1.3 ml./Kg./Hr.)	300-840	330-1000	165-100
Child (10-40 Kg.)	500-800	40-100	340-600	840-1500	100-1800	100-45
Adolescent or Adult (60 Kg.)	800-1000	100	600-1000 (0.5 ml./Kg./Hr.)	1500-2300	1800-2500	45-30

Additional Water Losses in Disease

In kidney disease, in which concentrating ability is limited, renal water loss may be twice as high as that listed in the table. Insensible losses may rise much higher than normal as a consequence of operations, in fever, or in the physically debilitated. When subjected to high environmental temperatures, patients will also sustain extremely high extrarenal water losses, as much as 2000 to 5000 ml. in some instances. Water losses from the intestine may be considerable, particularly in diarrhea and vomiting.

Factors Which Influence the Distribution of Body Water:

Water is retained in the body in a rather constant amount, but its distribution is continuously subject to change. Osmotic forces are the principal factors which control the location and the amount of fluid in the various compartments of the body. These osmotic forces are maintained by the solutes, the substances dissolved in the body water.

A. Solutes in the Body: The solutes in the body fluids are important not only in directing fluid distribution but also in maintaining acid-base balance (see p. 150). In a consideration of the various substances in solution in the fluids of the body and of their effect on water retention and distribution, it is convenient to divide them into three categories:
   1. Organic compounds of small molecular size (glucose, urea, amino acids, etc.) - Since these substances diffuse freely across cell membranes, they are not important in the distribution of water; if present in large quantities, however, they aid in retaining water and thus do influence **total** body water.
   2. Organic substances of large molecular size (mainly the proteins) - The importance of the plasma proteins in the exchange of fluid between the circulating blood and the interstitial fluid has been discussed (see p. 133). The effect of the protein fraction of the plasma and tissues is mainly on the **transfer** of fluid from one compartment to another, not on the total body water.
   3. The inorganic electrolytes - Because of the relatively large quantities of these materials in the body, the inorganic electrolytes are by far the most important, both in the **distribution** and in the **retention** of body water.

Plasma Electrolyte Concentrations (from Gamble)

Cations (+)	mEq. /L.	Anions (-)	mEq. /L.
Na^+	142	HCO_3^-	27
K^+	5	Cl^-	103
Ca^{++}	5	$HPO_4^=$	2
Mg^{++}	3	$SO_4^=$	1
		Organic acids	6
		Protein	16
Totals	155		155

B. Measurements of Solutes: In describing chemical reactivity, particularly acid-base balance, all reacting ions must be expressed in identical concentration units. Since one chemical equivalent of any substance is exactly equal in chemical reactivity to one equivalent of any other, this can be accomplished by converting the concentrations of each into **equivalents** per liter. Because of the small quantities involved in body fluids, the milliequivalent (mEq., 1/1000 Eq.) is preferred. Furthermore, when changes occur in the chemistry of the body fluid, there are usually compensatory shifts of one ion to make up for losses of another. For example, excessive losses of chloride over sodium in vomiting from the stomach result in a chloride deficit in the extracellular fluid. This is promptly compensated by an increase in bicarbonate to accompany the sodium left uncovered by the chloride loss. These changes can be readily understood and calculated when all reactants are expressed in the same units.
   1. Conversion of electrolyte concentrations to mEq. - For conversion of mg. per 100 ml. to mEq. per L. -
      a. Express the concentration on a per liter basis, i.e., multiply the number of mg. (per 100 ml.) by 10 to determine the number of mg. per L.

b. Divide the mg. per L. by the appropriate mEq. weight given below. The mEq. weight of an element is the millimolecular weight divided by the valence.
c. Milliequivalent weights -

Milliequivalent Weights

Na	23	Cl	35.5
K	39	Cl (as NaCl)	58.5
Ca	20	HPO_4 (as P)	17.2*
Mg	12	SO_4 (as S)	16.0

Examples: Plasma sodium: 322 mg. /100 ml. Multiply the mg. by 10 (to express on a per liter basis). Then divide by mEq. weight of sodium, 23.

$322 \times 10 = 3220$ mg. /L.; divided by 23 = 140 mEq. /L.

Chloride (reported as NaCl): 603 mg. /100 ml.

$603 \times 10 = 6030$ mg. /L.; divided by 58.5 = 103 mEq. /L.

Calcium: 10 mg. /100 ml.

$10 \times 10 = 100$ mg. /L.; divided by 20 = 5 mEq. /L.

2. Conversion of bicarbonate to milliequivalents - The bicarbonate of the plasma is measured by conversion to carbon dioxide and reported in volumes percent (Vol. %); to convert to mEq. of bicarbonate per liter, divide carbon dioxide combining power, expressed as Vol. %, by 2.3.†
3. Conversion of organic acids and proteins to mEq. - The organic acids and the proteins in the anion column of plasma are calculated from their combining power with base. The base equivalence of protein, in mEq. per L., is obtained by multiplying the grams of total protein per 100 ml. by 2.43.

C. Electrolyte Composition of blood plasma and of intracellular fluid may be graphically expressed as in the diagrams on p. 311.

*The inorganic phosphorus in the serum exists as a buffer mixture in which approximately 80% is in the form of $HPO_4^{=}$ and 20% as $H_2PO_4^{-}$. For this reason the mEq. weight is usually calculated by dividing the atomic weight of phosphorus by 1.8. Thus the mEq. weight for phosphorus in the serum is taken as 31/1.8, or 17.2. To avoid the problem which the two valences of the serum phosphorus present, some laboratories prefer to express the phosphorus as millimols (mM.) rather than mEq. One mM. of phosphorus is 31 mg. To convert mg. of phosphorus to mM., divide mg. per L. by 31, e.g., serum phosphorus = 3.1 mg. per 100 ml. = 31/31 = 1 mM. per L.

†The conversion of the carbon dioxide combining power to mEq. of bicarbonate is based on the following facts. One mol of a gas occupies 22.4 L. (at 0° C. and 760 mm. Hg), and therefore 1 mM. occupies 22.4 ml. or, what is the same thing, each 22.4 ml. of gas is equivalent to 1 mM.; 600 ml. of carbon dioxide per L. (a normal total blood carbon dioxide) thus equals 600/22.4 = 26.7 mM. total carbon dioxide per L. (1 mM. of carbon dioxide is the same as 1 mEq. of carbon dioxide).

The total carbon dioxide as determined in the blood includes carbonic acid, free carbon dioxide, and bicarbonate. The bicarbonate fraction alone can be calculated by assuming that a 1:20 ratio exists between carbonic acid and bicarbonate. Under these conditions, the plasma bicarbonate fraction is derived by dividing the total carbon dioxide (as carbon dioxide combining power in Vol. %) by 2.3.

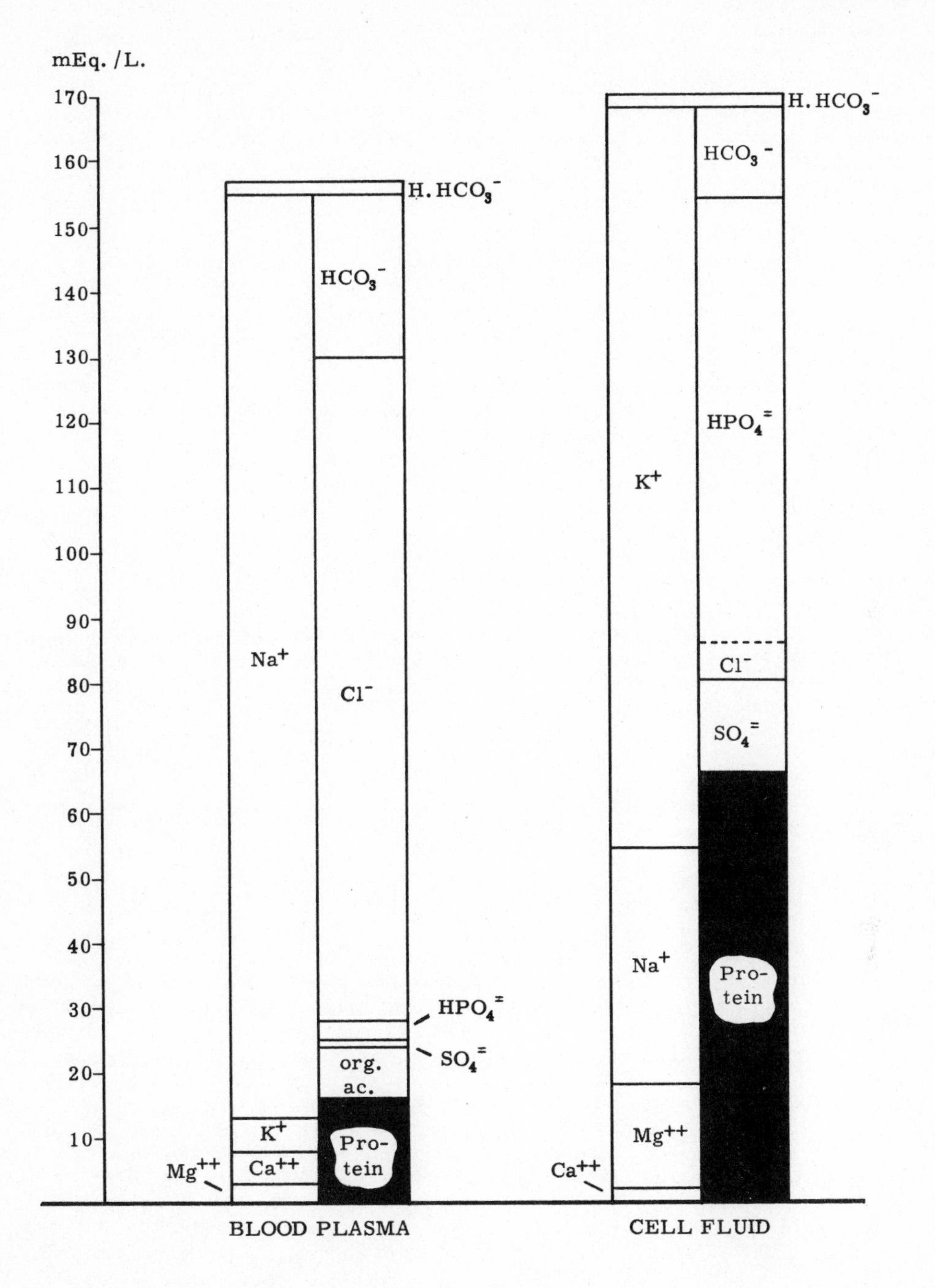

Electrolyte Composition of Blood Plasma and Intracellular Fluid. (Modified from Gamble.)

1. Composition of interstitial fluid - The electrolyte composition of the interstitial fluid is similar to that of the plasma except that chloride largely replaces protein in the anion column.
2. Composition of intracellular fluids - The intracellular fluid differs in electrolyte composition from that of the plasma in that potassium rather than sodium is the principal cation and, largely due to the presence of phosphorylated organic compounds, phosphate rather than chloride is the principal anion. The intracellular chloride content is variable in accordance with the metabolic circumstances. The amount of protein within the cell is also considerably larger than that in its extracellular environment.

Recent findings indicate that the intracellular concentration of sodium is higher than had been assumed. A normal average intracellular sodium of 37 mEq. per L. of intracellular water has been reported by Deane and Smith[3]. Furthermore, it is now clear that sodium may replace potassium within the cell when sodium salts are administered to potassium-deficient subjects. The adrenocortical steroids and ACTH also influence the concentration of sodium and potassium within the cell. Under the influence of these hormones, intracellular sodium may be increased.

D. Importance of Sodium and Potassium in Water Metabolism: Both from the standpoint of osmotic forces (directing the movement of water from one compartment to another in the body) and in controlling the total hydration of the body, sodium and potassium are the most important elements in the body fluids. As has been pointed out, in the normal individual sodium is largely confined to the extracellular space and potassium to the intracellular space.

1. Sodium - As Gamble has so well expressed it, sodium is the ''backbone'' of the extracellular fluid in that it, more than any other element, determines the quantity of extracellular fluid to be retained. This is the reason that sodium intake is restricted in order to control overhydration in various pathologic states.
2. Potassium - Under certain conditions, potassium leaves the cells. Important examples are found in prolonged gastrointestinal losses due to vomiting, diarrhea, or prolonged gastric suction. Replacement of lost electrolytes with only sodium salts leads to migration of sodium into cells to replace the potassium deficit. This produces profound alterations in cellular metabolism such as persistent alkalosis even after apparently adequate salt and water therapy. It is preventable by the prompt and concomitant replacement of potassium deficits as well as sodium deficits.
3. Electrolyte influence on water shifts - Since water is freely diffusible across the cell barrier, its movement is determined by the changes in concentration of the osmotically effective electrolytes (principally sodium and potassium) on either side. Changes in extracellular electrolyte concentration are most commonly the basis for these shifts of water.

Dehydration:

This term should not imply only changes in **water** balance. Almost always there must also be accompanying changes in electrolytes.

A. Water Loss or Restriction Causing Dehydration: When the supply of water is restricted for any reason, or when the losses are excessive, the rate of water loss exceeds the rate of electrolyte loss. The extracellular fluid becomes concentrated and hypertonic to the cells. Water then shifts from the cells to the extracellular space to compensate.

The symptoms of this intracellular dehydration are severe thirst, nausea and vomiting, a hot and dry body, a dry tongue, loss of coordination, and a concentrated urine of small volume. Intracellular dehydration is corrected by giving water by mouth, or dextrose and water parenterally, until symptoms are alleviated and the urine volume is restored.

B. Electrolyte Deficit: A relative deficit of electrolytes may occur when an excess of water is ingested. This condition of overhydration is commonly observed when large amounts of electrolyte-free solutions are administered to patients. More frequently, however, water and electrolytes are both lost, and replacement with only water leads to a deficiency of electrolytes in the presence of normal or excess total body water. The deficiency of sodium in the extracellular fluid is mainly responsible for the resulting hypotonicity of this fluid compartment. Some water passes into the cells, which are hypertonic to the extracellular fluid, causing the so-called intracellular edema. There follows a diminution in extracellular fluid volume which is very damaging. The resulting decrease in blood volume is conducive to a fall in blood pressure, slowing of circulation, and consequent impairment in renal function. Since the kidney is an essential aid in restoring the normal equilibrium, this latter complication is a serious one.

The patient becomes progressively weaker, but he does not complain of thirst and his urine volume is not notably changed. There is, however, reliable evidence of this type of dehydration in the elevated hematocrit or plasma total protein and the lowered sodium and chloride concentration in the plasma.

C. Correction of Dehydration: Because of the high content of electrolytes in the gastrointestinal secretions, loss of fluid from the gastrointestinal tract will readily produce serious fluid and electrolyte deficits if prompt and accurate replacement of the losses does not take place. In the table below the volume and composition of gastrointestinal fluids and of sweat are shown. Loss of chloride in excess of sodium will be expected when fluid is withdrawn from the upper gastrointestinal tract, as may occur in high intestinal obstruction, pyloric stenosis, gastric vomiting, or in continuous gastric suction. Ordinarily, sodium chloride solutions may be given parenterally to repair the losses since, in the presence of adequate kidney function, a proper adjustment of the electrolyte imbalance will occur. The importance of simultaneous replacement of potassium must also be recalled (see p. 321).

Fluid and electrolyte losses originating from the intestinal tract (as in prolonged diarrheas, pancreatic or biliary fistulas, etc.) are characterized by the removal of a fluid high in sodium and bicarbonate. This leads to a relative chloride excess and a bicarbonate deficit. This condition might best be repaired initially by the intravenous administration of a mixture of two-thirds isotonic saline solution and one-third sodium lactate solution (1/6 molar).

Volume and Composition of Gastrointestinal Secretions and Sweat[4 a,b]

Fluid	Average Volume (ml. /24 hrs.)	Electrolyte Concentrations (mEq. /L.)			
		Na^+	K^+	Cl^-	HCO_3^-
Blood plasma		135-150	3.6-5.5	100-105	24.6-28.8
Gastric juice	2500	31-90	4.3-12	52-124	0
Bile	700-1000	134-156	3.9-6.3	83-110	38
Pancreatic juice	> 1000	113-153	2.6-7.4	54-95	110
Small bowel (Miller-Abbott suction)	3000	72-120	3.5-6.8	69-127	30
Ileostomy					
Recent	100-4000	112-142	4.5-14	93-122	30
Adapted	100-500	50	3	20	15-30
Cecostomy	100-3000	48-116	11.1-28.3	35-70	15
Feces	100	< 10	< 10	< 15	< 15
Sweat	500-4000	30-70	0-5	30-70	0

Dehydration is frequently a complication of gastrointestinal tract disturbances, but it is not confined to these conditions. Other disorders in which dehydration is a problem include diabetes mellitus, Addison's disease, uremia, extensive burns, and shock.

Clinically, change in body weight during short periods is a reliable criterion of changes in hydration. When a patient is properly nourished and hydrated, his body weight remains relatively constant, with only a slight variation. Rapid daily gain in weight indicates overhydration. Loss of 8 to 12% in body weight represents a significant degree of dehydration if it is due to loss of fluids.

MINERAL METABOLISM

Although the mineral elements constitute a relatively small amount of the total body tissues, they are essential to many vital processes. The function of individual minerals has been mentioned at various points in this book: for example, blood calcium and its role in neuromuscular irritability and in the clotting of blood, the effect of various ions on activation of enzymes, and the activities of electrolytes in acid-base regulation.

The balance of ions in the tissues is often of importance. For example, normal ossification demands a proper ratio of calcium to phosphorus; the normal ratio between potassium and calcium in the extracellular fluid must be maintained to insure normal activity of the muscle.

Certain mineral elements, principally sodium and potassium, are the major factors in osmotic control of water metabolism, as described in the previous section of this chapter. Other minerals are an integral part of important physiologic compounds such as iodine in thyroxine, iron in hemoglobin, zinc in insulin, cobalt in vitamin B_{12}, sulfur in thiamine, biotin, Co A, and lipoic acid.

The animal body requires seven principal minerals: calcium, magnesium, sodium, potassium, phosphorus, sulfur, and chlorine. These minerals constitute 60 to 80% of all the inorganic material in the body. At least seven other minerals are utilized in trace quantities: iron, copper, iodine, manganese, cobalt, zinc, and molybdenum. Several other elements are present in the tissues, but their functions, if any, are not known. These include fluorine, aluminum, and boron.

CALCIUM

Functions:

Calcium is present in the body in larger amounts than any other cation. Almost all of it is in the bones and teeth. The very small quantity not in the skeletal structures is in the body fluids and is in part ionized. Ionized calcium is of great importance in blood coagulation, in the function of the heart, muscles, and nerves, and in the permeability of membranes.

Sources:

Dietary sources of calcium include milk, cheese, egg yolk, beans, lentils, nuts, figs, cabbage, turnip greens, cauliflower, and asparagus.

Requirements:

Men and women: 800 mg. daily.
During latter half of pregnancy and lactation: 1.5 to 2 Gm. daily.
Children: 1 to 1.4 Gm. daily.

To supply additional calcium, the carbonate, lactate, or gluconate salts as well as dicalcium phosphate may be administered.

The requirements for calcium listed above are thought to be excessive by some nutritional authorities[5]. There are also reports[6] of the occurrence of metastatic calcification associated with high intakes of calcium and of alkali in connection with dietary supplements prescribed for patients with peptic ulcer (the milk-alkali syndrome). A high intake of calcium in the presence of a high intake of vitamin D such as may occur in children is also a potential source of hypercalcemia and possibly of widespread excessive calcification. These considerations may indicate a need for revision of the presently recommended allowances for calcium, particularly in the case of adults other than pregnant and lactating women.

Absorption:

The ability of different individuals to utilize the calcium in foods varies considerably. On a high-protein diet, 15% of the dietary calcium is absorbed; on a low-protein diet, 5%. Phytic acid in cereal grains interferes with calcium absorption by forming insoluble calcium phytate in the intestine. Oxalates in foods (e.g., spinach) may have a similar effect. Other intestinal factors which influence absorption of calcium include:

A. pH: The more alkaline the intestinal contents, the less soluble the calcium salts. An increase in acidophilic flora (e.g., the lactobacilli) is recommended to lower the pH, which favors calcium absorption.

B. Phosphate: If the P:Ca ratio is high, much $Ca_3(PO_4)_2$ will be formed and absorption diminished.

C. Presence of Free Fatty Acids: When fat absorption is impaired, much free fatty acid is present. These free fatty acids react with free calcium to form insoluble calcium soaps.

D. Vitamin D promotes the absorption of calcium from the intestine.

Distribution:

The calcium other than that in the bones and teeth is distributed as follows:

Fluid or Tissue	mg. per 100 ml. or 100 Gm.
Serum	9.0-11 (5 mEq./1.)
Cerebrospinal fluid	4.5-5 (2 mEq./1.)
Muscle	70
Nerve	15

Metabolism:

The blood cells contain very little calcium. Most blood calcium is therefore in blood plasma, where it exists as two fractions: the diffusible and nondiffusible components. About 60% is in the diffusible form; the remainder, probably because it is attached to serum albumin, is nondiffusible. There is therefore a relationship between the plasma albumin and the serum calcium concentrations. When this protein fraction is increased, plasma calcium is also increased; a fall in protein demands an increase in diffusible calcium to compensate for the deficit in protein-bound calcium.

A decrease in the ionized fraction of serum calcium causes tetany. This may be due to an increase in the pH of the blood (alkalotic tetany; gastric tetany) or to lack of calcium because of poor absorption from the intestine, decreased dietary intake, increased renal excretion as in nephritis, or parathyroid deficiency. Increased retention of phosphorus, as in renal tubular disease, also predisposes to low serum calcium levels.

The Ca:P ratio is important in ossification. In the serum, the product of Ca × P (in mg. per 100 ml.) is normally in children about 50. In rickets, this product may be below 30.

Relatively small quantities of the total excreted calcium (200 mg. per day; 10 to 30% of the total) are eliminated in the urine; most of the calcium is excreted in the feces (70 to 90%). This represents unabsorbed dietary calcium almost entirely. The amounts of calcium re-excreted into the intestine are very small.

Disease States:

A. Relationship of Parathyroids: Calcium metabolism is profoundly influenced by the parathyroids.
   1. In hyperparathyroidism caused by hyperactive, hyperplastic, or adenomatous parathyroid glands, the following signs are noted: hypercalcemia (serum calcium 12 to 22 mg. per 100 ml.), decrease in serum phosphate, decreased renal tubular reabsorption of phosphate, increased phosphatase activity, rise in urinary calcium and phosphorus from bone decalcification, and dehydration and hemoconcentration. These signs are due to increased renal losses of phosphorus, causing a decrease in serum phosphate which elicits an increase in calcium to maintain the Ca:P product. The extra calcium and phosphorus is lost from soft tissues and from bone by increased osteoclastic (bone-destroying) activity.
   2. Hypocalcemia - The concentration of serum calcium may drop below 7 mg. per 100 ml. after operative removal of the parathyroids. There is a concomitant increase in serum phosphate and a decrease in urinary phosphates. The urinary calcium is extremely low as well.

B. Rickets is characterized by faulty calcification of bones due to a low vitamin D content of the body, a deficiency of calcium and phosphorus in the diet, or a combination of both. Usually the serum phosphate concentration is low or normal, except in renal disease (in which it may be elevated), and the serum calcium remains normal or may be lowered. There is an increase in fecal phosphate and calcium because of poor absorption of these elements, accompanied by a decrease in urine phosphate and calcium. An increase in alkaline phosphatase activity is also characteristic of rickets.

C. Renal rickets is caused by renal tubular defect (usually inherited) which interferes with reabsorption of phosphorus. This disease is not relieved by vitamin D in ordinary dosages.

D. In severe renal disease the serum calcium may decrease, in part because of increased losses in the urine but mainly because of increase in serum phosphorus, which causes a compensatory decrease in serum calcium.

PHOSPHORUS

Functions:

Phosphorus is found in every cell of the body, but most of it (about 80% of the total) is combined with calcium in the bones and teeth. About 10% is in combination with proteins, lipids, and carbohydrates, and in other compounds in blood and muscle. The remaining 10% is widely distributed in various chemical compounds. The great importance of the phosphate ester in energy transfer is discussed on p. 119. Phosphorylation is equally important in absorption of carbohydrate, and possibly fat, from the intestine, and in the reabsorption of glucose by the renal tubule.

Requirements and Sources:

It is recommended that the intake of phosphorus be at least equal to that of calcium in the diets of children, and of women during the latter half of pregnancy and in lactation. For other adults, one and one-half times the calcium intake is required. Since their distribution in foods is very similar, an adequate intake of calcium generally ensures an adequate intake of phosphorus also. Cow's milk is an exception since it contains more phosphorus than calcium.

Distribution:

The distribution of phosphorus in the body is as follows:

Fluid or Tissue	mg. per 100 ml. or 100 Gm.
Blood	40
Serum inorganic - Children	4.7
- Adults	3-4.5
Muscle	170-250
Nerve	360
Bones and teeth	22,000

Metabolism:

The metabolism of phosphorus is in large part related to that of calcium, as described heretofore. The ratio of Ca:P in the diet affects the absorption and excretion of these elements. When either element is given in excess, excretion of the other is increased. The optimal ratio is 1:1, if the intake of vitamin D is adequate.

An increase in carbohydrate metabolism, such as during absorption of carbohydrate, is accompanied by a temporary decrease in serum phosphate. A similar decrease may occur during absorption of some fats. In diabetes mellitus, there is a lower concentration of organic phosphorus but a higher concentration of inorganic phosphorus in the serum. In the correction of hyperglycemia and acidosis of diabetes with glucose and insulin, it is considered desirable to include phosphate in the repair solutions.

In rickets of the common low-phosphate variety, serum phosphate values may go as low as 1 to 2 mg. per 100 ml. (0.64 to 1.3 mEq./L.).

Phosphate retention is a prominent cause of the acidosis in severe renal disease, and the resultant elevated serum phosphorus also contributes to the lowered serum calcium. Blood phosphorus levels are also high in hypoparathyroidism. A relationship of phosphorus metabolism to growth hormone is indicated by the fact that growing children usually have high blood phosphorus levels and that in acromegaly an elevation of the blood phosphorus also occurs.

Blood phosphorus levels are low in hyperparathyroidism and in sprue and celiac disease. A low blood phosphorus together with an elevated alkaline phosphatase is also a characteristic finding in patients with an inherited or acquired renal tubular defect in the reabsorption of phosphate. Such cases include so-called vitamin D–resistant rickets, Milkman's syndrome, and the de Toni-Fanconi syndrome (see also p. 291). The greatly increased excretion of phosphate in the urine of these patients distinguishes them from those in which a deficiency of vitamin D is the cause of the low blood phosphorus and the accompanying defects in calcification of the bone.

MAGNESIUM

Functions and Distribution:

Seventy percent of the magnesium in the body is combined with calcium and phosphorus in the complex salts of bone. The remainder is in the soft tissues and body fluids. Magnesium is one of the principal cations of soft tissue. The blood contains 2 to 4 mg. per 100 ml. (1.7 to 3.4 mEq./L.). The serum contains less than half that in the blood cells (1.94 mEq./L.). This is in contrast to calcium, almost all of which is in the serum. Cerebrospinal fluid is reported to contain about 3 mg. per 100 ml. (2.40 mEq./L.). The magnesium content of muscle is about 21 mg. per 100 Gm. Here it probably functions in carbohydrate metabolism as an activator for many of the enzymes of the glycolytic systems.

Requirements:[7]

The daily requirement for magnesium is not known. It is estimated that about 300 mg. is consumed in a day. This is probably adequate.

Metabolism:

The metabolism of magnesium is similar to that of calcium and phosphorus. From 20 to 50% of the total excreted magnesium is eliminated in the urine; the remainder is excreted in the feces because of poor absorption from the intestine.

Magnesium replaces calcium in bone salts when calcium is deficient, but magnesium in excess inhibits calcification. Magnesium salts are diuretic and cathartic.

Magnesium deficiency is not definitely known to occur in man. However, one case has been reported of a child whose tetany was thought to be due to a magnesium deficiency and who was relieved by 0.3 Gm. of magnesium sulfate, administered orally three times daily; the concentration of plasma magnesium in this case simultaneously increased from 0.6 to 2.6 mg. per 100 ml. A decrease in the concentration of serum magnesium has also been noted in clinical hyperparathyroidism. It is possible that prolonged hyperparathyroidism could deplete the body stores of magnesium. After surgical correction of hyperparathyroidism, the development of tetany which is refractory to the administration of large amounts of calcium may indicate the need for magnesium.

In rats on a very low magnesium diet (0.18 mg. per 100 Gm. of food) vasodilatation and hyperemia, hyperirritability, cardiac arrhythmia, and convulsions developed which were subsequently fatal. The tetany which developed when the diet was low in magnesium was probably due to the low magnesium content of the serum since the calcium levels remained normal.

An antagonism between magnesium and calcium has been noted in certain experiments. The intravenous injection of magnesium in a quantity sufficient to raise the magnesium ion concentration in the serum to about 20 mg. per 100 ml. (normal, 2.4 mg. per 100 ml.) results in immediate and profound anesthesia together with paralysis of voluntary muscles. The intravenous injection of a corresponding amount of calcium results in an instantaneous reversal of this effect. It is suggested that these two cations are exerting differing effects on cell permeability. In the case of magnesium, there is about ten times as much of this element in the cells as in the extracellular fluid. For example, in plasma there is an average of 2.4 mg. per 100 ml.; in muscle cells, 23 mg. per 100 Gm. This differential distribution between plasma and muscle cells is not observed with calcium, but it is particularly prominent in the case of sodium and potassium as well as magnesium. Apparently magnesium and potassium are normally concentrated within the cell and sodium without. An alteration in this relationship is followed by profound physiologic changes.

SODIUM

Functions:

This element is the largest component of the extracellular total base. It is largely associated with chloride and bicarbonate in regulation of acid-base equilibrium. The other important function of sodium is the maintenance of the osmotic pressure of body fluid and thus protection of the body against excessive fluid loss. It also functions in the preservation of normal irritability of muscle and the permeability of the cells.

Requirements and Sources:

Daily requirements of 5 to as much as 15 Gm. of sodium chloride have been recommended for adults by various authorities. These requirements were established from observations on urinary losses in subjects who were not on controlled low intakes of sodium chloride, and much of the salt in the urine therefore represented merely the excretion of the excess intake. Dahl[8] has recently appraised the need for sodium chloride under conditions of controlled intakes. In his experiments, adults maintained on daily intakes of only 100-150 mg. sodium lost a total of less than 25 mg. of sodium per day, which probably represents the minimum losses in the sweat. Dahl estimates the normal obligatory (irreducible) daily losses of sodium as follows: urine, 5-35 mg.; stool, 10-125 mg.; skin (not sweating), 25 mg.; total 40-185 mg.

The most variable source of salt loss is the sweat, but even this route of salt loss can be minimized during prolonged exposure to high temperatures if a period of a few days is allowed for adaptation. It is concluded that a maximum sodium chloride intake of about 5 Gm. per day may be recommended for adults without a family history of hypertension. This is about half the daily amount which is ordinarily voluntarily consumed. Furthermore, an intake of 5 Gm. of sodium chloride per day is ten times the amount at which adequate sodium chloride balance can apparently be maintained. For persons with a family history of hypertension, Dahl recommends a diet containing no more than 1 Gm. of sodium chloride per day.

The main source of sodium is the sodium chloride used in cooking and seasoning; ingested foods contain additional sodium.* It is estimated that about 10 Gm. of sodium chloride (4 Gm. of sodium) is thus ingested each day.

About 95% of the sodium which leaves the body is excreted in the urine. Sodium is readily absorbed, so that the feces contain very little except in diarrhea, when much of the sodium excreted into the intestine in the course of digestion escapes reabsorption.

Distribution:

About one-third of the total sodium content of the body is present in the inorganic portion of the skeleton. However, most of the sodium is found in the extracellular fluids of the body, as shown below:

Fluid or Tissue	mg. per 100 ml. or 100 Gm.
Whole blood	160 (70 mEq. /L.)
Plasma	330 (143 mEq. /L.)
Cells	85 (37 mEq. /L.)
Muscle Tissue	60-160
Nerve Tissue	312

Metabolism:

The metabolism of sodium is influenced by the adrenocortical steroids. In adrenocortical insufficiency, a decrease of serum sodium and an increase in sodium excretion occur.

In chronic renal disease, particularly when acidosis coexists, sodium depletion may occur due to poor tubular reabsorption of sodium as well as to loss of sodium in neutralization of acids.

*A detailed compilation of the sodium and potassium content of foods and of water obtained from the drinking supply of many cities has been prepared by Bills et al.[9]

Unless the individual is well adapted to a high environmental temperature, extreme sweating may cause the loss of considerable sodium in the sweat; muscular cramps of the extremities and abdomen, headaches, nausea, and diarrhea may develop. Low serum sodium (hyponatremia) may also develop if patients are given large quantities of salt-free fluids, particularly after they have sustained large electrolyte losses.

Increased serum sodium (hypernatremia) is rare. It may occur in Cushing's disease; after the administration of corticotropin (ACTH), cortisone, or desoxycorticosterone; in nephrosis, and in congestive heart failure. Some of the sex hormones also cause a rise in the concentration of serum sodium. However, in all of these situations the concomitant retention of water may mask the sodium elevation. Rapid loss of water, such as in the dehydration associated with diabetes insipidus, on the other hand, may lead to a considerable rise in the serum sodium concentration.

Addison's disease, which is characterized by an increased sodium loss because of adreno-cortical insufficiency, is ameliorated during pregnancy presumably because of the production of steroidal hormones, which cause sodium retention. It has also recently been shown that the placenta elaborates hormones with sodium-retaining effects, and it is believed that these hormonal substances are responsible for sodium and water retention, accompanied by rapid gains in weight, commonly observed in certain stages of pregnancy.

Meneely and Ball[10] have made a study in rats of the effects of chronic ingestion of large amounts of sodium chloride on an otherwise standardized diet. Their test diets contained varying amounts of sodium chloride ranging from 2.8 to 9.8%. Among the animals eating a diet with 7% sodium chloride or more there occurred a syndrome resembling nephrosis, characterized by the sudden onset of massive edema and by hypertension, anemia, pronounced lipemia, severe hypoproteinemia, and azotemia. All of the affected animals died, and at autopsy showed evidence of severe arteriolar disease. Significant hypertension was uniformly observed at all levels of sodium chloride (from 2.8 to 9.8%), and there was a tendency for the degree of elevation in blood pressure to parallel the amount of salt in the diet. At the higher levels of salt intake, there was also a significant decrease in the survival time of the experimental animals. However, the addition of potassium chloride to the high sodium chloride diets produced a striking increase in the survival times on the various diets, although a moderating effect of potassium on the blood pressure was observed only on the high levels of sodium chloride intake.

The clinical aspects of hyponatremic states and of hypernatremia have recently been reviewed[11,12,13].

POTASSIUM

Functions:

Potassium is the principal cation of the **intracellular** fluid; but it is also a very important constituent of the extracellular fluid because it influences muscle activity, notably cardiac muscle. Within the cells it functions, like sodium in the extracellular fluid, by influencing acid-base balance and osmotic pressure, including water retention.

Requirements and Sources:

The normal intake of potassium in food is about 4 Gm. per day (see p. 318, note). It is so widely distributed that a deficiency is unlikely except in the pathologic states discussed below.

Distribution:

The predominantly intracellular distribution of potassium is illustrated by the following figures:

Fluid or Tissue	mg. per 100 ml. or 100 Gm.
Whole blood	200 (50 mEq. /L.)
Plasma	20 (5 mEq. /L.)
Cells	440 (112 mEq. /L.)
Muscle tissue	250-400
Nerve tissue	530

Metabolism:

Variations in extracellular potassium influence the activity of striated muscles so that paralysis of skeletal muscle and abnormalities in conduction and activity of cardiac muscle occur. Although potassium is excreted into the intestine in the digestive fluids (see table on p. 313), much of this is later reabsorbed. The kidney is the principal organ of excretion for potassium. Not only does the kidney filter potassium in the glomeruli but it is also secreted by the tubules. The excretion of potassium is markedly influenced by changes in acid-base balance as well as by the activity of the adrenal cortex. The renal mechanisms for potassium excretion are discussed on p. 296. The capacity of the kidney to excrete potassium is so great that hyperkalemia will not occur, even after the ingestion or intravenous injection at a moderate rate of relatively large quantities of potassium, if kidney function is unimpaired. This, however, is not the case when urine production is inadequate. It is important to emphasize that potassium should **not** be given intravenously until circulatory collapse, dehydration, and renal insufficiency have been corrected.

A. Elevated Serum Potassium (Hyperkalemia):

1. Etiology - Toxic elevation of serum potassium is confined for the most part to patients with renal failure, advanced dehydration, or shock. A high serum potassium, accompanied by a high intracellular potassium, also occurs characteristically in adrenal insufficiency (Addison's disease). This elevated serum potassium is corrected by the administration of desoxycorticosterone (D. O. C. A.). Hyperkalemia may also occur if potassium is administered intravenously at an excessive rate.
2. Symptoms - The symptoms of hyperkalemia are chiefly cardiac and central nervous system depression; they are related to the elevated plasma potassium, not to increases in intracellular levels. The heart signs include bradycardia and poor heart sounds, followed by peripheral vascular collapse and, ultimately, cardiac arrest. Electrocardiographic changes are characteristic and include elevated T waves, widening of the QRS complex, progressive lengthening of the P-R interval, and then disappearance of the P wave. Other symptoms commonly associated with elevated extracellular potassium include mental confusion; weakness, numbness, and tingling of the extremities; weakness of respiratory muscles; and a flaccid paralysis of the extremities.

B. Low Serum Potassium (Hypokalemia):

1. Etiology - Potassium deficiency is likely to develop in any illness, particularly in postoperative states when intravenous administration of solutions which do not contain potassium is prolonged. Potassium deficits are likewise to be expected in chronic wasting diseases with malnutrition, prolonged negative nitrogen balance, gastrointestinal losses (including those incurred in all types of diarrheas and gastrointestinal fistulas, and in continuous suction), and in metabolic alkalosis. In most of these cases intracellular potassium is transferred to the extracellular fluid, and this potassium is quickly removed by the kidney. Because adrenocortical hormones, particularly aldosterone, increase the excretion of potassium, overactivity of the adrenal cortex (Cushing's syndrome or primary aldosteronism), or injection of excessive quantities of D. O. C. A., cortisone, or corticotropin (ACTH) may induce a deficit.

   The excretion of potassium in the urine is increased by the activity of certain diuretic agents, particularly acetazolamide (Diamox®) and chlorothiazide (Diuril®). It is therefore recommended that potassium supplementation be provided when these drugs are used for more than a few days.

   A prolonged deficiency of potassium may produce severe damage to the kidney[14]. This may be associated secondarily with the development of chronic pyelonephritis. There is evidence that the initial damage to the kidney in potassium-depleted animals affects particularly the mitochondria in the collecting tubule.

   During heart failure, the potassium content of the myocardium becomes depleted; with recovery, intracellular repletion of potassium occurs. However, intracellular deficits of potassium increase the sensitivity of the myocardium to digitalis intoxication and to arrhythmias. This fact is of importance in patients who have been fully digitalized and are then given diuretic agents which may produce potassium depletion. Administration of potassium may prevent or relieve such manifestations of digitalis toxicity.

   Potassium deficits often become apparent only when water and sodium have been replenished in an attempt to correct dehydration and acidosis or alkalosis. Darrow states that changes in acid-base balance involve alterations in both **intracellular and extracellular** fluids and that the normal reaction of the blood cannot be maintained without a suitable relation between the body contents of sodium, potassium, chloride, and water.

When 1 Gm. of glycogen is stored, 0.36 mM. of potassium is simultaneously retained. In treatment of diabetic coma with insulin and glucose, glycogenesis is rapid and potassium is quickly withdrawn from the extracellular fluid. The resultant hypokalemia may be fatal.

Familial periodic paralysis is a rare disease in which potassium is rapidly transferred into cells, lowering extracellular concentration.

2. The symptoms of low serum potassium concentrations include muscle weakness, irritability, and paralysis; tachycardia and dilatation of the heart with gallop rhythm are also noted. Changes in the EKG record are also a prominent feature of hypokalemia; including first a flattened T wave and a prolonged QT interval, later inverted T waves with sagging ST segment and A-V block, and finally cardiac arrest.

   It is important to point out that a potassium deficit may not be reflected in lowered (less than 3.5 mEq. per liter) extracellular fluid concentrations until late in the process. This is confirmed by the finding of low intracellular potassium concentrations in muscle biopsy when serum potassium is normal. Thus the serum potassium is not an accurate indicator of the true status of potassium balance.

3. Treatment - In parenteral repair of a potassium deficit, a solution containing 25 mEq. of potassium (KCl, 1.8 Gm.) per liter may be safely given intravenously after adequate urine flow has been established. A daily maintenance dose of at least 50 mEq. of potassium (KCl, 3.6 Gm.) intravenously is probably necessary for most patients, with additional amounts to cover excessive losses, as from gastrointestinal drainage, up to 150 mEq. of potassium per day. When these larger doses are required, as much as 50 mEq. of potassium may be added to a liter of intravenous solution, although in this concentration a slower rate of injection is required (2 1/2 to 3 hours). The potassium salts may be added to saline solutions or to dextrose solutions. Some prefer to add also magnesium and calcium in order to provide a better ionic balance, suggesting 10 mEq. each of calcium and magnesium for each 25 mEq. of potassium. The following formula contains these three cations in that proportion:

KCl	1.8 Gm.	(25 mEq. K)
$MgCl_2$	0.5 Gm.	(10 mEq. Mg)
$CaCl_2$	0.6 Gm.	(10 mEq. Ca)

Whenever possible, the correction of a potassium deficit by the oral route is preferred. For adults, 4 to 12 Gm. of KCl (as 1 to 2% solution) per day in divided doses is recommended.

In muscle, the proportion of potassium to nitrogen is 3 mM. to each Gm. Storage of nitrogen as muscle protein therefore demands additional potassium. According to Moore, a loss of 5 Kg. of muscle protein requires 600 mEq. of potassium together with the protein nitrogen necessary for its replacement. For this reason, the administration of potassium along with parenterally-administered amino acids has been recommended. According to Frost and Smith[15], 5 mEq. of potassium per Gm. of amino acid nitrogen infused is required for optimal nitrogen retention.

CHLORINE

Function:

As a component of sodium chloride, this element is essential in water balance and osmotic pressure regulation as well as in acid-base equilibrium. In this latter function, chloride plays a special role in the blood by the action of the chloride shift (see p. 149). In gastric juice, chloride is also of special importance in the production of hydrochloric acid (see p. 155).

Requirement and Metabolism:

In the diet, the chloride occurs almost entirely as sodium chloride, and therefore the intake of chloride is satisfactory as long as sodium intake is adequate. In general, both the intake and output of this element are, in fact, inseparable from those of sodium. On low-salt diets, both the chloride and sodium in the urine drop to low levels.

Abnormalities of sodium metabolism are generally accompanied by abnormalities in chloride metabolism. When losses of sodium are excessive, as in diarrhea, profuse sweating, and certain endocrine disturbances, chloride deficit is likewise observed.

In loss of gastric juice by vomiting or in pyloric or duodenal obstructions, there is a loss of chloride in excess of sodium. This leads to a decrease in plasma chloride, with a compensatory increase in bicarbonate and a resultant hypochloremic alkalosis (see p. 151). In Cushing's disease or after the administration of an excess of corticotropin (ACTH) or cortisone, hypokalemia with an accompanying hypochloremic alkalosis may also be observed. Some chloride is lost in diarrhea because the reabsorption of chloride in the intestinal secretions is impaired.

Distribution in the Body:

The chloride concentration in cerebrospinal fluid is higher than that in other body fluids, including gastrointestinal secretions.

Fluid or Tissue	mg. per 100 ml. or 100 Gm.	mg. per 100 ml. (as NaCl)*
Whole blood	250 (70 mEq. /L.)	(450-500)
Plasma	365 (103 mEq. /L.)	(570-620)
Cells	190 (53 mEq. /L.)	
Cerebrospinal fluid	440 (124 mEq. /L.)	(720-750)
Muscle tissue	40	
Nerve tissue	171	

*It should be noted that the sodium and chloride contents of plasma are not equal. For this reason, there is no justification for expressing chloride concentrations as sodium chloride since they do not measure total sodium. This custom arose before the chemistry of body fluids was properly understood, when it was believed that all of the chloride in these fluids existed as sodium chloride. The concept is erroneous and should be discarded.

SULFUR

Function:

Sulfur is present in all of the cells of the body, primarily in the cell protein. The metabolism of sulfur and nitrogen thus tend to be associated.

The importance of sulfur-containing compounds in detoxication mechanisms (see p. 167) and of the SH group in tissue respiration has been noted (see p. 249). A high-energy sulfur bond similar to that of phosphate plays an important role in metabolism (see p. 86).

Sources and Metabolism:

The main (if not the only) sources of sulfur for the body are the two sulfur-containing amino acids, cystine and methionine. Elemental sulfur or sulfate sulfur is not known to be utilized. Organic sulfur is mainly oxidized to sulfate and excreted as inorganic or ethereal sulfate. The various forms of urinary sulfur are described on p. 303.

Utilization of sulfate in organic combination requires ''activation'' as described on p. 47.

Distribution in the Body:

In addition to cystine and methionine, other organic compounds of sulfur are heparin, glutathione, insulin, thiamine, biotin, coenzyme A, lipoic acid, ergothioneine, taurocholic acid, the sulfocyanides; sulfur conjugates, like phenol esters and indoxyl sulfate; and the chondroitin sulsuric acid in cartilage, tendon, and bone matrix. Small amounts of inorganic sulfates, with sodium and potassium, are present in blood and other tissues.

Keratin, the protein of hair, hoofs, etc., is rich in sulfur-containing amino acids. The sulfur (methionine and cystine) requirement of hairy animals, such as the rat and the dog, is higher than that of human beings, possibly because of their additional hair.

IRON

Function and Distribution:

The role of iron in the body is almost exclusively confined to the processes of cellular respiration. Iron is a component of hemoglobin, myoglobin, and cytochrome, as well as the enzymes catalase and peroxidase. In all of these compounds the iron is a component of a porphyrin. The remainder of the iron in the body is almost entirely protein-bound; these forms include the storage and transport forms of the mineral. The approximate distribution of iron-containing compounds in the normal adult human subject is as follows[16]:

	Total in Body (Gm.)	Iron Content (Gm.)	Percent of Total Iron in Body
Iron porphyrins (heme compounds)			
Hemoglobin	900	3.0	60-70
Myoglobin	40	0.13	3-5
Heme enzymes			
Cytochrome c	0.8	0.004	0.1
Catalase	5.0	0.004	0.1
Other cytochromes	-	-	-
Peroxidase	-	-	-
Nonporphyrin iron compounds			
Siderophilin (transferrin)	10	0.004	0.1
Ferritin	2-4	0.4-0.8	15.0
Hemosiderin	-	-	-
Total available iron stores		1.2-1.5	
Total Iron		4.0-5.0	

Requirements and Sources:

The need for iron in the human diet varies greatly at different ages and under different circumstances. During growth, pregnancy, and lactation, or after blood loss (including menstrual hemorrhages), when the demand for hemoglobin formation is increased, additional iron is needed in the diet. In the healthy adult male, or in healthy women after the menopause, the iron requirement is almost negligible.

Traces of copper are required for utilization of iron in hemoglobin formation. The required amounts of iron and copper currently suggested by nutritional authorities are as follows:

A. Infants: 5-7 mg. per day.

B. Young Children: 7-12 mg. per day.

C. Adolescents and Pregnant and Lactating Women: 15 mg. of iron and 2 mg. of copper per day.

D. Other Adults: Men - 10 mg. of iron and 2 mg. of copper per day; women - 12 mg. of iron and 2 mg. of copper per day. (Male adults and postmenopausal women probably do not require this much iron; 2 to 3 mg. per day are probably adequate.)

The best sources of iron are "organ meats": liver, heart, kidney, and spleen. Other good sources are egg yolk, whole wheat, fish, oysters, clams, nuts, dates, figs, beans, asparagus, spinach, molasses, and oatmeal.

Absorption From the Stomach and Intestine:

A peculiar and possibly unique feature of the metabolism of iron is that it occurs in what is virtually a closed system. Under normal conditions very little dietary iron is absorbed, and the amounts excreted are minimal. Because there is no way to excrete excess iron, its absorption from the intestine must be controlled if it is not to accumulate in the tissues in toxic amounts. In the ordinary diet, 10 to 20 mg. of iron are taken each day, but less than 10% of this is absorbed. The results of a study with labeled iron (Fe^{59}) illustrate this fact. After the administration of a test dose of the isotopic iron, only 5% was found in the blood, 87% was eliminated in the feces, and 8% remained unaccounted for. Infants and children absorb a higher percentage of iron from foods than do adults. Iron-deficient children absorb twice as much from foods as normal children; therefore, iron deficiency in infants can usually be attributed to dietary inadequacy.

In the male, excretion from the body is less than 1 mg. per day; in the female during the childbearing years, 1.5 to 2.0 mg. per day.

Most of the iron in foods occurs in the ferric (Fe^{+++}) state either as ferric hydroxide or as ferric organic compounds. In an acid medium, these compounds are broken down into free ferric ions or loosely bound organic iron. The gastric hydrochloric acid as well as the organic acids of the foods are both important for this purpose. Reducing substances in foods, SH groups (e.g., cysteine), or ascorbic acid convert ferric iron to the reduced (ferrous) state. In this form, iron is more soluble and should therefore be more readily absorbed.

Most of the absorption of iron occurs in the stomach and duodenum. Impaired absorption of iron is therefore observed in patients who have had subtotal or total removal of the stomach or in patients who have sustained surgical removal of a considerable amount of the small bowel. Iron absorption is also diminished in various malabsorption syndromes, such as steatorrhea.

In iron-deficiency anemias the absorption of iron may be **increased** to 2 to 10 times normal. It is also increased in pernicious anemia and in hypoplastic anemia.

A diet high in phosphate causes a decrease in the absorption of iron since compounds of iron and phosphate are insoluble; conversely, a diet very low in phosphates markedly increases iron absorption. Phytic acid (found in cereals) and oxalates also interfere with the absorption of iron.

Not all of the iron in foods is available to the body. An ordinary chemical determination for total iron of foodstuffs is thus not an accurate measure of nutritionally available iron. This can be determined by measuring the amount that will react with the α, α-dipyridyl reagent.

The intestine itself controls the absorption of iron. At the border of the intestinal mucosal cell there seems to be a "mucosal block" which regulates the amount of ferrous iron entering the cell. The previous administration of iron hinders the subsequent absorption of iron for 12 to 24 hours. Once within the cell, the ferrous iron is oxidized to the ferric state and combines there with a protein, **apoferritin**, to form an iron-containing derivative, **ferritin**, containing 23% iron by weight. The binding capacity of apoferritin for iron further limits iron absorption; when saturated with iron, no further storage of iron may occur in this form. Normally, the apoferritin content of the mucosal cells is low, but there is evidence for a rapid increase in its formation by the mucosal cells as a result of the administration of iron[17].

Transport in the Plasma:

Iron released from storage as ferritin is reduced again to ferrous iron and leaves the intestine to enter the plasma. There is a possibility that diminished oxygen tension in the blood may increase the activity of the intestinal mucosal cells transferring iron to the plasma. In the presence of carbon dioxide the iron from the plasma forms a complex with a metal-binding β-globulin known as **transferrin** or **siderophilin** which occurs in Cohn Fraction IV-7 (see p. 131). The normal content of protein-bound iron (BI) in the plasma of males is 120 to 140 μg. per 100 mg.; in females, 90 to 120 μg. per 100 ml. However, the total iron binding capacity (TIBC) is about the same in both sexes: 300 to 360 μg. per 100 ml. This indicates that normally only about 30 to 40% of the iron-binding capacity of the serum is utilized for iron transport and that the iron-free siderophilin, i.e., the unsaturated iron-binding capacity (UIBC), is therefore about 60 to 70% of the total.

In iron deficiency anemias the plasma-bound iron is low, whereas the total iron binding capacity tends to rise, resulting in an unsaturated iron-binding capacity which is higher than normal. In hepatic disease, both the bound iron and the total iron-binding capacity of the plasma may be low, so that the percentage of the total iron-binding capacity which is unsaturated is not significantly altered from normal.

The amount of bound iron in the plasma is reported to exhibit a diurnal variation[18] which can be as much as 60 μg. per 100 ml. over a 24-hour period. The lowest values were found two hours following retirement for sleep; the highest values were found five to seven hours later.

The failure of the kidney to excrete iron is probably due to the presence of iron in the plasma as a protein-bound compound which is not filtrable by the glomerulus. By the same token, losses of iron into the urine may occur in proteinuria. In nephrosis, for example, as much as 1.5 mg. of iron per day may be excreted with protein in the urine.

Metabolism:

The storage form of iron, ferritin, is found not only in the intestine but also in liver (about 700 mg.), spleen, and bone marrow. If iron is administered parenterally in amounts which exceed the capacity of the body to store it as ferritin, it accumulates in the liver as microscopically visible **hemosiderin**, a form of colloidal iron oxide in association with protein. The iron content of hemosiderin is 35% by weight.

The level of iron in the plasma is the result of a dynamic equilibrium; the factors which influence it include the rate of breakdown of hemoglobin, uptake by the bone marrow in connection with red blood cell synthesis, removal and storage by the tissues, absorption from the gastrointestinal tract, and the rate of formation and decomposition of siderophilin (transferrin). Studies of the turnover of iron, using isotopic Fe^{59}, indicate that about 27 mg. are utilized each day, 75% of this for the formation of hemoglobin. About 20 mg. are obtained from the breakdown of red blood cells, a very small amount from newly absorbed iron, and the remainder from the iron stores. Normally there is a rather slow exchange of iron between the plasma and the storage iron; in fact, following an acute hemorrhage in a normal individual, the level of iron in the plasma may remain low for weeks, a further indication that mobilization of iron from the storage depots is a slow process.

Iron deficiency anemias are of the hypochromic microcytic type. In experiments with rats made iron-deficient, it was found that cytochrome c levels were reduced even in the absence of anemia. This suggests that some of the symptoms in anemia may be due to decreased activity of intracellular enzymes rather than to low levels of hemoglobin.

A deficiency may result from inadequate intake (e.g., a high cereal diet, low in meat) or inadequate absorption (e.g., gastrointestinal disturbances such as diarrhea, achlorhydria, steatorrhea, or intestinal disease, after surgical removal of the stomach, or after extensive intestinal resection) as well as from excessive loss of blood. If absorption is adequate, the daily addition of ferrous sulfate to the diet will successfully treat the iron-deficiency type of anemia. Satisfactory preparations of iron (iron dextran) for intramuscular injection in patients who cannot tolerate or absorb orally administered iron are available. These must be used with caution because of the possibility of oversaturation of the tissues with resultant production of hemosiderosis.

Studies with isotopic iron have been used to determine the rate of red blood cell production. Fe^{59} is given intravenously in tracer doses and the rate of disappearance of the label is measured. Normally one-half of the radioactivity disappears exponentially from the circulating blood in 90 minutes. In hemolytic anemias, where there is hyperplasia of the erythroid tissue, and in polycythemia vera, one-half of the activity disappears in 11 to 30 minutes. In aplastic anemia the opposite situation prevails; the disappearance time is prolonged to as long as 250 minutes. The reappearance of the label in newly-formed blood cells is then noted. In an iron-deficiency type of anemia, the uptake of iron in the erythrocytes is accelerated; in aplastic anemia, it is diminished.

Because of the absence of an excretory pathway for iron, excess amounts may accumulate in the tissues. This is observed in patients with aplastic or hemolytic anemia who have received many blood transfusions over a period of years. The existence in some individuals of an excessive capacity for the absorption of iron from the intestine has been detected by studies with radioactive iron. Such individuals absorb 20 to 45% of an administered dose of the labeled iron; a normal subject absorbs 1.5 to 6.5%. The anomaly in iron absorption may be inherited. In such patients, a very large excess (as much as 40 to 50 Gm.) of iron accumulates in the tissues after many years. This hemosiderosis may be accompanied by a bronzed pigmentation of the skin, **hemochromatosis**, and, presumably because of the toxic effect of the unbound iron in the tissues, there may be liver damage with signs of cirrhosis, diabetes, and a pancreatic fibrosis. The condition is sometimes referred to as bronze diabetes. It is of interest that the unsaturated iron-binding capacity of the serum of the patient with hemochromatosis is very low. Thus, whereas the iron-binding proteins of the serum in a normal individual are only about 30% saturated, in patients with hemochromatosis they are about 90% saturated. This is doubtless due to the excess absorption of iron from the intestine.

An acquired siderosis of dietary origin has been reported to occur with great frequency among the Bantu peoples of Africa. This condition has been termed "Bantu siderosis" and is believed to be caused by the fact that the natives consume a diet which is very high in corn and thus low in phosphorus and that their foods are cooked in iron pots. The combination of a low-phosphate diet and a high intake of iron enhances absorption of iron sufficiently to produce siderosis with accompanying organ damage as described above. It is of further interest that iron-deficiency anemias, common among pregnant women in other areas of the world, are virtually unknown among the Bantu.

COPPER

Functions and Distribution:

The functions of this essential element are not well understood. Copper is a constituent of certain enzymes or is essential in their activity; these include cytochrome, cytochrome oxidase, catalase, tyrosinase, and ascorbic acid oxidase as well as uricase, which contains 550 μg. of copper per Gm. of enzyme protein. Along with iron, copper is necessary for the synthesis of hemoglobin. A copper-containing protein (hemocuprin) has been found in the red blood cells of mammals. This copper in the red cell may explain the need for copper in the formation of hemoglobin. Hemocyanin is a copper-protein complex in the blood of certain invertebrates, where it functions like hemoglobin as an oxygen carrier. Copper is also important in bone formation and in maintenance of myelin within the nervous system.

The adult human body contains 100 to 150 mg. of copper; about 64 mg. are found in the muscles, 23 mg. in the bones, and 18 mg. in the liver, which contains a higher concentration of copper than any of the other organs studied. It is of interest that the concentration of copper in the fetal liver is five to ten times higher than that in the liver of an adult. Both the blood cells and serum contain copper; but the copper content of the red blood cell is constant, while that of the serum is highly variable, averaging about 90 μg. per 100 ml.

The serum copper is present in two distinct fractions. The so-called "direct-reacting" copper is that fraction which reacts directly with diethyldithiocarbamate, the reagent used to determine copper colorimetrically[19]. This copper is loosely bound to albumin and may represent copper in transport. Relatively little of the serum copper is present in this form. Most of the copper in the serum (96%) is bound to an α globulin contained in the Cohn Fraction IV-1. Consequently, in determination of serum copper it is necessary first to treat the serum with hydrochloric acid to free the copper from the globulin-bound form so that it can react with the color reagent.

The copper-binding protein is called **ceruloplasmin.** It has a molecular weight of about 151,000 and contains 0.34% copper, or about 8 atoms of copper per mol. Normal plasma contains about 30 mg. of this protein per 100 ml. The absorption peak for the protein is at 6100 Å. According to Holmberg and Laurell[20], this copper-containing protein acts as an enzyme - a polyphenol oxidase.

Metabolism:

Experiments have been carried out with labeled copper (Cu^{64}). The copper was found largely associated with the albumin fraction of the plasma immediately after its ingestion; 24 hours later, most of it was in the globulin fraction, associated with ceruloplasmin. Since the copper in the plasma is largely bound to protein, it is not readily excreted in the urine. Apparently most of it is lost through the intestine.

Experimental animals on a copper-deficient diet lose weight and die; the severe hypochromic microcytic anemia which they exhibit is not the cause of death since an iron-deficiency anemia of equal proportions is not fatal. This suggests that copper has a role in the body in addition to its function in the metabolism of red cells. This additional role of copper may be related to the activity of oxidation-reduction enzymes of the tissues, such as the cytochrome system. A relation between copper and iron metabolism has been detected. In the presence of a deficiency of copper, the movement of iron from the tissues to the plasma is decreased and hypoferremia results. Copper favors the absorption of iron from the gastrointestinal tract.

A bone disorder associated with a deficiency of copper in the diet of young dogs has been described[21]. The bones of these animals were characterized by abnormally thin cortices, deficient trabeculae, and wide epiphyses. Fractures and deformities occurred in many of the animals. Anemia was present, and the hair turned gray. The disorder did not occur in any of the control animals, and was relieved by the administration of copper.

Wilson's disease (hepatolenticular degeneration) is associated with abnormalities in the metabolism of copper. In this disease the liver and the lenticular nucleus of the brain contain abnormally large amounts of copper, and there is excessive urinary excretion of copper and low levels of copper and of ceruloplasmin in the plasma. A generalized aminoaciduria also occurs in this disease.

According to Bearn and Kunkel[22] the absorption of copper from the intestine is considerably increased in the patient with Wilson's disease. As a result copper accumulates in the tissues and appears in the urine. If the deposition of copper in the liver becomes excessive, cirrhosis may develop. Accumulation of copper in the kidney may give rise to renal tubular damage which leads to increased urinary excretion of amino acids and peptides and, occasionally, glucose as well.

Others[23] feel that Wilson's disease is characterized by a failure to synthesize ceruloplasmin at a normal rate (although administration of ceruloplasmin does not ameliorate the disease) and/or a defect at the stage of incorporation of copper into the copper-binding globulin. As a result there is present in these patients a quantity of "unattached" copper which is free to combine in an abnormal manner, such as with the proteins of the brain or liver. The excessive excretion of copper in the urine is also explained as a result of the presence of abnormal amounts of unbound copper. In the normal individual, copper is bound to ceruloplasmin within 24 hours of its ingestion; in the patient with Wilson's disease, copper is still associated with the albumin fraction at this time. As a result the so-called "direct-reacting" fraction of the serum copper (see above) is not reduced in these patients; in fact, it may actually be increased. Thus the total serum copper may appear normal or only slightly reduced.

There have been some suggestions that inadequate excretion of copper via the intestine may be a factor in the genesis of Wilson's disease.

Hypercupremia occurs in a variety of circumstances. It does not seem to have any diagnostic significance.

Requirements:

The human requirement for copper has been studied by balance experiments. A daily allowance of 2.5 mg. has been suggested for adults; infants and children require about 0.05 mg./Kg. body weight. This is easily supplied in average diets, which contain 2.5 to 5 mg. of copper.

A nutritional deficiency of copper has never been positively demonstrated in man, although it has been suspected in cases of sprue or in nephrosis. However, there are recent reports[24] of a syndrome in infants which is characterized by low levels of serum copper and iron and by edema and a hypochromic microcytic anemia. Therapy with iron easily cures the disease; and "spontaneous" cures are also reported.

IODINE

Function:

The only known function of iodine is in the thyroid mechanism. Its metabolism is discussed under that subject (see p. 332).

Requirements and Source:

The human requirement for iodine, determined by iodine balance studies in normal persons under basal conditions of bed rest on a hospital diet, is 1 μg. per Kg. body weight per day. An optimal requirement has been estimated at 2 to 4 μg. per Kg. The use of iodized salt regularly will provide more than this.

The need for iodine is increased in adolescence and in pregnancy. Thyroid hypertrophy occurs if iodine deficiency is prolonged.

MANGANESE

Ordinary diets yield about 4 mg. per day of manganese. This amount is in excess of that suggested as required by man. Animal experiments have indicated that this element is essential, but there is no evidence of a manganese deficiency in man. Perosis (slipped tendon disease) in chickens and other fowl may be produced by a manganese deficiency.

The kidney and the liver are the chief storage organs for manganese. In blood, values of 4 to 20 μg. per 100 ml. have been reported. Most of the manganese is excreted in the intestine by way of the bile (determined by experiments with isotopic manganese).

The functions of manganese are not known. In vitro, manganese activates several enzymes: blood and bone phosphatases, yeast, intestine and liver phosphatases, arginase, carboxylase, cozymase, and cholinesterase.

When isotopic manganese (Mn^{56}) was injected intraperitoneally, a correlation was found between the mitochondrial content of a given organ and its ability to concentrate manganese[25]. Fractionation studies of liver cells confirmed the assumption that the mitochondria are the principal intracellular sites of manganese uptake. Since the mitochondria are also the structures with which the majority of the intracellular respiratory enzyme systems are associated, a role of manganese as a coenzyme for these enzymes is strongly suggested.

COBALT

This element is a constituent of vitamin B_{12} (see p. 95). Cobalt affects blood formation. A nutritional anemia in cattle and sheep living in cobalt-poor soil areas has been successfully treated with cobalt. There are reports of its favorable use in the anemias of children.

An excess of red cells, **polycythemia**, has been produced in rats when the element is fed or injected. Cobaltous chloride administration to human subjects also produced an increase in red blood cells.

Isotopic cobalt is quickly eliminated almost completely via the kidneys.

ZINC

When experimental animals (rats) are maintained on a diet which is very low in zinc, impaired growth and poor development of their coats are noted.

Zinc is a structural and functional component of the enzyme carboxypeptidase, and this element participates directly in the catalytic action of the enzyme[26]. Zinc is also closely associated with several other enzymes, including carbonic anhydrase and alcohol dehydrogenase, which contains one atom of zinc per mol of enzyme protein.

Insulin is known to contain zinc and it is therefore of interest to note that the pancreas of diabetics contains only about one-half the normal amount of zinc.

The zinc content of leukocytes from the blood of normal human subjects is reported to be $3.2 \pm 1.3 \times 10^{-10}$ μg. per million cells. However, the zinc content of white blood cells in human leukemia patients is reduced to 10% of the normal amount[27]. Temporary therapeutic amelioration of the leukemic process is accompanied by a rise in the zinc content of these cells to normal. Because there is no carboniç anhydrase in white blood cells, the function of the intracellular zinc in these cells is not known. The metal is firmly bound to protein within the cell; in fact, it is found in a constant ratio to the protein: 82 to 117 μg. of zinc per Gm. of protein.

Although it is believed that zinc is an essential element, little is known concerning the human requirement. As a result of balance studies, a daily intake of 0.3 mg. per Kg. has been recommended. This amount is easily obtained in the diet since the zinc content of natural foods approximates that of iron.

FLUORINE

This element is found in certain tissues of the body, particularly in bones and teeth.

Fluorine is a poison for some enzyme systems. Specifically, it inhibits the conversion of glyceric acid to pyruvic acid by enolase (see p. 182) in anaerobic glycolysis.

In very small amounts, it seems to improve tooth development, but a slight excess causes mottling of the enamel of the tooth. Mild mottling of the teeth occurs in less than 2% of the children living in areas where the fluorine content of the water is between 0.6 and 1.2 p.p.m. Severity and incidence of mottling increase when the fluorine in the water exceeds these amounts. However, at levels above 1 p.p.m., the incidence of dental caries in children is much lower. The topical application of fluorine to the teeth during the developmental stage or the addition of 1 p.p.m. fluoride to drinking water in areas where it is normally low, are now utilized in an effort to reduce the incidence of dental caries.

Studies with animals on low-fluorine diets have not provided evidence of its essentiality. In large quantities, it is definitely toxic.

ALUMINUM

While this element is widely distributed in plant and animal tissues, there is still no evidence that it is essential. Its physiologic role, if any, remains obscure. Rats subsisting on a diet which supplied as little as 1.0μg. per day showed no abnormalities. Large amounts fed to rats produced rickets by interfering with the absorption of phosphates.

The daily intake of aluminum in the human diet ranges from less than 10 mg. to over 100 mg. In addition to the very small amounts naturally present in foods and that derived from cooking utensils, aluminum may be added to the diet as sodium aluminum sulfate in baking powder and as alum, sometimes added to foods to preserve firmness. However, absorption of aluminum from the intestine is very poor. From measurements of urinary excretion it is estimated that only about 100 μg. of aluminum are absorbed per day, most of the ingested aluminum being excreted in the feces even on a reasonably high intake. The total amount of aluminum in the body is about 50 to 150 mg.

BORON

Boron is essential for the growth of plants, and traces are found in animal tissues. On very low boron diets, the growth of rats is not impaired.

MOLYBDENUM

In studies of the activity of the flavoprotein enzyme, xanthine oxidase (see p. 269), it was found that traces of molybdenum are required for the deposition and maintenance of normal levels of this enzyme in the intestine and liver of the rat. Highly purified preparations of the enzyme as obtained from milk contain molybdenum in a form which indicates that it is actually a part of the enzyme molecule. Liver aldehyde oxidase, a flavoprotein which catalyzes the oxidation of aldehydes, also contains molybdenum[28].

SELENIUM

Selenium is usually thought of as an element with pronounced toxic properties[29]. It is therefore of great interest that in very small quantities it may prove to be an essential factor in tissue respiration. This suggestion emanates from the work of Schwartz and co-workers on the identity of "Factor 3," which is protective against a hepatic necrosis produced in the rat by dietary means (see p. 205). The preventive factor is an organic compound containing selenium as the active ingredient. It is extremely active, as evidenced by the finding that four parts of selenite selenium per 100, 000, 000 parts of diet were found to be protective against dietary liver necrosis. Vitamin E is also protective against the hepatic lesion produced when the animals were maintained on the necrogenic diet. This suggests that the action of the selenium compound may be that of a participant in an electron transfer system involved in cellular respiration[30].

• • •

References:

1. Forbes, R.M., Cooper, A.R., and Mitchell, H.H.: J.Biol.Chem. **203**:359, 1953.
2. Butler, A.M., and Talbot, N.B.: New Eng.J.Med. **231**:585, 1944.
3. Deane, N., and Smith, H.W.: J.Clin.Invest. **31**:197, 1952.
4. a. Lockwood, J.S., and Randall, H.T.: Bull. New York Acad.Med. **25**:228, 1949.
   b. Randall, H.T.: Surg.Clin.N.Amer. **32**:3, 1952.
5. Hegsted, D.M.: Nutrition Reviews **15**:257, 1957.
6. Burnett, C.H., Commons, R.R., Albright, F., and Howard, J.E.: New Eng.J.Med. **240**:787, 1949.
7. Wacker, W.E.C., and Vallee, B.L.: New Eng.J.Med. **259**:431, 1958.
8. Dahl, L.K.: New Eng.J.Med. **258**:1152;1205, 1958.
9. Bills, C.E., McDonald, F.G., Niedermeier, W., and Schwartz, M.C.: J.Am.Dietet.A. **25**:304, 1949.
10. Meneely, G.R., and Ball, C.O.T.: Am.J.Med. **25**:713, 1958.
11. Edelman, I.S.: Metabolism **5**:500, 1956.
12. Wynn, V.: Metabolism **5**:490, 1956.
13. Knowles, H.C., Jr.: Metabolism **5**:508, 1956.
14. Kark, R.M.: Am.J.Med. **25**:698, 1958.
15. Frost, P.M., and Smith, J.L.: Metabolism **2**:529, 1953.
16. Granick, S.: Bull.New York Acad.Med. **30**:81, 1954.
17. Fineberg, R.A., and Greenberg, D.M.: J.Biol.Chem. **214**:91, 1955.
18. Hoyer, K.: Acta med.Scandinav. **119**:577, 1944.
19. Gubler, C.J., Lahey, M.E., Ashenbrucker, H., Cartwright, G.E., and Wintrobe, M.M.: J.Biol.Chem. **196**:209, 1952.
20. Holmberg, C.G., and Laurell, C.B.: Acta chem.Scandinav. **2**:550, 1948.
21. Baxter, J.H., and Van Wyk, J.J.: Bull.Johns Hopkins Hosp. **93**:1:25, 1953.
22. Bearn, A.G., and Kunkel, H.G.: J.Clin.Invest. **33**:400, 1954.
23. Earl, C.J., Moulton, M.J., and Selverstone, B.: Am.J.Med. **17**:205, 1954.
24. Gitlin, D., and Janeway, C.A.: Pediatrics **21**:1034, 1958.
25. Maynard, L.S., and Cotzias, G.C.: J.Biol.Chem. **214**:489, 1955.
26. Vallee, B.L., and Neurath, H.: J.Biol.Chem. **217**:253, 1955.
27. Hoch, F.L., and Vallee, B.L.: J.Biol.Chem. **195**:531, 1952.
28. Mahler, H.R., Mackler, B., and Green, D.E.: J.Biol.Chem. **210**:465, 1954.
29. Moxon, A.L., and Rhian, M.: Physiol.Rev. **23**:305, 1943.
30. Schwartz, K., in Brauer, R.W., Ed. "Liver Function," p. 509; Publication No. 4, Am.Inst.Biol.Sci., Washington, D.C., 1958.

Bibliography:

Bland, J.H.: Clinical Recognition and Management of Disturbances of Body Fluids. Saunders, 2nd Ed., 1956.
Elkinton, J.R., and Danowski, T.S.: The Body Fluids. Williams and Wilkins, 1955.
Gamble, J.L.: Chemical Anatomy, Physiology and Pathology of Extracellular Fluid. Harvard University Press, 1949.
Hardy, J.D.: Fluid Therapy. Lea and Febiger, 1954.
Statland, H.: Fluid and Electrolytes in Practice. Lippincott, 1954.
Weisberg, H.F.: Water, Electrolyte and Acid-Base Balance. Williams and Wilkins, 1953.

20 . . .

The Chemistry and Functions of the Hormones

Most glands of the body deliver their secretions by means of ducts. These are exocrine glands. Other glands manufacture chemical substances or hormones which they secrete into the blood stream for transmission to various tissues. These are the endocrine glands, or ductless glands. The hormonal secretions of these glands, often in minute quantities, exert profound effects on certain tissues. In general, the hormones have a regulatory effect on the diverse metabolic processes of the body. It is characteristic of the endocrine system that a state of balance is normally maintained among the various glands. Furthermore, a reciprocal interaction among them is often demonstrable. This is particularly notable in the relationship of the anterior pituitary tropic principles to the various target glands which they affect.

The chemical structure and functions of each hormone will be discussed in connection with the gland responsible for its production.

THE THYROID

The thyroid gland consists of two lobes, one on each side of the trachea, with a connecting portion making the entire gland more or less H-shaped in appearance. In the adult, the gland weighs about 25 to 30 Gm.

Function:

The principal function of thyroid hormone is that of a catalyst, a sort of "spark," for the oxidative reactions of the body cells. Thus it regulates the rate of metabolism in the body. The tissues from hypothyroid animals exhibit a low rate of oxygen consumption, and, conversely, those from hyperthyroid animals take up oxygen at an accelerated rate. In the absence of adequate quantities of the hormone, there is a notable slowing of bodily processes manifested by a slow pulse, lowered systolic blood pressure, decreased mental and physical vigor, and, generally, obesity. In hyperthyroid states, the reverse occurs. Some of the noteworthy symptoms are rapid pulse rate (tachycardia), irritability and nervousness, and, usually, loss of weight.

An effect of thyroid hormone on oxidative phosphorylation has been observed by several investigators. This is discussed on p. 122.

Chemistry and Normal Physiology:[1,2]

The normal daily intake of iodide is 100-200 μg. This iodide, absorbed mainly from the small intestine, is transported in the plasma in loose attachment to proteins. Small amounts of iodide are excreted by the salivary glands, stomach, and small intestine, and traces occur in milk. However, about two-thirds of the ingested iodide is excreted by the kidney and the remaining one-third is taken up by the thyroid gland. Thyroid-stimulating hormone (TSH, thyrotropic hormone) of the pituitary (see p. 364) stimulates iodide uptake by the gland.

Within the thyroid, iodide is oxidized to an active form of "iodine," a reaction which is catalyzed by an oxidizing enzyme not yet identified. TSH is also active in stimulating this reaction. Oxidation of iodide to iodine is followed rapidly by iodination of tyrosine, first in position 3 of the aromatic nucleus and then also at position 5, forming monoiodo- and diiodotyrosine, respectively.

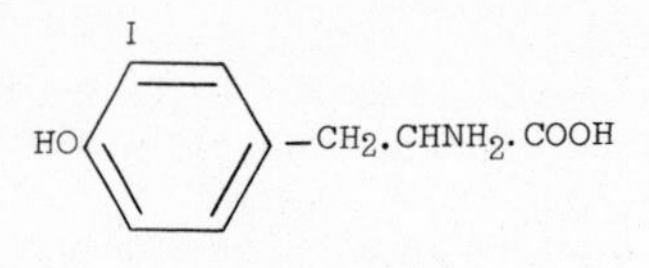

3-Monoiodotyrosine (MIT)

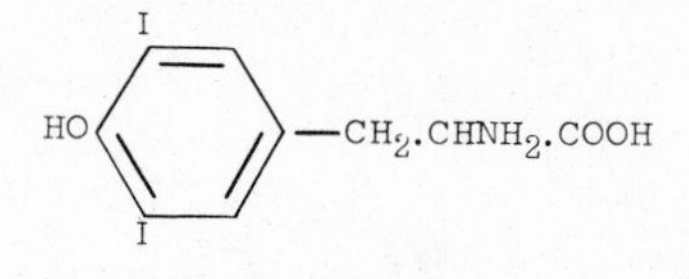

3, 5-Diiodotyrosine (DIT)

It is assumed that coupling of two molecules of diiodotyrosine (DIT) then occurs to form tetraiodothyronine, or thyroxine. Coupling of monoiodotyrosine (MIT) with DIT probably also occurs to form triiodothyronine (TIT), which occurs in the thyroid gland along with thyroxine.

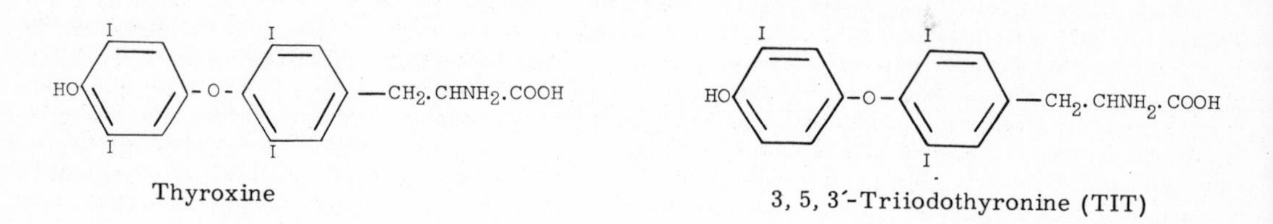

Thyroxine

3, 5, 3′-Triiodothyronine (TIT)

These various iodinated compounds may actually be formed in the amino acid chain of the protein, thyroglobulin, within the gland. In any event, the hormones, thyroxine and triiodothyronine, are stored within the colloid of the thyroid and released and made available to the body only after the enzymatic action of a protease in the colloid that breaks down thyroglobulin. MIT and DIT are also liberated, but these compounds are deiodinated by another enzyme and the freed iodine is reutilized within the gland.

Release of thyroxine and of triiodothyronine into the plasma is accelerated by pituitary TSH or by exposure to a cold environment. Administration of thyroxine impairs release of the hormones, presumably by suppression of pituitary TSH production. Within the plasma, thyroxine is transported in loose attachment to a glycoprotein (thyroxine-binding protein) which migrates electrophoretically in a region between the α_1 and α_2 globulins. Butanol readily removes thyroxine from this protein by a simple extraction procedure. Triiodothyronine binds preferentially to albumin rather than to thyroxine-binding protein. The total circulating thyroid hormone is measured as the so-called protein-bound iodine (PBI). In euthyroid individuals its concentration is between 4 and 8 μg./100 ml. About 80% of the organic iodine released by the thyroid is thyroxine, and 20% is probably triiodothyronine.

Triiodothyronine is not only almost twice as active on a weight basis as is thyroxine, but its onset of action is more rapid. This may be due to the fact that it penetrates into the cells more rapidly than thyroxine; it is also more rapidly degraded in the body and more rapidly cleared from the blood.

Both thyroxine and triiodothyronine are further metabolized, probably in the tissues, by deamination and decarboxylation to tetraiodothyroacetic acid (Tetrac) or to triiodothyroacetic acid (Triac). These metabolites are about one-fourth as active on a weight basis as their hormonal precursors, although their onset of action, according to some observers, is much more rapid.

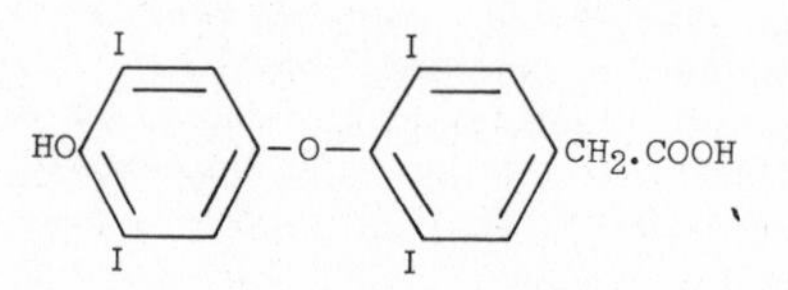

Tetraiodothyroacetic Acid (Tetrac)

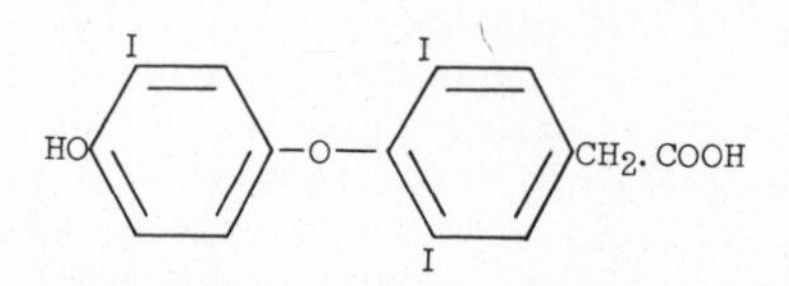

Triiodothyroacetic Acid (Triac)

Other demonstrated metabolic pathways for the iodothyronines include conjugation of thyroxine in the liver with glucuronic acid, followed by excretion of the conjugate in the bile and deiodination, to return iodide to the plasma for excretion into the urine or for reuse by the thyroid.

Iodine Metabolism:

Since iodine is essential to the production of thyroid hormone and since this is the only known function of iodine in the body, iodine metabolism and thyroid function are closely related.

The inorganic iodine in the body is largely taken up by the thyroid in connection with the synthesis of thyroid hormone. Of a total of 50 mg. of iodine in the body, about 10 to 15 mg. is in the thyroid. Thyroid activity appears to be related to the total amount of iodine in the gland rather than to the amount of thyroxine. For example, in the normal gland, an iodine content of 2 mg. per Gm. of dried tissue is found, whereas in hyperthyroidism the content may fall to 0.25 mg. per Gm. Some believe that a fall in the iodine content of the gland stimulates the production of more TSH, which therefore prolongs and exacerbates the hyperthyroid state.

The so-called **protein-bound iodine** of the blood is considered to be hormonal iodine in transit. There is some additional inorganic iodine in the blood as well. The blood iodine is measured as an adjunct in the diagnosis of thyroid disease. Normally, the **total** blood iodine varies from 5 to 15 μg. per 100 ml. However, the protein-bound (hormonal) iodine, normally 4 to 8 μg. per 100 ml., is physiologically more significant and is therefore of diagnostic value. It is increased in hyperthyroidism, but an increase may also be found in pregnancy, in malignancy with excessive tissue destruction, or in burns. If iodine has been given by mouth or if iodine-containing compounds have been used therapeutically or diagnostically (e.g., Diodrast®), the blood iodine may remain high for some time. The interference of these iodine compounds with accurate measurement of hormonal iodine may be lessened by using the "butanol-extractable iodine" (BEI) method[3]. In hypothyroidism the total blood iodine may be normal, but the protein-bound fraction is decreased to about 50% of normal.

It is of interest that many proteins, e.g., casein or soybean protein, may be "iodinated" by treatment with an alkaline solution of iodine at reasonably high temperatures. Such proteins show thyroid activity presumably due to the formation of diiodotyrosine and thyroxine from the tyrosine in the protein. Furthermore, the fact that protein-bound iodine is found in the blood of completely thyroidectomized animals suggests that compounds indistinguishable from thyroid hormone may be formed in vivo by iodination in the absence of glandular tissue.

Radioiodine (I^{131}), the radioisotope of iodine, is promptly concentrated in organic combination in the thyroid after its administration. The initial step, the accumulation phase, is believed to represent free iodine in the epithelial cells of the follicles of the gland. Later, the iodine is found in diiodotyrosine and, still later, in thyroxine. Finally, the radioiodine is detected in the protein-bound form, i.e., in the active hormone. The accumulation of inorganic iodine and its conversion to diiodotyrosine and thyroxine is completed over about a 48-hour period, but the labeled protein-bound iodine does not appear for several days after the original administration of the isotope.

The quantity of the administered radioiodine taken up by the thyroid gland is a valuable diagnostic tool. Normally, 20 to 30% is taken up by the gland. About 40% is suggestive of hyperthyroidism; less than 20%, of hypothyroidism.

Abnormalities of Thyroid Function:

A. Hypothyroid States: A deficiency of thyroid hormone produces a number of clinical states depending upon the degree of the deficiency and the age at which it occurs.
   1. Cretinism follows the incomplete development or congenital absence of the thyroid gland. It is characterized by a low basal metabolic rate and body temperature, almost complete cessation of mental and physical development; obesity, accompanied by a protruding abdomen; dry, coarse skin and hair; and a characteristic facial expression marked by a protruding tongue and open mouth from which the saliva dribbles.
   2. Childhood hypothyroidism (juvenile myxedema) appears later in life than cretinism. It is generally less severe, and some of the typical cretinoid symptoms are absent. The most important signs of juvenile myxedema are stunting of growth because of a lack of growth of bone, cessation of mental development, and, in some cases, changes in the skin as noted in cretinism.
   3. Myxedema is caused by complete hypothyroidism in the adult; it is the adult analogue of cretinism. The basal metabolic rate and body temperature are lowered, and the patient complains of undue sensitivity to cold. Other characteristic findings include puffiness of the face and extremities, thickening and drying of the skin, falling hair, and, in some patients, obesity. Anemia and retardation of physical and mental reactions are also present.

B. Hyperthyroid States and Goiter: Enlargement of the thyroid is called **goiter**. With the exception of malignant goiter and inflammatory disease of the thyroid, as in thyroiditis, goiter involves simple enlargement of the gland (without hyperactivity) or enlargement of the gland with associated hyperthyroidism.

1. Simple (endemic or colloid) goiter is a deficiency disease caused by an inadequate supply of iodine in the diet. Treatment with iodine or sodium iodide is successful. The disease is common where the soil and water are low in iodine (e.g., Great Lakes area). The use of iodized salt has done much to reduce the incidence of simple goiter.
2. Toxic goiter (hyperthyroidism) differs from simple goiter in that enlargement of the gland is accompanied by the secretion of excessive amounts of thyroid hormone, i.e., hyperthyroidism together with enlargement (goiter). The term "toxic" does not refer to the secretion of the gland but to the toxic symptoms incident to the hyperthyroidism. Hyperthyroidism occasionally occurs without goiter.

   The most common form of hyperthyroidism is exophthalmic goiter (Graves's disease, Basedow's disease). The enlargement of the gland may be diffuse or nodular. In the latter case the terms **nodular toxic goiter** or **toxic adenoma** (Plummer's disease) are sometimes used, but the symptoms are the same since hyperthyroidism is, of course, common to both.

   The most important signs and symptoms of hyperthyroidism, in addition to the goiter, are nervousness, easy fatigability, loss of weight in spite of adequate dietary intake, increased body temperature with excessive sweating, and an increase in the heart rate. A characteristic protrusion of the eyeballs (exophthalmos) usually accompanies hyperthyroidism, but this may be absent.

Laboratory Diagnosis of Thyroid Abnormalities:

A. Hypothyroid States: A basal metabolic rate (BMR) below −30% suggests hypothyroidism, although this is not necessarily diagnostic of hypothyroidism alone; a protein-bound iodine of less than 4 μg. per 100 ml. and an I^{131} uptake of less than 20% are also of diagnostic value. Blood cholesterol may be elevated (see p. 217). Urinary excretion of neutral 17-ketosteroids (see p. 354) is usually reduced in myxedema.

B. Hyperthyroid States: Laboratory signs in hyperthyroidism include elevated BMR, blood hormonal iodine (PBI) exceeding 8 μg. per 100 ml., and increased I^{131} uptake. Lowered blood cholesterol (see p. 217) and elevated urinary creatine may occasionally be observed.

Treatment:

A. Hypothyroid States: "Endemic cretinism," which is due to a lack of iodine in the geographical area involved, is treated prophylactically, i.e., by administration of iodine to pregnant women.

Desiccated thyroid gland is a satisfactory replacement for thyroid hormone in cretinism, but it must be given in early infancy. It will maintain normal growth and physical development, but mental development usually remains inadequate, probably because of irreversible changes which occur during intrauterine life.

Desiccated thyroid is also used in the treatment of childhood hypothyroidism and of myxedema. When rapid response is necessary, sodium lyothyronine (Cytomel®) may be used.

B. Hyperthyroid States and Goiter:

1. Subtotal thyroidectomy, after preliminary treatment with iodine and propylthiouracil, is probably still the most often used treatment of toxic hyperthyroidism.
2. Iodine (Lugol's solution) sometimes benefits hyperthyroid patients, at least briefly. The iodine depresses the activity of the gland, probably by decreasing TSH output. However, after a time the gland seems to "escape" from its influence since iodine no longer retards the production of TSH.
3. Radioactive iodine (I^{131}) therapy is now available in most medical centers. It consists of concentrating radioiodine solution in the gland, permitting the beta and gamma emissions of the solution to irradiate the tissue in the same way as x-rays, thus decreasing thyroid activity. This treatment may be particularly valuable in malignant thyroid disease if the cancer tissue can take up iodine.
4. Antithyroid drugs - Certain compounds act as antithyroid agents, inhibiting the production of thyroxine by preventing the gland from incorporating inorganic iodine into the organic form. Examples of these antithyroid drugs are thiouracil, propylthiouracil, methylthiouracil, thiourea, and methimazole (Tapazole®, 1-methyl-2-mercaptoimidazole).

Thiouracil is relatively toxic; the other compounds less so.

The antithyroid compounds have an almost immediate effect since they operate during the first stage of iodine uptake by the gland when the iodine is still in the inorganic form. Under their influence the thyroid gland does not manufacture the hormone; the protein-bound iodine gradually falls, and the symptoms of hyperthyroidism gradually decrease.

Studies of thyroid glands which have been "blocked" with thiouracil show that iodine is trapped in the gland but not incorporated into the organic form. In the normal thyroid which has been blocked with thiouracil, the accumulation of iodine is equal to the amount in about 500 ml. of plasma. However, in a thyrotoxic patient in which excess pituitary thyroid–stimulating hormone (TSH) activity accelerates the activity of the gland, the iodine accumulation within the blocked gland may be equivalent to that contained in as much as 35 liters of plasma, a gradient of iodine in the gland as compared to plasma of 500:1.

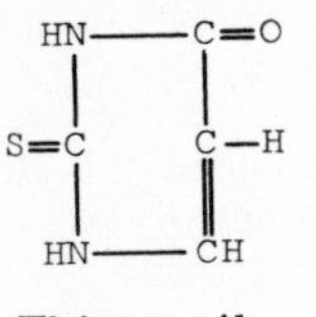

Thiouracil

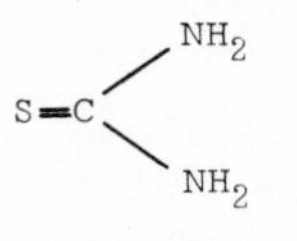

Thiourea

THE PARATHYROIDS

The parathyroid glands are four small glands so closely associated with the thyroid that they remained unrecognized for some time and were often removed in thyroidectomy operations. In man, the parathyroids are reddish or yellowish-brown, egg-shaped bodies 2 to 5 mm. wide by 3 to 8 mm. long and 0.5 to 2 mm. in thickness. The four glands together weigh about 0.05 to 0.3 Gm.

Function:

Parathyroid hormone exerts a profound effect on the metabolism of calcium and phosphorus. There may be other metabolic effects which have not yet been clarified. The metabolism of calcium and phosphorus is described on pp. 314 and 316.

Chemistry:

Parathyroid extract is prepared from bovine parathyroid glands extracted with aqueous hydrochloric acid solution. The hormone is precipitated from these solutions with trinitrophenol and the precipitate extracted with acid-acetone and alcohol, with final precipitation from this solution by excess acetone. This final acetone precipitate, dissolved in water, is marketed as parathyroid extract. The hormone has not yet been isolated in the pure state. It is believed to be a protein.

Parathyroid extracts are assayed by their ability to increase the serum calcium in normal dogs after subcutaneous injection. One unit is 0.01 of the amount necessary to raise the serum calcium by 1 mg. per 100 ml. within 16 to 18 hours after injection.

Physiology: (See also p. 315.)

The administration of parathyroid hormone (1) raises the serum calcium and lowers the serum phosphorus, (2) increases the elimination of calcium and phosphorus in the urine; (3) removes calcium from the bone, particularly if the dietary intake of calcium is inadequate; and (4) increases serum alkaline phosphatase if changes in bone have been produced.

There is increasing evidence that there are probably two (and perhaps more than two) types of parathyroid hormones[4]. The two primary hormones are as follows: (1) A calcium mobilizing principle, which by direct action on the bones stimulates a breakdown of this tissue (osteoclastic activity increased). A shift of calcium from the bones then occurs which is evidenced by a rise in serum calcium (hypercalcemia) and increased excretion of calcium by overflow into the urine (hypercalciuria). (2) A phosphaturic substance with a direct effect on the kidney, decreasing the renal tubular reabsorption of phosphate and thus increasing the excretion of phosphate into the urine and lowering the concentration of phosphate in the serum.

In the course of purifying crude parathyroid extracts, the phosphaturic activity may be lost. As noted above, parathyroid extracts are usually assayed by their ability to elevate the serum calcium. However, it has been found that a crude preparation of parathyroid extract as compared to a purified one may have three to four times as much phosphaturic activity although both are equal in calcium-mobilizing activity. It is also possible that signs of hyperparathyroidism may vary in different patients in accordance with the type of hormone which is being produced in the larger quantity. For example, hypercalcemia and hypercalciuria may be only transient features in some cases of hyperparathyroidism, whereas concurrent hypophosphatemia and hyperphosphaturia are consistently noted. In other cases the reverse is true, and hypercalcemia and renal stone formation are especially prominent.

Removal of the parathyroids causes tetany, a pronounced fall in the ionized fraction of the blood calcium, which in turn leads to increased neuromuscular irritability and tetania parathyreopriva. Extracts of the parathyroid relieve this tetany, and these extracts substitute completely for the glands.

The parathyroid hormone may affect organs other than bones and kidneys; indeed, patients with hyperparathyroidism may present without prominent renal or bone symptoms but with involvement of the central nervous system, the gastrointestinal tract, or the peripheral vascular system. Repeated doses of parathyroid extract cause acute symptoms such as oliguria or anuria, anorexia, gastrointestinal hemorrhage, nausea and vomiting, and finally loss of consciousness and death. These symptoms are probably due to acute water and electrolyte depletion as well as to other changes in the body which are not clearly understood.

Abnormalities of Parathyroid Function:

A. Hypoparathyroidism: Usually this is the result of surgical removal of the parathyroids, although an idiopathic hypoparathyroidism of unknown etiology has been reported. The symptoms of hypoparathyroidism include muscle weakness, tetany, and irritability.

B. Hyperparathyroidism:[5,6] An increase in parathyroid hormone production is usually due to a tumor of the gland. Decalcification of the bones, which results from hyperparathyroid activity, causes pain and deformities in the bones as well as spontaneous fractures. Anorexia, nausea, and polyuria are other symptoms. Deposits of calcium may form in the soft tissues, and renal stones may also occur. In fact, hyperparathyroidism should always be suspected in patients with chronic renal lithiasis. Deficiencies of magnesium may also result from long-continued hyperparathyroidism[7] (see p. 317).

Enlargement of the parathyroid glands (secondary hyperparathyroidism), probably as a result of increased serum phosphate, is often a feature of chronic renal disease.

Laboratory Diagnosis of Parathyroid Abnormalities:

A. Hypoparathyroidism: Low serum calcium (less than 5 to 6 mg. per 100 ml.), elevated blood phosphorus (4 to 6 mg. per 100 ml.), normal serum alkaline phosphatase, and reduced blood magnesium (1.5 to 1.8 mg. per 100 ml.) are the chemical findings in hypoparathyroidism.

B. Hyperparathyroidism: High serum calcium, low phosphorus, and high or normal alkaline phosphatase are characteristic chemical findings in hyperparathyroidism. In chronic renal disease, with secondary hyperactivity of the parathyroid, the serum alkaline phosphatase is either normal or elevated (because of associated bone abnormalities); on the other hand, calcium levels may be reduced because of increased loss in the urine.

Treatment:

A. Hypoparathyroidism: In the presence of tetany, prompt intravenous or intramuscular injection of calcium salts (calcium gluconate) and intramuscular injection of parathyroid hormone are required.

Parathyroid hormone (parathormone) is effective for only a short time, presumably because of the formation of an antihormone; therefore, dihydrotachysterol (A. T. 10) is now used in the maintenance management of hypoparathyroidism. A high-calcium, low-phosphorus diet, supplemented with oral calcium salts, is also used, together with A. T. 10. Dihydrotachysterol is prepared in the irradiation of ergosterol to produce vitamin D_2. Its action, similar to that of vitamin D, is to increase the absorption of calcium from the intestine. It also increases phosphate excretion by the kidney. Vitamin D_2 (calciferol) may be used in place of dihydrotachysterol and may, in fact, be preferred since it is less dangerous and easier to control.

The objective in treatment of hypoparathyroidism with A. T. 10 or calciferol is to raise the serum calcium to a normal level of about 9 mg. per 100 ml. Above this level, increasing amounts of calcium are excreted in the urine. A positive test for urinary calcium thus provides a simple indication of adequate treatment or overdosage. **Sulkowitch's reagent** (2.5 Gm. oxalic acid, 2.5 Gm. ammonium oxalate, and 5 ml. glacial acetic acid dissolved in water and diluted to 150 ml.) is used for conducting a simple qualitative analysis for urine calcium. The reagent and urine are mixed in equal proportions; if no precipitate forms, the blood calcium is between 5 and 7.5 mg. per 100 ml.; the formation of a fine, white cloud indicates that the blood calcium is probably normal (9 to 11 mg. per 100 ml.); a heavy precipitate suggests hypercalcemia.

There is a possibility that vitamin D at some dosage levels may decrease renal tubular reabsorption of calcium. This is indicated by reports of a number of observers who have found **increased** excretion of calcium in patients with hypoparathyroidism being treated with vitamin D[8]. The calciuria occurred only in patients with true hypoparathyroidism and not in those with so-called pseudohyperparathyroidism, in which the disease is presumably caused by a failure of the renal tubules to respond to the action of parathyroid hormone rather than to a deficiency of the hormone itself. Occasionally the calciuria during vitamin D therapy persisted even when the serum levels of calcium were still low and the serum phosphorus was still elevated. As noted above, when hypoparathyroid patients are being treated with vitamin D, the dosage may be regulated by the results of the Sulkowitch test; but under the circumstances described, a positive Sulkowitch test might be obtained when the serum calcium was still abnormally low.

B. Hyperparathyroidism: Parathyroid tumors should be removed surgically. Large quantities of fluids should be given to produce a dilute urine and lessen the opportunity for the formation of calcium phosphate stones.

THE PANCREAS

The pancreas has both exocrine and endocrine functions. Its exocrine functions are concerned with digestive processes. The endocrine function of the pancreas is centered in the **islets of Langerhans**, epithelial cells that are dispersed throughout the entire organ between the acinar tissue, which manufactures the digestive secretions. Two hormones which affect carbohydrate metabolism are produced by the pancreas: **insulin**, by the beta cells; and **glucagon**, by the alpha cells.

INSULIN

Insulin plays an important role in the metabolism of carbohydrate by its effect on the storage of glycogen in the liver (hepatic glycogenesis; see p. 174), the conversion of carbohydrate to fat, and the oxidation of glucose by the peripheral tissues; thus insulin also participates in the mechanisms for the regulation of the blood sugar (see p. 172).

Chemistry:

Insulin is a protein hormone which is secreted by the beta cells of the islets of Langerhans. The hormone has been isolated from the pancreas and prepared in crystalline form. Crystallization of insulin requires traces of zinc, which may be a constituent of insulin since normal pancreatic tissue is relatively rich in zinc.

Digestion of insulin protein with proteolytic enzymes inactivates the hormone, and for this reason it cannot be administered orally. Insulin may also be irreversibly denatured with alkali and reversibly with acid. It is inactivated by aldehydes and by other reducing agents. Incubation of insulin with cystine inactivates insulin.

The structure of insulin has been elucidated by Sanger and his co-workers. It is shown on p. 40. When insulin is treated with ketene so as to acetylate the free amino groups but not the free hydroxyl groups (of tyrosine), the hormone is not inactivated; however, its activity is reduced when the hydroxyl groups in the molecule are acetylated. Chemical treatment to reduce the S-S (cystine) linkage to SH (cysteine) also inactivates the hormone.

Protamine zinc insulin is a combination of insulin with protamine. It is absorbed more slowly than ordinary insulin; one injection of protamine zinc insulin may lower the blood sugar for more than 24 hours, whereas two or three injections of regular insulin might be required for the same effect. In practice, mixtures of regular and protamine zinc insulin in a ratio of 3:1 or 2:1 are frequently used.

Globin insulin, another combination of insulin with a protein - in this case globin - has an effect somewhere between those of regular and protamine insulin (a 12- to 15-hour duration of action).

Insulin preparations are standardized by measuring their effect on the blood sugar of experimental animals, usually rabbits. One unit of insulin is the amount required to reduce the blood sugar level of a normal 2-Kg. rabbit after a 24-hour fast from 120 to 45 mg. per 100 ml. An international standard preparation of zinc insulin crystals is maintained at the National Institute for Medical Research, London. One unit is 1/22 mg. of this preparation. Commercial insulin is usually sold in three concentrations: U-20, U-40, and U-80, the numbers referring to units per ml.

Physiology:

The injection of large doses of insulin into normal subjects causes a reduction in the blood sugar (hypoglycemia) accompanied by typical symptoms of weakness, hunger, irritability, faintness, tremors, and convulsions, all of which are relieved by the administration of glucose. The hormone exerts a catalytic effect on the metabolism of carbohydrates and, secondarily, on the metabolism of lipids and on amino acids. The exact manner of action of insulin is not known, but certain facts seem clear. According to Goldstein et al.[9], one action of insulin is to facilitate the transfer of glucose from the extracellular fluid into the intracellular space in certain organs and tissues. Increased oxidation and storage of glucose then follows as a result of the increase in the amount of glucose that has entered the cells. Insulin promotes the entry into the cells of only those sugars possessing the same configuration as D-glucose at carbons 1, 2, and 3. This essential structure is shown below.

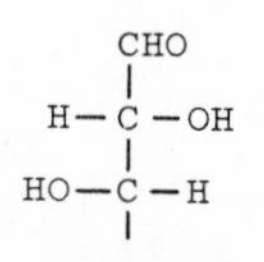

Such sugars would include D-glucose, D-galactose, D-xylose, and L-arabinose. Although the last sugar does not occur in the body, insulin - at least experimentally - influences its penetration into the cells. The failure of fructose to depend upon insulin for its entrance into the cells is readily understood in terms of the above hypothesis since in carbon No. 2 fructose has a ketone group and thus differs from the key structure which is necessary for insulin action.

A second action of insulin is related to its role as an antagonist to the effects of pituitary and adrenocortical extracts. This relationship as it affects the activity of glucokinase has been described on p. 179.

It is now well established that the formation of fat from carbohydrate (lipogenesis) is a major route for the disposal of the carbohydrate derived from the diet. Insulin plays a major role in the process of lipogenesis, probably by facilitating the entrance of glucose into glycolytic pathways and thus making available the precursors for fatty acid synthesis as well as energy for this process. Another important aspect of the relationship of insulin to lipogenesis is its role in facilitating entry of glucose into the direct oxidative pathway, whereby reduced TPN (TPN.H) can be made available for the reductive steps in fatty acid synthesis (see p. 187). The marked effect of insulin added in vitro on lipogenesis from glucose in adipose tissue has emphasized the possibility that adipose tissue, in addition to liver and muscle, is a major site of insulin action[10]. All of these effects of insulin on lipogenesis obviously contribute substantially to the reduction in the blood sufar levels which insulin accomplishes.

Insulin causes a reduction in the levels of amino acids in the blood and acts to accelerate the rate at which proteins are synthesized from free amino acids. These effects are probably also secondary to the primary role of insulin in facilitating the utilization of glucose, thus making energy available for protein synthesis.

An effect of insulin in connection with oxidative phosphorylation and the maintenance of the total supply of ATP has been suggested by Stadie. The details of the present status of concepts on the action of insulin have been reviewed by Stadie[11].

The level of sugar in the blood is the resultant of the rate at which carbohydrate is formed to the rate at which it is oxidized or deposited in the tissues as glycogen. It has been noted that when there is a deficiency of insulin the amount of sugar excreted in the urine exceeds the amount ingested. This suggests that there is an overproduction of carbohydrate under conditions of insulin deficiency; according to the proponents of this ''overproduction theory,'' this is the cause of the hyperglycemia and glycosuria of diabetes. According to this theory, insulin functions in maintaining a state of balance between glycogenesis and lipogenesis, and glycogenolysis and gluconeogenesis. The latter two metabolic activities are increased by the activity of pituitary and adrenal hormones.

Destruction of Insulin; Activity of Insulinase:

The activity of insulin is destroyed after incubation with homogenates, extracts, or slices of liver. Two systems appear to be involved in this process: (1) an insulin-inactivating enzyme, **insulinase,*** and (2) an inhibitor of this enzyme (**insulinase inhibitor**), which appears to be a heat-stable, dialyzable, nonprotein factor. Mirsky and his co-workers[13] have studied the insulinase system with the aid of I^{131}-labeled insulin. They have shown that liver slices can accomplish the degradation of insulin and that a nonprotein fraction of liver (insulinase inhibitor) which inhibits the action of insulinase also prevents the degradation of the labeled insulin by liver slices.

Insulinase is responsible for the very short life span of insulin in vivo; in man, half of the circulating insulin is degraded in about 35 minutes.

Not only the liver, but kidney and muscle also possess active insulinase systems. In plasma, insulin destruction proceeds only at a very slow rate.

Factors which are inhibitory to insulin (insulin antagonists) have been detected in the α globulin fraction of the serum of patients with diabetic ketosis. There is also a great increase in the insulin-inhibitory activity of the serum of patients with clinically demonstrated insulin resistance. The phenomenon of insulin resistance in these cases seems to be attributable to an increase in an insulin-binding ''antibody''[14]. The bound insulin is not available to the cells and is only slowly degraded; thus, much of the insulin administered is actually wasted. However, it is slowly released, so that after an acute episode of insulin resistance which has been treated with massive doses of insulin, the released insulin may cause repeated bouts of hypoglycemia.

Insulinoid Substances:

It has been noted that insulin is not effective when administered orally. However, several compounds have recently been discovered which under certain circumstances will produce a drop

*It has been pointed out that the insulin-inactivating system of the liver is not absolutely specific to insulin[12]. It seems rather to be a proteolytic enzyme which can also attack ACTH, casein, glucagon, and growth hormone. The term ''insulinase'' is, therefore, perhaps not entirely accurate since it connotes absolute specificity for insulin.

in the blood sugar when given by mouth. These hypoglycemic agents do not lower the blood sugar in alloxanized diabetic animals nor in animals or patients who have undergone total removal of the pancreas. They are also ineffective in ''juvenile'' diabetics whose pancreas contains little or no insulin, but they are valuable in the treatment of diabetics of the ''adult'' type. More than 75% of patients who have developed diabetes after reaching 40 years of age have shown marked and continued benefit from the use of these agents.

The orally effective hypoglycemic agents most extensively studied so far are sulfonamide derivatives, i.e., arylsulfonylureas. Examples are tolbutamide (Orinase®) and chlorpropamide (Diabinese®). The exact mode of action of these compounds has not yet been established, but most authorities incline to the view that they act to stimulate the beta cells of the pancreas to produce insulin.

Abnormalities:

A. Diabetes mellitus in man is characterized by a disturbance in carbohydrate metabolism, with hyperglycemia and glycosuria. Secondary disturbances in the metabolism of protein and fat (ketosis; see p. 211) lead to acidosis (see p. 150) and to dehydration because of the water required to excrete the excess glucose in the urine (see p. 288). The administration of insulin corrects these metabolic abnormalities. There is no doubt that diabetes mellitus in man is due to an insufficiency of insulin relative to the requirements by the tissues. This is supported by the fact that pancreatectomy or destruction of the islet tissue by specific drugs (alloxan) or infection produces typical diabetes. However, in only a small proportion of cases can diabetes in man be attributed only to decreased production of insulin. The demonstration that insulin can be destroyed by liver and kidney suggests that the insulin insufficiency of some diabetic patients may be due to an increase in the rate of insulin destruction, either because of increased insulinase activity or a decrease in the insulinase inhibitor. Since liver and kidney are involved in the degradation of insulin, it seems likely that in diseases affecting these organs a diminution in this function might occur. This would result in an apparent increase in the potency of administered insulin and a decrease in the requirement for insulin, which has actually been observed in certain instances in diabetics who have also associated kidney or liver disease.

Another puzzling clinical observation is the occurrence in certain patients of resistance to insulin. Furthermore, as noted on p. 172, overproduction of ''diabetogenic'' hormones such as those of the pituitary and adrenal cortex in the presence of normal amounts of insulin might also lead to diabetic symptoms. Thus diabetes may be not only of pancreatic origin, presumably due to a relative insulin deficiency; pituitary and steroid diabetes must also be considered as factors in the etiology of the disease process.

In the light of present knowledge it is believed that the abnormal metabolism observed in absolute or relative insulin deficiency is attributable to increase of hepatic glycogenolysis, and a decrease in glucose oxidation and in lipogenesis from glucose. There is also increased formation of sugar from amino acids and other precursors (gluconeogenesis) and accelerated mobilization of fat from the depots to the liver. The effects on amino acid and fatty acid metabolism are doubtless due to the predominance of pituitary and adrenocortical hormonal activity as a result of a decline in the antagonism to these effects normally exerted by insulin.

A summary of the abnormalities of metabolism which occur in uncontrolled diabetes is shown on the next page.

B. Experimental diabetes may be produced by total pancreatectomy. Destruction of the insulin-producing beta cells of the pancreas may be brought about in experimental animals by a single injection of alloxan, a substance related to the pyrimidines (see p. 44). This is a simpler method of producing permanent diabetes than surgical removal of the pancreas, although it is not equally effective in all animals.

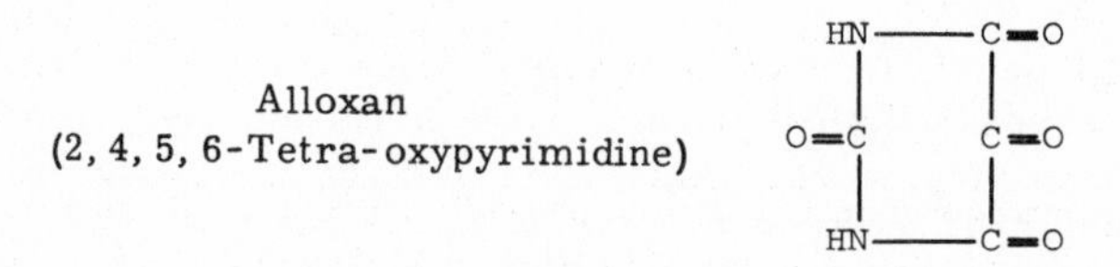

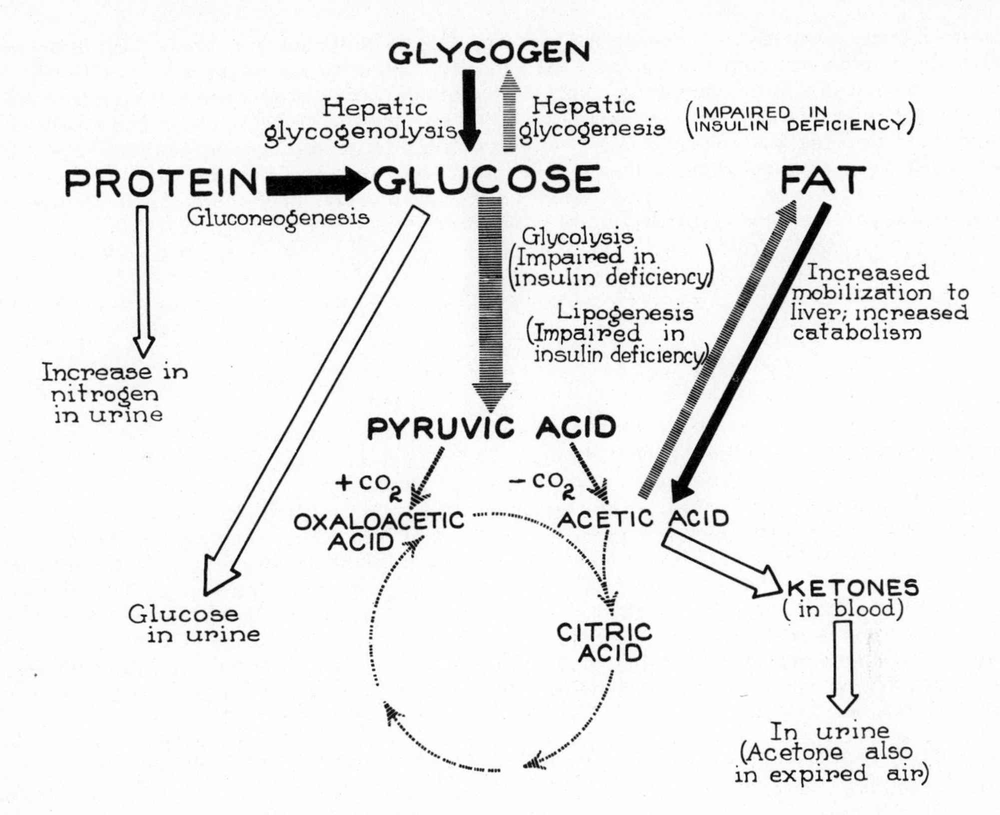

Abnormal Metabolism in Uncontrolled Diabetes. The black arrows indicate normal pathways and the broken lines represent impairment resulting from insulin deficiency. The white arrows show changes also resulting from insulin deficiency.

Diabetes may also be produced by repeated injection of the so-called diabetogenic hormone of the pituitary. This type of diabetes is considered to be due to overstimulation and exhaustion (?) of the beta cells rather than to their destruction, as in the case of alloxan poisoning.

Phlorhizin diabetes is actually a renal diabetes in which glycosuria is produced by failure of reabsorption of glucose by the renal tubules rather than by virtue of any endocrine abnormality.

Laboratory Diagnosis of Diabetes:

Test for sugar in the urine. A positive test suggests diabetes. An elevated fasting blood sugar and/or an abnormal sugar level two hours after a meal containing 100 Gm. carbohydrate or a dose of 100 Gm. of glucose is usually confirmatory if other disease states which affect carbohydrate metabolism have been ruled out. Glucose tolerance tests (see p. 173) may be used in the diagnosis of diabetes in doubtful cases.

Treatment of Hypo- and Hyperinsulinism:

A. Diabetes mellitus may usually be controlled with dietary regimens with or without insulin; insulin cannot effect a cure.

B. Hyperinsulinism results from excessive production of insulin or overdosage of the hormone. A pancreatic tumor affecting islet tissue is a frequent cause of hyperinsulinism; however, most cases of hyperinsulinism are not associated with a demonstrable tumor but seem rather to be due to "overshooting" or to increased reactivity to glucose. The condition is characterized by nervousness, weakness, depression, cold sweats, anxiety, confusion, delirium, and, in some cases, convulsions and final collapse. Blood sugar levels are considerably reduced, sometimes to the vanishing point. The administration of glucose will alleviate the condition. In the patient with hyperinsulinism, a diet high in protein and fat and low in carbohydrate is necessary in order to avoid excessive stimulation of the pancreas to produce insulin. If the hyperinsulinism is due to hyperplasia or a tumor of the islet tissue, surgical removal is necessary to correct the condition; however, cortisone (see p. 347) has been used successfully in treating hyperinsulinism in children.

GLUCAGON

Kimball and Murlin[15] were the first to note the occurrence of a rise in the blood sugar after the administration of certain pancreatic extracts. They postulated the existence in the pancreas of a hyperglycemic factor to which they gave the name "**glucagon.**" The existence of this factor, now often referred to as the hyperglycemic-glycogenolytic factor (HGF), has been confirmed. It has been purified and crystallized from pancreatic extracts and shown to be a polypeptide with a molecular weight of about 3485. It contains 15 different amino acids with a total of 29 amino acid residues, arranged in a straight chain. The sequence of the amino acids in the chain has also been determined. Histidine is the N-terminal amino acid and threonine is the C-terminal amino acid. In distinction to insulin. glucagon contains no cystine, proline, or isoleucine but does contain considerable amounts of methionine and tryptophan which are absent from insulin. Further, it can be crystallized in the absence of zinc or other metals. The structure of the glucagon polypeptide is as follows:[16]

His-Ser-Glu(NH_2)-Gly-Thr-Phe-Thr-Ser-Asp-Tyr-Ser-Lys-Tyr-Leu-Asp-Ser-Arg-

Arg-Ala-Glu(NH_2)-Asp-Phe-Val-Glu(NH_2)-Try-Leu-Met-Asp(NH_2)-Thr

Glucagon is believed to originate in the alpha cells of the pancreas and possibly also in other cells located in the gastric and duodenal mucosa which resemble alpha cells. There is evidence that administration of growth hormone may bring about secretion of a hyperglycemic factor from the pancreas.

The primary physiologic action of glucagon is to stimulate hepatic glycogenolysis and thus to raise the blood sugar. This is accomplished by the selective action of glucagon in the liver, where it causes an increase in the activity of liver phosphorylase[17]. In distinction to epinephrine, which also enhances the activity of liver phosphorylase, glucagon does not increase glycogenolysis in muscle nor increase blood lactic acid levels. It also fails to inhibit peripheral utilization of glucose.

An effect of glucagon on fat metabolism was reported by Haugaard[18]. An average decrease of 30% in the synthesis of fatty acids by liver slices was caused by the addition of glucagon. The hormone also increased ketosis by the liver slices. These effects of glucagon are directly opposite to the effects of insulin.

Totally depancreatized patients do not require as much insulin as do some diabetics with an intact pancreas. This may be explained by the loss of the anti-insulin action of glucagon which would accompany pancreatectomy.

It is currently believed that glucagon plays a role in metabolism, but its importance in this connection has not been established. In diabetes, the present view seems to be that glucagon may exert a modifying influence on the disease but that it is not important as a causative factor.

The physiologic and clinical aspects of glucagon have been reviewed by Van Itallie[19].

THE ADRENALS

The adrenal glands, located just superior to the kidneys, consist of an inner part, the medulla, and an outer part, the cortex. The endocrine functions of each area are quite different.

THE ADRENAL MEDULLA

Function:

The adrenal medulla is a derivative of the sympathetic portion of the autonomic nervous system; the hormone of the adrenal medulla, epinephrine (adrenalin) in general duplicates the effect of sympathetic stimulation of an organ. Despite its diverse physiologic functions, the adrenal medulla is not essential to life. According to Cannon, epinephrine is necessary to provide a rapid physiologic response to emergencies such as cold, fatigue, shock, etc. In this sense, it mobilizes what has been termed the ''fight or flight'' mechanism, a cooperative effort of the adrenal medulla and the sympathetic nervous system.

Chemistry:

The adrenal medulla contains a rather high concentration of epinephrine, 1 to 3 mg. per Gm. of tissue. The chemical structure of epinephrine is shown below. It is undoubtedly a derivative of tyrosine and phenylalanine. After the feeding of isotopically labeled phenylalanine, the C^{14} label was detected in the epinephrine molecule.

Naturally-occurring epinephrine is the L-isomer; the D-form is only 1/15 as active as the L-form. Epinephrine produced synthetically is racemic. When resolved, the L-form is identical in activity with the natural product.

Until recently it was believed that the adrenal medulla secretes but one hormone; however, it has now been shown that while 80% of the hormonal activity is due to epinephrine, another closely related hormone, norepinephrine (arterenol) is also present in the extracts of bovine adrenal medulla. It has also been shown that stimulation of the hepatic nerve of cats causes release of arterenol. Pure arterenol has been isolated from several commercial lots of natural epinephrine. Assays of the arterenol content of epinephrine from natural sources indicate concentrations of 10.5 to 18.5%. The physiologic action of arterenol is somewhat different from that of epinephrine.

From the chemical structures of epinephrine and norepinephrine, shown below, it can be seen that norepinephrine is a precursor of epinephrine which is formed from norepinephrine by methylation on the primary amino group.

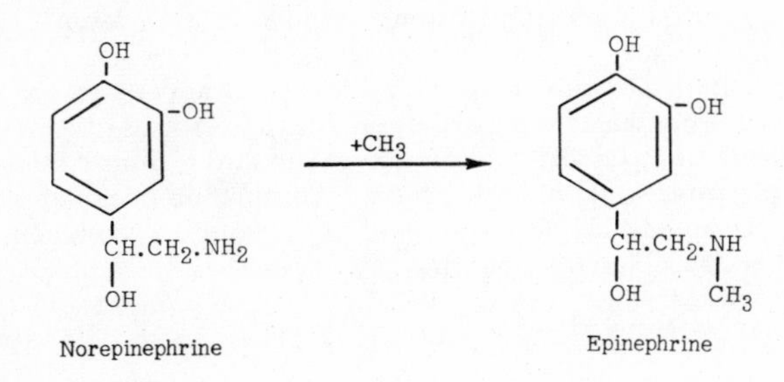

Normal Physiology:

A. Action on Cardiovascular System: Epinephrine brings about vasodilatation of the blood vessels of the skeletal muscles and vasoconstriction of the arterioles of the skin, mucous membranes, and splanchnic viscera. It is also effective as a stimulant of heart action, increasing the rate, irritability, strength of contraction, and cardiac output.

It has often been noted that the effect of natural epinephrine varies with the dosage. It is probable that some of the paradoxical effects can be attributed to the presence of norepinephrine. This hormone - sometimes called **excitor sympathin** or **sympathin N** - has less effect on the cardiac output than epinephrine although it has an excitatory effect on most areas of the cardiovascular system. Norepinephrine exerts an over-all vasoconstrictor effect, whereas epinephrine exerts, in general, an over-all vasodilator effect, with exceptions as noted above. Both hormones lead to an elevation of the blood pressure, more marked in the case of norepinephrine, as a result of their action on the heart and blood vessels. Norepinephrine has found important application in the treatment of hypotensive shock other than that caused by hemorrhage. Its marked action on the arterioles without producing tachycardia makes it particularly useful for this purpose.

B. Action on Smooth Muscle of the Viscera: Epinephrine causes relaxation of the smooth muscles of the stomach, intestine, bronchioles, and urinary bladder, together with contraction of the sphincters in the case of the stomach and bladder. Other smooth muscles may be contracted. The relaxing effect of epinephrine on bronchiolar smooth muscles makes this hormone particularly valuable in the treatment of asthmatic attacks.

C. Action on Blood Sugar: If the liver glycogen stores are adequate, epinephrine induces an elevation in blood sugar by promoting hepatic glycogenolysis (see pp. 171 and 174).

D. Stimulation of ACTH Production(?): Animal experiments had suggested that epinephrine stimulates ACTH production by the anterior pituitary gland. A depression in circulating eosinophils (see pp. 347 and 356) was used as the criterion for ACTH production. It is now believed that the eosinophil response was independent of the adrenal cortex; it is thus concluded that epinephrine does not stimulate ACTH production in man.

E. Mobilization of Nonesterified Fatty Acids: As noted on p. 199, epinephrine and norepinephrine cause a marked rise in plasma nonesterified fatty acids.

F. Other Effects: After the injection of epinephrine, skeletal muscle becomes fatigued at a slower rate than usual, the rate and depth of respiration are increased, and there is a stimulation of the metabolic rate with an increase in the respiratory quotient.

Abnormal Physiology:

The adrenal medulla is not often involved in disease. No clinical state directly attributable to a deficiency of the adrenal medulla is known. However, certain tumors of the medullary tissue (chromaphil cell tumors, ''pheochromocytoma'') produce a condition which simulates hyperactivity of the adrenal medulla. The symptoms of these tumors include intermittent hypertension which may progress into a permanent hypertension, leading eventually to death from complications such as coronary insufficiency, ventricular fibrillation, or pulmonary edema.

It has often been noted that the norepinephrine content of adrenal medullary tumors is much higher than their epinephrine content. For example, Holton[20] found, in the extract from one tumor he assayed, 3.7 mg. epinephrine per Gm. of tumor tissue and 6.3 mg. norepinephrine. In other cases of adrenal medullary tumors an even larger predominance of norepinephrine over epinephrine was reported. It is believed that the attacks of hypertension produced by these tumors are attributable to their high norepinephrine content.

THE ADRENAL CORTEX

The outer portion of the adrenal gland, the adrenal cortex, is essential to life. Its embryologic origin is quite different from that of the adrenal medulla. The adrenal cortex originates from the mesodermal tissue of the nephrotomic area; in a 20 mm. human embryo the chromaphil cells, which develop into medullary tissue, migrate into the developing cortical tissue so that eventually the compound gland is formed.

An extract of the adrenal cortex, **cortin**, contains a number of potent hormones all of which are steroid derivatives having the characteristic cyclopentanoperhydrophenanthrene nucleus. As will be noted later, the hormones of the gonads are also steroid hormones not remarkably different from those of the adrenal cortex. The similarity of embryologic origin of the adrenal cortex and of the gonads is of interest in connection with the close relationship of the chemistry of their respective hormones.

Function:

The hormones of the adrenal cortex exert well-defined effects on the metabolism of electrolytes and thus on water metabolism, and on the metabolism of protein, carbohydrate, and fat. Various other physiologic functions are also influenced by the cortical hormones. All of these observations suggest that these hormones influence basic physiologic processes, possibly cell permeability. There is also evidence of steroid hormone effects on cellular enzyme systems.

Chemistry:

About 30 crystalline compounds have been isolated from the adrenal gland, but only five are known to occur normally in human subjects and to possess physiologic activity. These include Compounds A, B, E (cortisone), and F (hydrocortisone, cortisol), and aldosterone. The major free circulating adrenocortical hormone in human plasma before and after administration of corticotropin is 17-hydroxycorticosterone (Compound F). Corticosterone (Compound B) and aldosterone are present in human plasma in relatively small concentrations.

Normal Physiology:

The well-defined effects of the corticosteroids can be classified under one of three groups.

A. Salt and Water Effects: All of the active corticosteroids increase the reabsorption of sodium and chloride by the renal tubules and decrease their excretion by the sweat glands, salivary glands, and the gastrointestinal tract. However, there is a considerable difference in the extent of this action by the various corticoids. Hydrocortisone (Compound F) shows the least sodium-retaining action, whereas aldosterone is extremely potent, being at least 1000 times as effective as hydrocortisone in electrolyte regulation. Accompanying the retention of sodium, there is increased excretion of potassium, caused by exchange of intracellular potassium with extracellular sodium and increased renal tubular excretion of potassium.

As a result of the shifts in electrolyte distribution there are accompanying changes in the volume of the fluid compartments by the body. For example, an increase in the extracellular fluid volume of as much as 20% may occur after the administration of large doses of cortisone or hydrocortisone as sodium is mobilized from connective tissue and transferred to the extracellular fluid. There will also be an increase in the volume of the circulating blood and in the urinary output. After removal of the adrenal glands, these effects on water and salt metabolism are quickly reversed. There is loss of water and of sodium into the cells, and of potassium into the extracellular fluid; dehydration is rapid, and there is a rise in the blood potassium and nonprotein nitrogen. Death occurs from circulatory collapse.

The naturally-occurring salt-regulating corticoid is aldosterone (electrocortin). It has been detected in extracts of the adrenal cortex and in the blood of the adrenal vein. In the peripheral blood, the content of aldosterone is reported to be very low (about 0.085 μg. per 100 ml.). It has also been found in the urine, which is its best source at the present time.

Unlike the other corticoids, the production of aldosterone by the adrenal is relatively uninfluenced by corticotropin (ACTH). However, aldosterone production in normal subjects is markedly increased by deprivation of sodium, as judged by the increases in urinary aldosterone which are observed. Conversely, the administration of sodium decreases aldosterone excretion. These changes in aldosterone production are accompanied by a fall or a gain, respectively, in body weight, presumably because of changes in the extracellular fluid volume of the body. It has therefore been suggested[21] that aldosterone secretion is mainly controlled by changes in extracellular fluid volume and is independent of total body sodium or of sodium concentration.

Although there is evidence that control of aldosterone secretion depends in part on intravascular volume, there remains the question how stimuli from changes in volume may affect the adrenal cortex. Mills, Casper, and Bartter[22] suggest that in view of the demonstrated modification of vagal nerve impulses by changes in intrathoracic volume, it may be that a similar mechanism is involved in mediating aldosterone secretion. In experiments designed to test this hypothesis they found that aldosterone secretion could be stimulated by constriction of the vena cava above the diaphragm so as to raise the femoral venous pressure by 10 cm. of water. Within 90 minutes after release of the constriction, secretion of aldosterone had fallen. Section of the vagi abolished the fall of aldosterone secretion which occurred on release of constriction but not that of an increase in secretion subsequent to constriction of the vena cava. Thus the effect of caval constriction on aldosterone secretion seems to depend on an intact vagus nerve.

A relation of aldosterone secretion to potassium has also been elicited. When potassium is withdrawn from the diet, aldosterone secretion is lowered; administration of large amounts of potassium brings about an increase in aldosterone production. It is believed that potassium has an effect on aldosterone secretion which is independent of that mediated by changes in fluid volume.

Secretion of aldosterone is increased in several diseases such as cirrhosis, nephrosis[23], and cardiac failure (see also p. 351). The cause of its increased production is not known, but the result is enhancement of sodium and water retention, which further aggravates the edema characteristic of these diseases. In nephritis the renal loss of sodium and water may directly increase aldosterone production, which then leads to greater losses of potassium and hydrogen ions.

The structure of aldosterone is shown below in the aldehyde form and in the hemi-acetal (see p. 6) form. It is believed that the hormone exists in solution in the hemi-acetal form. It will be noted that aldosterone has the same structure as corticosterone except that the methyl group at position 18 is replaced by an aldehyde group. Deoxycorticosterone (see p. (349)appears to be the precursor in the adrenal of both aldosterone and corticosterone.

11-Deoxycorticosterone (D.O.C.), usually employed as the acetate (D.O.C.A.), is essentially a synthetic steroid. It has a greater electrolyte-regulating effect than the other corticoids, with the important exception of aldosterone, which is far more potent. However, the fact that D.O.C.A. can be prepared synthetically and that aldosterone is not yet available for therapeutic use makes the former compound important in the treatment of Addison's disease (see p. 352). D.O.C.A. may be administered sublingually, since absorption from the buccal mucous membrane is possible.

Aldosterone is at least 25 times as potent as D.O.C. in its sodium-retaining effects but only about five times as potent in increasing the excretion of potassium. When used in excess, deoxycorticosterone leads to overhydration within the cells and a consequent water intoxication; this effect is much less prominent with aldosterone. Although aldosterone resembles deoxycorticosterone in many of its metabolic effects, it also possesses other physiologic properties which distinguish it from D.O.C. These include increasing the deposition of glyco-

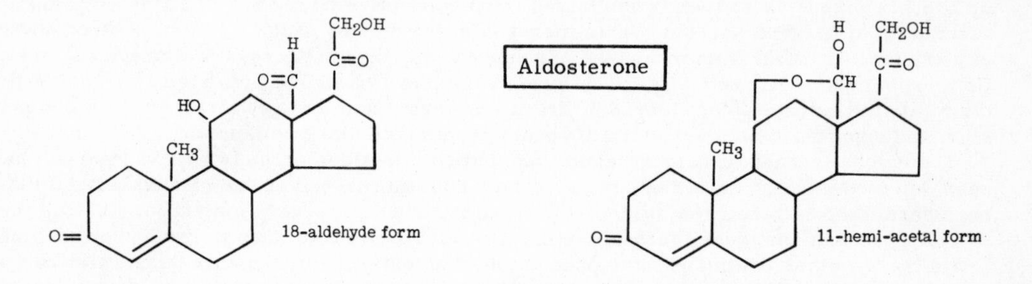

gen in the liver, decreasing the circulating eosinophils, and maintaining resistance to stress, such as exposure to low temperatures (so-called "cold stress test"). In maintaining the life of the adrenalectomized animal, aldosterone is more potent than any other known steroid.

B. Metabolic Effects on Protein, Carbohydrate, and Fat (C-21 Corticosteroids; Glycogenic or "S" Hormones): (See p. 349.) These metabolically active steroids are often grouped together as the 11-oxysteroids since they all possess an oxygen atom attached to C-11 of the nucleus. Those oxygenated at both 11 and 17 are the most potent in their effects on metabolism. Examples of the metabolic corticosteroids are corticosterone (Kendall's Compound B), 11-dehydrocorticosterone (Kendall's Compound A), 17-hydroxy-11-dehydrocorticosterone (Kendall's Compound E, also known as cortisone), and 17-hydroxycorticosterone (Kendall's Compound F, also known as hydrocortisone or cortisol).

The 11-oxysteroids produce the following effects (note in many cases antagonism to the effects of insulin): (1) Elevation of blood glucose. (2) Decrease in hepatic lipogenesis from carbohydrate. (3) Increase in gluconeogenesis (see p. 171) and a protein anti-anabolic effect which increases protein catabolism. This includes reduction of the osteoid matrix (see p. 384) of bone, with a resulting osteoporosis and excessive loss of calcium from the body. (4) A sparing action on carbohydrate by an increase in fat metabolism, including mobilization of depot fat to the liver. (5) Decrease in carbohydrate oxidation and an increase in hepatic glycogenesis. (6) Decrease in tubular reabsorption of urates, resulting in greater excretion of these compounds. (7) Involution of the thymus and other lymphatic tissues. (8) Effects on cellular elements of the blood: eosinopenia, lymphopenia, leukocytosis and erythremia. (9) Increase in the secretion of hydrochloric acid and pepsinogen by the stomach and of trypsinogen by the pancreas.

C. Sex Hormone Effects (C-19 Corticosteroids; Androgenic, Protein Anabolic, or "N" Hormones): (See p. 348.) The adrenal cortex also produces hormones similar in structure and action to the sex hormones, both male (androgenic) and female (estrogenic). An example is adrenosterone, which has been isolated from beef cortex. The androgenic activity of adrenosterone is about 1/5 that of androsterone (see p. 357), the androgenic steroid which is used as a standard of androgenic activity.

The adrenal origin of some sex hormones is supported further by the observation that the urine of castrates still contains sex hormones. These adrenocorticoids of the sex hormone type cause retention of nitrogen (a protein anabolic effect), phosphorus, potassium, sodium, and chloride. If present in excessive amounts, they also lead to masculinization in the female (see p. 352).

D. Other Effects (Anti-inflammatory Effects): The adrenocortical hormones exert many other effects which cannot at present be explained on the basis of their known metabolic activities. A wide variety of disease states have been modified by treatment with cortisone or by injection of adrenocorticotropic hormone of the pituitary (corticotropin, ACTH), which increases the production of corticosteroids by the intact adrenal. The so-called collagen diseases, such as rheumatoid arthritis, are outstanding examples. These hormones also appear to modify cellular response to disease such as reaction of the body to infection and allergic states, notably antibody response. Often these effects may be brought about by relatively small doses of hormones; and if treatment is continued only for short periods, the metabolic effects are minimal.

Synthetic Analogues of Natural Steroids:

During recent years, adrenal hormones have been synthesized which are in many instances more potent than the naturally-occurring hormones and often more specific in their action.

The introduction of halogen atoms (e.g., fluorine) in the 9-α position of cortisone, hydrocortisone, or corticosterone has produced compounds which are much more potent than the parent compounds. However, the increase in salt-retaining activity which occurs as a result of 9-α halogenation is relatively greater than that of anti-inflammatory or metabolic activities. For this reason the clinical usefulness of these derivatives is limited. On the other hand, the introduction of a double bond between carbon atoms 1 and 2 results in the production of cortisone and hydrocortisone analogues which are relatively inert as far as salt-retaining properties are concerned although they retain the anti-inflammatory activity of the natural steroids. The cortisone analogue is known as prednisone (Meticorten®); the hydrocortisone analogue is prednisolone (Meticortelone®).

BIOSYNTHESIS OF ADRENAL HORMONES

17-KETOSTEROID AND CORTICOSTEROID PATHWAYS

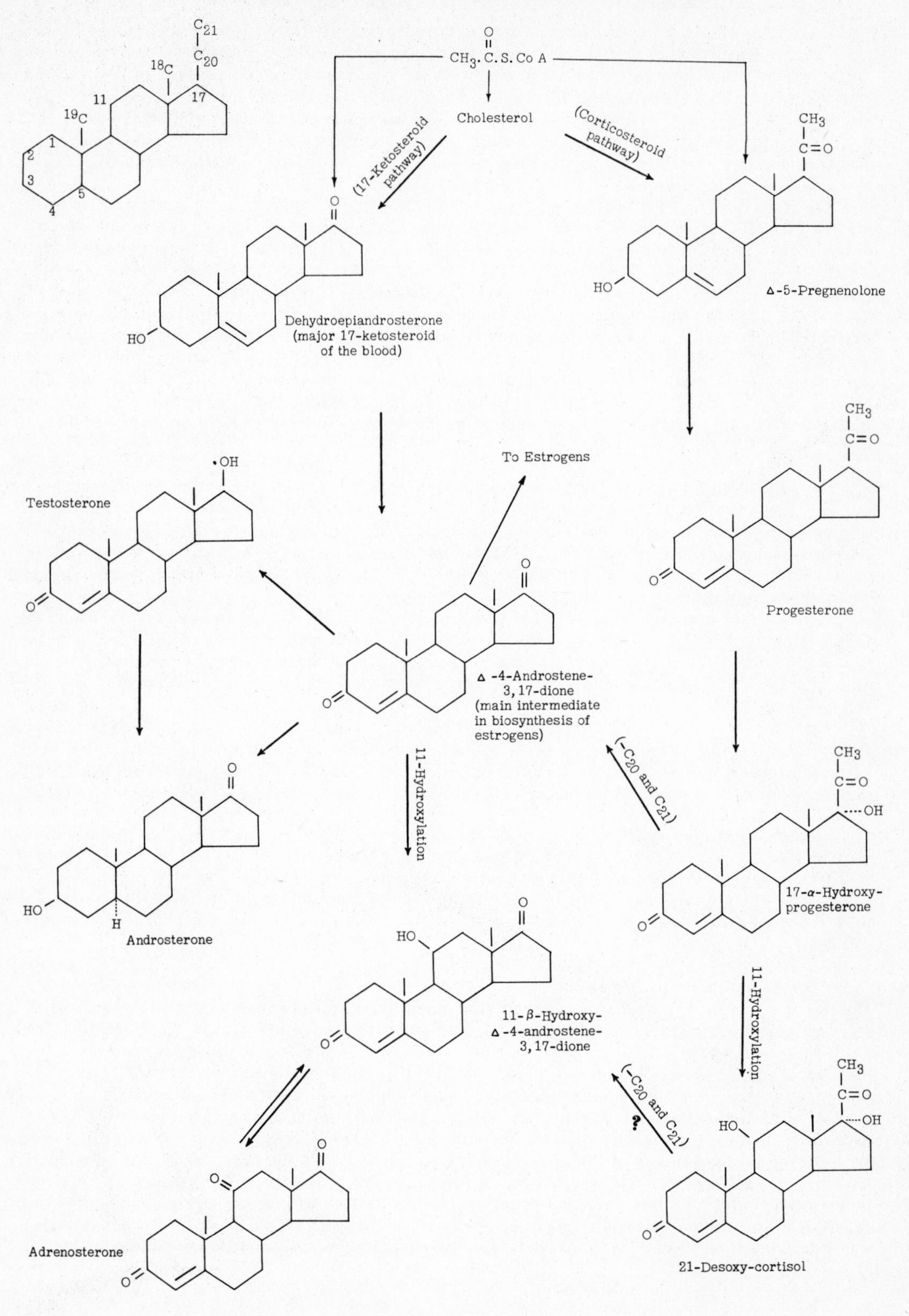

CORTICOSTEROID (C-21) PATHWAY

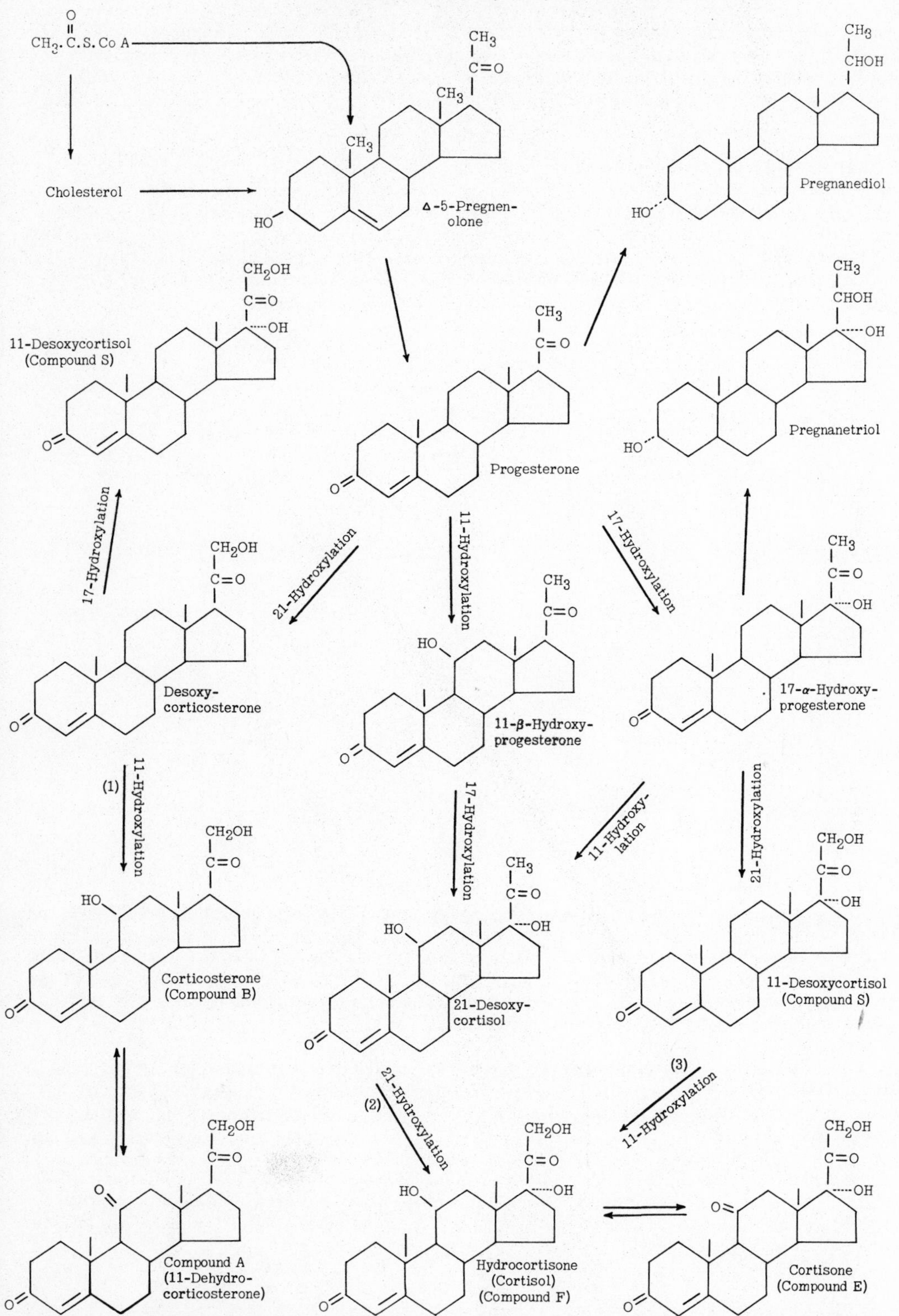

(1) ? Block in hypertensive form of adrenogenital syndrome (see p. 353).
(2) Block in uncomplicated adrenogenital syndrome (see p. 352).
(3) Block in hypertensive form of adrenogenital syndrome 353).

Synthetic steroids which contain a methyl group at position 2 have also been prepared. In steroids which have a hydroxy group on position 11 (e.g., hydrocortisone, 9-α-fluorohydrocortisone, or 11-β-hydroxyprogesterone), the addition of the two-methyl group markedly enhances the sodium-retaining and potassium-losing activity of the hormone. The most active derivative of this type which has been prepared is 2-methyl-9-α-fluorohydrocortisone. It is three times as potent as aldosterone, which makes it the most potent electrolyte-regulating corticoid yet discovered.

Another new, synthetic analogue of prednisolone, having similar but more potent anti-inflammatory action, is 9-α-fluoro-16-α-methylprednisolone (Decadron®, Deronil®). It is about 30 times more potent than hydrocortisone. Consequently, equivalent anti-inflammatory effects can be obtained with a dose of 0.75 mg. as compared to a requirement for 20 mg. of hydrocortisone. This lowered dosage markedly reduces the unwanted side effects of the drug. For example, it is claimed that abnormal salt and water retention and excess potassium excretion are rarely observed in patients receiving therapeutic dosages of this new compound.

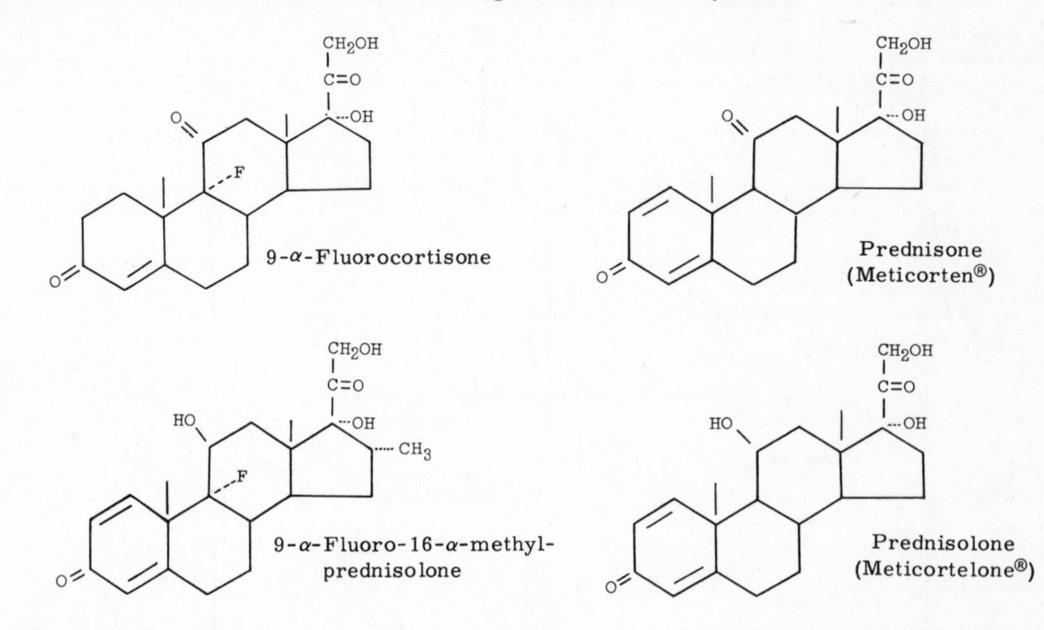

<u>Biosynthesis of Adrenal Hormones:</u> (See pp. 348 and 349.)

The adrenal cortex contains relatively large quantities of cholesterol and ascorbic acid. After stimulation of the gland by stress or by the tropic stimulus of ACTH, there is a rapid decline in the concentration of these two substances. It is believed that cholesterol is the precursor of the steroid hormones. The role of ascorbic acid is not clear. The measurement of cholesterol or ascorbic acid in the adrenal glands of hypophysectomized animals after the injection of ACTH serves as a method of assay for the tropic hormone.

The adrenal steroids may arise directly from C-2 (active acetate) fragments or cholesterol. From these intermediates, two pathways arise: (1) the 17-ketosteroid pathway, p. 348 (19 carbons, as exemplified by dehydroepiandrosterone), and (2) the corticosteroid pathway, pp. 348 and 349 (21 carbons, as exemplified by pregnenolone). At least three specific **hydroxylases** are present in the cortex. These enzymes catalyze reactions which add hydroxyl groups to the steroid derivatives at positions 11, 17, and 21. The action of these hydroxylases involves TPN.H and oxygen (see p. 114). Under normal conditions, in the human, the C-21 corticosteroids (Compounds A, B, E, and F) are synthesized in the largest amounts; the 17-ketosteroids and derivatives of progesterone and of estrogens are synthesized in smaller amounts.

<u>Adrenocortical Activity From Placenta:</u>

Placental extracts have been shown to possess adrenocortical activity. This may be why addisonian patients are improved during pregnancy and may explain the relief of arthritic symptoms which has also been noted during pregnancy.

Metabolism of Adrenal Corticosteroids:

About one-half of the 11- or of the 11, 17-hydroxycorticosteroids are transported in the blood stream bound loosely to the serum proteins. In the resting state, the plasma contains 5-15 μg. per 100 ml. of hydrocortisone, the major corticoid of the blood. In the course of their passage through the liver, the 17-hydroxycorticoids are conjugated, mainly with glucuronic acid (see p. 168), and reduced to the tetrahydro derivatives, which are completely inactive. Both free and conjugated corticoids are excreted into the intestine by way of the bile and, in part, reabsorbed from the intestine by the enterohepatic circulation. Excretion of free and, in particular, conjugated corticoids by the kidney also takes place, although there is some tubular reabsorption. A reversible combination occurs between the serum albumin and a number of steroids (except cholesterol). The binding of the steroid hormones with the serum albumin explains their low concentration in the urine. On the other hand, the metabolic products of the steroids which are largely conjugated with glucuronic and sulfuric acids are poorly bound to the serum proteins and also have a greater aqueous solubility. This undoubtedly explains their rapid excretion into the urine and the fact that most of the excreted steroidal metabolites are in the conjugated form.

The inactivating effect of the liver on steroids may decline during periods of prolonged malnutrition (protein and B vitamins) or in liver disease. A decreased excretion may also occur in renal insufficiency. Under any of these circumstances the levels of corticoids in the blood are markedly elevated. It has been suggested that in chronic liver disease (cirrhosis) as well as in congestive heart failure the liver does not completely inactivate salt-retaining steroids. This would lead to excessive salt retention, which may be an important cause of edema and ascites found in certain stages of these diseases.

Because of the effect of the liver on their metabolism, the dose of many steroids, when administered by mouth, must be increased, since after absorption from the gastrointestinal tract the steroids are carried by the portal circulation to the liver, where a portion of the administered dose may be inactivated.

The disappearance from the body of steroidal compounds such as hydrocortisone is normally very rapid. C^{14}-labeled hydrocortisone injected intravenously has a half-life of about four hours. Within 48 hours, 93% of the injected dose has disappeared from the body: 70% by way of the urine, 20% by the stool, and the remainder presumably through the skin. The steroid nucleus is eliminated in the intact form; no significant breakdown of the nucleus to carbon dioxide and water occurs.

Fate of Corticosteroids

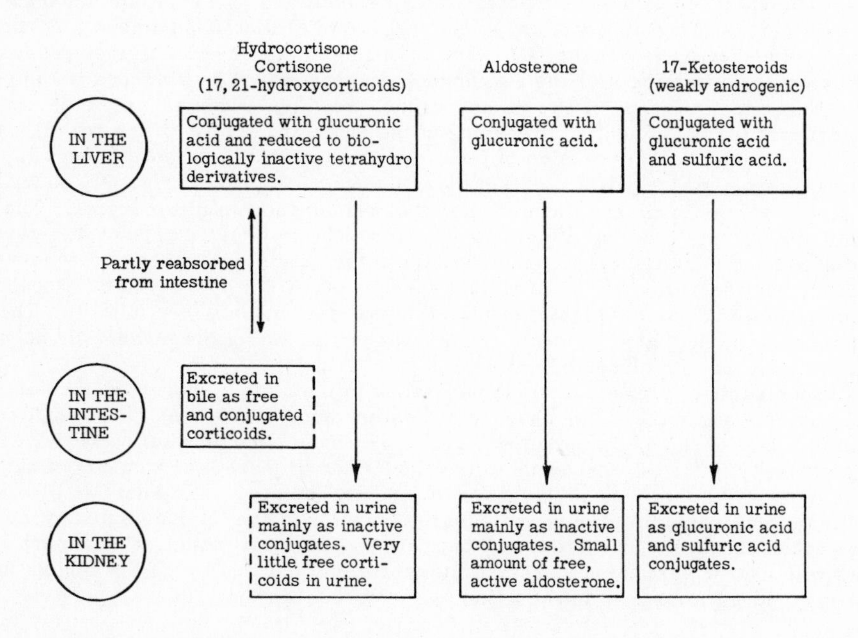

The blood also contains 17-ketosteroids which are carried both free and in conjugated form in association with the serum proteins. These consist mostly of dehydroepiandrosterone at a level of about 50 μg. per 100 ml. of plasma. About half as much is androsterone; it is found in both males and females. The conversion of small amounts of hydrocortisone and cortisone to 17-ketosteroids is believed to occur in the liver. In certain forms of liver disease it may be found that 17-hydroxycorticoids are elevated and 17-ketosteroids are reduced. Most of the 17-ketosteroids are excreted into the urine (see p. 354).

Abnormal Physiology:

A. Hypoadrenocorticalism:

1. Adrenalectomy - The effects of adrenalectomy on the metabolism of experimental animals are quite pronounced. Sodium is lost as a result of the failure of the renal tubules to reabsorb sodium ion. Sodium loss is accompanied by retention of potassium and losses of chloride and bicarbonate. Associated alterations in water metabolism lead to loss of water from the blood and the tissue spaces, resulting in severe dehydration and hemoconcentration. These effects on salt and water metabolism are promptly reversed by adrenocortical extract when adequate water and minerals are made available. In the adrenalectomized rat, merely providing sodium chloride suffices to keep the animal alive indefinitely.
2. Addison's disease - In humans, degeneration of the adrenal cortex often due to a tuberculous process, is the cause of Addison's disease. Depending on the extent of adrenocortical hypofunction, the effects are similar to those observed in the adrenalectomized animal. The symptoms of Addison's disease include low blood pressure, muscular weakness, gastrointestinal disturbances, low body temperature, hypoglycemia, and a progressive brownish pigmentation which increases over a period of months. These symptoms, with the exception of hypoglycemia and pigmentation, are mainly due to the lack of salt- and water-regulating hormones since addisonian patients may be maintained on deoxycorticosterone acetate (D.O.C.A.), although most are now maintained on small amounts of cortisone acetate by mouth together with a daily intake of 10 to 15 Gm. of sodium chloride. Addisonian patients withstand any shock or traumatic experience very poorly.

B. Hyperadrenocorticism: Adrenocortical hyperfunction may be caused by benign or malignant tumors of the cortex or by adrenocortical hyperplasia initiated by increased production of ACTH.

The continuous administration of steroid hormones or corticotropin (ACTH) may also induce signs of hyperadrenocorticism. These include (1) hyperglycemia and glycosuria (diabetic effect); (2) retention of sodium and water, followed by edema, increased blood volume, and hypertension; (3) negative nitrogen balance (protein antianabolic effect from gluconeogenesis); (4) potassium depletion and hypokalemic alkalosis; and (5) hirsutism and acne.

Certain tumors of the adrenals cause the production of increased amounts of androgenic (C-19) steroids (see p. 347, para. C). The resulting disturbance is termed **congenital virilizing hyperplasia** in infants or the **adrenogenital syndrome** when it occurs in the postnatal period. Under the influence of excess androgens, the female assumes male secondary sex characteristics (e.g., in adults, deepening of the voice and growth of the beard), enlargement of the clitoris, and repression of female characteristics. When it occurs in the male, there is excessive masculinization. Feminizing adrenal tumors may rarely occur in males.

Bongiovanni[24] distinguishes three varieties of the adrenogenital syndrome. The first and most common variety is the uncomplicated form which in either sex apparently causes little or no disturbance in salt and water metabolism in the child. Under these circumstances the metabolic defect seems to lie in the virtual absence of a C-21 hydroxylase. Consequently, hydrocortisone (or cortisone) is not produced in normal amounts (see p. 349). The urine of these patients contains large amounts of 17-ketosteroids and of pregnanetriol, but hydrocortisone is very low in both blood and urine.

It is assumed that hydrocortisone is necessary to maintain a proper balance between the pituitary and the adrenal. In the absence of this hormone, excess ACTH production occurs because of a lack of the "braking effect" which hydrocortisone normally exerts. The resulting hyperactivity of the adrenal leads to the production of increased amounts of androgenic steroids of the C-19 type (the 17-ketosteroid pathway; see p. 348). There is also increased production of pregnanetriol, which reflects the shunting of the intermediates of the C-21 pathway which cannot be converted to hydrocortisone (and cortisone). Treatment of these patients with hydrocortisone restores pituitary-adrenal balance by its braking action on ACTH production; androgenic steroid production by the adrenal is then reduced, as evidenced by a fall in the excretion of 17-ketosteroids to normal. The dose of hydrocortisone which is necessary for this purpose is regulated by keeping the level of urinary 17-ketosteroids between 1 and 5 mg. per day in children ranging in age from six months to ten years.

A second rather rare variety of the adrenogenital syndrome is the hypertensive form. It is characterized by high blood pressure and a large excretion of Compound S (11-deoxycortisol) which may not be distinguished in the ordinary assays for urine corticoids and hence measured as a corticosteroid of the 11-oxy type. The characteristic hypertension is apparently due to the production of deoxycorticosterone. The defect may lie in a failure of hydroxylation at C-11; but since there is no defect in C-21 hydroxylation, pregnanetriol excretion is not particularly high.

The third variety of the adrenogenital syndrome is the salt-losing form. Death may occur, as in addisonian crisis beginning shortly after birth, if the disease is not promptly recognized. So far, no biochemical defect has yet been found which is different from the uncomplicated form, but the metabolic effects are more intensive (e.g., pregnanetriol excretion and lack of hydrocortisone), suggesting that this third variant may actually be a more severe variant of the uncomplicated form.

The adrenogenital syndrome is caused by a mutation within a single autosomal-recessive gene having a heterozygous frequency of 1:128 in the general population.

Cushing's Syndrome:

This disease was first described by Cushing in 1932 and was attributed at that time to pituitary hyperactivity due to a basophilic tumor (adenoma) of the pituitary. Later evidence cast doubt on the pituitary origin of all instances of this disease; it has been suggested that the adrenal is directly responsible for most cases and that the disease should be considered as a sign of hyperadrenocorticism. This is supported by the fact that both cortisone and corticotropin (ACTH) induce Cushing's syndrome when administered for prolonged periods. The disease may be reversed when these agents are discontinued.

Cushing's syndrome is characterized by a rapidly increasing adiposity of the face, neck, and trunk (buffalo fat distribution), a tendency to purpura, purplish striae on the abdomen, hirsutism, sexual dystrophy, hypertension, osteoporosis, and low sugar tolerance and hyperglycemia (insulin resistance). There is also a tendency to depletion of potassium and to alkalosis (see p. 320). Cushing's syndrome is more common in females than in males.

Primary Aldosteronism:

There have been recent reports of the occurrence of tumors (aldosteromas) of the adrenals in which the hyperactivity of the adrenal cortex is apparently confined to excess production of aldosterone. This syndrome is termed "primary aldosteronism"[25a,b,c]. Patients with these tumors exhibit intermittent tetany, paresthesias, periodic muscle weakness, alkalosis, persistent low potassium levels, hypertension, and polyuria and polydipsia but no edema. The significant metabolic findings are hypokalemic alkalosis, hyperaldosteronuria, and hyposthenuria which is unresponsive to antidiuretic hormone (ADH). After sodium restriction, the tendency to alkalosis and hypokalemia is eliminated and sodium disappears from the urine. Measurement of aldosterone excretion in the urine, particularly after sodium loading, is also helpful in establishing the diagnosis of aldosteronism.

The lack of edema and the presence of polyuria in primary aldosteronism may be explained by failure of the renal tubule to respond to ADH because of a potassium depletion induced by the excess urinary potassium losses which the tumor engenders. This idea was suggested by Richter[26] on the basis of experiments in rats wherein severe potassium depletion was characterized by unresponsiveness of the renal tubule to exogenous ADH.

Laboratory Studies of Adrenocortical Function:

A. Urinary Excretion of Steroid Hormones or of Their Metabolites:

1. Determination of 17-hydroxycorticosteroids - Measurement of the levels of 17-hydroxycorticoids in the serum and in the urine is the most direct approach to the study of the function of the adrenal cortex. The best method for the diagnosis of adrenal insufficiency is the measurement of the response of the adrenal cortex to stimulation by ACTH. This is accomplished by intramuscular or intravenous injection of ACTH followed by quantitative

measurement of the increase in the serum and urinary levels of 17-hydroxycorticoids. In the urine, the normal 24-hour excretion of these corticoids (expressed as hydrocortisone) in adult males is 10 mg. ±4; in females, 7 mg. ±3. Most of the excreted corticoids are in the conjugated form with glucuronic acid, so that a preliminary hydrolysis with glucuronidase is necessary. After an eight-hour intravenous infusion with 20 units of corticotropin (ACTH), an increase in corticoid excretion of as much as 200% (e.g., from a normal of 10 mg. to a high of 30 mg. per day) may be seen in normal subjects. The levels of 17-hydroxycorticoids in the urine will probably be below 3 mg. in 24 hours in patients with adrenal insufficiency and, after administration of corticotropin (ACTH), no increase occurs. If the adrenal insufficiency is secondary to a lack of ACTH, a small response will occur which will increase if the test is soon repeated.

The normal values for hydrocortisone in the serum are 5 to 15 μg. per 100 ml. After the intramuscular injection of 25 units of corticotropin (ACTH), a normal individual may show a rise to as much as 45 μg. per 100 ml. In adrenal insufficiency, no response to corticotropin will occur.

An **elevation** of the 24-hour excretion of 17-hydroxycorticoids above 12 mg. strongly suggests that the patient may have **hyperactivity** of the adrenal cortex, such as Cushing's syndrome, unless the patient is under stress, has liver disease with hepatic insufficiency, or is taking corticotropin or cortisone. If the value is borderline, an eight-hour intravenous ACTH test can be used. If hyperadrenocorticism is present, there will be a rise in the 24-hour excretion of 17-hydroxycorticoids to values above 40 mg. The levels of 17-hydroxycorticoids in the blood will also be elevated in Cushing's syndrome.

The urinary excretion of 17-hydroxycorticoids is elevated in adrenocortical carcinoma as well as in adrenocortical adenoma. However, the increase in excretion is greater in carcinoma, and it is not elevated after an intravenous ACTH test as it is in adrenocortical adenoma.

2. Urinary 17-ketosteroids - The urine contains a number of steroidal compounds besides the 17-hydroxycorticoids described above. They may be subdivided into three general types: acidic, phenolic, and neutral. The acidic compounds are derived from the bile acids; the phenolic, from the estrogens; the neutral reflect mainly the excretion of metabolic derivatives from the endocrine secretions of the gonads and the adrenal cortex. The neutral steroids, which are further characterized by the occurrence of a carbonyl group at position 17 in the sterid nucleus, can be measured by chemical methods. The principal components of this so-called "neutral 17-ketosteroid" fraction of the urinary steroids are androsterone and its isomers, etiocholanolone and isoandrosterone, as well as dehydroepiandrosterone (see p. 357). The neutral 17-ketosteroids may be further subdivided into alpha and beta fractions. The beta fractions are those which are precipitated by digitonin; the alpha fractions are not precipitated by digitonin.
   a. Methods for determination of urinary 17-ketosteroids - In a determination of ketosteroids in the urine, the sample is first subjected to hydrolysis since the 17-ketosteroids are present as esters. This is accomplished with concentrated hydrochloric acid or sulfuric acid. Extraction with ether or carbon tetrachloride then follows. Acidic fractions are removed from the ether extract by washing with sodium bicarbonate; phenolic fractions by extracting with 2 N sodium hydroxide. The neutral fraction remains. It is estimated colorimetrically by means of the **Zimmermann reaction,** which involves coupling of the reactive 17-keto group of these compounds with **metadinitrobenzene**. A colored complex forms. The intensity of this color is measured in a spectrophotometer at 520 mμ wave length. The standard in the estimation of the urinary 17-ketosteroids is crystalline dehydroisoandrosterone, which is treated to produce a color in the same manner as the urinary extract. In some cases, the neutral fraction obtained as described above may be treated with **Girard's reagent T,** which separates it into ketonic and nonketonic components. The nonketonic fraction comprises only 10 to 15% of the original neutral fraction. The remaining 85 to 90% is the ketonic fraction; it is separable into alpha and beta components by digitonin as described above. A flow chart of these operations is shown on p. 355.
   b. Origin of urinary neutral 17-ketosteroids - These substances are a reflection of the androgenic function of the subject. In the female, 17-ketosteroids are produced entirely by the adrenal cortex; in the male, by the adrenal cortex and testes. The testes contribute one-third of the neutral urinary 17-ketosteroids. Although dehydroisoandrosterone is found in the urine of both normal men and women, it is greatly increased in some cases of hyperadrenocorticism, particularly those with excess production of androgens. This androgenic steroid is probably derived mainly from the adrenal. However, metabolites of some sex hormones also contribute to the neutral 17-ketosteroid fraction of

the urine since the administration of androgenic steroids, such as testosterone, increases 17-ketosteroid production and excretion; on the other hand, not all androgens increase the excretion of neutral 17-ketosteroids (e. g., methyltestosterone).

c. Normal values for neutral 17-ketosteroids in a 24-hour urine sample -

Children:	Birth to three years	None
	Three to eight years	0 to 2.5 mg.
	Eight to 12 years	1.5 to 5 mg.
	12 to 16 years	4 to 9 mg.
Adult Males		10 to 20 mg. (avg., 15 mg.)
Adult Females		5 to 15 mg. (avg., 10 mg.)

Ratio alpha to beta ketonic fractions, 9:1

d. Significance of ketosteroid excretion in abnormal states - Only a small amount of the 17-ketosteroid excretion is contributed by the breakdown of 17-hydroxycorticoids. Therefore, 17-ketosteroids do not represent a direct measurement of adrenocortical activity as do 17-hydroxycorticoids. Only a slight elevation occurs in Cushing's syndrome, acromegaly, simple hirsutism, pregnancy, and hyperthecosis. The degree of elevation may be so slight as to encompass the normal daily variation in a normal subject; thus the diagnostic value of these minimal changes is restricted. However, the simultaneous determination of 17-hydroxycorticoid excretion and 17-ketosteroid excretion is of value in differentiating Cushing's syndrome from the adrenogenital syndrome. Both types of adrenocortical hyperfunction would be expected to show increased 17-hydroxycorticoid excretion. However, whereas in Cushing's syndrome the 17-ketosteroid excretion barely exceeds the normal except in the presence of a carcinoma, in the adrenogenital syndrome 17-ketosteroid values above 30 mg. in 24 hours are found.

In adrenocortical carcinoma, as well as bilateral hyperplasia of the cortex, the 17-ketosteroid excretion is often markedly elevated. It is reported that the beta fraction may be elevated in cases of adrenocortical carcinoma, but in order for this to be diagnostically significant, the total 17-ketosteroid excretion should be over 50 mg. in 24 hours and over 50% of this should be β-ketonic neutral 17-ketosteroids. Testicular tumors (Leydig cell tumors) also produce noteworthy increases in neutral 17-ketosteroid excretion.

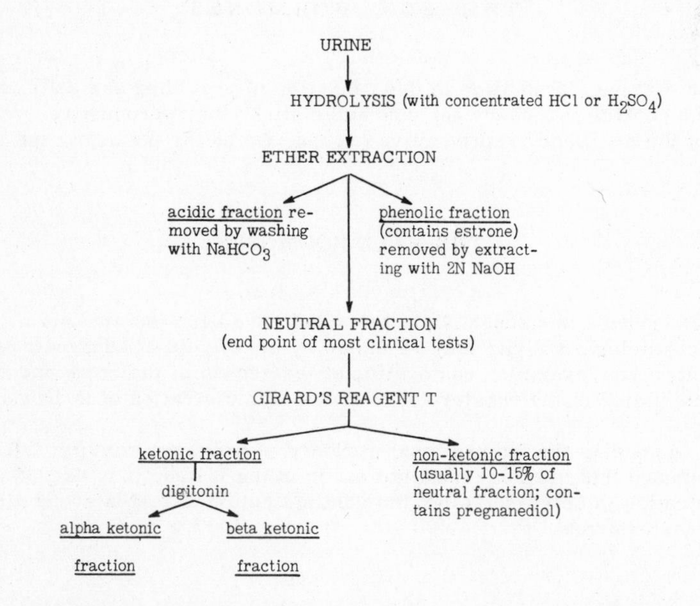

Flow Chart for Extraction of 17-Ketosteroids From Urine

During the administration of corticotropin, increases in urinary 17-ketosteroids may be noted if the adrenal is competent to produce steroid hormones. On the other hand, the administration of cortisone does not cause a significant increase in 17-ketosteroid excretion except possibly in addisonian patients. In fact, cortisone may effect a fall in 17-ketosteroid excretion because of inhibition of ACTH production.

Excretion of 17-ketosteroids is low in hypophyseal infantilism, Simmonds' disease, and occasionally in anorexia nervosa (see p. 366). In Addison's disease, the values for 17-ketosteroids in the urine drop to 1 to 5 mg. in 24 hours in the male, and to 0.5 to 1 mg. in the female. Excretion is also low in myxedema.

Slight reduction in normal values for urinary 17-ketosteroids may be noted in eunuchoidism, castrates, diabetes mellitus, the climacteric, and during any chronic debilitating disease. It is normal for the levels to decline somewhat in old age.

B. Eosinophil Count: A rapid and sensitive effect of cortisone or corticotropin (ACTH) administration is a decline in circulating eosinophils; consequently, counts of the circulating eosinophils before and after stimulation of the adrenal with corticotropin may be used as a simple method for the study of adrenocortical function. In the four-hour ACTH test, 25 units of corticotropin are injected intramuscularly, and the percent fall in the eosinophil count from the control level is ascertained at that time. A fall exceeding 50% excludes a diagnosis of adrenal insufficiency. A fall of less than 50% may be due to adrenocortical insufficiency, but this should be confirmed with a similar test after the eight-hour infusion of 20 units of corticotropin intravenously. In this case, failure to produce a fall in eosinophils of more than 50% is a definite indication of adrenal insufficiency.

C. The Modified Robinson-Power-Kepler Water Test: Addisonian patients do not respond to a water diuresis as do normal individuals. This fact is utilized in the simplified test devised by Soffer and Gabrilove. Fifteen hundred ml. of tap water are administered by mouth over a period of 15 to 20 minutes, and the urine is collected for five hours. The volume does not exceed 800 ml. in patients with adrenal insufficiency, as compared to an excretion of 1200 to 1900 ml. in normal subjects. The test is useful as a screening device but is not specific. A delayed diuresis may occur in diseases other than adrenal insufficiency if either the rate of absorption of water from the gastrointestinal tract or its elimination by the kidney is decreased, e.g., in patients with nephritis, cirrhosis, celiac disease, or cardiac failure.

THE SEX HORMONES

The testes and ovaries, in addition to their function of providing sex cells, manufacture steroid hormones which control secondary sex characteristics, the reproductive cycle, and the growth and development of the accessory reproductive organs, excluding the ovary and testis themselves.

THE MALE HORMONES

A number of androgenic hormones (C-19 steroids) have been isolated either from the testes or the urine. Their physiologic activity may be tested by the effects of administration in castrated animals. In the capon, for example, restoration of the growth of the comb and wattles, the secondary sex characteristics of the rooster, follows the administration of androgenic hormones.

Androsterone is used as the international standard of androgen activity; 1 I.U. = 0.1 mg. of androsterone. Although this hormone does not occur in the testes, it is excreted in the urine of males, and its excretion is increased after the administration of testosterone since androsterone is a metabolite of testosterone.

Testosterone:

The principal male hormone, testosterone, is synthesized by the activity of the interstitial (Leydig) cells of the testes from acetate and cholesterol through Δ-5-pregnenolone, progesterone, and 17-α-hydroxyprogesterone, which is then converted to the 17-ketosteroid, Δ-4-androsterone-3, 17-dione, the immediate precursor of testosterone. These reactions are shown on p. 348 since they are also a part of the biosynthetic pathway in the adrenal for the formation of the androgenic (C-19) steroids. It will also be noted that Δ-5-pregnenolone is a common precursor of the adrenocortical hormones and testosterone as well as of progesterone.

The principal metabolites of testosterone are androsterone and dehydroepiandrosterone, the major 17-ketosteroid of the blood. Their structures are shown below.

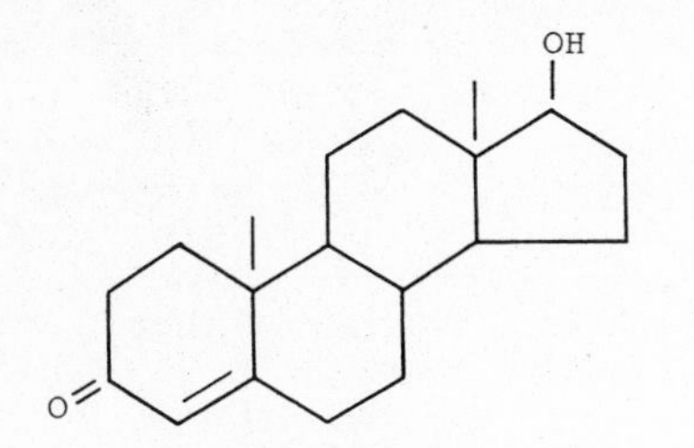

Testosterone

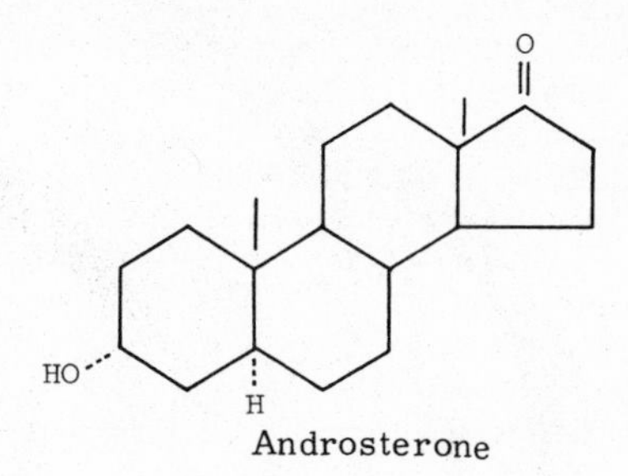

Androsterone

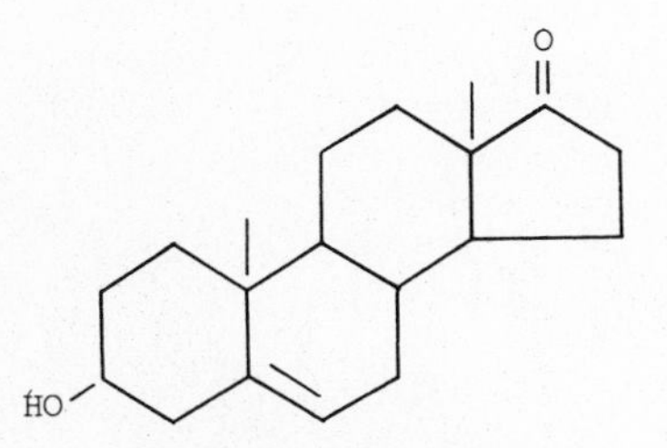

Dehydroepiandrosterone

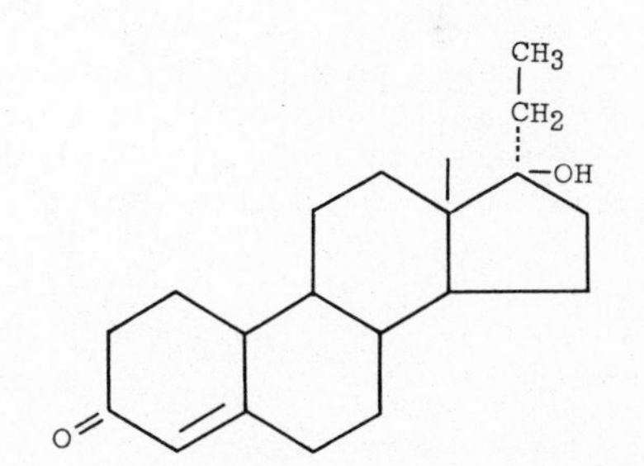

17-α-Ethyl-17-hydroxy-19-nor-4-androsten-3-one (Norethandrolone, Nilevar®)

Testosterone promotes the growth and function of the epididymis, vas deferens, prostate, seminal vesicle, and penis. It is of value as replacement therapy in eunuchoidism. Its metabolic effect as a protein anabolic steroid exceeds that of any other naturally-occurring steroid. By aiding the muscular and skeletal growth which accompanies puberty, this function is doubtless a necessary concomitant to its androgenic functions.

Promotion of nitrogen retention is of great importance in many clinical situations. Testosterone has been administered for this purpose, but the accompanying androgenic effects are often undesirable. A synthetic steroid which retains the protein anabolic activity of testosterone but is at the same time virtually free of its androgenic effects has recently been produced. This compound is 17-α-ethyl-17-hydroxy-19-nor-4-androsten-3-one (norethandrolone, Nilevar®). It resembles the estrogens in lacking a C-19 substituent and differs from both the androgens and estrogens by having C-20 and C-21. As determined in rats, the ratio of protein anabolic to androgenic effects was found to be 16:1.

Excretion of 17-ketosteroids, as noted above, is in part a reflection of testicular hormone production. The testis contributes about one-third of the urinary neutral 17-ketosteroid excretion. In normal children up to the eighth year of life, there is a gradual increase in 17-ketosteroid excretion up to 2.5 mg. per day; an increase up to 9 mg. occurs between the eighth year and puberty in both boys and girls; after puberty, the sex differences in ketosteroid excretion which were noted on p. 355 occur. In eunuchs, ketosteroids may be normal or lowered, whereas in testicular tumors with associated hyperactivity of the Leydig cells ketosteroid excretion may be considerably increased.

THE FEMALE HORMONES

Two main types of female hormones are secreted by the ovary: the **follicular** or **estrogenic** hormones produced by the cells of the developing graafian follicle; and the **progestational** hormones derived from the corpus luteum that is formed in the ovary from the ruptured follicle.

The Follicular Hormones:

The conversion of androgens to estrogens in the ovary, placenta, adrenal, and testes, has been demonstrated. The estrogenic hormones are C-18 steroids, differing from the androgens in lacking the C-19 position. The structures of the estrogens which are found in the tissues and their metabolites are shown below.

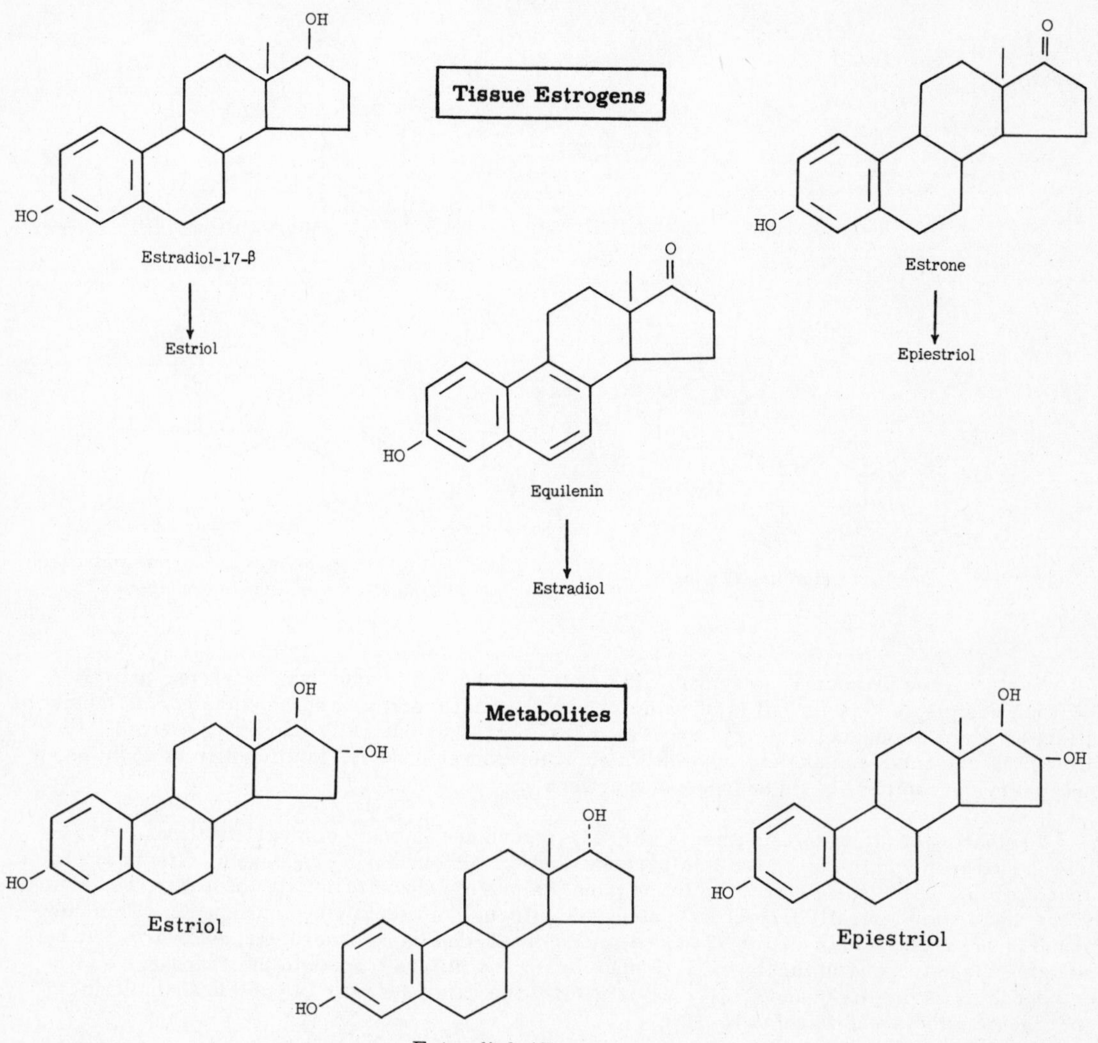

Estradiol is the principal hormone produced by the ovarian follicle. It occurs in two forms, α and β, varying in the position of the OH group on Carbon 17. The α estradiol is 30 times as potent as the β form.

Estriol is the principal estrogen found in the urine of pregnant women and in the placenta. This hormone and estrone are probably produced by oxidation of estradiol prior to its excretion.

Physiological effects of estrogenic hormones. The exact mechanism of estrone synthesis in endocrine tissue is not known, but the possibility that progesterone and testosterone may be precursors is supported by the fact that placental tissue can convert testosterone to Δ^4-androstene-3, 17-dione (see p. 348), which in turn goes to estrone. Ovarian tissue also forms Δ^4-androstene-3, 17-dione from testosterone as well as from progesterone. It is suggested that degradation of the androgens involves hydroxylation of the methyl group at C-19, followed by removal of the entire resultant CH_2OH group to produce a C_{18} steroid with an aromatic nucleus characteristic of the estrogens, as shown below.

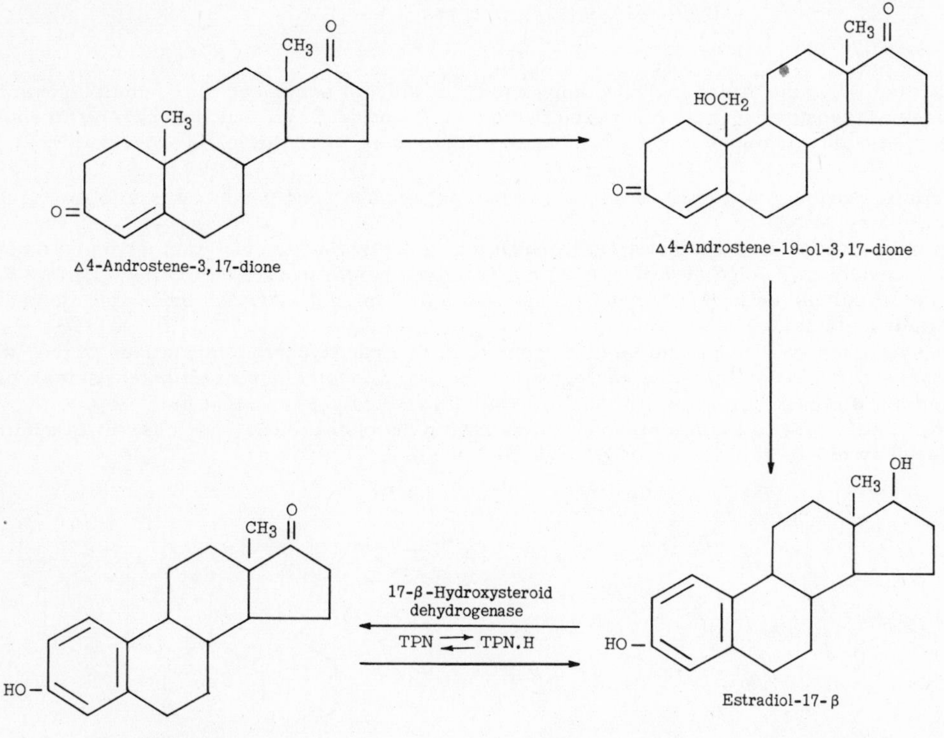

In the lower animals, the estrogenic hormones induce estrus, a series of changes in the female reproductive system associated with ovulation. These changes may be detected by the histologic appearance of the vaginal smear. The **Allen-Doisy test**, used for the detection of estrogenic activity, is based on the ability of a compound to produce estrus in ovariectomized, sexually mature rats.

In women, the follicular hormones prepare the uterine mucosa for the later action of the progestational hormones. The changes in the uterus include proliferative growth of the lining of the endometrium, deepening of glands, and increased vascularity; changes in the epithelium of the tubes and of the vagina also occur. All of these changes begin immediately after menstrual bleeding has ceased.

The estrogens also suppress the production of the pituitary hormone (the follicle-stimulating hormone, FSH) which initially started the development of the follicle. Estrogens are effective in maintenance of female secondary sex characteristics, acting antagonistically to testosterone.

Coenzyme functions. The hydroxysteroid dehydrogenases are a class of enzymes which catalyze the interconversion of specific hydroxyl and ketone groups of steroids, utilizing the pyridine nucleotides (DPN or TPN) as hydrogen carriers. These enzymes may actually be functioning as transhydrogenases (see p. 121) to accomplish the transfer of hydrogen between the two pyridine nucleotides. An example is the 17-β-hydroxysteroid dehydrogenase of human placenta, which interconverts estradiol-17-β and estrone as shown above. The reactions may be shown as follows:

$$\text{Estradiol} + \text{DPN} \rightleftharpoons \text{Estrone} + \text{DPN.H} + H^+$$

$$H^+ + \text{Estrone} + \text{TPN.H} \rightleftharpoons \text{Estradiol} + \text{TPN}$$

The net effect of the interconversions of estrone and estradiol is to accomplish the following reaction of transfer of hydrogen between DPN and TPN:

$$\text{TPN.H} + \text{DPN} \rightleftharpoons \text{TPN} + \text{DPN.H}$$

Talalay and Williams-Ashman[27] have suggested that steroid hormones may function generally as coenzymes of transhydrogenation between pyridine nucleotides* and that many of the biochemical effects of steroid hormones may be attributed to their control of this central mechanism.

Synthetic estrogens. A number of synthetic estrogens have been produced; the following two are clinically very valuable:

1. Ethinyl estradiol is a synthetic estrogen which is highly active even when given by mouth. For example, a dosage of 0.02 to 0.05 mg. daily by mouth is as effective as 1.25 to 2.5 mg. of water-soluble estrogenic preparations or 0.5 to 1 mg. of estradiol benzoate injected intramuscularly.
2. Diethylstilbestrol is an example of a group of para-hydroxyphenyl derivatives which, while apparently not resembling the estrogens chemically, nonetheless exert potent estrogenic effects. However, as shown in the formula below, it is possible that ring closure may occur in the body to form a structure resembling the phenanthrene nucleus. It is administered by mouth in a dosage of 0.5 to 2 mg. per day.

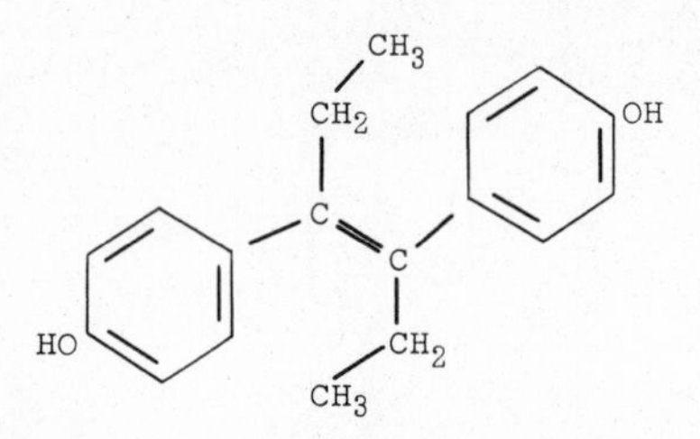

Diethylstilbestrol

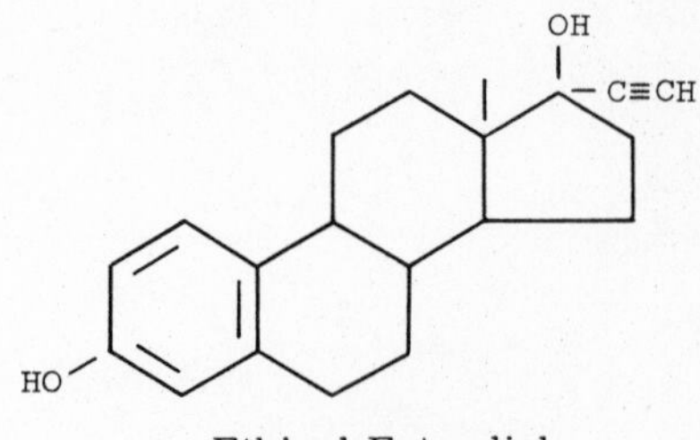

Ethinyl Estradiol

The Progestational Hormones (Luteal Hormones):

Progesterone (see p. 348) is the hormone of the corpus luteum, the structure which develops from the ruptured follicle. It is formed also by the placenta, which secretes progesterone during the latter part of pregnancy. Progesterone is also formed in the adrenal cortex as a precursor of both C-19 and C-21 corticoids.

Functions of progesterone. This hormone appears after ovulation and causes extensive development of the endometrium, preparing the uterus for the reception of the embryo and for its nutrition. The hormone also suppresses estrus, ovulation, and the production of pituitary luteinizing hormone (LH), which originally stimulated corpus luteum formation. Progesterone also stimulates the mammary glands. The Corner and Hisaw test for progesterone (progestin) is based on the specific secretory action of the hormone on the uterine endometrium. When pregnancy occurs, the corpus luteum is maintained and menstruation and ovulation are suspended. The concentration

*The ability of steroid hormones to function as coenzymes for transhydrogenation between pyridine nucleotides is not confined to hormones in the placenta. Testosterone has also been similarly implicated and a steroid-mediated pyridine transhydrogenation has recently been demonstrated with purified preparations of a 3-α-hydroxysteroid dehydrogenase in rat liver which reacts with 3-α-hydroxysteroids and 3-ketosteroids[28]. The transfer of hydrogen to DPN has been demonstrated with TPN.H generated either from oxidation of glucose-6-phosphate (in the direct oxidative pathway, p. 186) or from the isocitric dehydrogenase reaction (in the citric acid cycle, p. 181).

of progesterone decreases near term. If fertilization does not occur, the follicular and progestational hormones suddenly decrease on about the twenty-eighth day of the cycle; the new cycle then begins with menstrual bleeding and sloughing of the uterine wall (see p. 362).

The metabolic fate of C^{14}-labeled progesterone has been studied by the injection of C^{14}-labeled hormone. About 75% of injected progesterone (or its metabolites) is transported to the intestine by way of the bile and eliminated in the feces. Large amounts occur in the urine only if the biliary route of excretion is blocked.

Other progestational hormones. These are shown on p. 348. **Pregnanediol** is the chief excretory product of progesterone. It is present in quantities of 1 to 10 mg. per day during the latter half of the menstrual cycle (as the glucuronidate; see p. 168). Its presence in the urine signifies that the endometrium is progestational rather than follicular. **Δ-5-Pregnenolone** also has some progestational activity. It may be used therapeutically to produce a secretory endometrium or to inhibit uterine motility in threatened abortion.

The Pregnancy Hormone (Pregnancy Tests):

A gonadotropic hormone found in urine during pregnancy is used as the basis for a test for pregnancy. Injection of urine of a pregnant woman into immature female mice or rats causes rapid changes in the ovaries which can be easily seen as hemorrhagic spots and yellowish corpora lutea. This is the **Aschheim-Zondek test.** For the **Friedman test**, a virgin rabbit is used. The test urine is injected into an ear vein, and the ovaries are examined 24 hours later for ruptured or hemorrhagic follicles. The male frog (Rana pipiens), the female toad (Xenopus laevis), and the male toad (Bufo arenarum) are also used for pregnancy tests.

The hormone of pregnancy is a product of the very early placenta. It is also produced when there is abnormal proliferation of chorionic epithelial tissue such as hydatiform mole or chorionepithelioma; this would give rise to false results in pregnancy tests. Various terms have been used to describe this hormone - **Antuitrin S, anterior pituitary-like hormone (APL),** and **chorionic gonadotropin.** This last term is preferred since it is descriptive of the origin of the hormone.

The effects of the female hormones on the menstrual cycle are illustrated on p. 362.

THE PITUITARY GLAND (HYPOPHYSIS)

The human pituitary is a reddish-gray oval structure, about 10 mm. in diameter, located in the brain just behind the optic chiasm as an extension from the floor of the thalamus. The average weight of the gland in the male is 0.5 to 0.6 Gm.; in the female it is slightly larger, 0.6 to 0.7 Gm. The pituitary gland is composed of different types of tissue embryologically derived from two sources: a neural component, originating from the infundibulum (a downgrowth from the floor of the thalamus); and a buccal component which develops upward from the ectoderm of the primitive oral cavity (stomadaeum) to meet and surround the infundibular rudiment. The terms **adenohypophysis** and **neurohypophysis** are used to differentiate the buccal and neural components, respectively.

The adenohypophysis includes the anterior lobe and the intermediate or middle lobe of the developed endocrine organ, both of which are glandular in structure. The neurohypophysis includes the posterior lobe of the gland and the infundibular or neural stalk which attaches the gland to the floor of the brain at the hypothalamus.

Complete removal of the pituitary, hypophysectomy, in young animals (e.g., rats) causes a cessation of growth and a failure in maturation of the sex glands. Removal of the pituitary in the adult animal is followed by atrophy of the sex glands and organs, involution of thyroid, parathyroid, and adrenal cortex, and a depression of their functions. In addition, there are alterations in protein, fat, and carbohydrate metabolism. A notable characteristic of a hypophysectomized animal or of the human patient suffering from pituitary insufficiency is abnormal sensitivity to insulin and resistance to the glycogenolytic effect of epinephrine. These functions are undoubtedly due to a lack of the hormones of the anterior lobe.

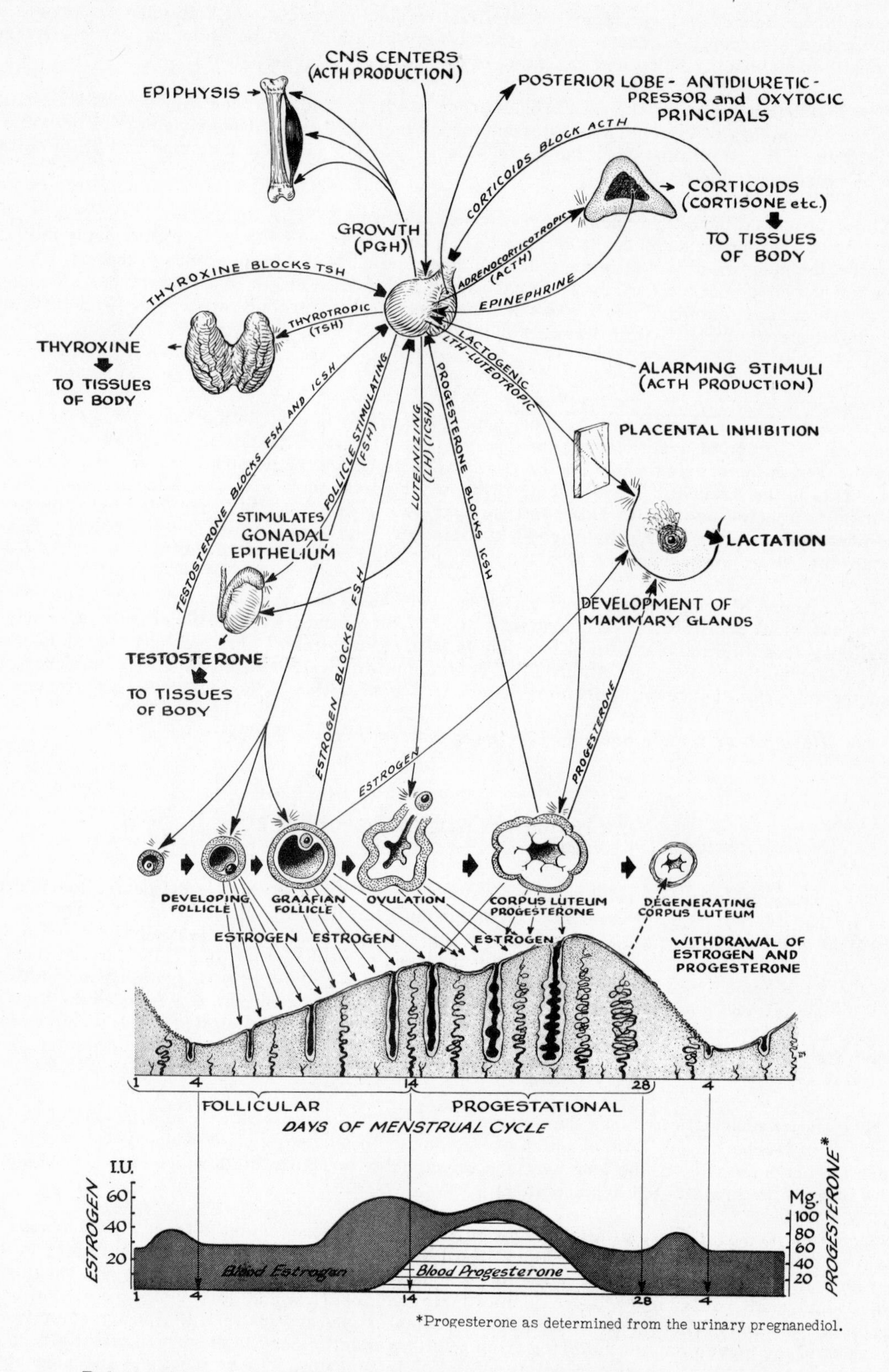

*Progesterone as determined from the urinary pregnanediol.

Relationships of the Pituitary Hormones to the Target Glands and Tissues

THE ANTERIOR PITUITARY GLAND

The anterior lobe is the largest and most essential portion of the pituitary. In man, this lobe comprises about 70% of the total weight of the gland. Histologically, the anterior lobe of the hypophysis consists of epithelial glandular cells of varying sizes and shapes. These cells are arranged in broad, circular columns, separated from one another by sinusoids containing blood. Three types of cells are differentiated by their staining qualities: (1) chromophobe cells (neutrophils), which stain poorly, located in the center of the columns of epithelium; (2 and 3) chromophil cells, located on the outside of the columns and therefore adjacent to the blood sinusoids; these stain well. Those chromophil cells which take the acid dyes are designated eosinophilic cells; those which stain with basic dyes, basophilic cells.

It is difficult to assign specific functions to each type of cell, although attempts have been made to do so. Most of the evidence is derived from association of physiologic disorders with the histologic changes observed at autopsy.

HORMONES OF THE ANTERIOR PITUITARY

The Growth Hormone (Somatotropin):

A. Chemistry: Pituitary growth hormone (PGH), also known as somatotropin, was originally obtained in quantity from the pituitary glands of cattle. It has recently been isolated by Li from human and monkey pituitary glands, and it has been found that while the hormones from the monkey and man are quite similar, both differ from that of cattle. The molecular weight of the hormone from cattle is about 46,000, whereas that of monkeys or man is about 27,000. The cattle hormone contains 400 amino acid residues as compared with 240 in that from man or monkeys. The latter hormones have a simple, straight-chain amino acid structure, whereas the cattle hormone has a branched, Y-type structure.

The growth hormone from cattle is effective in stimulating growth in rats but not in man. The monkey or human preparations are also active in rats. It has been hypothesized that activity resides in a fragment of the molecule since limited hydrolysis of cattle hormone with chymotrypsin does not abolish hormonal activity. The fragment of the molecule which is essential to activity may be common to all three hormones. It may be that the rat can degrade the cattle hormone to the active form, whereas man cannot.

The amino acid composition of growth hormone has been described by Li and Chung[29].

B. Function: Growth hormone stimulates the growth of the long bones at the epiphyses as well as the growth of soft tissue. At the same time, a considerable retention (positive balance) of nitrogen, potassium, and phosphorus occurs. There is evidence that growth hormone acts, in part, by suppressing amino acid catabolism, which may be the means by which it catalyzes nitrogen retention. Injection of growth hormone into hypophysectomized rats causes a significant rise in the concentration of plasma nonesterified fatty acids (see p. 199) derived from adipose tissue. It is suggested[30] that this rise in nonesterified fatty acids may be designed to provide a source of energy for the protein anabolic effect of growth hormone. As noted above, hypophysectomy in young animals results in complete cessation of growth; injection of pituitary extracts or of purified growth hormone restores the growth. The injection of growth hormone into normal animals causes an extensive increase in growth, or gigantism. The "diabetogenic" effect of growth hormone should also be recalled (see p. 172).

The Lactogenic Hormone (Prolactin; Mammotropin; Galactin):

This hormone was first identified by its ability to stimulate the enlargement and formation of "crop milk" in the crop glands of pigeons.

A hormone of the placenta in mammals also stimulates growth of the mammary glands and inhibits production of pituitary prolactin. At parturition, because the placental inhibition is removed, prolactin production begins; mild production then starts.

Prolactin has a molecular weight of about 25, 000. Its amino acid composition has also been determined[31].

The Pituitary Tropins:

The most characteristic function of the anterior pituitary is the elaboration of hormones which influence the activities of other glands, primarily those of the endocrine system. Such hormones are called **tropic** hormones. They are carried by the blood to other **target** glands and aid in maintaining these glands and stimulating production of their own hormones. For this reason, atrophy and decline in the function of many endocrine glands occur in pituitary hypofunction or after hypophysectomy. The reciprocal relationship among the various glands of the endocrine system is best illustrated in the activity of the tropins. A decline in the output of a given hormone, e.g., estrogen, is followed by an increase in the pituitary tropin, presumably to stimulate increased production of the target hormone; when estrogens are supplied in increased quantity, gonadotropic activity of the pituitary declines.

A. The Gonadotropins: These tropic substances influence the hormonal functions of the testis and ovary. The gonadotropins are protein hormones which also contain carbohydrates. They have not been as well studied or characterized as the other pituitary hormones.

1. Follicle-stimulating hormone (FSH, formerly termed Prolan A) affects the activity of the graafian follicle in producing estrogens. In the male, this hormone causes growth of the testis and induces spermatogenesis.
2. Luteinizing hormone (LH, formerly termed Prolan B) causes ovulation and development of the corpus luteum from the mature graafian follicle. A luteotropic hormone (LTH), possibly identical with the lactogenic hormone (see above), activates the corpus luteum and stimulates continued progesterone production by the developed corpus luteum.

   In the male, LH stimulates testosterone production by the testis, which in turn maintains spermatogenesis and provides for the development of accessory sex organs such as the vas deferens, prostate, and seminal vesicles. In this latter function it is also termed interstitial cell–stimulating hormone (ICSH).
3. Pituitary-like hormones from the placenta - Gonadotropic hormones not of pituitary origin are also found (cf. the pregnancy hormone, chorionic gonadotropin; see p. 361). The action of these hormones is almost identical with that of ICSH. They function not by themselves but only if the anterior pituitary is intact, and may act synergistically with FSH.

B. Thyrotropic Hormone; Thyroid-Stimulating Hormone (TSH): Pure preparations of TSH are not yet available. The hormone may be a mucoprotein with a molecular weight of around 10,000. Injections of TSH cause stimulation of thyroid growth and activity with all of the symptoms of hyperthyroidism (see p. 334). Pituitary deficiency, on the other hand, causes thyroid atrophy. The reciprocal relationship between the target gland and pituitary thyrotropin is demonstrated by the reduction in TSH after thyroxine (or iodine) administration.

C. Adrenocorticotropic Hormone (ACTH, Corticotropin): Hypophysectomy causes atrophy of the adrenal cortex. This can be prevented by injections of pituitary ACTH. Pituitary hyperfunction, as in Cushing's disease, is accompanied by adrenocortical hypertrophy.

1. Chemistry - When originally discovered, ACTH was thought to be a protein of high molecular weight. It has subsequently been found that, after careful hydrolysis, various peptides of lower molecular weight can be isolated which retain biologic activity. The structure of one of the most active components of ACTH, termed β-ACTH, has now been determined[32]. Its biologic activity is 150 U.S.P. units per mg. Eight equally active components of ACTH have been isolated in pure form from hog anterior pituitary, but β-ACTH occurs in the highest concentration.

   The structure of β-ACTH is said to be that of a single-chain polypeptide with a molecular weight of 4566 and containing 39 amino acids. The amino acid sequence, which has been determined by methods similar to those used in studies with insulin (see p. 39), is as follows: (N-terminal amino acid, serine; C-terminal amino acid, phenylalanine.)

Ser-Tyr-Ser-Met-Glu-His-Phe-Arg-Try-Gly-Lys-Pro-Val-Gly-Lys-Lys-Arg-Arg-Pro-Val-

Lys-Val-Tyr-Pro-Asp-Gly-Ala-Glu-Asp-Glu(NH_2)-Leu-Ala-Glu-Ala-Phe-Pro-Leu-Glu-Phe

The other peptides isolated from ACTH have amino acid sequences similar to β-ACTH; and it appears that the hormonal activity is dependent on the preservation of the N-terminal amino acid sequence, as evidenced by the fact that brief treatment with kidney aminopeptidase causes extensive inactivation of hog ACTH.

2. Physiologic effects - ACTH not only increases the synthesis of corticoids by the adrenal but also stimulates their release from the gland. It also increases the incorporation of C^{14}-labeled acetate into the adrenal tissue proteins, which indicates that it has both a **tropic** and a **trophic** effect. As noted on p. 345, ACTH does not significantly increase the output of aldosterone by the adrenal.

   The administration of ACTH (corticotropin) to normal human beings causes the following effects: (1) increased excretion of nitrogen, potassium, and phosphorus; (2) retention of sodium and chloride and secondary retention of water; (3) elevation of fasting blood sugar and a diabetic glucose tolerance curve; (4) increased excretion of uric acid; and (5) decline in circulating eosinophils and lymphocytes and elevation of polymorphonuclear leukocytes.

   A reciprocal relationship between corticoid production and ACTH is well-known. Exogenous cortisone therefore is observed to depress ACTH activity (see also p. 352).

3. Control of ACTH secretion - The control of the secretion of ACTH by the anterior pituitary gland is attributed to the action of chemical mediators in the brain which arise from the anterior portion of the hypothalamus. These substances reach the pituitary by way of the blood vessels which traverse the pituitary stalk; in the gland they excite the basophilic cells to secrete ACTH. The hypothalamic centers are themselves activated through the cerebral cortex by nonspecific stresses, such as cold or trauma, or by psychic reactions. Histamine or epinephrine may also stimulate the hypothalamic centers. The resulting increased production of ACTH leads to increased adrenocortical activity and thus to protective effects against the stress.

D. Other Tropic Hormones: A parathyrotropic hormone and a pancreatropic hormone affecting islet cell growth and insulin production have been described. Their existence is not established, and they have not been isolated in purified form.

The Metabolic Hormones:

An effect of the pituitary gland on carbohydrate metabolism is ascribed to a "diabetogenic" hormone, so-called because, in contrast to insulin, the blood sugar level is raised. Continued injection of pituitary extract in some animals (the dog, but not the cat) causes permanent diabetes with evidence of damage to the islet tissue of the pancreas.

As is well known, the complete removal of the pancreas leads to diabetes; however, the diabetes can be ameliorated by subsequent removal of the pituitary (Houssay). This observation, coupled with the fact that in hypophysectomized animals or in hypopituitarism there is abnormal sensitivity to insulin, suggests that normal metabolism of carbohydrate is to a large extent dependent on a check-and-balance system between the pituitary and the insulin activity of the pancreas (see also p. 171). Diabetes may then result not necessarily from **absolute** insulin lack but from **relative** insulin lack due to pituitary hyperfunction.

Whether or not a true diabetogenic hormone of the pituitary exists as a separate entity is not a settled question. Certain pituitary tropins, e.g., ACTH or TSH, can also cause a diabetic state through their action on the appropriate target gland. Furthermore, the growth hormone has been shown not only to abolish the insulin hypersensitivity of hypophysectomized animals but also to produce a diabetic state when an overdosage is used. It is now believed that while the growth hormone has diabetogenic properties, it is not solely responsible for all of the diabetogenic activity of pituitary extracts.

Abnormalities of Pituitary Function:

A. Hyperpituitarism:
   1. Excess production of growth hormone (eosinophilic adenoma).
      a. Gigantism results from hyperactivity of the gland during childhood or adolescence, i.e., before closure of the epiphyses. The long bones increase in length so that the patient reaches a height of six and one-half to eight feet or more. In addition, there are associated metabolic changes attributed to a generalized pituitary hyperfunction.
      b. Acromegaly results from hyperactivity that begins after epiphyseal closure has been completed and growth has ceased. The patient exhibits characteristic facial changes (growth and protrusion of the jaw, enlargement of the nose), growth and enlargement of the hands and feet and of the viscera, and a thickening of the skin.
   2. Excess production of ACTH (basophilic adenoma) produces Cushing's disease (see p. 353).

B. Hypopituitarism may occur as a result of certain types of tumors, hemorrhage (especially postpartum), infarct, or atrophy.

1. Dwarfism is a result of hypoactivity of the gland, sometimes caused by chromophobe tumors or craniopharyngioma. In either case, the underactivity is due to pressure of the tumor on the remainder of the gland. If the tumor begins early in life, dwarfism will result; if it occurs later, there will be a cessation of growth and metabolic abnormalities similar to those observed after total hypophysectomy.
2. Pituitary myxedema, due to a lack of TSH, produces symptoms similar to those described for primary hypothyroidism (see p. 334).
3. Panhypopituitarism refers to deficiency of function of the hypophysis which involves all of the hormonal functions of the gland. Simmonds' disease (hypopituitary cachexia) is a form of panhypopituitarism which occurs abruptly as a result of destruction of the gland because of hemorrhage or infarct of the gland. Simmonds' disease may be quite severe, producing extreme emaciation of the body as a result of the profound decline in the function of the endocrine system and of the metabolic processes which this system controls.

   Milder forms of long-standing panhypopituitarism may result from the pressure of a tumor on the pituitary gland. The symptoms in these cases are similar to those of the early stages of Simmonds' disease, including a tendency to hypoglycemia with sensitivity to insulin.

   The excretion of neutral 17-ketosteroids in the urine (see p. 354) is much reduced in hypopituitarism. The degree of reduction depends upon the severity of the disease.

   Hypopituitarism cannot at present be treated with extracts of the hypophysis, since no adequate preparation is yet available. Specific replacement therapy for the target glands involved, such as the thyroid, adrenal cortex, and gonads, is therefore usually resorted to.

THE MIDDLE LOBE OF THE PITUITARY

The middle lobe of the pituitary secretes a hormone, **intermedin**, which was first detected by its effect on the pigment cells in the skin of lower vertebrates. This hormone apparently also increases the deposition of melanin by the melanocytes of the human skin. In this role it is referred to as the melanocyte-stimulating hormone (MSH). Both hydrocortisone and cortisone inhibit the secretion of MSH; epinephrine and, even more strongly, norepinephrine inhibit its action. When production of the corticoids is inadequate, as in Addison's disease, the synthesis of melanin is increased and there is an accompanying brown pigmentation of the skin. In patients suffering from panhypopituitarism (see above), in which case there is a lack of MSH as well as corticoids, pigmentation does not occur. On the other hand, in totally adrenalectomized patients, as much as 30 to 50 mg. of orally administered cortisone must be supplied each day to prevent excess deposition of pigment in the skin.

Chemistry:

Two peptides (α-MSH and β-MSH) have been isolated from extracts of hog pituitaries. In α-MSH the amino acid sequence appears to be that of the first 13 amino acids of β-ACTH. β-MSH is a peptide containing 18 amino acids and has a minimum molecular weight of 2177. The amino acid sequence in β-MSH is as follows:[33, 34]

Asp-Glu-Gly-Pro-Tyr-Lys-[Met-Glu-His-Phe-Arg-Try-Gly]-Ser-Pro-Pro-Lys-Asp

The amino acid sequence between the brackets is identical with that between amino acids 4 and 10 of β-ACTH. This is of interest in view of the fact that ACTH has small but definite melanocyte-stimulating activity (about 1% that of MSH). This may be due to its content of the so-called intermedin sequence of amino acids, as described above. MSH, however, has no adrenocorticotropic activity.

THE POSTERIOR LOBE OF THE PITUITARY

Extracts of the posterior pituitary contain at least two active substances: a pressor-antidiuretic principle, **vasopressin**; and an oxytocic principle, **oxytocin.**

Function:

A. Vasopressin: This substance raises blood pressure by its vasopressor effect on the peripheral blood vessels. Vasopressin has been used in surgical shock as an adjuvant in elevating blood pressure. It may also be used occasionally in obstetrics in the case of delayed postpartum hemorrhage and, at delivery, for uterine inertia.

Vasopressin also exerts an antidiuretic effect as the so-called posterior pituitary antidiuretic hormone, ADH (see p. 289). The hormone affects the renal tubules and provides for the facultative reabsorption of water. In the absence of this hormone, **diabetes insipidus** occurs. This is characterized by extreme diuresis - up to 30 liters of urine per day. The disease may be controlled by subcutaneous administration of posterior pituitary extract or even by nasal instillation of the extract.

B. Oxytocin: This substance is the principal uterus-contracting hormone of the posterior pituitary. It is employed in obstetrics when induction of uterine contraction is desired.

The so-called "milk-ejecting hormone" of rabbits is also oxytocin.

Chemistry:

Du Vigneaud and his collaborators[35] determined the structure and accomplished the synthesis of the posterior pituitary hormones. The structure of oxytocin is shown below. It is a polypeptide containing eight different amino acids and has a molecular weight of about 1000. Note that five amino acids are arranged to form a cyclic disulfide (S-S) structure with cystine and the other three amino acids are attached as a side chain to the carboxyl group of half of the cystine.

Structure of Oxytocin

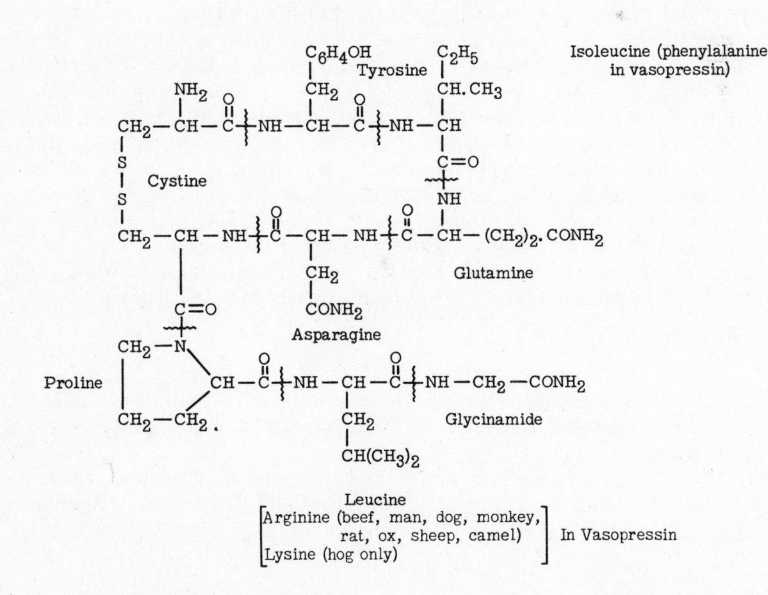

The structure of vasopressin is quite similar to that of oxytocin[36]. The differences are in two amino acids: (1) isoleucine of oxytocin is replaced in vasopressin by phenylalanine; and (2) leucine of oxytocin is replaced in vasopressin obtained from hog pituitary by lysine; in that from beef as well as many other animals, by arginine.

References:

1. Stanbury, J.B., and McGirr, E.M.: Am.J.Med. **22**:712, 1957.
2. Berson, S.A.: Am.J.Med. **20**:653, 1956.
3. Man, E.B., Kydd, D.M., and Peters, J.P.: J.Clin.Invest. **30**:531, 1951.
4. Munson, P.L.: Ann.N.Y.Acad.Sc. **60**:776, 1955.
5. Bogdonoff, M.D., Woods, A.H., White, J.E., and Engel, F.L.: Am.J.Med. **21**:583, 1956.
6. Janelli, D.F.: Surg., Gyn. and Obs. **102**:105, 1956.
7. Agna, J.W., and Goldsmith, R.E.: N.Eng.J.Med. **258**:222, 1958.
8. Litvak, J., Moldawer, M.P., Forbes, A.P., and Henneman, P.H.: J.Clin.Endocr. and Metab. **18**:246, 1958.
9. Goldstein, M.S., Henry, W.L., Huddlestun, B., and Levine, R.: Am.J.Physiol. **173**: 207, 1953.
10. Winegrad, A.I., and Renold, A.E.: J.Biol.Chem. **233**:267, 1958.
11. Stadie, W.C.: Am.J.Clin.Nutr. **5**:393, 1957.
12. Narahara, H.T., Tomizawa, H.H., Miller, R., and Williams, R.H.: J.Biol.Chem. **217**:675, 1955.
13. Mirsky, I.A., Perisutti, G., and Dixon, F.J.: Proc.Soc.Exper.Biol. and Med. **86**:228, 1954.
14. Berson, S.A., and Yalow, R.S.: Am.J.Med. **25**:155, 1958.
15. Kimball, C.P., and Murlin, J.R.: J.Biol.Chem. **58**:337, 1923.
16. Bromer, W.W., Sinn, L.G., and Behrens, O.K.: J.Am.Chem.Soc. **79**:2807, 1957.
17. Sutherland, E.W., and Cori, C.F.: J.Biol.Chem. **188**:531, 1951.
18. Haugaard, E.S., and Haugaard, N.: J.Biol.Chem. **206**:641, 1954.
19. Van Itallie, T.B.: New Eng.J.Med. **254**:794, 1956.
20. Holton, P.: Nature **163**:217, 1949.
21. Bartter, F.C.: Metabolism **5**:369, 1956.
22. Mills, I.H., Casper, A., and Bartter, F.C.: Science **128**:1140, 1958.
23. Luetscher, J.A., Jr., Dowdy, A., Harvey, J., Neher, R., and Wettstein, A.: J.Biol. Chem. **217**:505, 1955.
24. Bongiovanni, A.M.: Pediatrics **21**:103, 1958.
25. a. Conn, J.W.: J.Lab. and Clin.Med. **45**:3, 1955.
    b. Conn, J.W., and Louis, L.H.: Ann.Int.Med. **44**:1, 1956.
    c. Foye, L.V., and Feichtmeier, T.V.: Am.J.Med. **19**:966, 1955.
26. Richter, H.S.: Proc.Soc.Exp.Biol. and Med. **97**:141, 1958.
27. Talalay, P., and Williams-Ashman, H.G.: Proc.Natl.Acad.Sci. **44**:15, 1958.
28. Hurlock, B., and Talalay, P.: J.Biol.Chem. **233**:886, 1958.
29. Li, C.H., and Chung, D.: J.Biol.Chem. **218**:33, 1956.
30. Engel, H.R., Hallman, L., Siegel, S., and Bergenstal, D.M.: Proc.Soc.Exp.Biol. and Med. **98**:753, 1958.
31. Behrens, O.K., and Bromer, W.W.: Ann.Rev.Biochem. **27**:57, 1958.
32. Shepherd, R.G., Howard, K.S., Bell, P.H., Cacciola, A.R., Child, R.G., Davies, M.C., English, J.P., Finn, B.M., Meisenhelder, J.H., Moyer, A.W., and Vanden Scheer, J.: J.Am.Chem.Soc. **78**:5051; 5059; 5067, 1956.
33. Geschwind, I.I., Li, C.H., and Barnafi, L.: J.Am.Chem.Soc. **78**:4494, 1956.
34. Harris, J.I., and Roos, P.: Nature **178**:90, 1956.
35. Du Vigneaud, V., Ressler, C., and Trippett, S.: J.Biol.Chem. **205**:949, 1953.
36. Popenoe, E.A., and Du Vigneaud, V.: J.Biol.Chem. **206**:353, 1954.

Bibliography:

Williams, R.H., Ed.: Textbook of Endocrinology, 2nd Ed. Saunders, 1955.

Albright, F., and Reifenstein, E.C., Jr.: Parathyroid Glands and Metabolic Bone Disease. Williams and Wilkins, 1948.

Escamilla, R.F.: Laboratory Aids in Endocrine Diagnosis. Thomas, 1954.

Harris, R.S., and Thimann, K.V., Eds.: Vitamins and Hormones. Academic Press.

Pincus, G., Ed.: Recent Progress in Hormone Research. Academic Press.

21...

Calorimetry: Elements of Nutrition

Food is the source of the fuel which is converted by the metabolic processes of the body into the energy for vital activities. Calorimetry deals with the measurement of the energy requirements of the body under various physiologic conditions and of the fuel values of foods which supply this energy.

The Unit of Heat: Large Calorie or Kilocalorie (C):

Vital energy and the fuel value of foods are most conveniently measured in calories. A calorie is the amount of heat required to raise the temperature of 1 Gm. of water 1° C. (from 15° to 16° C.). This is a very small quantity of heat. In nutrition it is therefore more common to use the large calorie (written with a capital "C") or kilocalorie, 1000 times the small calorie. All references to caloric values will henceforth be to the kilocalorie.

Measurement of the Fuel Value of Foods:

The combustion of a foodstuff in the presence of oxygen results in the production of heat. The amount of heat thus produced can be measured in a bomb calorimeter. By this technic, the caloric value of a foodstuff can be determined.

The caloric content of the three principal foodstuffs, determined by burning in a bomb calorimeter, is given in the table below:

Fuel Values of Foods

	Kilocalories per Gm.	
	In Bomb Calorimeter	In the Body*
Carbohydrate	4.1	4
Fat	9.4	9
Protein	5.6	4

*Figures are expressed in round numbers.

These are average figures, since variations occur within each class, e.g., monosaccharides do not have exactly the same caloric content as polysaccharides.

When utilized in the body, carbohydrate and fat are completely oxidized to carbon dioxide and water, as they are in the bomb calorimeter. Proteins, however, are not burned completely since the major end product of protein metabolism, urea, still contains some energy which is not available to the body. For this reason, the energy value of protein in the body (4.1 C per Gm.) is less than that obtained in the bomb calorimeter.

It is customary to round off the energy value of foods as utilized in the body to the figures given in the table above. These figures also correct for the digestibility of the foods.

Control of Body Heat:

The heat generated by the body in the course of the metabolism of foodstuffs maintains the body temperature. Warm-blooded animals, such as birds and mammals, have heat-regulating mechanisms which either increase heat production or radiate or otherwise dissipate excess heat, depending on the temperature of their external environment. When the external temperature rises above the normal body temperature, 98.6° F. (37° C.), evaporation from the surface of the body becomes the only mechanism available for cooling the body.

Animal Calorimetry:

Since all of the energy produced in the body is ultimately dissipated as heat, measurement of the vital heat production of an animal is a way to estimate its energy expenditure. There are two methods of accomplishing this:

A. Direct Calorimetry: In the direct calorimeter, the subject is placed in an insulated chamber; his heat production is measured directly by recording the total amount of heat transferred to a weighed quantity of water circulating through the calorimeter. The oxygen intake, the carbon dioxide output, and the nitrogen excretion in the urine and feces are also measured during the entire period of observation. These data are used as described below.

B. Indirect Calorimetry: Direct calorimetry is attended by considerable technical difficulties. By measuring gas exchange and determining the respiratory quotient, energy metabolism studies are considerably simplified and thus rendered applicable to field studies and to clinical analysis.

Respiratory Quotients (R.Q.) of Foodstuffs:

The respiratory quotient is the ratio of the volume of carbon dioxide eliminated to the volume of oxygen utilized in oxidation.

A. Carbohydrates: The complete oxidation of glucose, for example, may be represented as follows:

$$C_6H_{12}O_6 + \boxed{6O_2} \longrightarrow \boxed{6CO_2} + 6H_2O$$

The R.Q. for carbohydrate is therefore: $\frac{CO_2}{O_2} = \frac{6}{6}$ or 1

B. Fats have a lower R.Q. because the oxygen content of their molecule in relation to the carbon content is quite low. Consequently they require more oxygen from the outside. The oxidation of tristearin will be used to exemplify the R.Q. for fat.

$$2C_{57}H_{110}O_6 + \boxed{163O_2} \longrightarrow \boxed{114CO_2} + 110H_2O$$

$$\frac{CO_2}{O_2} = \frac{114}{163} = 0.70$$

C. Proteins: The oxidation of proteins cannot be so readily expressed because their chemical structure is not known. By indirect methods the R.Q. of proteins has been calculated to be about 0.8.

D. R.Q. of Mixed Diets Under Varying Conditions: In mixed diets containing varying proportions of protein, fat, and carbohydrate, the R.Q. is about 0.85. As the proportion of carbohydrate metabolized is increased, the R.Q. approaches closer to 1. When carbohydrate metabolism is impaired, as in diabetes, the R.Q. is lowered. Therapy with insulin is followed by an elevation in the R.Q. A high carbohydrate intake, as used in fattening animals, will result in an R.Q. exceeding 1. This rise is caused by the conversion of much of the carbohydrate, an oxygen-rich substance, to fat, an oxygen-poor substance; a relatively small amount of oxygen is required from the outside, and the ratio of carbon dioxide eliminated to the oxygen taken in (the R.Q.) will be considerably elevated.

A reversal of the above process, i.e., the conversion of fat to carbohydrate, would lower the R.Q. below 0.7. This has been reported but has not been generally confirmed.

Performance of Indirect Calorimetry:

The use of the indirect method for calculating the total energy output and the proportions of various foodstuffs being burned may be illustrated by the following example.

The subject utilized oxygen at a rate of 414.6 L. per day and eliminated 353.3 L. of carbon dioxide in the same period. The urinary nitrogen for the day was 12.8 Gm.

Because of the incomplete metabolism of protein, the gas exchange is corrected for the amount of protein metabolized; a nonprotein R.Q. (nonprotein portion of the total R.Q.) is thus obtained.

One Gm. of urinary nitrogen represents the combustion of an amount of protein which would require 5.92 L. of oxygen and would eliminate 4.75 L. of carbon dioxide.

1. Calculating the nonprotein R.Q. -
   a. Multiply the amount of urinary nitrogen by the number of liters of oxygen required to oxidize that amount of protein represented by 1 Gm. of urinary nitrogen.

$$12.8 \text{ Gm.} \times 5.92 = 75.8 \text{ L.}$$

   b. Multiply the amount of urinary nitrogen by the number of liters of carbon dioxide which result from this oxidation.

$$12.8 \text{ Gm.} \times 4.75 = 60.8 \text{ L. of } CO_2$$

   Thus, 75.8 L. of the total oxygen intake was used to oxidize protein, and 60.8 L. of the carbon dioxide eliminated was the product of this oxidation. The remainder was used for the oxidation of carbohydrates and fats. Therefore, to determine the nonprotein R.Q., subtract these values for protein from the totals for the day.

$$\text{Oxygen: } 414.6 - 75.8 = 338.8 \text{ L.}$$
$$CO_2\text{: } 353.3 - 60.8 = 292.5 \text{ L.}$$

$$\frac{CO_2}{O_2} = \frac{292.5}{338.8} = 0.86 \text{ (nonprotein R.Q.)}$$

2. Convert the nonprotein R.Q. to grams of carbohydrate and fat metabolized - Reliable tables have been worked out which give the proportions of carbohydrate and of fat metabolized at various R.Q.'s. According to the tables of Zuntz and Shunberg (as modified by Lusk and later by McLendon), when the nonprotein R.Q. is 0.86, 0.622 Gm. of carbohydrate and 0.249 Gm. of fat are metabolized per liter of oxygen used. Therefore, to determine the total quantities of carbohydrate and fat used, multiply these figures by the number of liters of oxygen (derived from the nonprotein R.Q.) consumed during the combustion of carbohydrate and fat.

$$338.8 \times 0.622 \text{ Gm.} = 210.7 \text{ Gm. of carbohydrate}$$
$$338.8 \times 0.249 \text{ Gm.} = 84.4 \text{ Gm. of fat}$$

3. Determine the amount of protein metabolized - Each gram of urinary nitrogen represents the oxidation of 6.25 Gm. of protein. Therefore, to determine the amount of protein metabolized, multiply the total urinary nitrogen by 6.25.

$$12.8 \text{ Gm.} \times 6.25 = 80 \text{ Gm. of protein}$$

4. Calculate the total heat production - Multiply the quantity in grams of each foodstuff oxidized by the caloric value of that food to obtain the heat production due to its combustion. The sum of these caloric values equals the total heat production of the diet.

Carbohydrate:	210.7 Gm.	× 4 C	= 842.8 C
Fat:	84.4 Gm.	× 9 C	= 759.6 C
Protein:	80.0 Gm.	× 4 C	= 320.0 C

Total heat production = 1922.4 C

BASAL METABOLISM

The total heat production or energy expenditure of the body is the sum of that required merely to maintain life, together with such additional energy as may be expended for any additional activities.

This minimum level of activity is called **basal metabolism.** It represents the lowest level of energy production consonant with life. A measurement of the rate of heat production under basal conditions is called the **basal metabolic rate,** or BMR.

Conditions Necessary for Measurement of the BMR:

1. A post-absorptive state; patient should have had nothing by mouth for the past 12 hours.
2. Mental and physical relaxation immediately preceding the test; usually one-half hour of bed rest is used, although ideally the patient should not arise from bed.
3. Recumbent position during the test.
4. Patient awake.
5. Environmental temperature of between 20° and 25° C.

Factors Influencing Basal Metabolism:

A. Surface Area: The metabolic rates of different individuals, when expressed in terms of surface area (sq. M.), is remarkably constant. However, in general, smaller individuals have a higher rate of metabolism per unit of surface area than do larger individuals.

B. Age: In the newborn the rate is low; it rises to maximum at the age of five, after which the rate begins to decline, continuing into old age. There is, however, a relative rise just before puberty. Examples of influence of age on BMR: At six years of age, the normal BMR is between 50 and 53 C/sq. M./hr.; at age 21 it is between 36 and 41 C/sq. M./hr.

C. Sex: Women normally have a lower BMR than men. The BMR of females declines between the ages of five and 17 more rapidly than does that of males.

D. Climate: The BMR is lower in warm climates.

E. Racial Variations: When the BMR of various peoples is compared, certain variations are noted. For example, oriental female students living in America average 10% below the standard BMR for American women of the same age; the basal metabolism of adult Chinese is equal to or below the lower limit of normal for Occidentals; high values (33% above normal) have been reported for Eskimos living in the region of Baffin Bay.

F. State of Nutrition: In starvation and undernourishment, the BMR is lowered.

G. Disease: Infectious and febrile diseases raise the metabolic rate, usually in proportion to the elevation of the temperature. Diseases which are characterized by increased activity of cells also increase heat production because of this increased cellular activity. Thus, the metabolic rate may increase in such diseases as leukemia, polycythemia, some kinds of anemia, cardiac failure, hypertension, and dyspnea - all of which involve increased cellular activity. Perforation of an ear drum causes false high readings.

H. Effects of Hormones: The hormones also affect metabolism. Thyroxine is the most important of the hormones in this respect, and the principal use of calorimetry in clinical practice is in the diagnosis of thyroid disease. The rate is lowered in hypothyroidism and increased in hyperthyroidism. Changes in the BMR are also noted in pituitary disease.

Other than thyroxine, the only hormone which has a direct effect on the rate of heat production is epinephrine, although the effect of epinephrine is rapid in onset and brief in duration. Tumors of the adrenal (pheochromocytoma; see p. 344) cause an elevation in the BMR. In adrenal insufficiency (Addison's disease), the basal metabolism is subnormal, whereas adrenal tumors and Cushing's disease may produce a slight increase in the metabolic rate.

Measurement of Basal Metabolism:

In clinical practice, the BMR can be estimated with sufficient accuracy by merely measuring the oxygen consumption of the patient for two six-minute periods under basal conditions. This is corrected to standard conditions of temperature and barometric pressure. The average oxygen consumption for the two periods is multiplied by ten to convert it to an hourly basis and then multiplied by 4.825 C, the heat production represented by each liter of oxygen consumed. This gives the heat production of the patient in C per hour. This is corrected to C per square meter body surface per hour by dividing the C per hour by the patient's surface area. A simple formula for calculating the surface area is as follows:

$$\frac{\text{Circumference of thighs}}{\text{(or one thigh} \times 2)} \times \text{Height} = \text{Surface area}$$

The classic formula is that of Du Bois, as follows:

Du Bois' Surface Area Formula

$$A = H^{0.725} \times W^{0.425} \times 71.84$$

where **A** = surface area in sq. cm., **H** = height in cm., and **W** = weight in Kg.

(Surface area in sq. cm. divided by 10,000 = surface area in sq. M.)

In practice, a nomogram which relates height and weight to surface area is used. It is based on Du Bois' formula.

Calculation of Basal Metabolic Rate:

The normal BMR for an individual of the patient's age and sex is obtained from standard tables. The actual rate of the patient is then compared to the normal and his rate expressed as a plus or minus percentage of the normal.

Example - A male, 35 years of age, 170 cm. in height and 70 Kg. in weight, consumed an average of 1.2 L. of oxygen (corrected to normal temperature and pressure: 0° C., 760 mm. Hg) in a six-minute period.

1.2 × 10 = 12 L. of oxygen per hour
12 × 4.825 = 58 C per hour
Surface area = 1.8 sq. M. (from Du Bois' formula)
BMR = 58 C/1.8 = 32 C per sq. M. per hour

The normal BMR for this patient, by reference to the Du Bois standards, is 39.5 C per sq. M. per hour. His BMR, which is below normal, is then reported as:

$$\frac{39.5 - 32}{39.5} \times 100 = 18.5\% \text{ or } \underline{\text{minus } 18.5}$$

A BMR between −15 and +20% is considered normal. In hyperthyroidism, the BMR may exceed +50 to +75%. The BMR may be −30 to −60% in hypothyroidism.

MEASUREMENT OF ENERGY REQUIREMENTS

The metabolic rate increases when the subject engages in activity. Maximal increases occur during exercise - as much as 600 to 800% over basal. The energy requirement for the day will therefore vary considerably in accordance with the amount of physical activity. It is sometimes important to know how much energy a given subject uses in performing various tasks in order to recommend an appropriate caloric intake. The gas exchange methods of indirect calorimetry, discussed above, are used for this purpose. The **Douglas bag,** which may be carried on the back of the subject like a knapsack, is filled with a measured quantity of oxygen; after the task is completed, the oxygen used and the carbon dioxide eliminated during the study are measured. These data, together with the urinary nitrogen, are then used to calculate the energy expenditure. (Consolazio et al.)

Such studies indicate, for example, that for a 70 Kg. man, 65 C per hour are expended in sleeping, 100 C per hour sitting at rest, 200 C per hour walking slowly, 570 C per hour running, and as much as 1100 C per hour walking up a flight of stairs.

SPECIFIC DYNAMIC ACTION (S. D. A.)

The specific dynamic action of a foodstuff is the extra heat production, over and above the caloric value of a given amount of food, which is produced when this food is used by the body. For example, when an amount of protein which contains 100 C (25 Gm.) is metabolized, the heat production in the body is not 100 C but 130 C. This extra 30 C is the product of the specific dynamic action of the protein. In the body, a 100-C portion of fat produces 113 C, and a 100-C portion of carbohydrate produces 105 C. The origin of this extra heat is not clear, but it is attributable to the activity of the tissues which are metabolizing these foodstuffs.

The specific dynamic action of each foodstuff, as given above, is obtained when each foodstuff is fed separately; but when these foods are taken in a mixed diet, the dynamic effect of the whole diet cannot be predicted by merely adding the individual effects of each foodstuff in accordance with its contribution to the diet, e.g., 30% of the caloric value of protein, 13% of the caloric value of fat, and 5% of the caloric value of carbohydrate. According to Forbes, the dynamic effects of beef muscle protein, glucose, and lard, when each was fed separately, were 32, 20 and 16%, respectively, of their caloric content. But when the glucose and protein were combined, the dynamic effect was 12.5% less than predicted from the sum of their individual effects; combining lard, glucose, and protein produced 22% less dynamic effect than predicted; a glucose-lard combination was 35% less, and a protein-lard mixture 54% less than calculated from the individual percentages of each foodstuff.

These observations are of importance because they indicate that the high specific dynamic action of protein can be reduced depending on the quantities of other foodstuffs in the diet. Forbes's data show that fat (lard) has a greater influence on specific dynamic action than does any other nutrient, i.e., fat decreases the S. D. A. more than any other nutrient.

In calculating the total energy requirement for the day, it is customary to add 10% to the total required C to provide energy for the S. D. A., i.e., expense of utilization of the foods consumed. This figure may be too large, particularly if the fat content of the diet is high.

THE ELEMENTS OF NUTRITION

The Components of an Adequate Diet:

There are six major components of the diet. **Carbohydrate, fat,** and **protein** yield energy, provide for growth, and maintain tissue subjected to wear and tear. **Vitamins, minerals,** and **water,** although they do not yield energy, are essential parts of the chemical mechanisms for the utilization of energy and for the synthesis of various necessary metabolites such as hormones and enzymes. The minerals are also incorporated into the structure of the tissue and, in solution, play an important role in acid-base equilibrium.

The Energy Aspect of the Diet:

Energy for physiologic processes is provided by the combustion of carbohydrate, fat, and protein. The daily energy requirement or the daily caloric need is the sum of the basal energy demands plus that required for the additional work of the day. During periods of growth, pregnancy, or convalescence, extra calories must be provided.

While all three major nutrients yield energy to the body, carbohydrate and, to a lesser extent, fat are physiologically the most economical sources. Protein serves primarily to provide for tissue growth and repair; but if the caloric intake from other foods is inadequate, it is burned for energy.

The caloric requirements of persons of varying sizes and under various physiologic conditions are tabulated on p. 378.

Obesity is almost always the result of excess consumption of calories. Treatment is therefore directed at reducing the caloric intake from fat and carbohydrate but maintaining the protein, vitamin, and mineral intake at normal levels. An adequate diet containing not more than 800 to 1000 C should be maintained and normal or increased amounts of energy expended until the proper weight is reached.

"Protein-Sparing Action" of Carbohydrates and Fats:

Carbohydrate and fat "spare" protein and thus make it available for anabolic purposes. This is particularly important in the nutrition of patients, especially those being fed parenterally, when it is difficult or even impossible for them to take in enough calories. If the caloric intake is inadequate, giving proteins orally or amino acids intravenously is a relatively inefficient way of supplying energy because the primary function of proteins is tissue synthesis and repair and not energy production.

Distribution of Calories in the Diet:

In a well-balanced diet, 10 to 15% of the total calories is usually derived from protein, 55 to 70% from carbohydrate, and 20 to 30% from fat. These proportions vary under different physiologic conditions or in various environmental temperatures. For example, the need for calories is increased by the need to retain a constant body temperature; in extreme cold, the caloric intake of the diet must therefore increase and this requirement is usually met by increasing the fat content of the diet.

A. The Carbohydrate Intake: Carbohydrate is the first and most efficient source of energy for vital processes. Cereal grains, potatoes, and rice, the staple foods of most countries, are the principal sources of carbohydrate in the diet. Usually 50% or more of the daily caloric intake is supplied by carbohydrate; this is equivalent to 250 to 500 Gm. per day in the average diet, but it varies within wide limits. Carbohydrate intake is often the principal variable in gain or loss of weight. A minimum of 5 Gm. of carbohydrate per 100 C of the total diet is required to prevent ketosis.

B. The Fat Intake: Because of its high fuel value, fat is an important component of the diet. Furthermore, the palatability of foods is generally increased by their content of fat. As a form of energy storage in the body, fat has more than twice the value of protein or carbohydrate.

The human requirement for fat is not precisely known. An important aspect of the contribution of fats to nutrition may be their content of the so-called "essential fatty acids," linoleic and arachidonic acids (see p. 203). These unsaturated fatty acids have been shown to be essential in the diet of experimental animals, and although they seem to exert important effects on lipid metabolism in man, a deficiency of these substances has not yet been unequivocally demonstrated in human subjects. However, it is now clear from studies on the metabolic effects of the polyunsaturated fatty acids that the requirement for fats in the diet must be considered from a qualitative as well as from a quantitative standpoint.

Up to a certain point, the isocaloric replacement of carbohydrate or protein by fat results in better growth in animals. This may be due to the effect of fat in reducing the S. D. A. of the ration (see p. 374), thus improving the caloric efficiency of the diet.

At ordinary levels of intake, 20 to 25% of the calories of the diet should probably be derived from fats (for a 3000-C intake, 66 to 83 Gm. of fat), including 1% of the total calories from "essential" fatty acids. At higher caloric levels, 30 to 35% will probably come from fats.

C. The Protein Requirement: A minimal amount of protein is indispensable in the diet to provide for the replacement of tissue protein, which constantly undergoes destruction and resynthesis. This is often spoken of as the **wear and tear quota.** The protein requirement is considerably increased, however, by the demands of growth, increased metabolism (as in infection with fever), in burns, and after trauma. The recommended intake of protein for normal adults is 1 Gm. per Kg. body weight, or about 70 Gm. per day. As already noted, this presumes that the caloric demand is adequately supplied by other foods so that the ingested protein is available for tissue growth and repair.

The requirement for protein in the diet is, however, not only quantitative; there is also an important qualitative aspect since the metabolism of protein is inextricably connected with that of its constituent amino acids. Certain amino acids are called indispensable in the diet in the sense that they must be obtained preformed and cannot be synthesized by the animal organism. The remainder, the so-called dispensable amino acids, are also required by the organism since they are found in the protein of the tissues, but they can apparently be synthesized, presumably from alpha-keto acids, by amination (see p. 226). The list of indispensable or "essential" amino acids varies with the animal species tested (e.g., the chick requires glycine). It may also vary with the physiologic state of the animal. For some amino acids (e.g., arginine) the rate of synthesis may be too slow to supply fully the needs of the animal. Histidine in the diet is probably also necessary to maintain growth during childhood. Such amino acids are said to be **relatively** indispensable. The nutritive value of a protein is now known to be dependent on its content of essential amino acids. Examples of incomplete proteins are gelatin, which lacks tryptophan; and zein of corn, which is low in both tryptophan and lysine. Such incomplete proteins are unable to support growth if given as the sole source of protein in the diet.

The experiments of Rose have supplied for normal adult human subjects the requirements for the eight amino acids which are essential to the maintenance of nitrogen balance (see table on the following page). The diet must also furnish sufficient nitrogen for the synthesis of other amino acids. Normal adult male subjects were maintained in nitrogen balance only when all eight amino acids listed below were simultaneously present in their diets. The elimination of any one promptly produced a negative nitrogen balance. Although tyrosine and cystine are not listed among the essential amino acids, tyrosine can spare 70 to 75% of the phenylalanine requirement and cystine can spare 80 to 89% of the methionine requirement in humans. This is undoubtedly because the normal metabolism of phenylalanine and methionine involves their conversion, in whole or in part, to tyrosine or cystine, respectively.

Some of the unnatural (D-) forms of the amino acids have been found to fulfill in whole or in part the requirement for a given amino acid. Thus DL-methionine was as effective as L-methionine in supplying the requirement for this amino acid[1]. Significant amounts of D-phenylalanine are utilized by the human organism, perhaps as much as 0.5 Gm. per day[2]. On the other hand, D-valine, isoleucine, and threonine individually do not exert a measurable effect upon nitrogen balance when the corresponding L- form is absent from the diet.

In summarizing his observations on the amino acid requirements of man, Rose has stated[3] that when the diet furnishes the eight "essential" amino acids at their "recommended" levels of intake (see the table on the following page) and extra nitrogen is provided as glycine to provide a total daily intake of only 3 to 5 Gm., nitrogen balance could be maintained in his subjects. Expressed as protein ($N \times 6.25$), this quantity of nitrogen represents about 22 Gm. of protein, an amount far below the recommended daily intake for adults. This finding emphasizes the importance of a consideration of the individual amino acid intake in assessing protein requirements rather than whole proteins of varying essential amino acid content.

The caloric intake necessary for nitrogen balance in human subjects receiving amino acids (e.g., casein hydrolysate) is higher than for subjects receiving whole protein (e.g., whole casein, by mouth). For example, with whole protein, nitrogen balance could be attained with 35 C per Kg., whereas as much as 53 to 60 C per Kg. were required when amino acids were used as the sole source of nitrogen.

Proteins differ in "**biologic value**" depending on their content of essential amino acids. The proteins of eggs, dairy products, kidney, and liver have high biologic values because they contain all of the essential amino acids. Good quality proteins, which are somewhat less efficient in supplying amino acids, include shellfish, soybeans, peanuts, potatoes, and the muscle tissue of meats, poultry, and fish. Fair proteins are those of cereals and most root vegetables. The proteins of most nuts and legumes are of poor biologic value. It is important to point out, however, that two or more proteins in themselves only "poor" or "fair" in quality may have a "good" biologic value when taken together because they may complement one another in supplying the necessary amino acids.

It is of interest to recall that proteins, in addition to their other important functions, also constitute the most important sources of nitrogen, sulfur, and phosphorus for the body.

Amino Acids Essential to Maintenance of Nitrogen Balance in Human Subjects

	Minimum Daily Requirement (Gm.)	Recommended Daily Intake* (Gm.)
L-Tryptophan	†0.25	0.5
L-Phenylalanine	1.10	2.2
L-Lysine	0.80	1.6
L-Threonine	0.50	1.0
L-Valine	0.80	1.6
L-Methionine	1.10	2.2
L-Leucine	1.10	2.2
L-Isoleucine	0.70	1.4

*Note that the recommended daily intake is twice the minimum Daily requirement.
†The tryptophan requirement may vary with the niacin intake (see p. 82).

Many amino acids have specific functions in metabolism in addition to their general role as constituents of the tissue proteins. Examples have been cited in Chapter 15. They include the role of methionine as a methyl donor, cystine as a source of SH groups, the dicarboxylic acids in transamination, tryptophan as a precursor of niacin, arginine and the urea cycle, etc.

Vitamins:

The chemistry and physiologic functions of the vitamins have been discussed in Chapter 6. Normal individuals on an adequate diet can secure all of the required vitamins from natural foods; no supplementation with vitamin concentrates is necessary. In disease states in which digestion and assimilation are impaired or the normal requirement for the vitamins is increased, they must of course be supplied in appropriate quantities from other sources.

Many of the vitamins are destroyed by improper cooking. Some of the water-soluble vitamins, for example, are partially lost in the cooking water. Overcooking of meats also contributes to vitamin loss. Vitamin C is particularly labile both in cooking and storage. In fact, one can hardly depend on an adequate vitamin C intake unless a certain quantity of fresh fruits and vegetables is taken each day.

The refinement of cereal grains is attended by a loss of B vitamins. Enrichment of these products with thiamine, riboflavin, niacin, and iron is now used to restore these nutrients. Other foods are often improved by the addition of vitamins, e.g., the addition of vitamin D to milk and of vitamin A to oleomargarine.

Minerals:

The minerals, while forming only a small portion of the total body weight, are nonetheless of great importance in the vital economy. Their functions and the requirements for each as far as now known are discussed in Chapter 19. Fruits, vegetables, and cereals are the principal sources of the mineral elements in the diet. Certain foods are particularly outstanding for their contribution of particular minerals, e.g., milk products, which are depended on to supply the majority of the calcium and phosphorus in the diet.

Water:

Water is not a food, but since it is ordinarily consumed as an article of the diet it is included as one of its components. The water requirements and the functions of water in the body are discussed in Chapter 19.

RECOMMENDED DIETARY ALLOWANCES

The Food and Nutrition Board of the National Research Council has collected the best available data on the quantities of various nutrients needed by normal persons of varying sizes and in different physiologic states. These are not minimum requirements but are purposely made somewhat high to provide a safe level of intake against losses due to variation in food content or in storage and preparation of the foods (see table below).

RECOMMENDED DAILY DIETARY ALLOWANCES† (1)

Designed for the Maintenance of Good Nutrition of Healthy Persons in the U. S. A.
(Allowances are intended for persons normally active in a temperate climate.)

	Age Years	Weight Kg. (lb.)	Height cm. (in.)	Calories	Protein Gm.	Calcium Gm.	Iron mg.	Vitamin A I. U.	Thiamine mg.	Riboflavin mg.	Niacin‡ mg. equiv.	Ascorbic Acid mg.	Vitamin D I. U.
Men	25	70 (154)	175 (69)	3200 §	70	0.8	10	5000	1.6	1.8	21	75	
	45	70 (154)	175 (69)	3000	70	0.8	10	5000	1.5	1.8	20	75	
	65	70 (154)	175 (69)	2550	70	0.8	10	5000	1.3	1.8	18	75	
Women	25	58 (128)	163 (64)	2300	58	0.8	12	5000	1.2	1.5	17	70	
	45	58 (128)	163 (64)	2200	58	0.8	12	5000	1.1	1.5	17	70	
	65	58 (128)	163 (64)	1800	58	0.8	12	5000	1.0	1.5	17	70	
	Pregnant (second half)			+ 300	+20	1.5	15	6000	1.3	2.0	+ 3	100	400
	Lactating (850 ml. daily)			+1000	+40	2.0	15	8000	1.7	2.5	+ 2	150	400
Infants**	0-1/12**				See								
	2/12-6/12	6 (13)	60 (24)	Kg.×120	Footnote	0.6	5	1500	0.4	0.5	6	30	400
	7/12-12/12	9 (20)	70 (28)	Kg.×100	**	0.8	7	1500	0.5	0.8	7	30	400
Children	1 - 3	12 (27)	87 (34)	1300	40	1.0	7	2000	0.7	1.0	8	35	400
	4 - 6	18 (40)	109 (43)	1700	50	1.0	8	2500	0.9	1.3	11	50	400
	7 - 9	27 (60)	129 (51)	2100	60	1.0	10	3500	1.1	1.5	14	60	400
	10-12	36 (79)	144 (57)	2500	70	1.2	12	4500	1.3	1.8	17	75	400
Boys	13-15	49 (108)	163 (64)	3100	85	1.4	15	5000	1.6	2.1	21	90	400
	16-19	63 (139)	175 (69)	3600	100	1.4	15	5000	1.8	2.5	25	100	400
Girls	13-15	49 (108)	160 (63)	2600	80	1.3	15	5000	1.3	2.0	17	80	400
	16-19	54 (120)	162 (64)	2400	75	1.3	15	5000	1.2	1.9	16	80	400

†The allowance levels are intended to cover individual variations among most normal persons as they live in the United States under usual environ mental stresses. The recommended allowances can be attained with a variety of common foods, providing other nutrients for which human requirements have been less well defined. See original reference for more detailed discussion of allowances and of nutrients not tabulated.
‡Niacin equivalents include dietary sources of the preformed vitamin and the precursor, tryptophan. 60 milligrams tryptophan equals 1 milligram niacin.
§Calorie allowances apply to individuals usually engaged in moderate physical activity. For office workers or others in sedentary occupations they are excessive. Adjustments must be made for variations in body size, age, physical activity, and environmental temperature.
**See original reference for discussion of infant allowances. The Board recognizes that human milk is the natural food for infants and feels that breast feeding is the best and desired procedure for meeting nutrient requirements in the first months of life. No allowances are stated for the first month of life. Breast feeding is particularly indicated during the first month when infants show handicaps in homeostasis due to different rates of maturation of digestive, excretory, and endocrine functions. Recommendations as listed pertain to nutrient intake as afforded by cow's milk formulas and supplementary foods given the infant when breast feeding is terminated. Allowances are not given for protein during infancy.
(1)Reproduced, with permission, from Publication 589, National Academy of Sciences–National Research Council, 1958.

The ''Basic Seven'':

In order to simplify the concept of an adequate diet, foodstuffs have been arranged into seven groups each of which makes a major contribution to the diet. It is recommended that some food from each of the basic seven groups be taken each day.

1. Milk - Two or more glasses daily for adults; three or four or more glasses daily for children.
2. Vegetables - Two or more servings daily (in addition to potatoes); one vegetable, raw; green and yellow vegetables served often.
3. Fruits - Two or more servings daily; one a citrus fruit or tomato.
4. Eggs - Three to five a week; one daily preferred.
5. Meats, cheese, fish, or legumes - One or more servings daily.
6. Cereal or bread - Most of it whole grain or ''enriched.''
7. Butter - Two or more tablespoons daily.

Parenteral Nutrition:

In the nutrition of patients unable to take food by mouth, all nutrients must be given by a parenteral (usually the intravenous) route. However, it is difficult to supply parenterally the requirements of complete nutrition. Dextrose solutions are used to supply calories, but the quantity of dextrose which can be administered is limited. In most individuals it is not desirable to give dextrose intravenously at a rate exceeding 0.5 Gm. per Kg. body weight per hour. At faster rates, there is likely to be considerable glycosuria and accompanying diuresis. One hundred Gm. of dextrose is required to prevent ketosis and to spare protein. Additional amounts of dextrose improve this sparing action somewhat, but, considering the maximal rate of administration, it is inconvenient to administer dextrose in significantly larger quantities.

A major defect in parenteral nutrition is caloric inadequacy. This can be minimized by the provision of fat because of its high caloric content. A 15% emulsion of fat suitable for intravenous administration is now available, and it is likely to find increased use in those cases where prolonged intravenous feeding is necessary. The added calories from fat should also improve the nutritive efficiency of intravenously administered amino acids which otherwise are largely catabolized to supply needed calories.

The protein requirements of sick and injured persons are quite variable, but about 100 Gm. a day of protein is considered the minimal desired intake. This may be supplied intravenously by the use of amino acid preparations, usually made by hydrolyzing proteins so that they are not antigenic. Plasma, including concentrated solutions of albumin, and blood may also contribute to the protein nutrition of the body, but these substances are not used primarily for this purpose.

The requirements for water and sodium chloride have been discussed on pp. 308 and 318. Here also there are considerable variations depending upon the clinical situation.

A. Short-term Parenteral Nutrition: Assuming no unusual losses, a patient can be satisfactorily maintained for a few days by the use of the following regimen every 24 hours:
   1. Sodium chloride solution - 500 ml. of isotonic (0.9%) sodium chloride solution.
   2. Dextrose solution - 2000 to 2500 ml. of dextrose, 5 or 10%, in water. This should not be given faster than 0.5 Gm. per Kg. per hour.

B. In More Prolonged Parenteral Nutrition:
   1. Dextrose solution with amino acids - If the patient must be deprived of any other dietary source of protein for more than a few days some of the dextrose solutions (as above) should contain 5% amino acids. The rate of utilization of dextrose by the tissues will also be increased by the added amino acids.
   2. Potassium salts - After the third or fourth day of parenteral nutrition, potassium must be added to the regimen. The quantities necessary and the precautions in administering it parenterally are discussed on p. 321.
   3. Vitamins are not required during short illnesses unless the patient was previously undernourished. In a prolonged disability they may be added to the parenteral regimen*. The most important vitamins and the quantities recommended to be given each day are:

a. Thiamine	5-10 mg.	e. Pyridoxine	1-2 mg.
b. Riboflavin	5-10 mg.	f. Folic acid	1.0-1.5 mg.
c. Niacin	50-100 mg.	g. Vitamin B_{12}	3-4 mcg.
d. Calcium pantothenate	10-20 mg.	h. Ascorbic acid	100-250 mg.

*Vitamins added to fluids given intravenously may be lost in considerable quantity because of excretion in the urine. Oral administration or intramuscular injection are therefore the preferred routes for most efficient utilization of vitamin supplements.

References:

1. Rose, W. C., Coon, M. J., Lockhart, H. B., and Lambert, G. F.: J. Biol. Chem. **215**:101, 1955.
2. Rose, W. C., Leach, B. E., Coon, M. J., and Lambert, G. F.: J. Biol. Chem. **213**:913, 1955.
3. Rose, W. C., Wixom, R. L., Lockhart, H. B., and Lambert, G. F.: J. Biol. Chem. **217**: 987, 1955.

Bibliography:

Block, R. J., and Weiss, K. W.: Amino Acid Handbook. Thomas, 1956.

Best, C. H., and Taylor, N. B.: Physiological Basis of Medical Practice, 6th Ed. Williams and Wilkins, 1955.

Consolazio, C. F., Johnson, R. E., and Marek, E.: Metabolic Methods. Mosby, 1951.

Albritton, E. C., Ed.: Standard Values in Nutrition and Metabolism. Saunders, 1954.

Bourne, G. H., and Kidder, G. W., Eds.: Biochemistry and Physiology of Nutrition. 2 Vols. Academic Press, 1953.

Pollack, H., and Halpern, S. L.: Therapeutic Nutrition, Publication 234. National Research Council, 1952.

22...

The Chemistry of the Tissues

The tissues of the body are usually divided into five categories: vascular (including the blood), epithelial, connective, muscle, and nerve tissue. The chemistry of blood has been discussed in Chapter 9.

EPITHELIAL TISSUE

Epithelial tissue covers the surface of the body and lines hollow organs such as those of the respiratory, digestive, and urinary tracts.

Keratin:

A major constituent of the epidermal portion of the skin and of other epidermal derivatives, such as hair, horn, hoof, feathers, and nails, is the albuminoid protein, keratin. This protein is notable for its great insolubility and resistance to attack by proteolytic enzymes of the stomach and intestine, and for its high content of the sulfur-containing amino acid, cystine.

The composition of human hair differs in accordance with its color and with the race, sex, age, and genetic origin of the individual. The amino acid, cystine, accounts for about 20% of the amino acid content of the protein in human hair; this very high cystine content differentiates human hair from all other types of hair.

The keratin of wool can be rendered more soluble and digestible by grinding it in a ball mill. Such finely ground keratin, when supplemented with tryptophan, histidine, and methionine, can serve as a source of dietary protein in animals.

Melanin:

The color of the skin is due to a variety of pigments, of which melanin (see p. 253), a tyrosine derivative, is the most important. It is said that racial differences in skin color are due entirely to the amount of melanin present and that there is no qualitative difference in skin pigmentation of different races.

CONNECTIVE TISSUE

Connective tissue comprises all of that which supports or binds the other tissues of the body. It includes the bones and teeth, cartilage, and fibrous tissue.

White Fibrous Tissue:

The Achilles tendon of the ox is generally used as a characteristic example of white fibrous tissue. It contains about 63% water and 37% solids, of which only 0.5% is inorganic matter. The three distinct proteins identified in the tendon are:

Collagen	- 31.6%
Elastin (see p. 382)	- 1.6%
Tendomucoid	- 0.5%

The remainder of the organic material consists of fatty substances (1%) and extractives (about 0.98%).

A. Collagen: An albuminoid, the protein **collagen**, is the principal solid substance in white fibrous connective tissue. Collagen is difficult to dissolve and somewhat resistant to chemical attack, although not to the same extent as keratin. The amino acid distribution of collagen is, however, quite different from that of keratin: glycine replaces cystine as the principal amino acid, and in fact accounts for as much as one-third of all the amino acids present (the prolines constitute another third). Collagen can be slowly digested by pepsin and hydrochloric acid; it can be digested by trypsin only after pepsin treatment or at temperatures over 40° C.

An important property of collagen is its conversion to **gelatin** by boiling with water or acid. This seems to involve only a physical change, since there is no chemical evidence of hydrolysis. Gelatin contains no tryptophan and very small amounts of tyrosine and cystine. It differs from collagen and keratin in being easily soluble and digestible. It may therefore be used as a source of protein in the diet, but only in a supplementary capacity because of its amino acid deficiencies.

The collagen of bone, skin, cartilage, and ligaments differs in chemical composition from that of white fibrous tissue.

B. Tendomucoid: This is a glycoprotein similar to salivary mucin. There are similar mucoids in bone (osseomucoid) and in cartilage (chondromucoid). The chemistry of these glycoproteins is discussed below.

Yellow Elastic Tissue:

The nuchal ligaments exemplify yellow elastic tissue. This tissue is composed of about 40% solid matter, of which 31.7% is **elastin**, 7.2% is collagen, and about 0.5% is mucoid.

The albuminoid, elastin, the characteristic protein of this type of connective tissue, is insoluble in water but digestible by enzymes. It is not converted to gelatin by boiling. The sulfur content of elastin is low. Ninety percent of the amino acid content of the protein is accounted for by only five acids: leucine, isoleucine, glycine, proline, and valine.

Cartilage:

The three principal solids in the organic matrix of cartilage are **chondromucoid**, **chondroalbumoid**, and **collagen** (see above).

A. Chondromucoid: When chondromucoid is split, it yields protein and the prosthetic group, chondroitin sulfuric acid. The structure of chondroitin sulfuric acid is as follows:

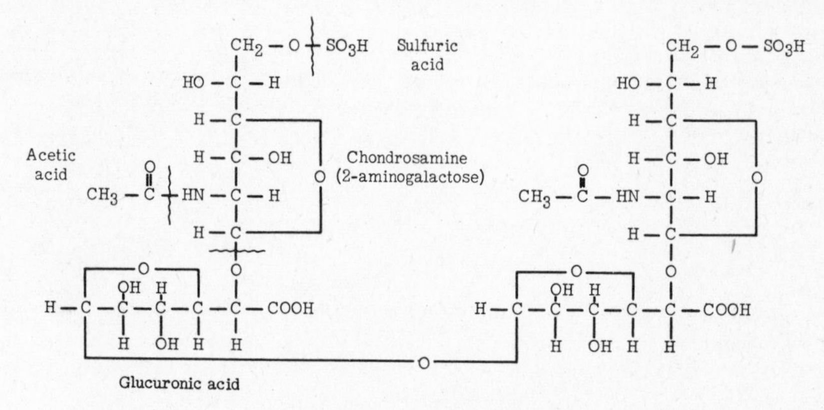

Chondroitin Sulfuric Acid

The hydrolytic breakdown of chondroitin sulfuric acid can be summarized as follows:

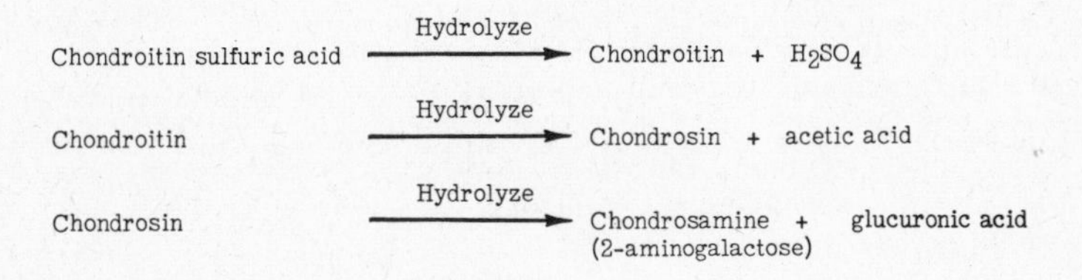

The chondromucoids of other tissues (e.g., skin, umbilical cord, spongy bone) differ only in the nature of the protein attached to the chondroitin sulfuric acid prosthetic group.

B. Chondroalbumoid is somewhat similar to elastin and keratin. In contrast to keratin, however, the sulfur content of chondroalbumoid is low and it is soluble in gastric juice.

Other Mucopolysaccharides:

Other polysaccharides containing hexosamines are found in the body. Such compounds, of which chondroitin sulfuric acid is an example, are called mucopolysaccharides. Other examples are heparin (see p. 126), mucoitin sulfuric acid, and hyaluronic acid.

A. Mucoitin sulfuric acid of the saliva contains a glucosamine (chitosamine, see p. 11), which is probably similar to that in the chitin of the exoskeleton of lower animals.

B. Hyaluronic acid contains glucosamine acetylated at the amino group (N-acetyl-D-glucosamine), and D-glucuronic acid. Its composition is therefore similar to chondroitin, the essential difference between the two structures being the presence of galactosamine in chondroitin instead of glucosamine, as is the case with hyaluronic acid. Hyaluronic acid is a component of the capsules of certain strains of pneumococci, streptococci, and certain other organisms, as well as of vitreous humor, synovial fluid, and umbilical cord (Wharton's jelly). The hyaluronic acid of the tissues acts as a lubricant in the joints, as a jelly-like cementing substance, and as a means of holding water in the interstitial spaces.

An enzyme, **hyaluronidase**, is present in certain tissues, notably testicular tissue and spleen, as well as in several types of pneumococci and the hemolytic streptococci. This enzyme, by destroying tissue hyaluronic acid, reduces viscosity and thus permits greater spreading of materials in the tissue spaces. Hyaluronidase is therefore sometimes designated the **spreading factor**. Its activity may be measured by the extent of spread of injected India ink as indicator. The invasive power of some pathogenic organisms may be increased because they secrete hyaluronidase. In the testicular secretions, it may dissolve the viscid substances surrounding the ova to permit penetration of the sperm cell. Hyaluronidase is used clinically to increase the efficiency of absorption of solutions administered by clysis.

C. Glycoproteins: The carbohydrates of the glycoproteins (see p. 132) probably have mucopolysaccharides as prosthetic groups. Such proteins are not coagulated by heat nor are they digested by the enzymes of the gastrointestinal tract. They therefore act as protective substances for the mucosal tissue.

Bone:

A. Chemistry: The water content of bone varies from 14 to 44%. From 30 to 35% of the fat-free dry material is organic. In some cases as much as 25% is fat. The organic material in bone, the bone matrix, is similar to that of cartilage i• that it contains collagen, which can be converted to gelatin (ossein gelatin). There are also a glycoprotein named **osseomucoid** and an **osseo-albumoid.** The presence of citrate (about 1%) in bone has been reported recently.

The inorganic material of bone consists mainly of phosphate and carbonate salts of calcium. There are also small amounts of magnesium, hydroxide, fluoride, and sulfate. A study of the x-ray pattern of the bone salts indicates a similarity to the naturally occurring mineral hydroxyapatite. The formula of hydroxyapatite is said to be:

$$Ca(OH)_2 . 3Ca(PO_4)_2 \text{ or } Ca_{10}(OH)_2(PO_4)_6$$

It is believed that the crystal lattices of bone are similar to the lattices of these apatite minerals but that elements may be substituted in bone without disturbing the structure. For example, calcium and phosphorus atoms may be replaced by carbon; magnesium, sodium, and potassium may replace calcium; and fluorine may replace hydroxide. This probably accounts for the alterations in bone composition which occur with increasing age, in rickets, as a result of dietary factors, or subsequent to changes in acid-base equilibrium (e.g., the acidosis of chronic renal disease).

B. Metabolism: Like all of the tissues of the body, the constituents of bone are constantly in exchange with those of the plasma. Demineralization of bone occurs when the intake of minerals necessary for bone formation is inadequate or when their loss is excessive.

The calcium and phosphorus content of the diet (see pp. 314 to 316) is obviously an important factor in ossification. Vitamin D (see p. 69) raises the level of blood phosphate and calcium, and this may in turn raise the calcium-phosphorus product to the point where calcium phosphate is precipitated in the bone. There is some evidence that the vitamin not only acts to promote better absorption of the minerals but also acts locally in the bone. In rickets, osteitis deformans, and other bone disorders the blood alkaline phosphatase rises, possibly in an effort to supply more phosphate. Treatment with vitamin D is accompanied by a reduction in phosphatase.

The influence of endocrines on calcification has been described in Chapter 20. The parathyroids, thyroid, anterior pituitary, adrenal, and sex glands are all important in this respect. They act either at the site of calcification or by altering absorption or excretion of calcium and phosphorus.

Ossification supposedly involves precipitation of bone salts in the matrix by means of a physicochemical equilibrium involving Ca^{++}, $HPO_4^{=}$, and $PO_4^{\equiv}$. The enzyme alkaline phosphatase, which liberates phosphate from organic phosphate esters, may produce the inorganic phosphate; this phosphate then reacts with the calcium to form insoluble calcium phosphate. Phosphatase is not found in the matrix but only in the osteoblasts of the growing bone.

The deposition of bone salt is not entirely explainable as due simply to physicochemical laws governing the solubility of the inorganic components of bone. The deposition of the bone salt in the presence of concentrations of inorganic phosphate and calcium similar to those of normal plasma may require the expenditure of energy from associated metabolic systems. In cartilage, calcification in vitro can be blocked with iodoacetate, fluoride, or cyanide, substances which are known to inhibit various enzymes involved in glycolysis. Furthermore, the deposition of bone salts in calcifying cartilage is preceded by swelling of the cartilage cells due to intracellular deposition of glycogen. Just prior to or simultaneously with the appearance of bone salt in the matrix of the cartilage, the stores of glycogen seem to disappear, which suggests that the breakdown of glycogen is necessary for the calcification of cartilage. All of the enzymes and intermediate compounds involved in glycolysis have been identified in calcifying cartilage, and, as noted above, enzyme inhibitors interfere with calcification in vitro.

It must be remembered that there are two important components in bone: the matrix, which is rich in proteins, and the mineral or inorganic component. Demineralization may result from effects on either. Steroids aid in the maintenance of osteoblasts and matrix; thus, in the absence of these hormones, osteoporosis may occur even though alkaline phosphatase is normal and, presumably, a favorable mineral environment exists. Changes in the concentration of alkaline phosphatase reflect activity of the osteoblasts which are simulated by stress on a weakened skeleton as may be found in rickets.

Teeth:

Enamel, dentin, and cementum of the teeth are all calcified tissues containing both organic and inorganic matter. In the center of the tooth is the **pulp**, a soft uncalcified organic mass containing also the blood vessels and nerves.

A. Composition and Structure: The average composition of human enamel and of dentin is shown in the accompanying table.

According to x-ray studies, the inorganic matter in the enamel and dentin of the teeth is arranged similarly to that in bone; it consists mainly of hydroxyapatite salts.

Keratin is the principal organic constituent of the enamel. There are also small amounts of cholesterol and phospholipid. In the dentin, collagen and elastin occur together with a glycoprotein and the lipids of the enamel.

Average Composition of Human Enamel and of Dentin*

	(% dry weight)	
	Enamel	Dentin
Calcium	35.8	26.5
Magnesium	0.27	0.79
Sodium	0.25	0.19
Potassium	0.05	0.07
Phosphorus	17.4	12.7
Carbon dioxide (from carbonate)	2.97	3.06
Chlorine	0.3	0.0
Fluorine	0.0112	0.0204
Iron	0.0218	0.0072
Organic matter	1.0	25.0

*From Hawk, Oser, and Summerson, "Practical Physiological Chemistry," 12th edition, The Blakiston Company, 1947.

Collagen is a major organic constituent in the cementum. Both dentin and, to a lesser extent, enamel contain citrate.

B. Metabolism of Teeth: Studies with radioactive isotopes (radiophosphorus) indicate that the enamel and especially the dentin undergo constant turnover; this is slow in adult teeth. The diet must contain adequate calcium and phosphorus and also vitamins A, C, and D to insure proper calcification. However, when the diet is low in calcium and phosphorus, the demineralization of bone exceeds that of the teeth, which may actually calcify during the restricted period but at a slower than normal rate. This and other data suggest that the mineral metabolism of teeth and that of bones is not necessarily parallel.

The relation of fluoride to teeth has been discussed on p. 329.

MUSCLE TISSUE

There are three types of muscle tissue in the body - striated (voluntary) or skeletal muscle, nonstriated (involuntary) or smooth muscle, and cardiac muscle.

The chemical constitution of the skeletal muscle has been most completely studied: 75% is water, 20% is protein, and the remaining 5% is composed of inorganic material, certain organic "extractives," and carbohydrate (glycogen and its derivatives).

The Proteins in Muscle:

The muscle fibrils are mainly protein. These fibrillar proteins are characterized by their elasticity or contractile power (see p. 34).

A. Myosin: This is the most abundant protein in the muscle. Myosin is probably the only protein in the fibril; the other muscle proteins are considered to be either constituents of the sarcoplasm, the material which imbeds the myofibrils, or at least extracellular in origin. Myosin is a globulin, soluble in dilute salt solutions and insoluble in water. In the presence of adenosine triphosphate (ATP, see p. 119), a complex of actin and myosin (actomyosin, see below) dissociates into actin (see below) and myosin A, accompanied by muscular contraction; the energy of the contraction is supplied by the breakdown of ATP to ADP. This latter reaction is catalyzed by **adenosine triphosphatase (ATPase).** The enzyme is closely associated with myosin itself. Some workers feel that ATPase is identical with myosin, but this is denied by others who have separated a magnesium-activated ATPase from myosin (Meyerhof). Other enzymes are also closely associated with myosin; it seems, therefore, that this protein readily absorbs various associated enzymes.

B. Actin: Another protein, actin, has been isolated from muscle. It now seems that the "myosin" usually isolated from muscle is actually a complex of actin plus myosin, which is designated **actomyosin** (myosin B of Szent-Györgyi).

C. Globulin X: This globulin remains after myosin is removed from a saline extract of muscle. It is probably a protein of the sarcoplasm. It coagulates at 50° C., and its molecular weight is about 160, 000.

D. Myogen is another protein of the sarcoplasm, probably an albumin. One myogen fraction is said to have enzymatic properties.

E. Myoglobin: This conjugated protein, often called muscle hemoglobin, is similar to hemoglobin and may function as an oxygen carrier. The molecular weight of myoglobin is one-fourth that of hemoglobin. In crush injuries, myoglobin is liberated and may appear in the urine since it is filtrable by the glomerulus. It may also precipitate in the renal tubules, obstructing the tubules resulting in anuria - a feature of the **crush syndrome.** The precipitation of hemoglobin in the tubule is said to cause a similar anuria after the hemolytic episodes of a transfusion reaction.

Enzymes in Muscle:

The considerable metabolic activity of muscle, particularly in carbohydrate metabolism, requires the presence of a number of enzymes. Some have been well studied, but many more have not. **Phosphorylase** (see p. 176), which catalyzes the reversible conversion of glucose-1-phosphate to glycogen, is an example. Other examples include the group of enzymes in anaerobic and aerobic glycolysis described in Chapter 13.

Muscle Extractives:

Certain compounds which can be easily extracted from muscle with water, alcohol, or ether are designated muscle extractives. Those containing nitrogen include creatine and phosphocreatine; the purines, adenine, guanine, xanthine, hypoxanthine, uric acid, and adenylic acid (from ATP); carnosine, anserine, and the betaine, carnitine. A vitamin-like substance (vitamin B_t) necessary for the growth of the meal worm, Tenebrio molitor, was recently identified as carnitine. This is the first demonstration of a specific requirement for this substance in animal tissue. The presence in carnitine of methyl groups in a linkage similar to choline and betaine suggests that carnitine may function as a methyl donor. This has not been definitely established.

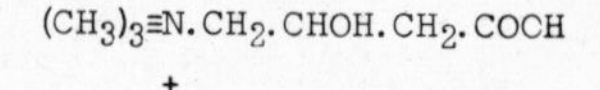

Carnitine

It is believed that these extractives are responsible for the flavor of meat and of broths made from meat such as "beef tea" or bouillon. The so-called Liebig extract of beef used in bacteriologic media is rich in these extractives.

The other extractives of muscle, which do not contain nitrogen, are mainly the muscle carbohydrate, glycogen, and all of its derivatives formed in glycolysis.

Inositol (see pp. 22 and 95) is also found in combination in fresh muscle.

Inorganic Constituents of Muscle:

The cations of muscle (potassium, sodium, magnesium, and calcium) are the same as in extracellular fluids except that, in muscle, potassium predominates. The anions include phosphate, chloride, and small amounts of sulfate. The apparent high inorganic phosphate content may actually be an artefact produced by the breakdown of the organic phosphates of ATP and phosphocreatine in the course of the analysis.

Intracellular potassium plays an important role in muscle metabolism. When glycogen is deposited in muscle and when protein is being synthesized, a considerable amount of potassium is also incorporated into the tissue. Muscle weakness is a cardinal sign of potassium deficiency (see p. 321). The calcium and magnesium of muscle appear to function as activators or inhibitors of intramuscular enzyme systems.

NERVE TISSUE

The tissues of the brain, spinal cord, the cranial and spinal nerves and their ganglia and plexuses, and those of the autonomic nervous system contain a considerable quantity of water. The gray matter, which represents a concentration of nerve cell bodies, always contains more water than the white matter, where the nerve fibers are found. In the adult brain, where gray and white matter are mixed, the water content averages 78%; in the cord the water content is slightly less, about 75%.

The solids of nervous tissue consist mainly of protein and lipids. There are also smaller amounts of organic extractives and of inorganic salts.

The Proteins of Nerve Tissue:

These constitute 38 to 40% of the total solids. They include various globulins, nucleoprotein, and a characteristic albuminoid called **neurokeratin.**

The Lipids of Nerve Tissue:

Over half (51 to 54%) of the solid content of nerve tissue is lipid material. In fact, this tissue is one of the highest in lipid content. It is noteworthy that very little if any simple lipid is present.

Representatives of all types of compound lipids are found. These include:

Compound Lipids in Nerve Tissue	Percent of Total Solids
Phospholipids (lecithins, cephalins and sphingomyelins)	28%
Cholesterol	10%
Cerebrosides or galactolipids (glycolipins)	7%
Sulfur-containing lipids, aminolipids, etc.	9%
	54%

The chemistry of these substances is discussed in Chapter 2.

The rate at which the lipids of the brain are exchanged is relatively slow in comparison to that in an active organ such as the liver. Tracer studies with deuterium indicate that while 50% of the liver fats may be exchanged in 24 hours, only 20% of the brain fat is replaced in seven days.

Inorganic Salts:

These substances in nervous tissue are components of the 1% ash produced as the result of combustion. The principal inorganic salts are potassium phosphate and chloride, with smaller amounts of sodium and other alkaline elements. The potassium of the nerve is thought to be important in the electrical nature of the nerve impulse, which depends on depolarization and repolarization at the membrane boundary of the nerve fiber.

Metabolism of Brain and Nerve:

When a nerve is stimulated to conduct an impulse, a small but measurable amount of heat is produced. The heat is produced in two stages, as it is in working muscle. This suggests that the rapidly released initial heat represents the energy involved in transmission of the impulse, and that the delayed or recovery heat (which may continue for 30 to 45 minutes) is related to restoration of the energy mechanisms. Similarly, the nerve may conduct impulses and develop heat under anaerobic conditions as, for example, in an atmosphere of nitrogen; but recovery depends on the admission of oxygen, as indicated by the extra consumption of readmitted oxygen.

The respiratory quotient of metabolizing nerve is very close to 1, which suggests that the nerve is utilizing carbohydrate almost exclusively. The metabolism of carbohydrate in nerve tissue seems to be similar to that in muscle, since lactic and pyruvic acid appear under anaerobic conditions (see p. 174). These end products disappear very slowly; oxygen does not accelerate the process.

The glycogen stores of brain and nerve are very small; hence a minute-to-minute supply of blood glucose is particularly important to the nervous system. This may be the major reason for the prominence of nervous symptoms in hypoglycemia. In contrast to muscle extracts, brain extracts act more readily on glucose than on glycogen; ATP does not accelerate glycolysis in the absence of phosphate in the brain, but reduced glutathione does.

Glutamic acid is apparently the only amino acid metabolized by brain tissue. Ammonia produced in brain metabolism, or delivered to the brain when the arterial blood ammonia is elevated, is taken up by the brain tissue presumably by conversion of ketoglutaric acid, produced in the citric acid cycle, to glutamic acid, and of glutamic acid, to glutamine.

The formation of γ-aminobutyric acid in central nervous system tissue from glutamic acid has been discussed on p. 243. The significance of γ-aminobutyric acid as an important regulatory factor in neuronal activity has also been mentioned.

Chemical Mediators of Nerve Activity:

At the junction of the nerve fiber and the effector organ, such as the muscle which is stimulated **(the myoneural junction)**, a chemical substance is elaborated by the action of the nerve impulse. This substance actually brings about the activity of the effector. Such a chemical substance is called a **chemical mediator** of the nerve impulse or a **neurohormone.**

A. The Neurohormones:

1. Acetylcholine - In parasympathetic and in voluntary nerves to the skeletal muscles, the chemical mediator is acetylcholine.
2. Sympathin - Stimulation of the sympathetic nerves produces sympathin, with an effect opposite to that of acetylcholine. Sympathin exerts two effects - one excitatory, referred to as sympathin E; the other inhibitor, referred to as sympathin I. The action of sympathin suggests that it may be identical with epinephrine, and, in fact, sympathetic stimulation does cause liberation of epinephrine from the adrenal medulla.

B. Chemistry of Acetylcholine:

1. Breakdown of acetylcholine; action of acetylcholine esterase (cholinesterase) - Acetylcholine is readily hydrolyzed to choline and acetic acid by the action of the enzyme acetylcholine esterase, found not only at the nerve endings but also within the nerve fiber. The reaction is:

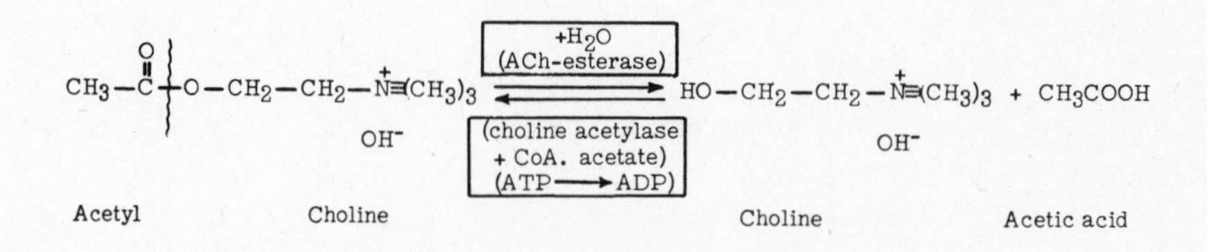

The action of acetylcholine in the body is controlled by the inactivating effect of acetylcholine esterase (designated ACh-esterase by Nachmansohn to distinguish it from a pseudocholinesterase found in the serum which hydrolyzes other esters).

2. Resynthesis of acetylcholine - The breakdown of acetylcholine is apparently an exothermic reaction since energy is required for its resynthesis. Active acetate (Co A-acetate; see p. 86) serves as acetate donor for the acetylation of choline. The enzyme, **choline acetylase**, which is activated by potassium and magnesium ions, catalyzes the transfer of acetyl from Co A-acetate to choline. The regeneration of ATP from ADP is accomplished by phosphocreatine, which, in turn, is resynthesized with the aid of the energy derived from glycolysis. All of these reactions are apparently similar to those in the muscle.
3. Anticholinesterases - Inhibition of acetylcholine esterase with resultant prolongation of parasympathetic activity is effected by physostigmine (eserine). The action is reversible.

   Neostigmine (Prostigmin®) is an alkaloid which is thought to function also as an inhibitor of cholinesterase and thus to prolong acetylcholine or parasympathetic action. It has been used in the treatment of myasthenia gravis, a chronic progressive muscular weakness with atrophy.

   A synthetic compound, diisopropylfluorophosphate (DFP), also inhibits the esterase activity but in an irreversible manner.

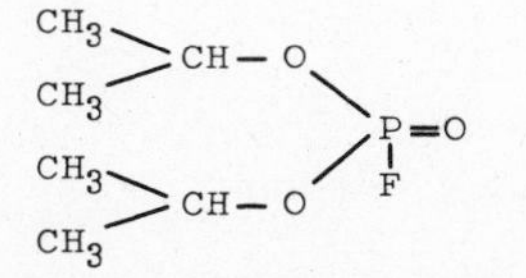

Diisopropylfluorophosphate (DFP)

This compound appears to be the most powerful and specific enzyme inhibitor yet discovered. It inhibits acetylcholine esterase when present in molar concentrations as low as 1×10^{-10} M. A mechanism for detoxifying DFP exists in the body in the action of an enzyme capable of bringing about the hydrolysis of the compound to fluoride and diisopropylphosphoric acid. This enzyme, diisopropylfluorophosphatase, has been identified in the kidney. The enzyme is activated by Mn^{++} or Co^{++} and specific co-factors such as imidazole and pyridine derivatives (e.g., proline or hydroxyproline).

DFP has also been used in the treatment of myasthenia gravis, although not with clinical results equal to those obtainable with Prostigmin®. It is a dangerous drug to use since the toxic dose is too close to the effective dose.

A number of **anticholinesterases** similar in their action to DFP have been investigated. They serve as the active principle of insecticides. The so-called "nerve gases" proposed for gas warfare, as well as many insecticides (e.g., parathion) are also anticholinesterases. These insecticides may produce toxic effects in individuals exposed to high concentrations when these are used as plant sprays.

Atropine is used as an antidote to the toxic effects of DFP and other anticholinesterases, together with a stimulant such as nikethamide (Coramine®), for treatment of respiratory paralysis.

• • •

Bibliography:

Elliott, K. A. C., Page, I. H., and Quastel, J. H., Eds.: Neurochemistry, The Chemical Dynamics of Brain and Nerve. Thomas, 1955.

Waelsch, H., Ed.: Biochemistry of the Developing Nervous System. Academic Press, 1955.

McLean, F. C., and Urist, M. R.: Bone, an Introduction to the Physiology of Skeletal Tissue. University Chicago Press, 1955.

Hawk, P. B., Oser, B. L., and Summerson, W. H.: Practical Physiological Chemistry. Blakiston, 13th Ed., 1954.

White, A., Handler, P., Smith, E., and Stetten, D., Jr.: Principles of Biochemistry. McGraw-Hill, 1954.

West, E. S., and Todd, W. R.: Textbook of Biochemistry. Macmillan, 2nd Ed., 1955.

Index